SINGULAR INTEGRAL EQUATIONS

Boundary Problems of Function Theory and Their Application to Mathematical Physics

N.I. MUSKHELISHVILI

SECOND EDITION

Translated from the Russian
by J.R.M. Radok

DOVER PUBLICATIONS, INC.
New York

This Dover edition, first published in 1992, is an unabridged and corrected republication of the edition published in 1953 by P. Noordhoff N.V., Groningen, Holland, based on the second Russian edition, Moscow, 1946.

Manufactured in the United States of America
Dover Publications, Inc., 31 East 2nd Street, Mineola, N.Y. 11501

Library of Congress Cataloging-in-Publication Data

Muskhelishvili, N. I. (Nikolaĭ Ivanovich), 1891–
[Singuli͡arnye integral'nye uravnenii͡a. English]
Singular integral equations : boundary problems of function theory and their application to mathematical physics / N.I. Muskhelishvili ; translated from the Russian by J.R.M. Radok.—2nd ed.
p. cm.
Translation of: Singuli͡arnye integral'nye uravnenii͡a.
"Unabridged and corrected republication of the edition published in 1953 by P. Noordhoff N.N., Groningen, Holland, based on the second Russian edition, Moscow, 1946"—Verso t.p.
Includes bibliographical references and index.
ISBN 0-486-66893-2 (pbk.)
1. Integral equations. 2. Boundary value problems. I. Radok, J. R. M. (Jens Rainer Maria) II. Title.
QA431.M813 1992
515'.45—dc20 92-14006
CIP

ANNOTATION

This monograph by N. I. Muskhelishvili systematically acquaints the reader with the mathematical apparatus of Cauchy type integrals and singular integral equations, in the study of which the author and his students took active interest. A considerable part of the book is devoted to applications to the solution of numerous problems of potential theory, the theory of elasticity and other sections of mathematical physics.

The book is intended for postgraduates and students of advanced courses of the physico-mathematical faculties, and likewise for research engineers.

EDITOR'S PREFACE

In preparing this translation for publication certain minor modifications and additions have been introduced into the original Russian text, in order to increase its readibility and usefulness. Thus, instead of the first person, the third person has been used throughout; wherever possible footnotes have been included with the main text. The chapters and their subsections of the Russian edition have been renamed parts and chapters respectively and the last have been numbered consecutively.

An authors and subject index has been added. In particular, the former has been combined with the list of references of the original text, in order to enable the reader to find quickly all information on any one reference in which he may be especially interested. This has been considered most important with a view to the difficulties experienced outside Russia in obtaining references, published in that country.

Russian names have been printed in Russian letters in the authors index, in order to overcome any possible confusion arising from transliteration.

Zürich, November 1952. J. R. M. Radok

CONTENTS

PART I

FUNDAMENTAL PROPERTIES OF CAUCHY INTEGRALS

Chapter 1

The Hölder Condition

Chapter 2

Integrals of the Cauchy type

Chapter 3

Some corollaries on Cauchy integrals

Chapter 4

Cauchy integrals near ends of the line of integration

PART II

THE HILBERT AND THE RIEMANN-HILBERT PROBLEMS AND SINGULAR INTEGRAL EQUATIONS (CASE OF CONTOURS)

Chapter 5

The Hilbert and Riemann-Hilbert boundary problems

Chapter 6

Singular integral equations with Cauchy type kernels (case of contours)

Contents

PART IV

THE HILBERT PROBLEM IN THE CASE OF ARCS OR DISCONTINUOUS BOUNDARY CONDITIONS AND SOME OF ITS APPLICATIONS

Chapter 10

The Hilbert problem in the case of arcs or discontinuous boundary conditions

Chapter 11

Inversion formulae for arcs

Chapter 12

Effective solution of some boundary problems of the theory of harmonic functions

Chapter 13

Effective solution of the principal problems of the static theory of elasticity for the half-plane, circle and analogous regions

PART V

SINGULAR INTEGRAL EQUATIONS FOR THE CASE OF ARCS OR DISCONTINUOUS COEFFICIENTS AND SOME OF THEIR APPLICATIONS

Chapter 14

Singular integral equations for the case of arcs and continuous coefficients

Chapter 15

Singular integral equations in the case of discontinuous coefficients

Chapter 16

Application to the Dirichlet problem and similar problems

Contents

Chapter 17

Solution of the integro-differential-equation of the theory of aircraft wings of finite span

PART VI

THE HILBERT PROBLEM FOR SEVERAL UNKNOWN FUNCTIONS AND SYSTEMS OF SINGULAR INTEGRAL EQUATIONS

Chapter 18

The Hilbert problem for several unknown functions

Chapter 19

Systems of singular integral equations with Cauchy type kernels and some supplements

Appendix 1

Appendix 2

Appendix 3

PREFACE

This book is intended for an extensive field of readers, in particular for those interested in applications to the theory of elasticity, hydromechanics and other branches of mathematical physics. The book is accessible to those acquainted with the basic theory of functions of a complex variable and the theory of Fredholm integral equations. To facilitate the reading of the book, theorems, the method of proof of which is not of essential independent interest, have been printed in italics, so that the proofs may be omitted without affecting an understanding of the nature of the matter. In addition, wherever possible, the Parts and their chapters devoted to different applications have been made independent of one another. It is hoped that the methods studied in this book may be effectively employed in the solution of many problems of an applied character. Some simple applications to potential theory, the theory of elasticity and hydromechanics are given in this book.

The idea of writing this book resulted from the Author's lectures in a seminar at the Mathematical Institute, Tiflis, in 1940--1942. Under the influence of a series of results obtained by members of the seminar (chiefly due to the excellent work of I. N. Vekua) the range of problems which the Author proposed to examine was essentially altered; the Author may state with much satisfaction that a large proportion of the contents of this book must be considered as the result of the collective work of the young members of the Tiflis Mathematical Institute of the Academy of Sciences of the Georgian S.S.R. together with I. N. Vekua and the Author himself.

Tiflis, Autumn 1944 N. I. MUSKHELISHVILI

In adding the last corrections to the book the Author desires to use this opportunity to express his deep gratitude to the publishing house which always willingly complied with the Author's suggestions. The difficult and responsible task of proof-reading was considerably eased for the Author by the extraordinary kindness of L. I. Bokshitski on the staff of the press, to whom he extends his sincere gratitude just as to all the other members of the staff of the 16th Printing Establishment of the State publishers for their prompt and efficient work.

Moscow, Autumn 1945 N. I. M.

INTRODUCTION

In recent years the theory of singular integral equations has assumed increasing importance in applied problems.

In this book only one-dimensional (i.e., where the range of integration is one-dimensional, i.e., a line) singular equations involving Cauchy principal values will be examined, since the theory of multi-dimensional equations of corresponding form is still far from completion. (Some references to the literature dealing with the latter are given in § 59).

The fundamentals of the theory of one-dimensional singular integral equations of the type described were included in the work of Poincaré and Hilbert, almost directly after the development of the classical theory of integral equations by Fredholm. However, the theory of singular integral equations did not receive the attention of mathematicians for some time. On the other hand many problems of an applied character naturally reduced to singular equations, e.g. problems of the theory of elasticity, etc.; thus often in practice these equations were arrived at by "ordinary methods" and this did not always lead to satisfactory results.

However, during recent years, the theory of one-dimensional singular integral equations has advanced considerably and it can now be presented in a finished form.

This theory appears to be particularly simple and effective, if the solution of a boundary problem of the theory of functions of a complex variable, to be called the Hilbert problem, is considered. Therefore the theory of singular equations is here closely linked with the above boundary problem. The solution of the latter will be used for the development of the theory of singular equations; afterwards this theory will be applied to the solution of other more complicated boundary problems, in particular, to problems encountered in potential theory, the theory of elasticity and in hydromechanics.

Having in mind the implications for different problems of mathematical physics, some restrictions will be imposed upon the unknown and the given functions appearing in the integral equations under consideration or in the boundary conditions of the problems considered, which will largely simplify the investigation, but not affect the final theory.

The fundamental tools of the investigation are Cauchy integrals, the elementary theory of which is given in Part I together with a number of direct simpler applications.

As regards the contents of the remaining Parts nothing will be said here, since a preliminary idea can be gained by means of the sufficiently detailed list of contents at the beginning of the book.

PART I

FUNDAMENTAL PROPERTIES OF CAUCHY INTEGRALS

CHAPTER 1

THE HÖLDER CONDITION

All functions considered will be defined on smooth lines and satisfy the Hölder condition. In this chapter some properties of such functions will be given.

§ 1. Smooth and piecewise smooth lines.

In the sequel only lines lying in one and the same plane will be considered. A coordinate system Oxy in this plane will always be a right-handed cartesian system.

Unless otherwise stated, lines are always assumed to be simple, i.e., not to intersect.

Curves will be called smooth arcs, when they can be represented in the form

$$x = x(s),\ y = y(s),\ s_a \leqq s \leqq s_b, \tag{1.1}$$

where s_a, s_b are finite constants, $x(s)$ and $y(s)$ functions, continuous in the interval of definition, with the following properties:

1°. They have continuous first derivatives $x'(s)$, $y'(s)$ within the interval (s_a, s_b), including the end points, and these derivatives are never simultaneously zero; $x'(s)$, $y'(s)$ at the ends of the interval are to be interpreted as $x'(s_a + 0,)$ $y'(s_a + 0)$ and $x'(s_b - 0)$, $y'(s_b - 0)$ respectively.

2°. The equalities $x(s_1) = x(s_2)$, $y(s_1) = y(s_2)$ are incompatible for

$$s_a \leqq s_1,\ s_2 \leqq s_b,\ s_1 \neq s_2.$$

The first condition shows that the arc L is smooth, i.e., that the direction of its tangent varies continuously with the position of its point of contact (see later). The second condition states that the arc is simple.

The points a and b, corresponding to the values s_a and s_b, are the ends of the arc. In agreement with (1.1) the ends a, b belong to the arc L; when it is necessary to stress this fact, L will be called a closed arc. When the end points do not belong to the arc, i.e., when $s_a < s < s_b$ in (1.1), the arc will be called open.

For each smooth arc considered a definite positive direction is chosen, namely that direction which corresponds to an increase of the parameter s. The arc with end points a, b will always be denoted by ab, where the order of letters is to indicate that the positive direction is from a to b.

The smooth arc L is obviously rectifiable; therefore it is possible to take as parameter s the length of arc from any fixed point of L, where the sign of s depends on the direction of displacement. This is done in all subsequent calculations. Hence

$$[x'(s)]^2 + [y'(s)]^2 = 1. \tag{1.2}$$

In this case the parameter s is called the arc coordinate of a point on the arc L. The point which corresponds to s will be denoted by $t(s)$ or simply t, and points corresponding to coordinates s_i will be denoted by $t(s_i)$ or simply t_i.

By the tangent to an arc L is always understood the positive tangent, i.e., the tangent in the increasing direction of s. If θ is the angle between the tangent at the point t and the axis Ox, measured from the latter counter-clockwise, then

$$\cos\theta = x'(s), \ \sin\theta = y'(s). \tag{1.3}$$

These formulae show that the angle between the tangent and any fixed direction is a continuous function of s. Conversely, it is clear that, if θ varies continuously with s, $x'(s)$ and $y'(s)$ are continuous functions of s.

Curves will be termed smooth contours L, if they differ from smooth arcs only in that, in condition 2°, the equalities $x(s_1) = x(s_2)$, $y(s_1) = y(s_2)$ are incompatible for $s_a < s_1$, $s_2 < s_b$, $s_1 \neq s_2$, but $x(s_b) = x(s_a)$, $y(s_b) = y(s_a)$ and $x'(s_b - 0) = x'(s_a + 0)$, $y'(s_b - 0) = y'(s_a + 0)$. The first of these equalities indicates that the curve L is closed, and the second that the direction of the tangent changes continuously on passing the point of contact corresponding to the values s_a, s_b of the arc coordinate. This point will sometimes be called the point of discontinuity of the arc-coordinate; it does not differ essentially from any other point of L, since it only depends on the choice of the parametric representation and clearly can be any point of the contour L.

In the following, without stating it explicitly in many cases, if any arc ab is under consideration consisting of portion of a contour L, the point of discontinuity of the arc coordinate is assumed moved on L so that it is outside or at one of the endpoints of the arc ab.

The union of a finite number of non-intersecting smooth arcs or contours will be called a smooth line L. Thus, by definition, a line may consist of several disconnected parts, while arcs or contours consist of one continuous portion. The main text of the book will refer to smooth lines only. However, many of the results obtained below may be applied directly or almost directly to piecewise smooth lines. (Some of the necessary modifications are given in Appendices 1 and 2.)

Arcs or contours will be called piecewise smooth, if they are continuous, simple, and consist of a finite number of separate smooth curves with contiguous ends. By a piece-wise smooth line is to be understood the union of a finite number of piece-wise smooth arcs (and contours) which do not intersect one another.

§ 2. Some properties of smooth lines.

In the sequel, the following easily proved properties of smooth lines are needed: (For the proofs see Appendix 1)

Let L be a given smooth line, α_0 any arbitrary acute angle (i.e., $0 < \alpha_0 < \frac{\pi}{2}$) and let $R_0 = R_0(\alpha_0)$ be a positive number depending on α_0, but not on the position of the point t on L, such that:

1°. The part of L, lying within a circle Γ with radius $R \leqq R_0$ with its centre at any point t on L, consists of a single arc ab. (If L consists only of a closed contour, the ends a, b of this arc lie always on the circumference of Γ. If an arc is part of L, one or both of its end points may be inside Γ.)

2°. The non-obtuse angle between the tangents at any two points of the arc ab does not exceed α_0.

From the above the following properties of the arc ab result immediately:

3°. The non-obtuse angle between the chord joining any two points of the arc ab and the tangent at an arbitrary point of this arc does not exceed α_0.

In fact, there is always a point on the arc ab at which the corresponding tangent is parallel to the chord, so that the statement is reduced to 2°.

4°. Let β_0 be any angle satisfying the condition $\alpha_0 < \beta_0 \leqq \frac{\pi}{2}$, also let Δ_a, Δ_b be two parallel lines through a and b which make a non-obtuse angle $\beta \geqq \beta_0$ with the tangent at any point t on ***ab***. Then every line Δ, parallel to the straight lines Δ_a, Δ_b and lying between them, cuts ab at one point only.

That Δ cuts ab in at least one point is obvious; that it does not intersect ab at any other point follows from 3° and from the condition $\beta \geqq \beta_0 > \alpha_0$.

5°. Let Δ be a straight line through any point t on ab, making a non-obtuse angle not less than $\beta_0 > \alpha_0$ with the tangent at t, and let t' be any other point on ab. Then the non-obtuse angle between Δ and the chord, connecting t and t', is not less than $\omega_0 = \beta_0 - \alpha_0 > 0$ [1]).

For the given angle α_0, $R_0 = R_0(\alpha_0)$ is called the standard radius, the circle Γ_0 of radius R_0 the standard circle, the arc ab, cut from the line L by the circle Γ_0 drawn at any point on it, the standard arc.

Let $L = ab$ be a standard arc, t_0 a fixed point on L (which may coincide with a or b), t a variable point on L and r the length of the chord joining t and t_0. Then

$$\frac{dr}{ds} = \pm \cos\alpha, \tag{2.1}$$

where s is the arc coordinate of the point t and α the acute angle between the chord tt_0 and the tangent at t, the upper sign referring to the part t_0b and the lower sign to the part at_0. Since from above $0 \leqq \alpha \leqq \alpha_0 < \frac{\pi}{2}$, it follows from (2.1) that r is a monotonic function of s over each of the parts at_0, t_0b (decreasing on the former, increasing on the latter part of ab), and consequently that the position of t on each of these parts is uniquely defined for any given r.

A frequently used inequality follows from (2.1), namely

$$|ds| \leqq K\,|dr|, \tag{2.2}$$

where K is a positive constant independent of the position of t_0 on ab.

Consider one of the arcs at_0, t_0b, integrate both sides of (2.1) between the limits s_1, s_2 and apply the mean value theorem; then for any pair of points t_1, t_2 on at_0 or t_0b

$$|r_2 - r_1| = k\,|s_2 - s_1|, \quad 0 < k_0 \leqq k \leqq 1, \tag{2.3}$$

where k_0 is a constant and r_1, r_2 are the distances of t_1, t_2 from t_0.

Taking t_1 as t_0 gives

$$r_{12} = k\sigma_{12}, \quad 0 < k_0 \leqq k \leqq 1, \tag{2.4}$$

---

[1]) If ω is the non-obtuse angle between Δ and the chord and α, β are the non-obtuse angles between the tangent at t and Δ and the chord respectively, then either $\omega = \beta - \alpha \geqq \beta_0 - \alpha_0$ or $\omega = \beta + \alpha$ (this may be so for $\beta + \alpha \leqq \frac{\pi}{2}$), and then $\omega \geqq \beta_0 > \beta_0 - \alpha_0$, or $\omega = \pi - \beta - \alpha$ (if $\beta + \alpha \geqq \frac{\pi}{2}$) and then $\omega \geqq \frac{\pi}{2} - \alpha \geqq \beta_0 - \alpha_0$.

where r_{12} is the distance between t_1 and t_2 and $\sigma_{12} = | s_1 - s_2 |$ is the length of that part of L which lies between t_1 and t_2.

The relation (2.4) is easily seen to remain valid in the case when L is an arbitrary arc or contour, provided that in the case of a contour the part of shorter length is taken as the part included between t_1 and t_2.

§ 3. The Hölder Condition (H condition).

Let there be given on the arc L a function of position $\varphi(t)$ (which is in general complex). In future let t denote both the point $t(x, y)$ and the corresponding complex number $t = x + iy$.

The function $\varphi(t)$ will be said to satisfy a Hölder condition (O. Hölder) on L, if for any two points t_1, t_2 of L

$$| \varphi(t_2) - \varphi(t_1) | \leqq A \, | t_2 - t_1 |^{\mu}, \tag{3.1}$$

where A and μ are positive constants. A is called the Hölder constant and μ the Hölder index.

A function which satisfies a Hölder condition will be said to obey the H condition or, when it is necessary to specify the index μ, the $H(\mu)$ condition. The value of the constant A is generally of no interest.

A function satisfying the H condition on the arc L is clearly continuous on L. Further, if the function $\varphi(t)$ satisfies the $H(\mu)$ condition with some constant A for any pair of points t_1, t_2 for which the distance r_{12} between them does not exceed some positive constant δ, then it will satisfy the $H(\mu)$ condition over the whole arc L.

If (3.1) holds when $r_{12} \leqq \delta$, then for the entire arc L

$$| \varphi(t_2) - \varphi(t_1) | \leqq A' r_{12}^{\mu},$$

where A' is the larger of the numbers A and $\dfrac{2M}{\delta^{\mu}}$, and M is the upper limit of $| \varphi(t) |$ on L.

From (2.4) the $H(\mu)$ condition is equivalent to

$$| \varphi(t_2) - \varphi(t_1) | \leqq A \, \sigma_{12}^{\mu}, \tag{3.2}$$

where $\sigma_{12} = | s_2 - s_1 |$ is the length of that part of the arc L included between t_1 and t_2. (In case L is a contour take the shorter of the two parts into which L is divided).

Therefore, because of (2.3), the $H(\mu)$ condition is equivalent to the condition that on any standard arc ab belonging to L

$$| \varphi(t_2) - \varphi(t_1) | \leqq A \, | r_2 - r_1 |^{\mu}, \tag{3.3}$$

where $r_1 = | t_1 - a |$, $r_2 = | t_2 - a |$.

In the inequalities (3.2), (3.3) as in (3.1), A is some positive constant; in all these three inequalities A may be considered as having the same value, for if necessary A may be changed to a higher value in some of the inequalities.

Further note that for $\mu > 1$ it follows from (3.2) that at every point on the arc $\frac{d\varphi}{ds} = 0$, and hence $\varphi = \text{const.}$ Since this case is of no interest, it will always be assumed that

$$0 < \mu \leqq 1.$$

Clearly, if $\varphi(t)$ satisfies the $H(\mu)$ condition, it satisfies the condition $H(\nu)$ for all $\nu \leqq \mu$.

§ 4. Generalization to the case of several variables.

Naturally the concept of the Hölder condition can be generalized to the case of functions of several variables. For example, the function $\varphi(u, v)$, where u and v lie in certain intervals, will be said to satisfy the $H(\mu, \nu)$ condition with $0 < \mu \leqq 1$, $0 < \nu \leqq 1$, if for any given pair of values (u_1, v_1), (u_2, v_2)

$$| \varphi(u_2, v_2) - \varphi(u_1, v_1) | \leqq A \, | u_2 - u_1 |^{\mu} + B \, | v_2 - v_1 |^{\nu}, \quad (4.1)$$

where A and B are constants.

It is clear that, if $\varphi(u, v)$ satisfies the $H(\mu, \nu)$ condition, then it also satisfies a condition denoted by $H(\lambda)$:

$$| \varphi(u_2, v_2) - \varphi(u_1, v_1) | \leqq C\{| u_2 - u_1|^{\lambda} + | v_2 - v_1 |^{\lambda}\}, \quad (4.2)$$

where λ is the lesser of the numbers μ, ν and C is a constant. Clearly, if $\varphi(u, v)$ satisfies the condition $H(\mu, \nu)$, then it satisfies the condition $H(\mu)$ for the variable u uniformly with respect to v, and the condition $H(\nu)$ for v uniformly with respect to u. (This shows that the Hölder constants in the conditions $H(\mu)$ and $H(\nu)$ may be chosen independently, respectively, of v and u.) That the converse is also true follows from the inequality

$$|\varphi(u_2, v_2) - \varphi(u_1, v_1)| \leqq |\varphi(u_2, v_2) - \varphi(u_1, v_2)| + |\varphi(u_1, v_2) - \varphi(u_1, v_1)|.$$

§ 5. Two auxiliary inequalities.

The two following well known inequalities for any positive numbers σ_1 and σ_2, $0 \leqq \mu \leqq 1$ will be needed:

$$\frac{\sigma_1{}^{\mu} + \sigma_2{}^{\mu}}{(\sigma_1 + \sigma_2)^{\mu}} \leqq 2^{1-\mu}, \quad (5.1)$$

$$\frac{| \sigma_1{}^{\mu} - \sigma_2{}^{\mu} |}{| \sigma_1 - \sigma_2 |^{\mu}} \leqq 1. \qquad (\sigma_1 \neq \sigma_2) \quad (5.2)$$

In proving these inequalities it is possible, without loss of generality, to suppose $\sigma_1 \geqq \sigma_2$; substitution of $\sigma = \frac{\sigma_2}{\sigma_1}$ gives the above inequalities the form:

$$\frac{1+\sigma^\mu}{(1+\sigma)^\mu} \leqq 2^{1-\mu} \ (0 \leqq \sigma \leqq 1), \quad \frac{1-\sigma^\mu}{(1-\sigma)^\mu} \leqq 1 \ (0 \leqq \sigma < 1).$$

These latter inequalities follow in a completely elementary manner by finding the maximum of the functions of σ on the left.

§ 6. Sufficient conditions for the H condition to be satisfied.

In this section some of the simplest criteria for fulfilment of the H condition will be given. Throughout this section L is taken to mean a smooth arc or contour.

1°. Let the point t_0 divide the arc $L = ab$ into two parts at_0 and t_0b. If a function $\varphi(t)$ is continuous on L and satisfies the $H(\mu)$ condition on both parts at_0 and t_0b separately, then it satisfies this condition for the whole of the arc L.

In fact, let t_1 and t_2 be two points on the arc L. It t_1 and t_2 lie on the same side of t_0, then the inequality (3.2) is satisfied from the data. If, however, t_1 and t_2 lie on different sides of t_0, then, denoting by σ_1 and σ_2 the lengths of the arcs t_1t_0 and t_0t_2 and using (5.1),

$$|\varphi(t_2) - \varphi(t_1)| \leqq |\varphi(t_2) - \varphi(t_0)| + |\varphi(t_1) - \varphi(t_0)| \leqq$$
$$\leqq A(\sigma_1{}^\mu + \sigma_2{}^\mu) \leqq 2^{1-\mu} A \, |\sigma_1 + \sigma_2|^\mu = 2^{1-\mu} A \, \sigma_{12}{}^\mu,$$

which proves the statement.

2°. If $\varphi(t)$ and $\psi(t)$ satisfy on L the $H(\mu)$ and $H(\nu)$ conditions respectively, then the functions $\varphi(t) + \psi(t)$ and $\varphi(t) \, . \, \psi(t)$ satisfy the $H(\lambda)$ condition on L, where λ is the smaller of the numbers μ, ν.

For example, for the product $\varphi(t) \, . \, \psi(t)$ the proof is as follows:

$$|\varphi(t_2)\,\psi(t_2) - \varphi(t_1)\,\psi(t_1)| \leqq$$
$$\leqq |\varphi(t_2)\,\psi(t_2) - \varphi(t_2)\,\psi(t_1)| + |\varphi(t_2)\,\psi(t_1) - \varphi(t_1)\,\psi(t_1)| \leqq$$
$$\leqq M\,|\psi(t_2) - \psi(t_1)| + N\,|\varphi(t_2) - \varphi(t_1)|,$$

where M and N are the upper bounds of $|\varphi(t)|$ and $|\psi(t)|$ on L. Hence the result follows.

3°. If $\varphi(t)$ satisfies the $H(\mu)$ condition on L and $\varphi(t) \neq 0$ everywhere on L, then $\frac{1}{\varphi(t)}$ also satisfies the $H(\mu)$ condition.

The proof is obvious.

4°. Let t and t_0 be variable and fixed points respectively on L.

The function of t

$$r^\mu = |t - t_0|^\mu, \quad 0 < \mu \leqq 1$$

satisfies the $H(\mu)$ condition on L, because by (5.2)

$$|r_2^\mu - r_1^\mu| \leqq |r_2 - r_1|^\mu.$$

A similar result is obtained for t fixed and t_0 variable. Hence the function $|t - t_0|^\mu$ of the two variables t and t_0 satisfies the $H(\mu)$ condition on L.

5°. Let $\varphi(t)$ satisfy the $H(\mu)$ condition on L, and let $0 \leqq \lambda < \mu \leqq 1$. Then the function of t

$$\psi(t) = \frac{\varphi(t) - \varphi(t_0)}{|t - t_0|^\lambda},$$

where t_0 is a fixed point on L, satisfies the $H(\mu - \lambda)$ condition on L.

Without loss of generality it is clearly possible to assume that t lies on a standard arc ab, cut from L by a standard circle with centre at t_0; also, from 1°, this may be restricted to the case in which t lies, for example, on the segment t_0b. The position t on t_0b is determined by $r = |t - t_0|$ (cf. § 2); sometimes $\varphi(r)$ and $\psi(r)$ will be written for $\varphi(t)$ and $\psi(t)$. Taking $h > 0$ (which does not affect the generality) and writing $\varphi(t) - \varphi(t_0) = \omega(r)$, one obtains

$$|\psi(r+h) - \psi(r)| = \left| \frac{\omega(r+h)}{(r+h)^\lambda} - \frac{\omega(r)}{r^\lambda} \right| =$$

$$= \left| \frac{\omega(r+h) - \omega(r)}{(r+h)^\lambda} + \omega(r) \left\{ \frac{1}{(r+h)^\lambda} - \frac{1}{r^\lambda} \right\} \right| \leqq$$

$$\leqq \frac{|\omega(r+h) - \omega(r)|}{(r+h)^\lambda} + |\omega(r)| \frac{(r+h)^\lambda - r^\lambda}{r^\lambda (r+h)^\lambda}.$$

Since

$$|\omega(r+h) - \omega(r)| \leqq Ah^\mu, \quad |\omega(r)| = |\varphi(t) - \varphi(t_0)| \leqq Ar^\mu,$$

it follows that

$$|\psi(r+h) - \psi(r)| \leqq \Delta_1 + \Delta_2,$$

where

$$\Delta_1 = \frac{Ah^\mu}{(r+h)^\lambda}, \quad \Delta_2 = Ar^{\mu-\lambda} \frac{(r+h)^\lambda - r^\lambda}{(r+h)^\lambda}.$$

Further

$$\Delta_1 = A \left[\frac{h}{r+h} \right]^\lambda . h^{\mu-\lambda} \leqq Ah^{\mu-\lambda},$$

so that Δ_1 satisfies the required condition.

For Δ_2 consider the two possible cases $r \leqq h$ and $r > h$. In the first case ($r \leqq h$), using the inequality

$$(r+h)^\lambda - r^\lambda \leqq h^\lambda$$

(cf. 4°), it is seen that

$$\Delta_2 \leqq A \cdot \frac{h^\mu}{(r+h)^\lambda} \leqq A\left[\frac{h}{r+h}\right]^\lambda h^{\mu-\lambda} \leqq Ah^{\mu-\lambda},$$

while in the second case ($r > h$), using the inequality [1])

$$(r+h)^\lambda - r^\lambda = r^\lambda\left[\left(1+\frac{h}{r}\right)^\lambda - 1\right] \leqq \lambda h r^{\lambda-1},$$

it is seen that

$$\Delta_2 \leqq A\lambda h r^{\mu-\lambda-1} = A\lambda\left(\frac{h}{r}\right)^{1-\mu+\lambda} \cdot h^{\mu-\lambda} \leqq A\lambda h^{\mu-\lambda},$$

and thus the statement is verified.

6°. It is clear that the above inequalities do not depend on the position of t_0 on L; in addition, it is possible to interchange the parts played by t and t_0. Therefore the function of the two variables t and t_0

$$\psi(t_0, t) = \frac{\varphi(t) - \varphi(t_0)}{|t - t_0|^\lambda}$$

satisfies the $H(\mu - \lambda)$ condition on L, if $\varphi(t)$ satisfies the $H(\mu)$ condition there and if $0 \leqq \lambda < \mu \leqq 1$.

7°. Now consider the function $\varphi(t, \tau)$, where t lies on L and τ is a parameter varying in some region T. Let $\varphi(t, \tau)$ satisfy the condition $H(\mu)$ (with respect to t and to τ), where t lies on L and τ in T (cf. § 4). It will be shown that the function of three variables

$$\psi(t_0, t, \tau) = \frac{\varphi(t, \tau) - \varphi(t_0, \tau)}{|t - t_0|^\lambda}$$

(where t_0 as well as t lie on L) satisfies the $H(\mu - \lambda)$ condition for all three variables.

With respect to t_0 and t this has been proved already. In order to prove the statement for τ, put

---

[1]) For $0 \leqq \mu \leqq 1$ and $x \geqq 0$

$$(1+x)^\mu - 1 \leqq \mu x;$$

in fact, if $f(x) = (1+x)^\mu - \mu x - 1$, then

$$f(0) = f'(0) = 0,\ f''(x) \leqq 0.$$

$$\Delta = \psi(t_0, t, \tau + h) - \psi(t_0, t, \tau) =$$
$$= \frac{\varphi(t, \tau + h) - \varphi(t_0, \tau + h)}{|t - t_0|^\lambda} - \frac{\varphi(t, \tau) - \varphi(t_0, \tau)}{|t - t_0|^\lambda} =$$
$$= \frac{\varphi(t, \tau + h) - \varphi(t, \tau)}{|t - t_0|^\lambda} - \frac{\varphi(t_0, \tau + h) - \varphi(t_0, \tau)}{|t - t_0|^\lambda}.$$

For $|t - t_0| \leqq |h|$, the first expression for Δ gives
$$|\Delta| \leqq 2A\,|t - t_0|^{\mu-\lambda} \leqq 2A\,|h|^{\mu-\lambda}.$$
For $|t - t_0| \geqq |h|$, the second expression gives
$$|\Delta| \leqq \frac{2A\,|h|^\mu}{|t - t_0|^\lambda} \leqq 2A\,|h|^{\mu-\lambda},$$
and the statement is proved. A direct consequence of the above is that, if the function $\varphi(t, t_0)$ of two points on L satisfies the $H(\mu)$ condition, then the function
$$\psi(t, t_0) = \frac{\varphi(t, t_0) - \varphi(t_0, t_0)}{|t - t_0|^\lambda} \quad (0 \leqq \lambda < \mu \leqq 1)$$
satisfies the $H(\mu - \lambda)$ condition.

8°. Finally, the following obvious property should be noted.

For simplicity suppose $u(s)$ is a real function which satisfies the $H(\mu)$ condition over some interval $s_1 \leqq s \leqq s_2$; let $f(u)$ be a function (in general complex) defined for values $u = u(s)$ with $s_1 \leqq s \leqq s_2$ and having a bounded derivative $f'(u)$. Then $f(u)$ satisfies the $H(\mu)$ condition.

§ 7. Sufficient conditions for the H condition to be satisfied (continued).

The following frequently used proposition will now be proved. Let $\varphi(t)$ satisfy the $H(\mu)$ condition on the smooth arc or contour L and let $\omega(t)$ be a bounded function on L having a derivative [1]) with respect to t, except possibly at $t = t_0$, such that
$$\left|\frac{d\omega}{dt}\right| < \frac{C}{|t - t_0|} \qquad (t \neq t_0), \tag{7.1}$$
where C is a constant and t_0 is some fixed point on L. Then the function of t
$$\psi(t) = [\varphi(t) - \varphi(t_0)]\,\omega(t)$$
also satisfies the $H(\mu)$ condition on L.

---

[1]) $\dfrac{d\omega}{dt}$ is defined by $\lim\limits_{t_1 \to t} \dfrac{\omega(t_1) - \omega(t)}{t_1 - t}$ (t_1, t points on L). Obviously $\left|\dfrac{d\omega}{dt}\right| = \left|\dfrac{d\omega}{ds}\right|$, where s is the arc coordinate.

It may be supposed, without loss of generality, that t lies on a standard arc having the point t_0 as one of its ends. Then the condition (7.1) is equivalent to

$$\left|\frac{d\omega}{ds}\right| < \frac{C_0}{|s - s_0|}, \tag{7.2}$$

where C_0 is a constant and s, s_0 are the arc coordinates corresponding to t and t_0. Further, it may be assumed that $\omega(t)$ takes only real values (otherwise the reasoning may be applied separately to the real and imaginary parts).

Replacing $\varphi(t)$, $\psi(t)$, $\omega(t)$ by $\varphi(s)$, $\psi(s)$, $\omega(s)$, one has

$$\begin{aligned}\psi(s+h)-\psi(s) &= [\varphi(s+h)-\varphi(s_0)]\omega(s+h)-[\varphi(s)-\varphi(s_0)]\omega(s) = \\ &= [\varphi(s+h)-\varphi(s)]\omega(s+h)+[\varphi(s)-\varphi(s_0)][\omega(s+h)-\omega(s)].\end{aligned}$$

The suppositions $s - s_0 \geqq 0$, $h \geqq 0$ do not affect the generality. Consequently, from the boundedness of $\omega(s)$, the first term in the last line does not exceed in magnitude C_1h^μ, where C_1 is a constant. The same is obviously true for the second term, provided $s - s_0 \leqq h$, while for $s - s_0 \geqq h$ and $0 < \theta < 1$

$$\begin{aligned}|\varphi(s)-\varphi(s_0)|\,.\,|\omega(s+h)-\omega(s)| &\leqq |\varphi(s)-\varphi(s_0)|\frac{C_0h}{s-s_0+\theta h} \leqq \\ \leqq \frac{AC_0(s-s_0)^\mu h}{s-s_0+\theta h} &\leqq AC_0\left[\frac{h}{s-s_0}\right]^{1-\mu}.\,h^\mu \leqq AC_0h^\mu,\end{aligned}$$

and the proposition is proved.

Some simple applications of this result will now be given:

1°. Let t be a variable, t_0 a fixed point on L. The function of t

$$\psi(t) = |t - t_0|^\mu \log|t - t_0|,\quad 0 < \mu \leqq 1$$

satisfies the $H(\mu - \varepsilon)$ condition on L, where ε is any positive number less than μ.

In fact, the last result may be applied to $\psi(t)$ by putting $\varphi(t) = |t - t_0|^{\mu-\varepsilon}$ and $\omega(t) = |t - t_0|^\varepsilon \log|t - t_0|$. It is clear that t and t_0 may interchange roles and that $|t - t_0|^\mu \log|t - t_0|$ satisfies the $H(\mu - \varepsilon)$ condition for both variables t and t_0.

2°. Let t be a variable, t_0 a fixed point on L. Denote by $\vartheta(t_0, t) = \arg(t - t_0) + \text{const.}$ the signed angle which the vector $\overrightarrow{t_0t}$ makes with any fixed direction ($\arg z$ stands for the argument of the complex number z). This angle is defined uniquely apart from an additive term of the form $2n\pi$, where n is an integer. Suppose that $\vartheta(t_0, t)$ varies continuously as t moves on L without passing through t_0. If t_0 is not an end of L, this angle must change discontinuously by an amount $(2n - 1)\pi$, n an integer, when t passes through t_0.

It will be shown that the function $\omega(t) = \vartheta(t_0, t)$ satisfies the condition (7.1). To prove this it is obviously sufficient to consider only the neighbourhood of the point t_0. Take as the axis Ox the tangent to L at t_0 and let the origin of the coordinate system be t_0. Let $x = x(s)$, $y = y(s)$ be the parametric representation of L and measure the arc coordinate from $t_0 = 0$. Then on either side in the neighbourhood of the point t_0

$$\vartheta(t_0, t) = \operatorname{artan} \frac{y(s)}{x(s)} + \text{const}, \quad \frac{d\vartheta}{ds} = \frac{y'(s)x(s) - x'(s)y(s)}{x^2(s) + y^2(s)}.$$

But from the choice of the coordinate system: $x(0) = y(0) = 0$, $x'(0) = 1$, $y'(0) = 0$, and hence

$$x(s) = s \,.\, x'(\theta_1 s), \quad y(s) = s \,.\, y'(\theta_2 s), \quad 0 < \theta_1, \ \theta_2 < 1.$$

When $|s|$ is sufficiently small $x'(\theta_1 s) > k$, where k is a positive constant. Hence it is seen that

$$\left| \frac{d\vartheta}{ds} \right| < \frac{C_0}{|s|},$$

and the proposition is proved.

Any function $f(\vartheta)$ satisfies an analogous inequality, provided $df/d\vartheta$ is bounded.

As $|t - t_0|^\mu$ satisfies the $H(\mu)$ condition for $0 < \mu \leqq 1$, it follows from the above that the function

$$\psi(t) = (t - t_0)^\mu = |t - t_0|^\mu e^{i\mu\vartheta}, \quad 0 < \mu \leqq 1, \tag{7.3}$$

where $\vartheta = \vartheta(t_0, t) = \arg(t - t_0)$, satisfies the $\mathrm{H}(\mu)$ condition on L. Similarly the function

$$\psi(t) = |t - t_0|^\varepsilon e^{\alpha\vartheta}, \quad 0 < \varepsilon \leqq 1, \ (\alpha \text{ any constant}) \tag{7.3a}$$

satisfies the $H(\varepsilon)$ condition on L. The function

$$\psi(t) = (t - t_0)^{\mu + i\nu} = e^{(\mu + i\nu)(\log|t - t_0| + i\vartheta)}, \tag{7.4}$$

where $0 < \mu \leqq 1$ and ν any real constant, likewise satisfies the $H(\mu)$ condition. For $\mu = 0$, $\nu \neq 0$, the function

$$\begin{aligned} \psi(t) = (t - t_0)^{i\nu} &= e^{i\nu \log|t - t_0| - \nu\vartheta} = \\ &= e^{-\nu\vartheta} \{\cos(\nu \log|t - t_0|) + i \sin(\nu \log|t - t_0|)\} \end{aligned} \tag{7.5}$$

obviously does not satisfy the H condition in the neighbourhood of $t = t_0$; it is not even continuous there. However, the function

$$\psi(t) = |t - t_0|^\varepsilon [t - t_0]^{i\nu} \tag{7.6}$$

satisfies the $H(\varepsilon)$ condition for all arbitrarily small values of $\varepsilon > 0$.

Clearly, as before, t and t_0 may interchange roles.

3°. Let t be a variable, t_0 a fixed point on L. Consider the function

$$\omega(t) = \frac{t - t_0}{s - s_0}$$

in the neighbourhood of t_0. It is easy to prove directly that $\omega(t)$ satisfies the condition (7.1) or, what is the same thing, (7.2).

Hence it may be stated, for example, that

$$| t - t_0 |^\varepsilon \frac{t - t_0}{s - s_0}, \quad | s - s_0 |^\varepsilon \frac{s - s_0}{t - t_0}, \quad 0 < \varepsilon \leqq 1$$

satisfy the $H(\varepsilon)$ condition. Similarly it is easy to prove that

$$| t - t_0 |^\varepsilon \frac{| t - t_0 |^\mu}{| s - s_0 |^\mu}, \quad | s - s_0 |^\varepsilon \frac{| s - s_0 |^\mu}{| t - t_0 |^\mu}, \quad 0 < \varepsilon \leqq 1,$$

μ an arbitrary constant, satisfy the $H(\varepsilon)$ condition.

§ 8. Sufficient conditions for the H condition to be satisfied (continued).

Finally another simple result will be given.

Let the function $f(s)$ of the real variable s, defined in the interval $s_1 \leqq s \leqq s_2$, have a continuous n-th derivative $f^{(n)}(s)$ in this interval. Put

$$F(s_0, s) = \frac{f(s) - f(s_0)}{s - s_0}, \quad s_1 \leqq s, \; s_0 \leqq s_2; \tag{8.1}$$

then the function of the two variables s_0, s

$$\frac{\partial^{n-1} F(s_0, s)}{\partial s^k \partial s_0{}^l}, \quad (k + l = n - 1),$$

is continuous for $s_1 \leqq s$, $s_0 \leqq s_2$. If, further, $f^{(n)}(s)$ satisfies the $H(\mu)$ condition, then the preceding function satisfies the $H(\mu)$ condition for both variables.

In fact, it follows from

$$f(s) - f(s_0) = \int_{s_0}^{s} f'(\sigma) d\sigma = (s - s_0) \int_0^1 f'[s_0 + u(s - s_0)] du \tag{8.2}$$

that

$$\frac{\partial^{n-1} F(s_0, s)}{\partial s^k \partial s_0{}^l} = \int_0^1 u^k (1 - u)^l f^{(n)}[s_0 + u(s - s_0)] du. \tag{8.3}$$

This will be applied to a simple example.

Let $L = ab$ be a smooth arc. In addition, assume that the signed angle $\theta(t)$ between the tangent to L at t and any fixed direction satisfies the H condition. Then clearly the derivatives $x'(s)$, $y'(s)$ satisfy the H condition.

Let $t = t(s)$ and $t_0 = t(s_0)$ respectively be variable and fixed points on L and let $\vartheta(t_0, t) = \arg(t - t_0) + \text{const}$, as in 2° of § 7, denote the signed angle between the vector $\overrightarrow{t_0 t}$ and some fixed direction and assume that ϑ varies continuously with t, provided t does not pass through t_0.

It will be proved that under the given conditions ϑ satisfies the H condition with respect to t (clearly an interchange of the roles played by t and t_0 is again possible) on each of the arcs at_0, t_0b separately and that

$$\frac{d\vartheta}{ds} = \frac{K(t_0, t)}{|t - t_0|^{\lambda}}, \tag{8.4}$$

where $K(t_0, t)$ satisfies the H condition (with respect to t and t_0) on L and λ is some real number less than 1.

Without loss of generality, $L = ab$ may be supposed to be a standard arc. If the Ox axis is taken parallel to the tangent at some point on this arc, $x'(s) \neq 0$ for $s \neq s_0$, $x(s) - x(s_0) \neq 0$.

Put

$$X(s_0, s) = \frac{x(s) - x(s_0)}{s - s_0}, \quad Y(s_0, s) = \frac{y(s) - y(s_0)}{s - s_0}.$$

From what has been proved at the beginning of this section the functions $X(s_0, s)$ and $Y(s_0, s)$ are continuous on L and satisfy the H condition; in addition, $X(s_0, s) \neq 0$. By

$$\operatorname{artan} \frac{Y(s_0, s)}{X(s_0, s)}$$

will be understood a branch which varies continuously on L. Then

$$\vartheta(t_0, t) = \operatorname{artan} \frac{y(s) - y(s_0)}{x(s) - x(s_0)} + C = \operatorname{artan} \frac{Y(s_0, s)}{X(s_0, s)} + C,$$

where C remains constant, while t lies on either of the arcs at_0, t_0b and changes discontinuously by some odd multiple of π when t passes through t_0.

From the above formula $\vartheta(t_0, t)$ satisfies the H condition with respect to t, if t lies on one of the arcs at_0, t_0b. Further, assuming for convenience that the angles θ and ϑ are reckoned from the Ox axis,

$$\frac{d\vartheta}{ds} = \frac{y'(s)[x(s) - x(s_0)] - x'(s)[y(s) - y(s_0)]}{[x(s) - x(s_0)]^2 + [y(s) - y(s_0)]^2} =$$
$$= \frac{\sin\theta\cos\vartheta - \cos\theta\sin\vartheta}{r} = \frac{\sin(\theta - \vartheta)}{r}, \tag{8.5}$$

where $r = |t - t_0|$, and (8.4) follows easily from § 6 (5^0), since obviously $\sin(\theta - \vartheta)$ satisfies the H condition and is zero at $t = t_0$.

Next assume that L has the continously changing curvature

$$\frac{1}{\varrho(s)} = \frac{d\theta}{ds}.$$

Then, clearly, the second derivatives $x''(s)$ and $y''(s)$ are continuous and it follows from the above that

$$\frac{\partial X}{\partial s}, \ \frac{\partial Y}{\partial s}$$

are also continuous. Hence

$$\frac{d\vartheta}{ds}$$

will also be a continuous function of s (and of s_0). $\left(\text{As before, } \frac{d\vartheta}{ds}\right.$ is written for $\left.\frac{\partial\vartheta}{\partial s}.\right)$ Further, if the curvature satisfies the H condition, $\frac{d\vartheta}{ds}$ will also satisfy the H condition.

CHAPTER 2

INTEGRALS OF THE CAUCHY TYPE

§ 9. Definitions.

In what follows, L is assumed to be a smooth line (for a generalization of the results to the case in which L is piecewise smooth see Appendix 2), i.e., the union of a finite number of contours or arcs which do not intersect one another and are of finite length.

A certain direction on L will be defined as positive. If arcs form parts of L, the ends of these arcs will be called ends of the line L.

Around each point t_0, lying on L and not coinciding with an end, a circle may be drawn with radius so small that it is divided by L into two parts, lying respectively to the right and left of L, when viewed in the positive direction of L. Accordingly they will be regarded as the left and right neighbourhoods of the point t_0.

If L consists of contours and is the boundary of some connected region, the positive direction of L will be selected so that, if L is described in this direction, that region is always on the left or on the right. That part of the plane which lies on the left will always be denoted by S^+ and that to the right by S^-.

A part of L is always to be understood as a portion, consisting of a finite number of contours or arcs.

§ 10. The Cauchy integral.

Let $\varphi(t)$ be a function of the point t on L, bounded everywhere on L with the possible exception of a finite number of points c_k where, however,

$$|\varphi(t)| \leqq \frac{C}{|t-c|^{\alpha}}; \tag{10.1}$$

c stands for any of the points c_k, and C, α are positive constants, $\alpha < 1$.

In addition, let the function $\varphi(t)$ be integrable with respect to s (the arc coordinate of t) in the Riemann sense on any part of L (cf. the end of § 9) which does not include points c_k; from the condition (10.1) $\varphi(t)$ and $|\varphi(t)|$ will be integrable everywhere on L.

Consider the integral

$$\Phi(z) = \frac{1}{2\pi i}\int_L \frac{\varphi(t)\,dt}{t-z}, \tag{10.2}$$

where z is any point of the plane; this integral is called a Cauchy integral. (Cauchy integrals may naturally be defined for much larger classes of lines L and functions $\varphi(t)$. Cf. for example I. I. Privalov [1], [2], [3].) $\varphi(t)$ will be called the density function.

Below (§ 12) a definite meaning will be given to this integral in certain cases where the point z lies on L (but not at one of its ends); meanwhile it will be assumed that the point z does not lie on L.

It is clear that the function $\Phi(z)$ is holomorphic in the entire region excluding L and that for large $|z|$

$$\Phi(z) = O\left(\frac{1}{|z|}\right)$$

(since L is of finite length).

§ 11. Connection with logarithmic potential. Historical remarks.

Cauchy integrals are closely related to the potentials of simple and double layers, distributed along the line of integration L. To show this connection, $\varphi(t)$ will be considered as a real function, since the general case follows directly from this.

Assume that z does not lie on L and put

$$\Phi(z) = U(x, y) + iV(x, y) = \frac{1}{2\pi i}\int_L \frac{\varphi(t)dt}{t-z}. \tag{11.1}$$

Further, put

$$t - z = re^{i\vartheta}, \tag{11.2}$$

where $r = |t-z|$ and $\vartheta = \vartheta(z, t) = \arg(t-z)$. Taking the logarithmic derivative of (11.2) (for variable t and constant z),

$$\frac{dt}{t-z} = d\log r + id\vartheta = \frac{dr}{r} + id\vartheta. \tag{11.3}$$

Substituting this in (11.1) and separating real and imaginary parts leads to

$$U(x, y) = \frac{1}{2\pi}\int_L \varphi d\vartheta = \frac{1}{2\pi}\int_L \varphi\frac{d\vartheta}{ds}ds = -\frac{1}{2\pi}\int_L \varphi\frac{d\log r}{dn}ds =$$

$$= \frac{1}{2\pi}\int_L \varphi\frac{\cos(r, n)}{r}ds, \tag{11.4}$$

$$V(x, y) = -\frac{1}{2\pi}\int_L \varphi d\log r = -\frac{1}{2\pi}\int_L \varphi\frac{dr}{r}, \tag{11.5}$$

where s is the arc coordinate of the point t, n the normal at this point, directed to the left of L, and (r, n) is the angle between the vector $\overrightarrow{tz}$ and n (Fig. 1). The transformation in (11.4) used the relation

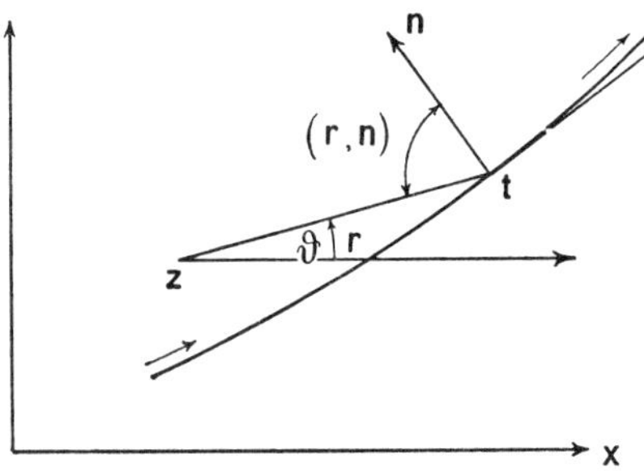

Fig. 1.

$$\frac{d\vartheta}{ds} = -\frac{d\log r}{dn} \tag{11.6}$$

which is one of the Cauchy Riemann equations for a system of axes consisting of the positive tangent T and the normal n (this system is right-handed as is the Oxy system) for the function $\log(t - z) = \log r + i\vartheta$ (for constant z and variable t).[1]

It follows from (11.4) that $U(x, y)$ represents the potential of a double layer with moment-density $\varphi/2\pi$.

After an integration by parts (assuming for simplicity that φ has an integrable derivative with respect to s; this assumption may be replaced by a more general one, e.g. that φ is only continuous, if use is made of the Stieltjes integral) (11.5) may be written in the form

$$V(x, y) = \frac{1}{2\pi}\int_L \frac{d\varphi}{ds}\log r\, ds + \frac{1}{2\pi}\Sigma \pm \varphi(c_k)\log r_k, \tag{11.7}$$

where the c_k are the ends of L (if it contains arcs) and the r_k are the distances between the points (x, y) and c_k. If L consists only of contours,

$$V(x, y) = \frac{1}{2\pi}\int_L \frac{d\varphi}{ds}\log r\, ds. \tag{11.7a}$$

The last formula shows that in the case of contours $V(x, y)$ represents the potential of a simple layer

---

[1]) In general, if $f(z) = u + iv$ is an analytic function, then the Cauchy-Riemann equations are: $\frac{\partial u}{\partial s} = \frac{\partial v}{\partial n}, \frac{\partial u}{\partial n} = -\frac{\partial v}{\partial s}$.

$$V(x, y) = \int_L \mu(s) \log \frac{1}{r}\, ds = -\int_L \mu(s) \log r\, ds \tag{11.8}$$

with the density

$$\mu(s) = -\frac{1}{2\pi} \frac{d\varphi}{ds}. \tag{11.9}$$

If L contains arcs, the potentials due to point masses, concentrated at the ends c_k, must be added to the potential (11.7a). However, it must be noted that, although on the one hand the function $V(x, y)$ defined by (11.5) is a generalization of the potential of a simple layer (for in that formula the function $\varphi(t)$ is not assumed differentiable), in the case of a differentiable $\varphi(t)$ it does not give the potential of a simple layer in its usual form. More specifically, the potential defined by (11.5) corresponds to the ordinary potential of a simple layer, if the masses

$$m_k = \int_{L_k} \mu(s)\, ds$$

distributed over the individual contours L_k which enter into the composition of L are zero; this is clear from (11.9). Hence, in future, an expression of the form (11.5) will be referred to as a modified potential of a simple layer.

From the above it is clear that the study of Cauchy integrals may lead to the consideration of logarithmic potentials of double and simple layers. A. Harnack's work [1] was essentially in this direction and it was one of the first important investigations devoted to Cauchy integrals (1885). However, the direct study of these integrals leads to more general results which are very important from the applied point of view. The first such investigation, which however did not attract due attention, came from G. Morera [1]. (1889). The next important results were those of J. Plemelj [1] published in a short note in 1908 which also did not receive much attention until later. Further very general and important results on Cauchy integrals are due to I. I. Privalov; they are set out in his books [2] and [3]; see also his textbook [1].

§ 12. The values of Cauchy integrals on the path of integration.

Returning to the direct investigation of Cauchy integrals the case will be considered in which in (10.2) the point z, now to be denoted by t_0, lies on L. Write purely formally

$$\Phi(t_0) = \frac{1}{2\pi i} \int_L \frac{\varphi(t)dt}{t - t_0}. \tag{12.1}$$

The integral on the right, in general, has no meaning from the ordinary point of view. However, for an extensive and important class of functions $\varphi(t)$, this integral may be given a definite sense, if the concept of the principal value of a Cauchy integral is introduced. This is as follows.

Let t_0 not coincide with either of the ends of L. Describe about t_0 as centre a circle with so small a radius ε that it intersects L in the two points t' and t'', and consider the integral

$$\frac{1}{2\pi i} \int_{L-l} \frac{\varphi(t)dt}{t - t_0}, \tag{12.2}$$

where l stands for the arc $t't''$. If for $\varepsilon \to 0$ the preceding integral tends to a definite limit, then this limit is called the principal value of the Cauchy integral. It is obvious that, if the integral (12.1) exists in the ordinary (i.e., Riemann) sense, then the principal value also exists (but not conversely). (The integral (12.1) exists in the ordinary sense, if the integral (12.2) tends to a definite limit whatever the arc l cut off around t_0 may be, as long as the length of this arc tends to zero; it is essential for the definition of the principal value that the ends t' and t'' of the arc l lie at equal distances from t_0. See below). Therefore the principal value of the integral will be denoted by the same symbol as is used for the ordinary integral, but if the integral has no meaning in the ordinary sense, its principal value is to be understood. (In contrast to this convention, many authors indicate the principal value of the integral by means of a special symbol, e.g. the integral sign is accented (′) or the letters VP (Valeur Principale) are put in front of it.)

No time will be spent on giving the most general conditions for the existence of the principal value, and only one important case will be treated. Namely, if the function $\varphi(t)$ satisfies the H condition in the neighbourhood of the point t_0, it will be proved that the principal value exists, and at the same time its value will be found in terms of an integral in the ordinary sense.

Naturally the case need only be considered in which L consists of a single smooth curve. First it will be assumed that the curve $L = ab$ is an arc.

Then

$$\int_{L-l} \frac{\varphi(t)dt}{t - t_0} = \int_{L-l} \frac{\varphi(t) - \varphi(t_0)}{t - t_0} dt + \varphi(t_0) \int_{L-l} \frac{dt}{t - t_0}. \tag{21.3}$$

The last integral becomes in finite form

$$\int_{L-l} \frac{dt}{t-t_0} = \Big[\log(t-t_0)\Big]_a^{t'} + \Big[\log(t-t_0)\Big]_{t''}^{b} =$$

$$= \log(b-t_0) - \log(a-t_0) - [\log(t''-t_0) - \log(t'-t_0)],$$

where $\log(t-t_0)$ on each of the arcs at_0 and t_0b is a branch which changes continuously on this arc. For definiteness, these branches will be connected by the following condition: the value $\log(t''-t_0)$ is obtained from the value $\log(t'-t_0)$ by means of a continuous change of $\log(t-t_0)$, while t varies on the arc of an infinitesimal circle, with centre at t_0, so that it passes the point t_0 on the left (with respect to L) (cf. Fig. 2).

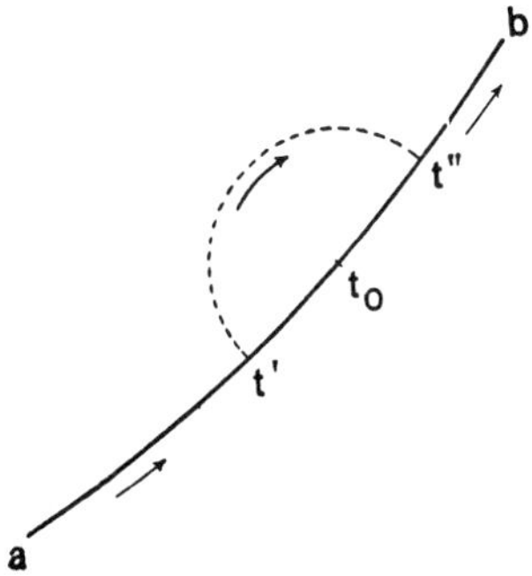

Fig. 2.

By the condition

$$|t''-t_0| = |t'-t_0|$$

one has obviously

$$\lim_{\varepsilon\to 0} [\log(t''-t_0) - \log(t'-t_0)] = -i\pi.$$

Further, since

$$|\varphi(t)-\varphi(t_0)| \leqq A\,|t-t_0|^{\mu},\ 0<\mu\leqq 1$$

near t_0, the limit of the first integral on the right side of (12.3) exists and is equal to the integral in the ordinary sense

$$\int_L \frac{\varphi(t)-\varphi(t_0)}{t-t_0}\,dt.$$

Therefore the principal value of the integral (12.1) also exists and is given by

$$\Phi(t_0) = \frac{1}{2\pi i}\int_L \frac{\varphi(t)dt}{t-t_0} = \tfrac{1}{2}\varphi(t_0) + \frac{\varphi(t_0)}{2\pi i}\log\frac{b-t_0}{a-t_0} + \frac{1}{2\pi i}\int_L \frac{\varphi(t)-\varphi(t_0)}{t-t_0}\,dt, \tag{12.4}$$

where

$$\log\frac{b-t_0}{a-t_0} = \log(b-t_0) - \log(a-t_0)$$

for the above values of the logarithms.

It is easy to pass from this to the case where L is a contour; namely, it is sufficient to assume that the point b coincides with the point a. Assuming for definiteness that the finite part of the plane, bounded by the contour L, lies to the left when L is described in the positive direction, it is seen that

$$\log(b - t_0) - \log(a - t_0) = 0,$$

adopting the previous convention for the meaning of $\log(t - t_0)$. Consequently in this case

$$\Phi(t_0) = \tfrac{1}{2}\varphi(t_0) + \frac{1}{2\pi i}\int_L \frac{\varphi(t) - \varphi(t_0)}{t - t_0}\, dt. \tag{12.5}$$

Yet another representation of the integral (12.1) when it exists in the above sense will be introduced; this will be used in some applications. Write (cf. last section)

$$t - t_0 = re^{i\vartheta},$$

where $r = r(t_0, t) = |t - t_0|$, $\vartheta = \vartheta(t_0, t) = \arg(t - t_0)$, and take the logarithmic derivative with respect to t leaving t_0 constant, then

$$\frac{dt}{t - t_0} = \frac{dr}{r} + i d\vartheta,$$

and hence

$$\Phi(t_0) = \frac{1}{2\pi i}\int_L \frac{\varphi(t)dt}{t - t_0} = \frac{1}{2\pi}\int_L \varphi(t)d\vartheta + \frac{1}{2\pi i}\int_L \varphi(t)\,\frac{dr}{r}. \tag{12.6}$$

Taking the arc coordinate as the integration variable and noting that

$$\frac{dr}{ds} = \cos\alpha(t_0, t), \quad \frac{d\vartheta}{ds} = \frac{-d\log r}{dn} = \frac{\cos(r, n)}{r(t_0, t)} = \frac{\sin\alpha\,(t_0, t)}{r(t_0, t)}, \tag{12.7}$$

where $\alpha(t_0, t)$ is the angle between the positive tangent T to L at the point t and the vector $\overrightarrow{t_0 t}$, measured from the latter in the positive direction, while (r, n) is the angle between the vector $\overrightarrow{t t_0}$ and the normal n at t directed to the left (Fig. 3), the required representation is

$$\Phi(t_0) = \frac{1}{2\pi i}\int_L \frac{\varphi(t)dt}{t - t_0} = \frac{1}{2\pi}\int_L \varphi(t)\frac{\sin\alpha(t_0, t)}{r(t_0, t)}ds + \frac{1}{2\pi i}\int_L \varphi(t)\frac{\cos\alpha(t_0, t)}{r(t_0, t)}ds. \tag{12.8}$$

[The second formula of (12.7) is identical with (8.5), deduced in a different way, for it is obvious that apart from multiples of 2π

$$\alpha(t_0, t) = \theta(t) - \vartheta(t_0, t),$$

where $\theta(t)$ is the angle between T and the Ox axis measured from the latter in the positive direction.]

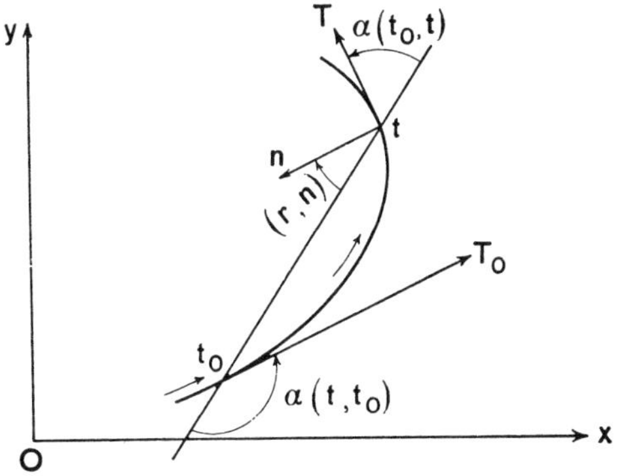

Fig. 3.

The first integral on the right side of (12.8) may be understood in the ordinary sense, if the angle made by T and Ox satisfies the H condition, for in this case, from § 8,

$$\frac{\sin \alpha(t_0, t)}{r(t_0, t)} = \frac{d\vartheta}{ds} = \frac{K(t_0, t)}{|t - t_0|^\lambda}, \ \lambda < 1,$$

where $K(t_0, t)$ is a continuous function (which moreover satisfies the H condition); but the second integral must be taken as Cauchy principal value.

NOTE 1. Consider any sufficiently small part t_1t_2 of the arc L which contains the point t_0. If the branch of the logarithm is taken as indicated above,

$$I = \int_{t_1t_2} \frac{dt}{t - t_0} = \log \frac{t_2 - t_0}{t_1 - t_0} + \pi i = \log \frac{|t_2 - t_0|}{|t_1 - t_0|} + \varepsilon i, \qquad (12.9)$$

where ε is the angle between the vectors $\overrightarrow{t_0t_2}$ and $\overrightarrow{t_1t_0}$ measured in the positive direction from the latter. (Fig. 4).

The expression (12.9) may also be written

$$I = \int_{t_1t_2} \frac{dt}{t - t_0} = \log (t_2 - t_0) - \\ - \log (t_0 - t_1) = \log \frac{t_2 - t_0}{t_0 - t_1}, \quad (12.9a)$$

if it is agreed to take

$$\log (t_0 - t_1) = \log (t_1 - t_0) - \pi i.$$

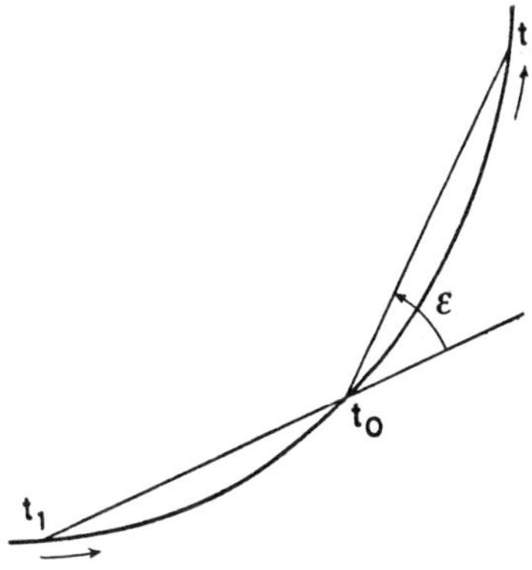

Fig. 4.

Now let $t_2 \to t_1$, then it is obvious that $\varepsilon \to 0$. Further, if $|t_2 - t_0|$ and $|t_1 - t_0|$ are infinitesimals of the same order, i.e.,

$$\frac{|t_2 - t_0|}{|t_1 - t_0|} \to 1, \tag{*}$$

then $I \to 0$. Clearly, if convergence to the limit in (*) takes place uniformly (with respect to t_0), $I \to 0$ uniformly.

NOTE 2. From the preceding Note and from the definition of the principal value of an integral it follows, in particular, that it is unnecessary for the definition of the principal value to assume that that part of the path of integration, which was denoted by $l = t't''$, satisfies strictly the condition $|t'' - t_0| = |t' - t_0|$; in fact, it is sufficient that

$$\frac{|t'' - t_0|}{|t' - t_0|} \to 1. \tag{**}$$

In particular, the points t' and t'' may be chosen in such a way that the two arcs $t't_0$ and t_0t'' are of equal length.

From the above it is clear that, if convergence to the limit in (**) is uniform (with respect to t_0), then

$$\frac{1}{2\pi i} \int_{L-l} \frac{\varphi(t)dt}{t - t_0}$$

will also converge uniformly for $|t' - t_0| \to 0$, $|t'' - t_0| \to 0$ to the principal value

$$\frac{1}{2\pi i} \int_{L} \frac{\varphi(t)dt}{t - t_0};$$

this will occur, for example, when $|t' - t_0| = |t'' - t_0|$ or when the arcs $t't_0$ and t_0t'' are of equal length.

§ 13. The tangential derivative of the potential of a simple layer.

Let $L = ab$ be a smooth arc with the additional property that the angle between its tangent at any point t and a fixed direction satisfies the H condition with respect to t or, what is the same thing, with respect to the arc coordinate s.

Further, let $\varphi(t)$ be a function defined on L and satisfying the H condition there. The potential of a simple layer

$$V(x, y) = \int_L \varphi(t) \log r \, ds, \tag{13.1}$$

where $r = |z - t|$ is the distance between the points $z = x + iy$ and $t(s)$, will now be discussed. The function $V(x, y)$ is known to be continuous everywhere in the finite part of the plane surrounded by L; its value $V(t_0)$ at the point t_0 on L is obtained by direct substitution of t_0 for z in (13.1) so that

$$V(t_0) = \int_L \varphi(t) \log r ds = \int_L \varphi(t) \log |t - t_0| ds. \qquad (13.2)$$

It will be shown that *under the assumed conditions the derivative dV/ds_0 exists for every point $t(s_0)$ which does not coincide with the ends a, b and that this derivative is given by*

$$\frac{dV}{ds_0} = \int_L \varphi(t) \frac{\cos \alpha (t, t_0)}{r} ds = \int_L \varphi(t) \frac{d \log r}{ds_0} ds, \qquad (13.3)$$

obtained by formal differentiation of (13.2); in this expression $\alpha(t, t_0)$ is the angle enclosed by the vector $\overrightarrow{tt_0}$ and the positive tangent T_0 at the point t_0 (see Fig. 3 of the preceding section).

In fact, let $a'b'$ be some fixed arc which is part of ab but has no end in common with it. It will be assumed that the point t_0 always lies on $a'b'$. Separate from L on either side of t_0 the equal arcs $t't_0$ and t_0t'' of sufficiently small length ε and denote by l the arc $t't''$ with centre at t_0. Put

$$V_\varepsilon (t_0) = \int_{L-l} \varphi(t) \log r ds = \int_{s_a}^{s_0-\varepsilon} \varphi(t) \log r ds + \int_{s_0+\varepsilon}^{s_b} \varphi(t) \log r ds, \qquad (13.4)$$

where s_a and s_b are the arc coordinates of the points a and b. Obviously

$$\lim_{\varepsilon \to 0} V_\varepsilon(t_0) = V(t_0).$$

Differentiate both sides of (13.4) with respect to s_0

$$\frac{dV_\varepsilon}{ds_0} = \int_{L-l} \varphi(t) \frac{d \log r}{ds_0} ds + \varphi(t') \log r' - \varphi(t'') \log r'', \qquad (13.5)$$

where $r' = |t' - t_0|$, $r'' = |t'' - t_0|$.

Since $\varphi(t)$ satisfies the H condition and

$$\varphi(t'') \log r'' - \varphi(t') \log r' = [\varphi(t'') - \varphi(t')] \log r'' + \varphi(t') \log \frac{r''}{r'},$$

it is easily seen that this difference converges uniformly to zero as $\varepsilon \to 0$.

Returning to the integral appearing on the right side of (13.5), it is possible to substitute

$$\frac{d \log r}{ds_0} = \frac{\cos \alpha\,(t, t_0)}{r}, \tag{13.6}$$

but, in order to give the integral the form studied in the preceding section, a somewhat different argument will at first be used. Since

$$r = (t - t_0)e^{-i\vartheta},$$

where ϑ is the argument of the difference $t - t_0$, one obtains, taking logarithms and differentiating with respect to s_0,

$$\frac{d \log r}{ds_0} = -\frac{1}{t - t_0}\frac{dt_0}{ds_0} - i\frac{d\vartheta}{ds_0} = -\frac{e^{i\theta(t_0)}}{t - t_0} - i\frac{d\vartheta}{ds_0},$$

where $\theta(t_0)$ is the angle between the Ox axis and the tangent to L at t_0.

In virtue of the assumed conditions (cf. § 8, example)

$$\left|\frac{d\vartheta}{ds_0}\right| < \frac{\text{const}}{r^\lambda}, \; \lambda < 1,$$

and therefore the integral

$$\int_{L-l} \varphi(t)\frac{d\vartheta}{ds_0}\,ds$$

converges uniformly, for $\varepsilon \to 0$, to

$$\int_L \varphi(t)\frac{d\vartheta}{ds_0}\,ds.$$

Further, from Note 2 at the end of § 12, the integral

$$\int_{L-l} \frac{\varphi(t)e^{i\theta(t_0)}\,ds}{t - t_0} = \int_{L-l} \frac{\varphi(t)e^{i[\theta(t_0)-\theta(t)]}\,dt}{t - t_0}$$

also tends uniformly, for $\varepsilon \to 0$, to

$$\int_L \frac{\varphi(t)e^{i[\theta(t_0)-\theta(t)]}\,dt}{t - t_0} = \int_L \frac{\varphi(t)e^{i\theta(t_0)}}{t - t_0}\,ds,$$

where now the principal value must be taken.

Thus it is seen that $\dfrac{dV_\varepsilon}{ds}$ converges uniformly to the limit

$$\lim_{\varepsilon \to 0} \frac{dV_\varepsilon}{ds_0} = \int_L \varphi(t) \frac{d \log r}{ds_0} ds.$$

Hence, by a well-known theorem of Analysis, it is concluded that the derivative dV/ds_0 exists and that

$$\frac{dV}{ds_0} = \int_L \varphi(t) \frac{d \log r}{ds_0} ds$$

from which the required result (13.3) follows by (13.6).

It is obvious that the result obtained remains valid, if L is the union of a finite number of arcs or contours with their tangents satisfying the above conditions.

The formula (13.3) was proved by G. Bertrand [2] for less general assumptions and by a more involved method; Bertrand's proof was reproduced by E. Picard [1]. The quite simple proof given here was communicated to me by A. V. Bitsadze.

§ 14. Sectionally continuous functions.

As before, let L be a smooth line, i.e., the union of a finite number of smooth arcs or contours which do not intersect one another; let the function $\Phi(z)$ of the point $z = x + iy$ be defined and continuous everywhere in the neighbourhood of the line L and perhaps also for points on the line L. Let t be a point on L which does not coincide with an end. It will be said that $\Phi(z)$ is continuous at the point t from the left (or from the right), if $\Phi(z)$ tends to a definite limit $\Phi^+(t)$ (or $\Phi^-(t)$) when z approaches t along any path, which remains, however, on the left (or on the right) of L. In other words, the range of values which z may take during the approach to t is only limited by the conditions $|z - t| \to 0$ and that z remains on the left (or on the right) of L. Then and only then it will be said that $\Phi(z)$ takes the boundary value $\Phi^+(t)$ on the left (or the boundary value $\Phi^-(t)$ on the right).

If the function $\Phi(z)$ is continuous from the left (right) at any point of some part L' of the line L, it will be said that $\Phi(z)$ is continuous from the left (right) on L'. In this case the function $\Phi^+(t)$ ($\Phi^-(t)$) is necessarily continuous on L'. In fact, for any given $\varepsilon > 0$, there is a $\delta > 0$, depending only on ε, such that

$$|\Phi(z) - \Phi^+(t)| < \varepsilon, \tag{*}$$

if $|z - t| < \delta$ and z lies towards the left from L. Now let t' be a second point on L' such that $|t' - t| < \delta$. If z, while remaining on the left of L and satisfying the condition $|z - t| < \delta$, tends to t',

$\Phi(z)$ will tend to $\Phi^+(t')$; but then it follows from (*) that

$$|\Phi^+(t') - \Phi^+(t)| \leqq \varepsilon,$$

which proves the statement [1]). (Similarly for $\Phi^-(t)$).

From the above statement it follows that, if $S^+(S^-)$ refers to the left (right) neighbourhood of the line L' and if $\Phi(z)$ is ascribed the value $\Phi^+(t)$ ($\Phi^-(t)$) on L', the function $\Phi(z)$ will be continuous in $S^+ + L'$ ($S^- + L'$).

The following corollary will often be used.

Let ab be a standard arc on L. Consider the family Π of parallel straight lines which make with the tangents to ab a non-obtuse angle, not less than some constant angle $\beta_0 > \alpha_0$; α_0 is here the acute angle corresponding to the standard arc (cf. § 2, 2°). Then, as is known (cf. § 2, 4°), each straight line Δ of the family Π, which lies between the two straight lines Δ_a, Δ_b of Π passing through a and b respectively, cuts ab in one and only one point. Assume now that $\Phi(z)$ tends uniformly to $\Phi^+(t)$ ($\Phi^-(t)$) as $z \to t$ along the straight line Δ of Π, z remaining on the left (right) of ab. Then the function $\Phi(z)$ is continuous from the left (right) on any part of the arc ab which does not include its ends.

In fact, from the uniformity of the approach to the limit it follows first of all that the function $\Phi^+(t)$ ($\Phi^-(t)$) is continuous on ab. Now let z tend to the point t (not an end point) of ab along some path which, say, stays on the left of ab. For sufficiently small $|z - t|$ the straight line of the family Π, which passes through z, cuts ab in some point t'. In this case $|z - t'|$ and $|t' - t|$ will be arbitrarily small. [This follows from examination of the triangle ztt'. Since the non-obtuse angle ω between the chord $t't$ and the segment $t'z$ is not less than some angle $\omega_0 > 0$ (§ 2. 5°), it is obvious that

$$|t' - t| \leqq \frac{|t - z|}{\sin\omega} \leqq \frac{|t - z|}{\sin\omega_0}, \quad |z - t'| \leqq \frac{|t - z|}{\sin\omega} \leqq \frac{|t - z|}{\sin\omega_0}.]$$

Consequently also the difference

$$\Phi(z) - \Phi^+(t) = [\Phi(z) - \Phi^+(t')] + [\Phi^+(t') - \Phi^+(t)]$$

will be arbitrarily small, which proves the statement.

Hitherto, when considering the limiting values, the ends of the line L were excluded. Now let t be one of the ends. It will be said that the function $\Phi(z)$ is continuous at the end t, if $\Phi(z)$ tends to a definite limit when z tends to t along any path which does not touch L. This limit will sometimes be denoted by $\Phi^{\pm}(t)$.

[1]) As can be seen from this proof, it is sufficient for the continuity of $\Phi^+(t)$ and $\Phi^-(t)$ to assume that $\Phi(z)$ is only continuous from the left (or from the right) at every point t of the portion L'. This result is due to P. Painlevé (cf. e.g. W. F. Osgood [1] p. 53).

Thus in the case of end points no distinction is made between the limits from the left and from the right. (Such a distinction can be introduced by extending L beyond the corresponding end.) However, it will sometimes be said of a function, continuous at a given end, that it is continuous at this end from the left and right.

A function $\Phi(z)$ which is continuous in some neighbourhood of the line L and which is continuous on L from the left and right, including the ends, will be said to be sectionally continuous in the neighbourhood of L (including its ends). If the continuity does not extend to some end, a corresponding reservation will be made.

The line L will be called the line of discontinuity when sectionally continuous functions are being considered.

NOTE. In future, when using the symbols $\Phi^+(t)$, $\Phi^-(t)$ for the limits from the left or right, it will always be assumed that these limits are attained along any paths lying to the left or to the right of L respectively.

§ 15. Sectionally holomorphic functions.

Let L have the same significance as in § 14 and $\Phi(z)$ be a function holomorphic in each finite region not containing points of the line L. Further, let the function $\Phi(z)$ be continuous on L from the left and from the right, with the possible exception of the ends, but near the ends satisfy the condition

$$|\Phi(z)| \leqq \frac{C}{|z-c|^\alpha}, \tag{15.1}$$

where c is the corresponding end, C and α are certain real constants, and $\alpha < 1$ (which is essential).

Such functions $\Phi(z)$ will be called sectionally holomorphic functions with the line of discontinuity L; the line L will sometimes be called the boundary.

If in the expansion of $\Phi(z)$ in the neighbourhood of the point at infinity

$$\Phi(z) = \sum_{j=-\infty}^{+\infty} a_j z^j \tag{15.2}$$

there are only a finite number of terms with positive powers of z, $\Phi(z)$ will be said to be of finite degree at infinity.

If a_k is the last coefficient in the expansion (15.2), which is different from zero, (excluding now the case when all $a_j = 0$, i.e., when $\Phi(z) \equiv 0$ in some region which includes the point $z = \infty$), then the degree of $\Phi(z)$ at infinity will be said to be equal to k. For $k > 0$, the point $z = \infty$ is a pole of order k of the function $\Phi(z)$, while for $k < 0$ it is a zero (or root) of order (or multiplicity) $-k$. For $k = 0$, i.e., when $\Phi(\infty) = a_0$ has a definite limit different from zero, $\Phi(z)$ may conveniently be said to have a pole or zero of zero-order at $z = \infty$. Finally, for $k \leq 0$, $\Phi(z)$ will be said to be sectionally holomorphic, including the point at infinity.

A well-known property of analytic functions, frequently used, will now be recalled.

Let S_1 and S_2 be two regions of the plane, bounded by smooth contours, having no interior points in common, but meeting along some arc L, a common part of their boundaries; the ends of L are not to be included in L. Further, let $\Phi_1(z)$ and $\Phi_2(z)$ be functions, holomorphic in S_1 and S_2 and continuous on L from S_1 and S_2 respectively, and let their boundary values along L be equal, i.e.,

$$\Phi_1(t) = \Phi_2(t), \tag{15.3}$$

where t denotes a point on L and $\Phi_1(t)$, $\Phi_2(t)$ are the boundary values of the functions $\Phi_1(z)$ and $\Phi_2(z)$. Then *the function, defined in the following manner,*

$$\Phi(z) = \Phi_1(z) \text{ for } z \in S_1, \quad \Phi(z) = \Phi_2(z) \text{ for } z \in S_2,$$
$$\Phi(t) = \Phi_1(t) = \Phi_2(t) \text{ for } t \in L, \tag{15.4}$$

is holomorphic in the region $S_1 + S_2 + L$.

For the proof it is obviously sufficient to show that the function defined above is holomorphic in the neighbourhood of any point of L not coinciding with its ends. Describe, with t_0 as centre, a circle γ with radius so small that it cuts L in exactly two points a and b. Let σ be the circular region bounded by γ and σ_1, σ_2 those parts of this region which lie respectively in S_1 and S_2. Also let γ_1, γ_2 be the boundaries of these portions described in the positive directions; γ_1 and γ_2 have the part ab in common which is described in opposite directions.

It is easily verified directly from Cauchy's theorem that

$$\Phi(z) = \frac{1}{2\pi i} \int_{\gamma_1} \frac{\Phi_1(t)dt}{t - z} + \frac{1}{2\pi i} \int_{\gamma_2} \frac{\Phi_2(t)dt}{t - z}$$

for all points z inside either of the regions σ_1, σ_2, since the first integral equals $\Phi_1(z)$ for $z \in \sigma_1$ and is zero for $z \in \sigma_2$, while the second equals $\Phi_2(z)$ for $z \in \sigma_2$ and is zero for $z \in \sigma_1$. But the sum of the two integrals reduces to a single integral taken along γ, since the parts, corresponding to the arc ab, cancel because of the condition (15.3) holding on L. Therefore the above formula may be written

$$\Phi(z) = \frac{1}{2\pi i}\int_{\gamma}\frac{\Phi(t)dt}{t-z}$$

after which the proposition becomes obvious, since clearly the right hand side of this formula represents a function, holomorphic inside γ.

To this the following remark may be added, if the above conditions are retained and if the functions $\Phi_1(z)$ and $\Phi_2(z)$ are assumed continuous on L from S_1 and from S_2 respectively, except possibly at a finite number of points $c_1, c_2, \ldots c_m$ on L near which, however,

$$|\Phi_1(z)| \leqq \frac{\text{const}}{|z-c_j|^\alpha},\ |\Phi_2(z)| \leqq \frac{\text{const}}{|z-c_j|^\alpha},\ \alpha < 1; \quad (15.5)$$

then *the function* $\Phi(z)$ *defined by the formula* (15.4) *will be holomorphic in* $S_1 + S_2 + L$, *if it is assigned suitable values at the points* $c_1, c_2, \ldots c_m$. In fact, it follows from the preceding argument that the function $\Phi(z)$ will be holomorphic in $S_1 + S_2 + L$ with the possible exception of the points c_j.But by (15.5) these points will clearly no longer be singular points and the statement is proved. [See for example I. I. Privalov [1] p. 200. The expansion of the function $\Phi(z)$ in the neighbourhood of the c_j in a Laurent series obviously cannot contain negative powers by (15.5).]

The following extension follows directly from what has been said. If $\Phi(z)$ is a sectionally holomorphic function with a line of discontinuity L and if on any part L' of L, with the possible exception of the end-points of L', $\Phi^+(t) = \Phi^-(t)$, then this part of L may be removed, i.e., the function $\Phi(z)$ will have the line of discontinuity $L—L'$.

Hitherto it was assumed that L lies entirely in a finite region of the plane. Unless the contrary is specifically stated, this assumption will hold in what follows. The definition of a sectionally holomorphic function is easily extended to the case when L contains parts extending to infinity, since in this case the definition remains the same as before with the only exception that the behaviour of $\Phi(z)$ near the point $t = \infty$ cannot be restricted (excluding special cases).

§ 16. The limiting value of a Cauchy integral.

The most important part of the problem under consideration is the investigation of the behaviour of the Cauchy integral

$$\Phi(z) = \frac{1}{2\pi i}\int_{L}\frac{\varphi(t)\,dt}{t-z} \quad (16.1)$$

near the line of integration L. As before, L is supposed to be a smooth line, i.e., the union of a finite number of non-intersecting smooth arcs or contours.

The following remark will often assist in simplifying the discussion.

Let it be required to investigate the behaviour of the integral $\Phi(z)$ near some part L_0 of L. If the integral $\Phi(z)$ is taken as the sum of two integrals, $\Phi_1(z)$ and $\Phi_2(z)$, one of which refers to that part L_1 of L which includes the portion L_0, while the other corresponds to the remaining part L_2 of L, and if this latter portion L_2 lies at a finite distance from L_0 to be investigated, then $\Phi_2(z)$ will be holomorphic in the neighbourhood of L_0, including L_0. Thus the problem is reduced to the study of $\Phi_1(z)$.

The principal result, to be proved in this section, may be formulated thus:

THEOREM. *If the density $\varphi(t)$ satisfies the H condition on L, then the function $\Phi(z)$ is continuous on L from the left and from the right, with the exception of those ends at which $\varphi(t) \neq 0$.*

Commencing with the case in which L consists of a single arc or contour, the behaviour of the integral

$$\Psi(z) = \frac{1}{2\pi i} \int_L \frac{\varphi(t) - \varphi(t_0)}{t - z}\, dt \tag{16.2}$$

when $z \to t_0$, where t_0 is an arbitrary point on L, will be investigated (the case in which t_0 coincides with one of the ends of L is now not excluded). The following Lemma will be proved:

LEMMA: *Let β_0 be an arbitrary non-obtuse angle (i.e., $0 < \beta_0 \leq \pi/2$) and let z approach t_0 in such a way that the non-obtuse angle β between the segment t_0z and the tangent to L at the point t_0 is not less than β_0. Then $\Psi(z)$ tends uniformly (with respect to the position t_0 on L) to the limit*

$$\Psi(t_0) = \frac{1}{2\pi i} \int_L \frac{\varphi(t) - \varphi(t_0)}{t - t_0}\, dt \tag{16.3}$$

(independently of whether z approaches t_0 from the left or from the right of the tangent.)

It is obviously sufficient to prove the lemma for the integral

$$\psi(z) = \frac{1}{2\pi i} \int_l \frac{\varphi(t) - \varphi(t_0)}{t - z}\, dt,$$

where l is a definite standard arc containing t_0 and corresponding to a standard radius $R(\alpha_0)$, $0 < \alpha_0 < \beta_0$ (cf. § 2).

Consider the difference

$$\psi(z) - \psi(t_0) = \frac{h}{2\pi i} \int_l \frac{\varphi(t) - \varphi(t_0)}{(t - t_0)(t - z)} \, dt,$$

where $h = z - t_0$. With t_0 as centre describe a circle γ with radius ϱ (for sufficiently small ϱ this circle will intersect l in one or two points); denote by $t_1 t_2$ that part of l which lies inside γ and by $l - t_1 t_2$ the remaining part. Then

$$\psi(z) - \psi(t_0) = I_1 + I_2,$$

where

$$I_1 = \frac{h}{2\pi i} \int_{t_1 t_2} \frac{\varphi(t) - \varphi(t_0)}{(t - t_0)(t - z)} \, dt, \quad I_2 = \frac{h}{2\pi i} \int_{l - t_1 t_2} \frac{\varphi(t) - \varphi(t_0)}{(t - t_0)(t - z)} \, dt.$$

I_1 will be considered first. Using the bounds for $\varphi(t)$ on $t_1 t_2$ given by the H condition $|\varphi(t) - \varphi(t_0)| \leqq A |t - t_0|^\mu$, [note that the H condition is only applied for any small part of l which includes t_0] and introducing the notation $r = |t - t_0|$, one obtains, since by (2.2) $|dt| = |ds| \leqq K |dr|$,

$$|I_1| \leqq \frac{\delta . A . K}{2\pi} \int_{t_1 t_2} \frac{r^{\mu - 1}}{|t - z|} |dr|,$$

where $\delta = |h|$. But if ω is the non-obtuse angle between the segments $t_0 z$ and $t_0 t$ (Fig. 5), then obviously

$$|t - z| \geqq \delta \sin \omega \geqq \delta \sin \omega_0,$$

where ω_0 is some constant such that $0 < \omega_0 \leqq \frac{\pi}{2}$ (cf. § 2, 5°). Therefore

$$|I_1| \leqq \frac{AK}{2\pi \sin \omega_0} \int_{t_1 t_2} r^{\mu - 1} |dr| \leqq \frac{AK}{\pi \sin \omega_0} \int_0^\varrho r^{\mu - 1} \, dr = \frac{AK\varrho^\mu}{\pi\mu \sin \omega_0}.$$

Take ϱ so small that $|I_1| < \frac{\varepsilon}{2}$; the choice of ϱ may obviously be made independently of the position of t_0 on l and of the position of z. Further, take $\delta \leqq \frac{\varrho}{2}$. Then for t on $l - t_1 t_2$, i.e., outside the circle γ, $|t - t_0| \geqq \varrho$, $|t - z| \geqq \varrho/2$, and therefore

$$|I_2| \leqq \frac{\delta}{\pi \varrho^2} \int_{l - t_1 t_2} |\varphi(t) - \varphi(t_0)| \, ds \leqq \frac{\delta . M}{\pi \varrho^2},$$

where M is a constant which depends neither on the position of t_0 on L nor on the position of z. Thus, for sufficiently small δ, $|I_2| < \varepsilon/2$, and the lemma is proved.

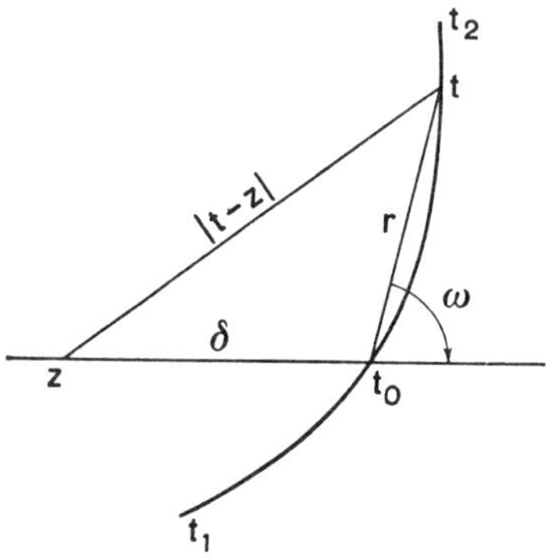

Fig. 5.

Before going further the following obvious remark may be added. The above remarks on the convergence of $\Psi(z)$ to the limit also remain true in the case, where $\varphi(t)$ does not satisfy the H condition on the entire arc L, but only over the part L', while it is integrable over the remaining portion, if the investigation is restricted to points t_0 of L' lying at finite distances from its ends.

The proof of the theorem formulated at the beginning of this section will now be given and the limiting values of the function $\Phi(z)$ found.

First consider the case when L is a contour and assume that the positive direction of L is selected in such a manner that the finite part of the plane, bounded by L, remains to the left for motion in this direction. Conforming to the convention in § 9, denote by S^+ that part of the plane lying to the left of L and by S^- the remainder, not including L in either S^+ or S^-.

Let z be any point of the plane not lying on L and t_0 some point on L. Then

$$\Phi(z) = \frac{1}{2\pi i}\int\limits_L \frac{\varphi(t)-\varphi(t_0)}{t-z}\,dt + \frac{\varphi(t_0)}{2\pi i}\int\limits_L \frac{dt}{t-z},$$

whence

$$\Phi(z) = \Psi(z) + \varphi(t_0) \quad \text{for } z \in S^+,$$
$$\Phi(z) = \Psi(z) \qquad\qquad \text{for } z \in S^-.$$

Now let z first tend to t_0, so that the non-obtuse angle between t_0z and the tangent to L at t_0 remains larger than some positive constant. Then, by the lemma proved above, $\Phi(z)$ tends uniformly to the limits

$$\begin{aligned}\Phi^+(t_0) &= \varphi(t_0) + \frac{1}{2\pi i}\int\limits_L \frac{\varphi(t)-\varphi(t_0)}{t-t_0}\,dt,\\ \Phi^-(t_0) &= \frac{1}{2\pi i}\int\limits_L \frac{\varphi(t)-\varphi(t_0)}{t-t_0}\,dt,\end{aligned} \tag{16.4}$$

depending on whether z during its approach to t_0 remains to the left or right of L.

But from the uniformity of the convergence to the limit it follows (see § 14) that $\Phi(z)$ is continuous on L from the left and right. Therefore, in particular, $\Phi(z)$ tends to the limits (16.4) when $z \to t_0$ on any path entirely on the left or on the right of L.

Now assume that L is an arc ab. In this case too the investigation is easily made directly, but the result may be obtained from the preceding one by the following simple method.

Extend $L = ab$ into the smooth contour L^* (so that the condition assumed above with regard to the positive direction is observed) and put $\varphi(t) = 0$ on the extension. The points a and b will, in general, be points of discontinuity of the function $\varphi(t)$ on L, but, if $\varphi(a) = 0$, the function $\varphi(t)$ will be continuous on L^* near a and will also satisfy the H condition; similarly for the case $\varphi(b) = 0$.

From what has been said above it is obvious that $\Phi(z)$ will be continuous from the left and from the right on L with the exclusion of those ends at which $\varphi(t) \neq 0$ and that it will take the limiting values given by the same formulae (16.4), if L is replaced in them by L^* and t_0 is instead any point on L including those ends at which $\varphi(t) = 0$.

From the above follows directly the theorem stated at the beginning of this section.

It is easy to show that this theorem also remains true in the case where L is a piecewise smooth line; the formulae (16.4) will hold as well in the case where L is a piecewise smooth contour (cf. Appendix 2). The formulae (16.4) were given by G. Morera [1] who did not state sufficiently clearly the conditions under which they are valid; he also did not give a strict proof. Morera assumed that z tends to t_0 at a finite angle with the tangent (or tangents in the case of a corner) and excluded cusps.

Next the behaviour of $\Phi(z)$ will be determined in the neighbourhood of the ends of L. Clearly it is sufficient to consider the case in which L consists of a single arc ab. Then one may write

$$\Phi(z) = \frac{1}{2\pi i}\int_L \frac{\varphi(t)-\varphi(a)}{t-z}\,dt + \frac{\varphi(a)}{2\pi i}\int_L \frac{dt}{t-z} = \frac{\varphi(a)}{2\pi i}\log\frac{z-b}{z-a} + \Phi_0(z), \tag{16.5}$$

where

$$\Phi_0(z) = \frac{1}{2\pi i}\int_L \frac{\varphi_0(t)\,dt}{t-z}, \qquad \varphi_0(t) = \varphi(t) - \varphi(a) \tag{16.6}$$

and where by $\log\dfrac{z-b}{z-a} = \log(z-b) - \log(z-a)$

is to be understood a definite branch, single-valued in the plane cut along $L = ab$. In particular, by $\log(z - a)$ near the point a one may understand an arbitrary branch, holomorphic in this neighbourhood (in the cut plane); then the function $\log(z - b)$ also has a definite meaning in the neighbourhood of the point a and it will already be holomorphic in this neighbourhood in the uncut plane. Since, further, $\varphi_0(a) = 0$, $\Phi_0(z)$ will be continuous on L from the left and from the right near and at a. Hence, finally,

$$\Phi(z) = \frac{\varphi(a)}{2\pi i} \log \frac{1}{z - a} + \Phi^*(z), \tag{16.7}$$

near a, where the function $\Phi^*(z)$ is holomorphic near a in the cut plane and continuous on L from the left and from the right near and at a.

Similarly near the end b

$$\Phi(z) = -\frac{\varphi(b)}{2\pi i} \log \frac{1}{z - b} + \Phi^{**}(z), \tag{16.8}$$

where $\Phi^{**}(z)$ is holomorphic near b in the plane cut along L and continuous on L from the left and from the right near and at b.

Thus, under the conditions assumed for the density $\varphi(t)$, the Cauchy integral (16.1) is seen to be a sectionally holomorphic function, vanishing at infinity, with the line of discontinuity L.

§ 17. The Plemelj formulae.

The formulae (16.4), giving the limiting value of a Cauchy integral, are inconvenient in that they are directly applicable only in the case in which L is a contour. If the principal value of the Cauchy integral is considered, these formulae may be put into a form applicable to any smooth line.

Since, if L is a contour,

$$\frac{1}{2\pi i} \int_L \frac{dt}{t - t_0} = \frac{1}{2}, \tag{17.1}$$

where it is assumed that the positive direction on L is selected as indicated in the preceding section, one has

$$\begin{aligned} \Phi^+(t_0) &= \frac{1}{2}\varphi(t_0) + \frac{1}{2\pi i} \int_L \frac{\varphi(t)dt}{t - t_0}, \\ \Phi^-(t_0) &= -\frac{1}{2}\varphi(t_0) + \frac{1}{2\pi i} \int_L \frac{\varphi(t)dt}{t - t_0}. \end{aligned} \tag{17.2}$$

Now it is easy to see that *these formulae are also valid for the case when L is an arbitrary smooth line*, provided $\varphi(t)$ satisfies the H condition on L and t_0 does not coincide with those end points at which $\varphi(t_0) \neq 0$. *$\Phi^+(t_0)$ and $\Phi^-(t_0)$ are the limiting values from the left and from the right of L respectively*; if t_0 coincides with an end, where $\varphi(t_0) = 0$, then $\Phi^+(t_0) = \Phi^-(t_0) = \Phi(t_0)$.

The proof of the preceding formulae is obvious in the case where t_0 lies on any contour which is part of L. [For the proof of (17.2) it is now unnecessary that the finite part of the plane, bounded by L, lies to the left of L; as always, one only has to understand by Φ^+ the limit from the left, and by Φ^- the limit from the right. One is immediately convinced of this, if in (17.2) the direction of the line of integration is reversed and the symbols $\Phi^+(t_0)$ and $\Phi^-(t_0)$ are interchanged correspondingly.]

In the case in which t_0 lies on an arc, it may be closed, as was done in the preceding section, and $\varphi(t_0)$ put equal to zero on the added part; in the final expressions (17.2) the integrals over this portion vanish, since $\varphi(t) = 0$ there.

The formulae (17.2) were given by I. Plemelj [1] in 1908. The same formulae, but under more general assumptions, were later obtained by I. I. Privalov (cf. his works [2] and [3]).

In principle, the proof given above is not different from that of Plemelj; but some simplifications have been introduced.

Two formulae which are equivalent to (17.2) and which will be often used should be noted, namely,

$$\Phi^+(t_0) - \Phi^-(t_0) = \varphi(t_0), \tag{17.3}$$

$$\Phi^+(t_0) + \Phi^-(t_0) = \frac{1}{\pi i}\int_L \frac{\varphi(t)dt}{t - t_0}. \tag{17.4}$$

§ 18. Generalization of the formulae for the difference in limiting values.

The expression

$$\Phi^+(t_0) - \Phi^-(t_0) = \varphi(t_0) \tag{18.1}$$

was obtained under the assumption that $\varphi(t)$ satisfies the H condition at least in the neighbourhood of t_0. However, this formula may also be given a definite meaning in the case where $\varphi(t)$ is only continuous.

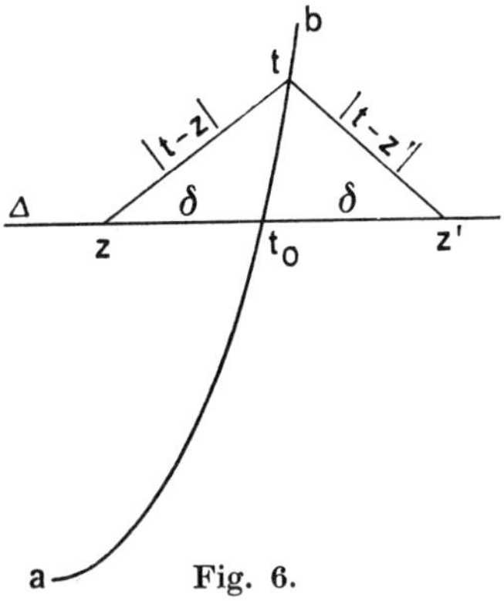

Fig. 6.

Take on the straight line Δ (Fig. 6), passing through t_0 and not tangential to L at t_0, two points z and z' at equal distances from t_0 to the left and right of L respectively, and define $\Phi^+(t_0) - \Phi^-(t_0)$ as

$$\lim\, [\Phi(z) - \Phi(z')] \text{ for } z, z' \to t_0. \tag{18.2}$$

It will be shown that, *if the function* $\varphi(t)$ *is continuous in the neighbourhood of* t_0, *the above limit exists and equals* $\varphi(t_0)$. *Further, this limit is attained uniformly on any part of* L *where* $\varphi(t)$ *is continuous (with the possible exception of the neighbourhoods of the ends of this portion), if the non-obtuse angle* β *between the straight line* Δ *and the tangent at* t_0 *is not less than some arbitrary non-obtuse angle* $\beta_0 > 0$.

Let $z = t_0 + h$, $z' = t_0 - h$.

Then

$$\Phi(z) - \Phi(z') = \frac{1}{2\pi i}\int_L \varphi(t)\left\{\frac{1}{t - t_0 - h} - \frac{1}{t - t_0 + h}\right\} dt =$$

$$= \frac{h}{\pi i}\int_L \frac{\varphi(t)dt}{(t - t_0)^2 - h^2}$$

or

$$\Phi(z) - \Phi(z') = \frac{h}{\pi i}\int_L \frac{\varphi(t) - \varphi(t_0)}{(t - t_0)^2 - h^2}\,dt + \frac{h\varphi(t_0)}{\pi i}\int_L \frac{dt}{(t - t_0)^2 - h^2}. \tag{*}$$

The last integral may be evaluated in an elementary manner, but to simplify its expression still more assume the line L to be a contour, which of course does not affect the generality. Since the positive direction on L is chosen so that the finite part of the plane bounded by L lies to the left and the point $t_0 + h = z$ lies in this part while the point $z' = t_0 - h$ is outside it, on applying the theory of residues, it is immediately seen that the second term in (*) equals $\varphi(t_0)$, and hence

$$\Phi(z) - \Phi(z') = \varphi(t_0) + I,$$

where

$$I = \frac{h}{\pi i}\int_L \frac{\varphi(t) - \varphi(t_0)}{(t - t_0)^2 - h^2}\,dt.$$

It remains to show that $I \to 0$ when $h \to 0$. With t_0 as centre describe a circle γ of sufficiently small radius ϱ intersecting L in two points a and b; ϱ will be assumed so small that $|\varphi(t) - \varphi(t_0)| < \eta$ for all points t on ab, where η is a given positive quantity. Also put $I = I_1 + I_2$, where I_1 is taken over ab and I_2 over $L - ab$.

Hence

$$|I_1| \leqq \frac{\delta \,.\, \eta}{\pi}\int_{ab} \frac{ds}{|t - z|\,.\,|t - z'|}, \quad \text{where } \delta = |h|.$$

By (2.2), putting $|t - t_0| = r$, one finds $ds \leqq K\,|dr|$, if $\varrho \leqq R_0(\alpha_0)$, where α_0 is an arbitrary positive quantity less than β_0 and $R_0(\alpha_0)$ is the radius of the corresponding standard circle.

Further, if ϑ is the angle between the vectors $\overrightarrow{t_0 z}$ and $\overrightarrow{t_0 t}$ and $\omega = \vartheta$ for $\vartheta \leqq \pi/2$, $\omega = \pi - \vartheta$ for $\vartheta > \pi/2$, then

$$|t - z|^2 = r^2 + \delta^2 - 2r\delta\cos\vartheta \geqq r^2 + \delta^2 - 2r\delta\cos\omega \geqq$$
$$\geqq r^2 + \delta^2 - - 2r\delta\cos\omega_0,$$

where $\omega_0 = \beta_0 - \alpha_0$ is a fixed acute angle (§ 2, 5°); similarly

$$|t - z'|^2 \geqq r^2 + \delta^2 - 2r\delta\cos\omega_0.$$

Consequently

$$|t - z|\,.\,|t - z'| \geqq r^2 + \delta^2 - 2r\delta\cos\omega_0 =$$
$$= (r - \delta\cos\omega_0)^2 + \delta^2\sin^2\omega_0,$$

$$|I_1| \leqq \frac{2K\delta\eta}{\pi}\int_0^{\varrho}\frac{dr}{(r - \delta\cos\omega_0)^2 + \delta^2\sin^2\omega_0} =$$

$$= \frac{2K\eta}{\pi\sin\omega_0}\left[\operatorname{artan}\left(\frac{r}{\delta\sin\omega_0} - \cot\omega_0\right)\right]_{r=0}^{r=\varrho} \leqq \frac{2K\eta}{\sin\omega_0},$$

and therefore, for sufficiently small ϱ, $|I_1| < \varepsilon/2$, where ε is an arbitrarily small positive quantity.

Further, it is obvious that, by choosing δ sufficiently small (for a fixed ϱ), $|I_2| < \varepsilon/2$, and the statement follows.

One important consequence of the proposition proved, which is almost obvious from what has been said in § 14, should be noted:

Let the function $\varphi(t)$ be continuous on some portion L' of L (L' may coincide with L) and let $\Phi(z)$ be continuous on L' from the left (right); then the function $\Phi(z)$ is continuous on L' from the right (left) as well, with the possible exception of the ends of L'.

The results, stated in this section, are also due to Plemelj [1]. The proof given here follows in detail his treatment (compare also Picard [1]).

§ 19. The continuity behaviour of the limiting values.

As before let L be smooth and $\varphi(t)$ satisfy the H condition on L. It has been seen that under these conditions the function

$$\Phi(z) = \frac{1}{2\pi i}\int_L \frac{\varphi(t)}{t - z}\,dt \tag{19.1}$$

is continuous on L from the left and from the right, with the exception of those ends at which $\varphi(t) \neq 0$. Therefore the limiting values $\Phi^+(t)$, $\Phi^-(t)$ are continuous functions of t everywhere on L, except at the above-mentioned ends. However, still more may be asserted, namely the following theorem (Plemelj—Privalov).

THEOREM: *If $\varphi(t)$ satisfies the $H(\mu)$ condition on L, then also $\Phi^+(t)$, $\Phi^-(t)$ satisfy everywhere, except in arbitrarily small neighbourhoods of those ends where $\varphi(t) \neq 0$, the $H(\mu)$ condition for $\mu < 1$ and the $H(1 - \epsilon)$ condition for $\mu = 1$, where ε is an arbitrarily small positive number.*

The following remark simplifies the discussion. If at any end c $\varphi(c) = 0$, then, by extending L beyond c, e.g. by a segment of the tangent, and by putting on the added part $\varphi(t) = 0$, one arrives at the case in which c is no longer the end of the path of integration. Therefore it is sufficient in proving the theorem to consider only points lying at finite distances from the ends. Clearly only the case need be considered in which L consists of a single smooth arc ab.

Let $L' = a'b'$ be any part of $L = ab$ the ends of which lie at finite distances from the ends a, b of L. It has to be shown that for any pair of points t_0, t_1 of L'

$$\begin{aligned} | \Phi^+(t_1) - \Phi^+(t_0) | &\leqq C \, | t_1 - t_0 |^\nu, \\ | \Phi^-(t_1) - \Phi^-(t_0) | &\leqq C \, | t_1 - t_0 |^\nu, \end{aligned} \tag{19.2}$$

where C is some constant and $\nu = \mu$, if $\mu < 1$, and $\nu = 1 - \varepsilon$, if $\mu = 1$, ε being an arbitrarily small positive quantity.

By (17.2)

$$\Phi^\pm(t_0) = \pm \frac{1}{2}\varphi(t_0) + \frac{1}{2\pi i} \int_L \frac{\varphi(t)dt}{t - t_0} =$$

$$= \pm \frac{1}{2}\varphi(t_0) + \frac{1}{2\pi i} \int_L \frac{\varphi(t) - \varphi(t_0)}{t - t_0} dt + \frac{\varphi(t_0)}{2\pi i} \int_L \frac{dt}{t - t_0},$$

where the upper or lower sign is to be taken throughout, or, noting that for a definite choice of a branch of the logarithm [see (12.9a)],

$$\int_L \frac{dt}{t - t_0} = \log \frac{b - t_0}{t_0 - a},$$

$$\Phi^\pm(t_0) = \pm \frac{1}{2}\varphi(t_0) + \frac{\varphi(t_0)}{2\pi i} \log \frac{b - t_0}{t_0 - a} + \Psi(t_0), \tag{19.3}$$

where

$$\Psi(t_0) = \frac{1}{2\pi i} \int_L \frac{\varphi(t) - \varphi(t_0)}{t - t_0} dt. \tag{19.4}$$

As the first two terms in (19.3) obviously satisfy the $H(\mu)$ condition on L', there remains to show that for any pair of points t_0, t_1 on L'

$$|\Psi(t_1)-\Psi(t_0)| \leqq C\,|t_1-t_0|^{\nu}, \tag{19.5}$$

where C is some constant and ν is defined above.

As usual let s, s_0, s_1 denote the arc coordinates corresponding to the points t, t_0, t_1 respectively and write

$$t_1-t_0=h, \quad s_1-s_0=\sigma;$$

it may obviously be assumed that $\sigma > 0$ and that 2σ is not greater than the distances of the points a', b' from the points a, b.

One has

$$\Psi(t_1)-\Psi(t_0)=\Psi(t_0+h)-\Psi(t_0)=$$
$$=\frac{1}{2\pi i}\int_L\left\{\frac{\varphi(t)-\varphi(t_0+h)}{t-t_0-h}-\frac{\varphi(t)-\varphi(t_0)}{t-t_0}\right\}dt. \tag{19.6}$$

Take on either side of t_0 on L arcs $t't_0$, t_0t'' of equal lengths 2σ and denote by $l=t't''$ the union of these two arcs; for the remaining part of L write $L-l$.

Correspondingly (19.6) is written as the sum of two integrals:

$$\Psi(t_0+h)-\Psi(t_0)=I_0+I,$$

where I_0 is taken over l and I over $L-l$.

From the $H(\mu)$ condition for $\varphi(t)$, i.e., from

$$|\varphi(t_2)-\varphi(t_1)| \leqq A\,|t_2-t_1|^{\mu},$$

where t_1, t_2 are any two points on L, it follows that

$$|I_0| \leqq \frac{A}{2\pi}\int_l |t-t_0-h|^{\mu-1}ds+\frac{A}{2\pi}\int_l |t-t_0|^{\mu-1}\,ds. \tag{19.7}$$

Since for any two points t_1, t_2 on L

$$0 < k_0 \leqq \frac{|t_2-t_1|}{|s_2-s_1|} \leqq 1, \tag{19.8}$$

where k_0 is a constant, then from (19.7), by an elementary calculation,

$$|I_0| \leqq A_0\left\{\int_{s_0-2\sigma}^{s_0+2\sigma} |s-s_0-\sigma|^{\mu-1}ds+\int_{s_0-2\sigma}^{s_0+2\sigma} |s-s_0|^{\mu-1}\,ds\right\} \leqq$$
$$\leqq B_0\sigma^{\mu} \leqq C_0\,|h|^{\mu},$$

where A_0, B_0, C_0 are some constants.

Now consider the integral I which may be written

$$I = I_1 + I_2,$$

where

$$I_1 = \frac{1}{2\pi i}\int_{L-l} \frac{\varphi(t_0) - \varphi(t_0 + h)}{t - t_0}\,dt =$$

$$= \frac{1}{2\pi i}[\varphi(t_0) - \varphi(t_0 + h)]\left\{\log\frac{b - t_0}{a - t_0} - \log\frac{t'' - t_0}{t' - t_0}\right\} \quad \text{(cf. § 12)},$$

$$I_2 = \frac{1}{2\pi i}\int_{L-l} [\varphi(t) - \varphi(t_0 + h)]\left\{\frac{1}{t - t_0 - h} - \frac{1}{t - t_0}\right\}dt =$$

$$= \frac{h}{2\pi i}\int_{L-l} \frac{\varphi(t) - \varphi(t_0 + h)}{(t - t_0 - h)(t - t_0)}\,dt.$$

The second factor in brackets in the expression for I_1 is obviously bounded for t_0 on $a'b'$. Therefore

$$|I_1| \leqq C_1 |h|^\mu,$$

where C_1 is a constant. The integral I_2 remains to be considered.

Using (19.8) and noting that $|\tau| \leqq \frac{1}{2}$, where $\tau = \sigma/(s - s_0)$, one obtains

$$|I_2| \leqq A_2 |h| \int_{L-l} \frac{ds}{|s - s_0| \cdot |s - s_0 - \sigma|^{1-\mu}} =$$

$$= A_2 |h| \int_{L-l} \frac{ds}{|s - s_0|^{2-\mu} |1 - \tau|^{1-\mu}} \leqq B_2 |h| \int_{L-l} \frac{ds}{|s - s_0|^{2-\mu}} =$$

$$= B_2 |h| \left\{\int_{s_a}^{s_0 - s\sigma} \frac{ds}{(s - s_0)^{2-\mu}} + \int_{s_0 + s\sigma}^{s_b} \frac{ds}{(s - s_0)^{2-\mu}}\right\},$$

where s_a and s_b are the arc coordinates of the ends a and b of $L = ab$ and A_2, B_2 are constants. Evaluating the integrals leads easily to

$$|I_2| \leqq C_2 |h|^\mu \text{ for } \mu < 1, \quad |I_2| \leqq C_2 |h| \cdot |\log |h|| \text{ for } \mu = 1,$$

where C_2 is a constant.

The above inequalities prove the theorem. This theorem is easily generalized to the case in which L is a piecewise smooth line (which may also have cusps) (cf. Appendix 2).

The theorem proved in this section is due to Plemelj [1]; the author did not state sufficiently clearly the conditions referring to the path of integration; judging by the context, he had the case of a smooth line in mind. Plemelj made no reservation except in the case $\mu = 1$, apparently implying $\mu < 1$. In 1918, I. I. Privalov [2] proved, seemingly independently of Plemelj, this theorem for the case in which L is a circle. Then (cf. I. I. Privalov [3] and also his textbook [1]) he gave a proof also for the case of any piecewise smooth line without cusps. The proof given here follows in principle that of Plemelj; in form, however, it agrees (except for some minor alterations) with that of I. I. Privalov.

§ 20. **The continuity behaviour of the limiting values** (continued).

From the theorem of the preceding section, by (17.2) or else directly, it follows that, *if $\varphi(t)$ satisfies the $H(\mu)$ condition on a smooth line L, then*

$$\Phi(t_0) = \frac{1}{2\pi i} \int_L \frac{\varphi(t)dt}{t - t_0} \tag{20.1}$$

also satisfies everywhere on L, except in the neighbourhood of those of its ends at which $\varphi(t) \neq 0$, the $H(\mu)$ condition for $\mu < 1$ and the $H(1-\varepsilon)$ condition for $\mu = 1$, where ε is an arbitrarily small positive quantity.

For the case in which the density depends on some parameter τ

$$\Phi(t_0, \tau) = \frac{1}{2\pi i} \int_L \frac{\varphi(t, \tau)}{t - t_0}\, dt \tag{20.2}$$

will be considered, assuming that $\varphi(t, \tau)$ satisfies the H condition for t and for τ, when $t \in L$ and $\tau \in T$. It will be shown that under these conditions $\Phi(t_0, \tau)$ satisfies the H condition when $\tau \in T$ and $t_0 \in L'$, where L' is any portion of L having no ends in common with those of L. It is sufficient to prove that $\Phi(t_0, \tau)$ satisfies the H condition with respect to τ for t_0 fixed, as it is already known from above that $\Phi(t_0, \tau)$ satisfies the H condition for t_0 when τ is fixed. Without loss of generality L may be assumed to be a smooth arc. Then

$$\Phi(t_0, \tau + h) - \Phi(t_0, \tau) = \frac{1}{2\pi i} \int_L \frac{\varphi(t, \tau + h) - \varphi(t, \tau)}{t - t_0} dt =$$

$$= \frac{1}{2\pi i} \int_L \frac{[\varphi(t, \tau + h) - \varphi(t_0, \tau + h)] - [\varphi(t, \tau) - \varphi(t_0, \tau)]}{t - t_0} dt +$$

$$+ [\varphi(t_0, \tau + h) - \varphi(t_0, \tau)] \frac{1}{2\pi i} \int_L \frac{dt}{t - t_0}.$$

Obviously the second term is not greater in magnitude than $B \mid h \mid^\nu$, where ν is the Hölder index for the function $\varphi(t, \tau)$ for variable τ and B is a constant. There remains to investigate the integral

$$I = \frac{1}{2\pi i} \int_L \frac{[\varphi(t, \tau + h) - \varphi(t_0, \tau + h)] - [\varphi(t, \tau) - \varphi(t_0, \tau)]}{t - t_0} dt.$$

Let l be an arc of length $\sigma = \mid h \mid$ with t_0 at its centre; assume $\mid h \mid$ so small that l lies inside L. Write I as the sum of two integrals

$$I = I_1 + I_2,$$

where I_1 is taken over l and I_2 over $L - l$. Obviously

$$\mid I_1 \mid \leqq A_1 \int_l \frac{ds}{\mid s - s_0 \mid^{1-\mu}} \leqq B_1 \sigma^\mu,$$

where μ is the Hölder index of $\varphi(t, \tau)$ with t variable, and A_1 and B_1 are constants. Further

$$I_2 = \frac{1}{2\pi i} \int_{L-l} \frac{\varphi(t, \tau + h) - \varphi(t, \tau)}{t - t_0} dt -$$

$$- [\varphi(t_0, \tau + h) - \varphi(t_0, \tau)] \frac{1}{2\pi i} \int_{L-l} \frac{dt}{t - t_0}.$$

The modulus of the second term is not greater than $B'\sigma^\nu$ where B' is a constant. The absolute value of the first term is not greater than

$$A_2 \mid h \mid^\nu \int_{L-l} \frac{ds}{\mid s - s_0 \mid} \leqq B_2 \sigma^\nu \mid \log \sigma \mid,$$

where A_2, B_2 are constants. Thus the statement is proved.

In particular, consider the integral

$$\Phi(t_0) = \frac{1}{2\pi i} \int_L \frac{\varphi(t, t_0)}{t - t_0} dt$$

in which $\varphi(t, t_0)$ satisfies the H condition for $t \in L$, $t_0 \in L$. It follows immediately from the above that $\Phi(t_0)$ satisfies the H condition on any part of L having none of its ends in common with those of L.

§ 21. On the behaviour of the derivative of a Cauchy integral near the boundary.

In the sequel a simple bound will be required for the modulus of the derivative of a Cauchy integral near the boundary.

As before, suppose L to be a smooth line and let $\varphi(t)$ satisfy the $H(\mu)$ condition on L.

Consider the derivative of the Cauchy integral

$$\Phi'(z) = \frac{1}{2\pi i} \int_L \frac{\varphi(t)dt}{(t-z)^2}. \tag{21.1}$$

Let t_0 be an arbitrary point on L such that the distance from t_0 to the nearest end of L, if the latter contains arcs, is not less than some arbitrarily fixed number R.

Let $R_0 = R_0(\alpha)$ be the standard radius for L, corresponding to some arbitrarily fixed acute angle α, and ϱ any positive constant such that

$$\varrho < R_0, \ \varrho < R; \tag{*}$$

in the case in which L consists only of contours the second condition is omitted.

In all that follows it will be assumed that the distance δ of the point z from the point t_0 does not exceed ϱ, i.e.,

$$\delta = |z - t_0| \leqq \varrho, \tag{**}$$

and that the non-obtuse angle between t_0z and the tangent to L at t_0 is not less than some fixed quantity $\beta_0 > \alpha_0$.

Then *the following inequalities hold:*

$$|\Phi'(z)| < C\delta^{\mu-1} \text{ when } \mu < 1, \ |\Phi'(z)| < C|\log \delta| \text{ when } \mu = 1, \tag{21.2}$$

where C is a constant.

In fact, describing about t_0 as centre the circle Γ_0 with radius R_0 and denoting by $l = ab$ the part of L included in Γ_0, the integral (21.1) can be divided into two parts

$$\Phi'(z) = \Psi_1(z) + \Psi_2(z)$$

the first of which is taken over l and the second over $L - l$. As by the condition (**) $\Psi_2(z)$ is obviously bounded for the values of z considered, it is sufficient to examine the integral

$$\Psi_1(z) = \frac{1}{2\pi i}\int\limits_l \frac{\varphi(t)dt}{(t-z)^2} =$$

$$= \frac{1}{2\pi i}\int\limits_l \frac{\varphi(t)-\varphi(t_0)}{(t-z)^2}dt + \frac{\varphi(t_0)}{2\pi i}\left\{\frac{1}{z-b}-\frac{1}{z-a}\right\}. \tag{21.3}$$

Since the second term on the right side is bounded (by the condition restricting the position of z), the first term only need be considered. Write $|t-t_0| = r$ and

$$I = \frac{1}{2\pi i}\int\limits_l \frac{\varphi(t)-\varphi(t_0)}{(t-z)^2}dt,$$

then, since $|ds| \leqq K\,|dr|$, $|\varphi(t)-\varphi(t_0)| \leqq A\,|t-t_0|^\mu$, $|t-z|^2 \geqq (r-\delta\cos\omega_0)^2+\delta^2\sin^2\omega_0$, where $\omega_0=\beta_0-\alpha_0$, (cf. § 18),

$$|I| \leqq \frac{AK}{2\pi}\int\limits_{ab}\frac{r^\mu\,|dr|}{(r-\delta\cos\omega_0)^2+\delta^2\sin^2\omega_0} \leqq$$

$$\leqq \frac{AK}{\pi}\int\limits_0^{R_0}\frac{r^\mu dr}{(r-\delta\cos\omega_0)^2+\delta^2\sin^2\omega_0}.$$

On writing $r = \delta\,.\,u$, for $\mu < 1$, one obtains

$$|I| < \frac{AK}{\pi}\delta^{\mu-1}\int\limits_0^\infty\frac{u^\mu du}{(u-\cos\omega_0)^2+\sin^2\omega_0} = C\delta^{\mu-1},$$

where C is a constant. For $\mu = 1$ the bound for $|I|$ can be evaluated in finite form and reduced to the estimate $|I| \leqq C\,|\log\delta|$. Hence the proposition is proved.

If L contains arcs and if the condition that t_0 lies at a finite distance from the ends does not hold and correspondingly the second of the conditions (*) is omitted, it can be easily seen from (21.3), retaining all other conditions on the position of z, that

$$|\Phi'(z)| < C\delta^{-1}, \tag{21.4}$$

where C is a constant.

A further result may be drawn from (21.2). Let z_1 and z_2 be two points on a straight line through t_0 making a non-obtuse angle, not less than β_0, with the tangent at t_0. Assuming that z_1 and z_2 lie on the same side of L, one has

$$\Phi(z_2)-\Phi(z_1) = \int\limits_{z_1}^{z_2}\Phi'(z)dz,$$

where it may be assumed that the integral is taken over the straight segment z_1z_2. Now, using (21.2), at first assuming $\mu < 1$ and denoting by δ_1 and δ_2 the distances of the points z_1 and z_2 from t_0,

$$| \Phi(z_2) - \Phi(z_1) | \leqq \frac{C}{\mu} | \delta_2{}^\mu - \delta_1{}^\mu | \leqq C_0 | \delta_2 - \delta_1 |^\mu = C_0 | z_2 - z_1 |^\mu, \tag{21.5}$$

where C_0 is a constant.

Similarly, for $\mu = 1$,

$$| \Phi(z_2) - \Phi(z_1) | \leqq C_0 | \delta_2 - \delta_1 |^{1-\varepsilon} = C_0 | z_2 - z_1 |^{1-\varepsilon}, \tag{21.6}$$

where ε is an arbitrarily fixed positive number. Replacing z_2 by z, δ_2 by δ and supposing that $z_1 \to t_0$, then, provided z lies to the left of L,

$$| \Phi(z) - \Phi^+(t_0) | \leqq C_0\delta^\mu \qquad (\text{if } \mu < 1), \tag{21.7}$$

$$| \Phi(z) - \Phi^+(t_0) | \leqq C_0\delta^{1-\varepsilon} \qquad (\text{if } \mu = 1). \tag{21.8}$$

A similar estimate holds for z when it lies to the right of L.

§ 22. On the behaviour of a Cauchy integral near the boundary [1].

Now a more detailed study of the behaviour of a Cauchy integral near the boundary L will be given and the following theorem will be proved.

THEOREM: *Let L be a contour and let $\varphi(t)$ satisfy the $H(\mu)$ condition on L; let S^+, S^- be the regions of the plane bounded by L. Then in each of the regions $S^+ + L$, $S^- + L$ the function*

$$\Phi(z) = \frac{1}{2\pi i} \int_L \frac{\varphi(t)dt}{t - z} \tag{22.1}$$

satisfies the condition

$$| \Phi(z_2) - \Phi(z_1) | \leqq C | z_2 - z_1 |^\mu \qquad (\text{for } \mu < 1) \tag{22.2}$$

or

$$| \Phi(z_2) - \Phi(z_1) | \leqq C | z_2 - z_1 |^{1-\varepsilon} \tag{22.2a}$$

(for $\mu = 1$, ε being an arbitrary positive number),

where C is a constant and by $\Phi(z)$ with $z \in L$ must be understood the corresponding limiting value (Φ^+ or Φ^-).

J. L. Walsh and W. E. Sewell [1], generalizing a result obtained by S. Warschawski [2], proved the following theorem.

Let R be the interior of the closed Jordan curve C and let $f(z)$ be a function,

---

1) This section is of no importance for the understanding of the later work.

analytic in R, continuous in $R + C$. Let α be a constant such that $0 < \alpha \leqq 1$ and let for all z and z_0 on C $(z \neq z_0)$

$$\left| \frac{f(z) - f(z_0)}{(z - z_0)^\alpha} \right| \leqq K,$$

where K does not depend on z and z_0. Then this inequality also holds in $R + C$.

The theorem, stated at the beginning of this section, follows directly from this result and the theorem proved in § 19. Nevertheless a direct proof is given as it is incomparably simpler than Walsh and Sewell's proof of the above theorem; this is, of course, related to the fact that only smooth contours are considered here (the case of piecewise smooth contours also does not present any difficulties; cf. Appendix 2).

For the proof assume $\mu < 1$, because, if $\varphi(t)$ satisfies the $H(1)$ condition, it satisfies a fortiori the $H(1 - \varepsilon)$ condition. The symbols S^+ and S^- will refer respectively to the finite and infinite parts of the plane, and the positive direction on L will be chosen correspondingly.

Let t_0 be a point on the contour L and z a variable point of the region S^+. Consider a branch of the function

$$\Psi(z) = \frac{\Phi(z) - \Phi^+(t_0)}{(z - t_0)^\nu}, \quad 0 \leqq \nu < \mu \tag{22.3}$$

which is one-valued in S^+. The limiting value of this function

$$\Psi^+(t) = \frac{\Phi^+(t) - \Phi^+(t_0)}{(t - t_0)^\nu} \tag{22.4}$$

satisfies the H condition on L by the theorem, proved in § 19 and § 6, 5°. In order to apply the Maximum Modulus theorem, it has to be shown that the function $\Psi(z)$ is continuous in $S^+ + L$. This is a consequence of the following considerations. An infinitesimal portion σ, contained in a circle described about t_0 as centre, is removed from the region S^+. Then in the region $S^+ - \sigma$ the function $\Psi(z)$ is represented by the Cauchy integral

$$\Psi(z) = \frac{1}{2\pi i} \int_{L'} \frac{\Psi^+(t)dt}{t - z},$$

where L' is the boundary of $S^+ - \sigma$. But since in the neighbourhood of t_0

$$\Psi(z) = O\left(\frac{1}{|z - t_0|^\nu}\right), \quad \nu < 1,$$

the integral over an arc of the circle about t_0 tends to zero and therefore in the entire region S^+

$$\Psi(z) = \frac{1}{2\pi i} \int_L \frac{\Psi^+(t)}{t - z}\, dt.$$

But then, by the theorem proved in § 16, the function $\Psi(z)$ is continuous in $S^+ + L$, if it is given the value $\Psi^+(t)$ on L, and consequently

$$\frac{|\,\Phi(z) - \Phi^+(t_0)\,|}{|\,z - t_0\,|^\nu} = |\,\Psi(z)\,| \leqq \max |\,\Psi^+(t)\,| \leqq C,$$

where C does not depend on the position of t_0 on L or on $\nu < \mu$.

This follows from the fact that $\nu < \mu$ and that $\Phi^+(t)$ satisfies the $H(\mu)$ condition.

Therefore the above inequality also holds in the case when $\nu = \mu$, and hence

$$|\,\Phi(z) - \Phi^+(t_0)\,| \leqq C\,|\,z - t_0\,|^\mu. \tag{22.5}$$

A similar inequality can be established for the point z when it lies in S^-.

The function $\Psi(z)$ is not one-valued in S^-. However, only those points z lying near t_0 are of interest here. Hence it is sufficient to consider a finite part of S^- bounded by some part of L, containing t_0 and extended to a smooth contour, so that the region bounded by this contour lies in S^-. This reduces the problem to the preceding one.

Thus the inequality (22.2) is established when at least one of the points z_1, z_2 lies on L.

Now both points z_1, z_2 will be assumed to lie in S^+. As it is easily seen, it is sufficient to consider the case in which at least one of these points, say z_1, lies at a distance from the boundary which is less than some constant $\varrho < R_0$, where R_0 is some standard radius.

If the distances of both points from the boundary are $\geqq \varrho$, where ϱ is some fixed number, then (22.2) necessarily holds, since $\Phi(z)$ is holomorphic.

Temporarily denote z_1 by z_0 and z_2 by z. Let t_0 be the point on the contour L nearest to z_0.

Cut the region S^+ by a straight line $t_0 z_0$ (the segment $t_0 z_0$ is obviously normal to L at the point t_0) and consider

$$\Psi_0(z) = \frac{\Phi(z) - \Phi(z_0)}{(z - z_0)^\mu} \tag{22.6}$$

which is holomorphic in the cut region and continuous from the left on its entire boundary. By (22.5), the limiting value of this function satisfies $|\,\Psi_0^+(t)\,| \leqq C$ on L. But the limiting value of $\Psi_0(z)$ on either side of the cut $z_0 t_0$ satisfies the same condition by (21.5). Hence, by the Maximum Modulus theorem, $|\,\Psi_0(z)\,| \leqq C$ everywhere in S^+. A similar inequality may be established for S^-. Thus the theorem is proved.

CHAPTER 3

SOME COROLLARIES ON CAUCHY INTEGRALS

The various results given in this chapter follow almost directly from the preceding work and are essential in future applications (except for the results of § 24).

§ 23. Poincaré-Bertrand transformation formula.

Let L be a smooth arc or contour, $\varphi(t, t_1)$ a function of the two points t, t_1 of this line, satisfying the H condition with respect to t and t_1, and let t_0 be a fixed point on L not coinciding with one of its ends.

Consider the repeated integrals

$$A = \int_L \frac{dt}{t - t_0} \int_L \frac{\varphi(t, t_1)dt_1}{t_1 - t}, \quad B = \int_L dt_1 \int_L \frac{\varphi(t, t_1)dt}{(t - t_0)(t_1 - t)} \tag{23.1}$$

differing only in the order of integration. Both integrals exist, because the function

$$\chi(t) = \int_L \frac{\varphi(t, t_1)dt_1}{t_1 - t} \tag{23.2}$$

satisfies the H condition on L, with the possible exception of the ends; it is easily seen from

$$\chi(t) = \int_L \frac{\varphi(t, t_1) - \varphi(t, c)}{t_1 - t} dt_1 + \varphi(t, c) \int_L \frac{dt_1}{t_1 - t}$$

that near the end c

$$\chi(t) = O(\log | t - c |). \tag{23.3}$$

Therefore the integral

$$A = \int_L \frac{\chi(t)dt}{t - t_0} \tag{23.4}$$

has a definite value (because the point t_0 is not an end).

Further

$$B = \int_L \frac{dt_1}{t_1 - t_0} \int_L \left\{ \frac{1}{t - t_0} - \frac{1}{t - t_1} \right\} \varphi(t, t_1) dt =$$

$$= \int_L \frac{\omega(t_0, t_1) - \omega(t_1, t_1)}{t_1 - t_0} dt_1, \tag{23.5}$$

where

$$\omega(\tau, t_1) = \int_L \frac{\varphi(t, t_1)}{t - \tau} dt, \tag{23.6}$$

τ being some point on L. The function $\omega(\tau, t_1)$ satisfies for variable τ the H condition on L, except possibly in the neighbourhood of the ends near which, however,

$$\omega(\tau, t_1) = O(\log |\tau - c|)$$

which is obtained analogously to the bound (23.3). Therefore the integral (23.5) exists in the ordinary (Riemann) sense.

However, as was first proved by Poincaré, the integrals A and B are not, in general, equal to one another. The following very important formula of Poincaré-Bertrand holds:

$$\int_L \frac{dt}{t - t_0} \int_L \frac{\varphi(t, t_1)}{t_1 - t} dt_1 = -\pi^2 \varphi(t_0, t_0) + \int_L dt_1 \int_L \frac{\varphi(t, t_1) dt}{(t - t_0)(t_1 - t)}. \tag{23.7}$$

A simple proof of this formula will now be given. Let

$$\Phi(z) = \int_L \frac{dt}{t - z} \int_L \frac{\varphi(t, t_1)}{t_1 - t} dt_1, \quad \Psi(z) = \int_L dt_1 \int_L \frac{\varphi(t, t_1)}{(t - z)(t_1 - t)} dt, \tag{23.8}$$

where z is a point of the plane, not on L.

It is easily seen that in the repeated integrals (23.8) inversion of the order of integration is legitimate, since one of the singularities of the integrand (namely for $t = t_0$) has been removed. For the present assume this, so that

$$\Phi(z) = \Psi(z). \tag{23.9}$$

A proof of (23.9) is given at the end of the present section (1°).

The functions $\Phi(z)$ and $\Psi(z)$ are closely associated with the integrals A and B. By the Plemelj formula (17.4)

$$\Phi^+(t_0) + \Phi^-(t_0) = 2 \int_L \frac{dt}{t - t_0} \int_L \frac{\varphi(t, t_1) dt_1}{t_1 - t}. \tag{23.10}$$

Further

$$\Psi(z) = \int_L \frac{\psi(t_1, z)dt_1}{t_1 - z}, \tag{23.11}$$

where

$$\psi(t_1, z) = \int_L \left\{ \frac{1}{t - z} - \frac{1}{t - t_1} \right\} \varphi(t, t_1)dt. \tag{23.12}$$

Denote by $\psi^+(t_1, t_0)$, $\psi^-(t_1, t_0)$ the limits of $\psi(t_1, z)$ as $z \to t_0$ from the left and from the right of L. By Plemelj's formulae (17.3), (17.4)

$$\psi^+(t_1, t_0) - \psi^-(t_1, t_0) = 2\pi i \varphi(t_0, t_1), \tag{23.13}$$

$$\psi^+(t_1, t_0) + \psi^-(t_1, t_0) = 2 \int_L \left\{ \frac{1}{t - t_0} - \frac{1}{t - t_1} \right\} \varphi(t, t_1)dt =$$

$$= 2(t_1 - t_0) \int_L \frac{\varphi(t, t_1)}{(t - t_0)(t_1 - t)} dt.$$

Besides

$$\begin{aligned} \psi(t_1, z) &= \psi^+(t_1, t_0) + \varepsilon^+ \quad (\text{if } z \text{ is to the left of } L), \\ \psi(t_1, z) &= \psi^-(t_1, t_0) + \varepsilon^- \quad (\text{if } z \text{ is to the right of } L), \end{aligned} \tag{23.14}$$

where $\varepsilon^+ \to 0$, $\varepsilon^- \to 0$ as $z \to t_0$. It is proved later (2° at the end of this section) that

$$\int_L \frac{\varepsilon^+ dt_1}{t_1 - z} \to 0, \qquad \int_L \frac{\varepsilon^- dt_1}{t_1 - z} \to 0 \tag{23.15}$$

when $z \to t_0$ along a straight line forming a finite angle with the tangent at t_0. Replacing $\psi(t_1, z)$ in (23.11) by the expression (23.14), using (23.15) and Plemelj's formulae, one obtains

$$\Psi^+(t_0) = \pi i \psi^+(t_0, t_0) + \int_L \frac{\psi^+(t_1, t_0)dt_1}{t_1 - t_0},$$

$$\Psi^-(t_0) = -\pi i \psi^-(t_0, t_0) + \int_L \frac{\psi^-(t_1, t_0)dt_1}{t_1 - t_0},$$

and hence by (23.13)

$$\Psi^+(t_0) + \Psi^-(t_0) = -2\pi^2 \varphi(t_0, t_0) + 2\int_L dt_1 \int_L \frac{\varphi(t, t_1)dt}{(t - t_0)(t_1 - t)}. \tag{23.16}$$

But from (23.9) the left sides of (23.10) and (23.16) are equal. Thus comparison of the right sides gives the required formula (23.7). The formulae (23.9), (23.15) remain to be proved.

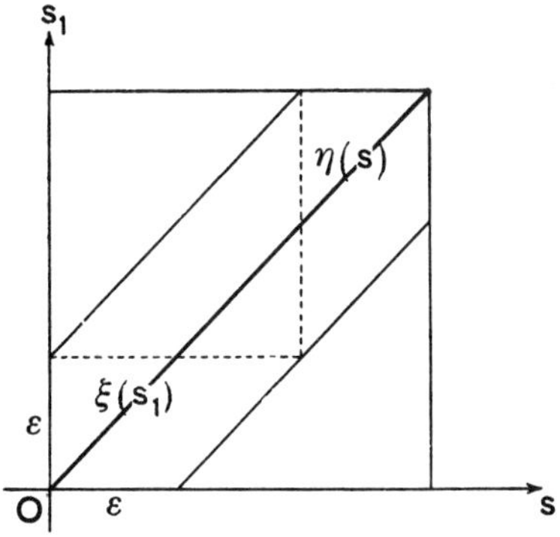

Fig. 7.

1° It will now be shown that $\Phi(z) = \Psi(z)$. For clarity the positions of the points t, t_1 on L will be defined by the arc coordinates $s, s_1 (0 \leqq s \leqq s^*, 0 \leqq s_1 \leqq s^*$, where s^* is the length of L) and s, s_1 will be considered as rectangular coordinates of points in the auxiliary plane Oss_1. The region of variation of s and s_1 in this plane is the square Q with sides equal to s^* (see Fig. 7). The singularities of the integrand

$$\frac{\varphi(t, t_1)}{(t - z)(t_1 - t)}$$

are concentrated along the diagonal $s = s_1$ of Q. Using the straight lines $s_1 = s \pm \varepsilon$, $\varepsilon > 0$, cut from Q the strip q, as shown in Fig. 7. Obviously

$$\Phi(z) = I_0 + I_1,\ \Psi(z) = I_0 + I_2,$$

where

$$I_0 = \iint\limits_{Q-q} \frac{\varphi(t, t_1) dt\, dt_1}{(t - z)(t_1 - t)} = \iint\limits_{Q-q} \frac{\varphi(t, t_1)}{(t - z)(t_1 - t)} \frac{dt}{ds} \frac{dt_1}{ds_1} ds\, ds_1,$$

$$I_1 = \int\limits_L \frac{dt(s)}{t - z} \int\limits_{\eta(s)} \frac{\varphi(t, t_1) dt_1(s_1)}{t_1 - t},\quad I_2 = \int\limits_L dt_1(s_1) \int\limits_{\xi(s_1)} \frac{\varphi(t, t_1)}{(t - z)(t_1 - t)} dt(s).$$

$\eta(s)$ denotes the segment of the straight line, parallel to the Os_1 axis lying at a distance s from it, which is inside the strip q, while $\xi(s_1)$ is the segment of the straight line parallel to the Os axis and at a distance s_1 from it, contained in the same strip.

The statement will be proved, if it is shown that

$$I_1 \to 0,\ I_2 \to 0 \text{ as } \varepsilon \to 0.$$

But this can be seen by the use of simple bounds, e.g.:

$$\chi_1(t) = \int\limits_{\eta(s)} \frac{\varphi(t, t_1) dt_1}{t_1 - t} = \int\limits_{\eta(s)} \frac{\varphi(t, t_1) - \varphi(t, t)}{t_1 - t} dt_1 + \varphi(t, t) \int\limits_{\eta(s)} \frac{dt_1}{t_1 - t}. \quad (*)$$

If $\varepsilon \leqq s \leqq s^* - \varepsilon$, the last term in (*) tends uniformly to zero when $\varepsilon \to 0$ (see § 12, Note 1); for $0 < s < \varepsilon$ or $s^* - \varepsilon < s < s^*$ this last term will be of order $\log s$ or $\log (s^* - s)$. Clearly therefore

$$\int_L \frac{\varphi(t, t)dt}{t - z} \int_{\eta(s)} \frac{dt_1}{t_1 - t} \to 0.$$

The first term on the right side of (*) obviously tends uniformly to 0. Consequently $I_1 \to 0$. Similarly it can be shown that $I_2 \to 0$.

2°. The proof of (23.15) will now be given.

For example, let $z \to t_0$ from the left of L. Let $R_0 = R_0(\alpha)$ be the standard radius and let l be the arc cut from L by the circle of radius R_0 with centre at t_0. Then it is obviously sufficient to show that

$$I = \int_l \frac{\varepsilon^+ dt_1}{t_1 - z} \to 0$$

as $z \to t_0$ along a straight line making a non-obtuse angle, not less than $\beta_0 > \alpha_0$, with the tangent at t_0.

By (21.7)

$$| \varepsilon^+ | = | \psi(t_1, z) - \psi^+(t_1, t_0) | \leqq C\delta^\mu,$$

where C is a constant, $\delta = | z - t_0 |$ and μ is the Hölder index of $\varphi(t, t_1)$ for variable t; assume $\mu < 1$ which, of course, does not affect the generality. Further (see § 18)

$$| t_1 - z |^2 \geqq (r - \delta \cos \omega_0)^2 + \delta^2 \sin^2 \omega_0,$$

where $r = | t_1 - t_0 |$, $0 < \omega_0 < \frac{\pi}{2}$; therefore

$$| I | \leqq B\delta^\mu \int_0^{R_0} \frac{dr}{\sqrt{(r - \delta \cos \omega_0)^2 + \delta^2 \sin^2 \omega_0}} =$$

$$= B\delta^\mu \Big[\log \Big\{ r - \delta \cos \omega_0 + \sqrt{(r - \delta \cos \omega_0)^2 + \delta^2 \sin^2 \omega_0} \Big\} \Big]_0^{R_0},$$

where B is some constant. Consequently $I \to 0$ as $\delta \to 0$. The case when $z \to t_0$ from the right of L can be treated in exactly the same manner.

Thus the proof of the transformation formula (23.7) is complete. This formula was first given by H. Poincaré [1] under very restrictive assumptions on the function $\varphi(t, t_1)$ and the line L; under wider conditions this formula was proved by G. Bertrand [1], [2], though with less generality than here. The rigorous proof, under conditions essentially those assumed here, was given by G. Giraud [1]. Giraud considered multi-dimensional regions of integration and the one-dimensional region as a special case. The proof given in this section is much simpler than the other proofs known to the Author. (This refers of course to strict proofs.)

Its simplicity is due to the use of the Plemelj formula proved earlier and the bounds obtained in the preceding section which permitted the point t_0, lying on the path of integration, to be replaced by the point z, not on this path.

NOTE. Let a function $K(t, t_1)$, given on L, be of the form

$$K(t, t_1) = \frac{\psi(t, t_1)}{|t_1 - t|^{\alpha}}, \quad 0 \leqq \alpha < 1, \tag{23.17}$$

where $\psi(t, t_1)$ satisfies the H condition. Then

$$\int_L \frac{dt}{t - t_0} \int_L K(t, t_1) dt_1 = \int_L dt_1 \int_L \frac{K(t, t_1)}{t - t_0} dt, \tag{23.18}$$

i.e., in this case the order of integration may be inverted. This follows directly from the Poincaré-Bertrand transformation formula, if it is noted that by (23.17)

$$K(t, t_1) = \frac{\varphi(t, t_1)}{t_1 - t},$$

where

$$\varphi(t, t_1) = \psi(t, t_1)\, |t_1 - t|^{1-\alpha} e^{i\vartheta}, \quad \vartheta = \arg(t_1 - t)$$

satisfies the H condition (cf. § 7) and $\varphi(t, t) = 0$.

§ 24. On analytic continuation of a function given on the boundary of a region.

Let S^+ be a connected region bounded by one or several smooth non-intersecting contours L_0, $L_1, \ldots L_p$, of which L_0 contains all the others (Fig. 8). The contour L_0 may be absent (i.e., it may be at infinity); then the region S^+ will be infinite (i.e., it will be an infinite region with certain contours as internal boundaries). By L denote the union of the contours L_0, $L_1 \ldots L_p$, by S^- the complement of $S^+ + L$. Thus S^- consists of the finite regions $S_1^- \ldots S_p^-$, bounded by $L_1 \ldots L_p$, and, if L_0 exists, of the infinite region S_0^-, bounded by L_0.

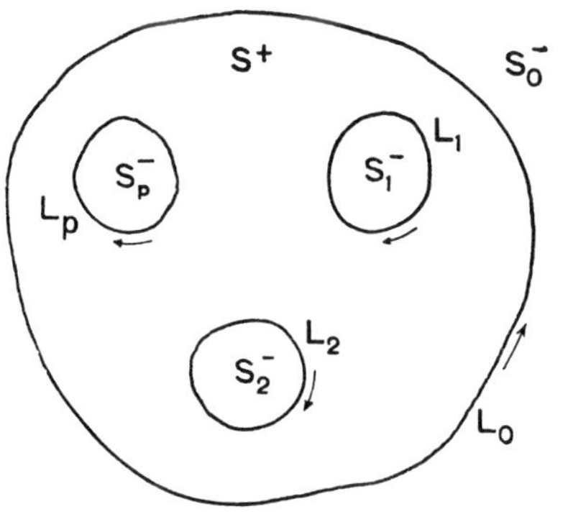

Fig. 8.

From the condition, laid down in § 9, the positive direction on L will be assumed such that the region S^+ lies to the left of L.

Let $\varphi(t)$ be a continuous function given on L. Consider the following question: What conditions must be satisfied by the function $\varphi(t)$, so that it will be the limiting value of some function $\varphi(z)$ holomorphic in S^+ and continuous in $S^+ + L$? In this case the function $\varphi(t)$ will sometimes be said to continue analytically into S^+.

If $\varphi(t)$ is the boundary value of the function $\varphi(z)$, holomorphic in S^+ and continuous in $S^+ + L$, and if (in the absence of L_0) it is zero at infinity, then by Cauchy's theorem

$$\frac{1}{2\pi i}\int_L \frac{\varphi(t)dt}{t-z} = 0 \text{ for all } z \in S^-. \tag{24.1}$$

Conversely, if (24.1) holds, $\varphi(t)$ is the boundary value of the function $\varphi(z)$, defined by

$$\varphi(z) = \frac{1}{2\pi i}\int_L \frac{\varphi(t)dt}{t-z} \text{ for } z \in S^+,$$

which is holomorphic in S^+ and continuous from the left on L. Indeed, for any point z not on L,

$$\Phi(z) = \frac{1}{2\pi i}\int_L \frac{\varphi(t)dt}{t-z}. \tag{24.2}$$

By (24.1), $\Phi(z) \equiv 0$ in S^-; therefore $\Phi(z)$, taken on the right of L, has the boundary value $\Phi^-(t) = 0$. Consequently, by § 18, the function $\Phi(z)$ is also continuous on L from the left and

$$\varphi(t) = \Phi^+(t) - \Phi^-(t) = \Phi^+(t),$$

which proves the statement.

Thus *the condition* (24.1) *is necessary and sufficient for the continuous function $\varphi(t)$, given on L, to be the boundary value of a function, holomorphic in S^+, continuous in $S^+ + L$ and vanishing at infinity in the absence of L_0.*

If $\varphi(t)$ satisfies on L the H condition, these assumptions may be given the following form. Namely, in (24.1) let $z \to t_0$, where t_0 is any point on L, then

$$-\tfrac{1}{2}\varphi(t_0) + \frac{1}{2\pi i}\int_L \frac{\varphi(t)dt}{t-t_0} = 0 \text{ for all } t_0 \in L. \tag{24.3}$$

Conversely, it is also clear that (24.1) follows from (24.3) [1]), *so that* (24.3) *is a necessary and sufficient condition for* $\varphi(t)$, *satisfying the H condition, to be the boundary value of a function, holomorphic in* S^+, *continuous in* $S^+ + L$ *and vanishing at infinity in the absence of* L_0.

Analogously one obtains *the necessary and sufficient condition for a continuous* $\varphi(t)$ *to be the boundary value of a function* [2]), *holomorphic in* S^-, *continuous in* $S^- + L$ *and vanishing at infinity in the absence of* L_0. This condition is

$$\frac{1}{2\pi i}\int_L \frac{\varphi(t)dt}{t-z} = 0 \text{ for all } z \in S^+. \tag{24.4}$$

If $\varphi(t)$ satisfies the H condition, (24.4) is equivalent to

$$\frac{1}{2}\varphi(t_0) + \frac{1}{2\pi i}\int_L \frac{\varphi(t)dt}{t-t_0} = 0 \text{ for all } t_0 \in L. \tag{24.5}$$

The above condition is easily generalized to the case in which $\varphi(z)$ is not required to vanish at infinity. For example, suppose L_0 not at infinity and let it be required to find *a necessary and sufficient condition for* $\varphi(t)$ *to be on* L *the boundary value of a function* $\varphi(z)$, *holomorphic in* S^-, *continuous in* $S^- + L$ *and having at infinity a pole with a given principal part, i.e.*, $\varphi(z)$ has the form

$$\varphi(z) = \gamma(z) + O\left(\frac{1}{z}\right),$$

where $\gamma(z)$ is a given polynomial.

Then (24.4) is obviously replaced by

$$\frac{1}{2\pi i}\int_L \frac{\varphi(t)-\gamma(t)}{t-z}dt = 0 \text{ for all } z \in S^+$$

or, what is the same thing, by

$$\frac{1}{2\pi i}\int_L \frac{\varphi(t)dt}{t-z} - \gamma(z) = 0 \text{ for all } z \in S^+. \tag{24.4a}$$

---

[1]) In fact, it follows from (24.3) that $\Phi^-(z)$, defined by (24.2), holomorphic in S^- and continuous from the right on L, assumes the boundary value $\Phi^-(t)=0$ on L; consequently $\Phi(z) = 0$ everywhere in S^-.

[2]) Here a "holomorphic function" is the union of the separate functions $\varphi_0(z), \varphi_1(z), \ldots \varphi_p(z)$, holomorphic respectively in the regions, $S_0^-, S_1^-, \ldots S_p^-$.

Correspondingly (24.5) is replaced by

$$\frac{1}{2}\varphi(t_0) + \frac{1}{2\pi i}\int_L \frac{\varphi(t)dt}{t-t_0} - \gamma(t_0) = 0 \text{ for all } t_0 \in L. \quad (24.5a)$$

The conditions (24.3) and (24.5a) were proved by Plemelj [1] for the case of a single contour, under the special assumption $\gamma(z) = \text{const}$; F. Casorati [1] also confined himself to the case of a single contour L and a finite region S^+ and gave considerably earlier than Plemelj, but without proper derivation, the condition

$$\int_L \frac{\varphi(t)-\varphi(t_0)}{t-t_0}dt = 0,$$

equivalent to (24.3). Later on, Morera [1] gave this condition a stricter basis and also stated the other condition

$$\int_L \varphi(t)\log(t-t_0)dt = 0$$

which is (with some additional assumptions) equivalent to the preceding one.

From the more general point of view pursued here the conditions were studied by I. I. Privalov [2].

§ 25. Generalization of Harnack's theorem.

Results which are a direct consequence of § 24 will now be given.

Following the notation of the preceding section, let $\varphi(t)$ be a real function, continuous on L. As before, let

$$\Phi(z) = \frac{1}{2\pi i}\int_L \frac{\varphi(t)dt}{t-z}. \quad (25.1)$$

If $\Phi(z) \equiv 0$ in S^+, then $\varphi(t) = C_k$ on L_k $(k = 0, 1, \ldots p)$, where the C_k are constants and, in the presence of L_0, $C_0 = 0$ (when L_0 is at infinity, one has to take $k = 1, 2, \ldots p$); if $\Phi(z) \equiv 0$ in S^-, then $\varphi(t) = C = \text{const.}$ on L.

If $\Phi(z) \equiv 0$ in S^+, $\varphi(t)$ is the boundary value of the functions $\varphi_k(z)$, holomorphic in S_k^-, and, with L_0 present, $\varphi_0(\infty) = 0$. As the function $\varphi(t)$ is assumed real, it follows that the imaginary parts of the functions $\varphi_k(z)$ vanish on L_k; therefore these imaginary parts are zero in S_k^- and consequently $\varphi_k(z) = C_k'$ in S_k^-, (in fact, if in some region the imaginary (real) part of a holomorphic function is zero, this function is a constant) and, with the contour L_0 present, $C_0 = 0$. Hence the first part of the proposition follows. The second part is proved quite analogously.

The proposition proved is a generalization of one given by A. Harnack [1] who did not formulate it quite correctly.

NOTE 1. The result proved is still valid, if the requirement $\Phi(z)=0$ in $S^+(S^-)$ is replaced by $\Re\Phi(z) = 0$ in $S^+(S^-)$.

In fact, first let $\Re\Phi(z) = 0$ in S^+. Then $\Phi(z) = Ai$ in S^+, where A is a real constant. So $\Phi^+(t) = Ai$, $\Phi^-(t) = \Phi^+(t) - \varphi(t) = Ai - \varphi(t)$; therefore $\Im\Phi^-(t) = A$ on L_k, $\Im\Phi(z) = A$ in S_k^- and, consequently, $\Phi(z) = Ai - C_k$ in S_k^-, where the C_k are real constants and, if there is a contour L_0, $C_0 = 0$. Hence

$$\varphi(t) = \Phi^+(t) - \Phi^-(t) = C_k.$$

Now let $\Re\Phi(z) = 0$ in S^-. Then $\Phi(z) = A_k i$ in S_k^-, where the A_k are real constants and, in the presence of the contour L_0, $A_0 = 0$. Hence $\Phi^-(t) = A_k i$, $\Phi^+(t) = A_k i + \varphi(t)$ on L_k and therefore $\Im\Phi^+(t) = A_k$ on L_k. But under these conditions, as is known (the proof will be given below, in § 60.), $A_k = 0$, $k = 0, 1, 2, \ldots p$ (the assertion $A_0 = 0$ must be omitted when the contour L_0 is at infinity), $\Im\Phi(z) = 0$ in S^+ and hence $\Phi(z) = C$ in S^+, where C is a real constant. Therefore $\varphi(t) = \Phi^+(t) - \Phi^-(t) = C$ on L.

NOTE 2. If in the proved result the condition $\Phi(z) = 0$ in S^+ or in S^- is replaced by $\Im\Phi(z) = 0$ in S^+ or S^-, then it is still valid, the difference being that in this case C_0 is not necessarily zero. The proof is quite analogous to the preceding ones and is left to the reader.

§ 26. Determination of sectionally holomorphic functions with given discontinuities (case of contours).

The following problem, of great importance later, will now be solved. Let L, S^+, S^- be as in § 24 and let $\varphi(t)$ satisfy the H condition on L. It is required *to find a sectionally holomorphic function* $\Phi(z)$, *zero at infinity and satisfying the given boundary condition*

$$\Phi^+(t) - \Phi^-(t) = \varphi(t) \text{ on } L. \tag{26.1}$$

The problem is immediately solved by Plemelj's formula. In fact, it is clear that the function

$$\Phi(z) = \frac{1}{2\pi i}\int_L \frac{\varphi(t)dt}{t-z} \tag{26.2}$$

satisfies all the required conditions. The problem has no other solutions, because the difference $\Psi(z)$ of the two solutions would satisfy the condition $\Psi^+(t) - \Psi^-(t) = 0$; hence it follows that $\Psi(z)$ is a function holomorphic in the entire plane; but as it vanishes at infinity, necessarily $\Psi(z) \equiv 0$.

The solution (26.2) was indicated by Plemelj [1] in connection with other problems (cf. § 35).

If the more general problem is posed: To find a function, sectionally holomorphic everywhere, except possibly at the point $z = \infty$, and having finite non-negative degree there, i.e., a function such that for large $|z|$

$$\Phi(z) = O(z^k), \tag{26.3}$$

where k is a given non-negative integer, then the most general solution is obviously of the form

$$\Phi(z) = \frac{1}{2\pi i}\int_L \frac{\varphi(t)dt}{t-z} + P_k(z), \tag{26.4}$$

where $P_k(z)$ is an arbitrary polynomial of degree not higher than k.

So far it has been assumed that $\varphi(t)$ satisfies the H condition. If $\varphi(t)$ is simply continuous on L, then one may only assert:

If a sectionally holomorphic function exists, satisfying the conditions (26.1) and (26.3), then it necessarily has the form (26.4).

In order to verify this statement it is sufficient to multiply both sides of (26.1) by

$$\frac{1}{2\pi i}\frac{dt}{t-z},$$

to integrate along L and to apply Cauchy's formula in an appropriate manner.

Thus, *for the assumed form of L, any sectionally holomorphic function having finite degree at infinity can be represented in the form* (26.4).

§ 27. Inversion of the Cauchy integral (case of contours).

Let L, S^+, S^- be the same as in § 24. Consider the equation

$$\frac{1}{\pi i}\int_L \frac{\varphi(t)\,dt}{t-t_0} = \psi(t_0), \tag{27.1}$$

where t_0 is an arbitrary point on L, $\psi(t)$ is given satisfying the H condition on L, and $\varphi(t)$ is an unknown function also subject to the H condition.

Equation (27.1) represents the simplest singular integral equation and is of great importance. Several methods will be given for its solution.

First method. Introduce the sectionally holomorphic function

$$\Phi(z) = \frac{1}{2\pi i}\int_L \frac{\varphi(t)dt}{t-z}, \tag{27.2}$$

vanishing at infinity. Then (27.1), by (17.4), may be written in the form

$$\Phi^+(t_0) + \Phi^-(t_0) = \psi(t_0). \tag{27.3}$$

Consider a second sectionally holomorphic function which is defined by

$$\Psi(z) = \Phi(z) \text{ for } z \in S^+, \quad \Psi(z) = -\Phi(z) \text{ for } z \in S^-. \tag{27.4}$$

Then (27.3) may be rewritten

$$\Psi^+(t_0) - \Psi^-(t_0) = \psi(t_0),$$

and from the preceding section

$$\Psi(z) = \frac{1}{2\pi i}\int_L \frac{\psi(t)dt}{t-z}. \tag{27.5}$$

On the other hand, by (17.3), (27.2) and (27.4),

$$\varphi(t_0) = \Phi^+(t_0) - \Phi^-(t_0) = \Psi^+(t_0) + \Psi^-(t_0),$$

so that finally, by (27.5) and (17.4),

$$\varphi(t_0) = \frac{1}{\pi i}\int_L \frac{\psi(t)dt}{t-t_0}. \tag{27.6}$$

Thus (27.6) follows from (27.1). But in exactly the same way (27.1) follows from (27.6). Hence (27.6) is the only solution of the equation (27.1); in other words, *each of the equations*

$$\frac{1}{\pi i}\int_L \frac{\varphi(t)dt}{t-t_0} = \psi(t_0), \quad \frac{1}{\pi i}\int_L \frac{\psi(t)dt}{t-t_0} = \varphi(t_0) \tag{A}$$

follows from the other, i.e., these relations are reciprocal.

Second method. The inversion formulae (A) can be deduced by the following simple argument. Let $\omega(t)$ be any function satisfying the H condition on L. The function

$$\Omega(z) = \frac{1}{\pi i}\int_L \frac{\omega(t)dt}{t-z}$$

is holomorphic in S^+ and continuous in $S^+ + L$, if it is ascribed on L the value

$$\Omega^+(t) = \omega(t) + \frac{1}{\pi i}\int_L \frac{\omega(t_1)dt_1}{t_1-t}. \tag{*}$$

Therefore

$$\frac{1}{\pi i}\int_L \frac{\Omega^+(t)dt}{t-z} = 0 \text{ for all } z \in S^-;$$

proceeding to the limit $z \to t_0$ from the right, one obtains a relation which might have been written down directly from (24.3):

$$-\Omega^+(t_0) + \frac{1}{\pi i}\int_L \frac{\Omega^+(t)dt}{t-t_0} = 0.$$

Replacement of $\Omega^+(t)$ by (*) leads to

$$-\omega(t_0) - \frac{1}{\pi i}\int_L \frac{\omega(t_1)dt_1}{t_1-t_0} + \frac{1}{\pi i}\int_L \frac{dt}{t-t_0}\left\{\omega(t) + \frac{1}{\pi i}\int_L \frac{\omega(t_1)dt_1}{t_1-t}\right\} = 0,$$

and after simplification to

$$\frac{1}{\pi i}\int_L \frac{dt}{t-t_0}\frac{1}{\pi i}\int_L \frac{\omega(t_1)dt_1}{t_1-t} = \omega(t_0). \tag{B}$$

This holds for any function $\omega(t)$ satisfying the H condition on L. The formula (B) expresses the same as the inversion formulae (A), i.e., that each of the expressions (A) follows from the other.

THIRD METHOD. Rewrite (27.1) in the form

$$\psi(t) = \frac{1}{\pi i}\int_L \frac{\varphi(t_1)dt_1}{t_1-t};$$

multiplying both sides by $\dfrac{1}{\pi i}\dfrac{dt}{t-t_0}$, integrating with respect to t along L and applying the Poincaré-Bertrand formula (§ 23), it gives

$$\frac{1}{\pi i}\int_L \frac{\psi(t)dt}{t-t_0} = -\frac{1}{\pi^2}\int_L \frac{dt}{t-t_0}\int_L \frac{\varphi(t_1)dt_1}{t_1-t} =$$

$$= \varphi(t_0) - \frac{1}{\pi^2}\int_L \varphi(t_1)dt_1 \int_L \frac{dt}{(t-t_0)(t_1-t)}. \tag{**}$$

But, as is easily verified directly,

$$\int_L \frac{dt}{(t-t_0)(t_1-t)} = \frac{1}{t_1-t_0}\int_L \left\{\frac{1}{t-t_0} - \frac{1}{t-t_1}\right\} dt = 0.$$

Therefore the second term in (**) vanishes and (27.6) is obtained. By the same method (27.1) is deduced from (27.6).

So it is seen that the inversion formulae (A) are an almost trivial consequence of the Plemelj and of the Poincaré-Bertrand formula. In applications to particular contours (infinite straight lines, circles; for the case of circles cf. the next section) they are often, in one or another equivalent (or almost equivalent) form, encountered in the literature under the name of Hilbert formulae. The Author finds it difficult to determine who first applied these formulae to arbitrary smooth contours. I. N. Vekua [1] obtained them as the special case of the solution of one class of singular equations, and also informed the Author of the method of deduction which is given above as the second method.

§ 28. The Hilbert inversion formulae.

In particular, the results obtained above will be applied to the case when L is a circle of radius 1 with centre at the origin. Then

$$t = e^{i\vartheta},\ t_0 = e^{i\vartheta_0} \tag{28.1}$$

and

$$\frac{dt}{t-t_0} = \frac{ie^{i\vartheta}d\vartheta}{e^{i\vartheta}-e^{i\vartheta_0}} = \frac{ie^{i\frac{\vartheta-\vartheta_0}{2}}d\vartheta}{e^{i\frac{\vartheta-\vartheta_0}{2}}-e^{-i\frac{\vartheta-\vartheta_0}{2}}} = \frac{1}{2}\,\frac{\cos\dfrac{\vartheta-\vartheta_0}{2}+i\sin\dfrac{\vartheta-\vartheta_0}{2}}{\sin\dfrac{\vartheta-\vartheta_0}{2}}\,d\vartheta =$$

$$= \frac{1}{2}\cot\left(\frac{\vartheta-\vartheta_0}{2}\right)d\vartheta + \frac{i}{2}\,d\vartheta. \tag{28.2}$$

The functions $\varphi(t)$ and $\psi(t)$ will now be denoted by $\varphi(\vartheta)$, $\psi(\vartheta)$ [assuming that $\varphi(\vartheta+2k\pi)=\varphi(\vartheta)$, $\psi(\vartheta+2k\pi)=\psi(\vartheta)$ for integral k] and, further, $\psi(\vartheta)$ will be replaced by $\psi(\vartheta)/i$. Then the formulae (A) of the preceding section take the form

$$\frac{1}{2\pi}\int_0^{2\pi}\varphi(\vartheta)\cot\left(\frac{\vartheta-\vartheta_0}{2}\right)d\vartheta + \frac{i}{2\pi}\int_0^{2\pi}\varphi(\vartheta)d\vartheta = \psi(\vartheta_0), \tag{28.3}$$

$$\frac{1}{2\pi}\int_0^{2\pi}\psi(\vartheta)\cot\left(\frac{\vartheta-\vartheta_0}{2}\right)d\vartheta + \frac{i}{2\pi}\int_0^{2\pi}\psi(\vartheta)d\vartheta = -\varphi(\vartheta_0), \tag{28.4}$$

where the integrals involving $\cot\left(\frac{\vartheta-\vartheta_0}{2}\right)$ must, of course, be considered as Cauchy principal values.

The known inversion formula of Hilbert, which the latter obtained in a different (less direct) manner, is easily deduced from (28.3), (28.4).

Now examine the equation

$$\frac{1}{2\pi}\int_0^{2\pi}\psi(\vartheta)\cot\left(\frac{\vartheta-\vartheta_0}{2}\right)d\vartheta = \psi(\vartheta_0), \tag{28.5}$$

where $\varphi(\vartheta)$ is an unknown and $\psi(\vartheta)$ is a given function, satisfying the H condition. It is easily seen that this equation will only have a solution, if the following condition is met:

$$\int_0^{2\pi} \psi(\vartheta)d\vartheta = 0. \tag{28.6}$$

This condition is obtained, if both sides of (28.5) are integrated with respect to ϑ_0 from 0 to 2π, noting that

$$\int_0^{2\pi} \cot\left(\frac{\vartheta - \vartheta_0}{2}\right)d\vartheta_0 = 0. \tag{*}$$

Assume that (28.6) is satisfied and seek the solution of (28.5) satisfying, in addition, the condition

$$\int_0^{2\pi} \varphi(\vartheta)d\vartheta = 0. \tag{28.7}$$

But then (28.5) coincides with (28.3), the solution of which is given by (28.4), i.e., by (28.6),

$$\varphi(\vartheta_0) = -\frac{1}{2\pi}\int_0^{2\pi} \psi(\vartheta) \cot\left(\frac{\vartheta - \vartheta_0}{2}\right)d\vartheta. \tag{28.8}$$

The formulae (28.5) and (28.8) together with the conditions (28.6) and (28.7) represent the inversion formulae of Hilbert. [cf. Hilbert [2] Chapt. IX p. 75]

A solution of (28.5) has been found, satisfying the supplementary condition (28.7). It will be shown that other solutions result from it by addition of arbitrary constants, assuming of course that (28.6) is satisfied. First of all, obviously, $\varphi(\vartheta)$ + const. is also a solution; this follows from (*). Now let $\varphi_1(\vartheta)$ be any solution of (28.5); one obtains

$$\varphi^*(\vartheta) = \varphi_1(\vartheta) - \frac{1}{2\pi}\int_0^{2\pi} \varphi_1(\vartheta)d\vartheta.$$

Then $\varphi^*(\vartheta)$ is a solution, satisfying the supplementary condition (28.7); but such a solution is unique and consequently coincides with the solution (28.8) already found. Hence follows the statement.

From the above the following inversion formulae are easily deduced (cf. O. D. Kellogg [1] who quoted from Hilbert's lectures):

$$\frac{1}{2\pi}\int_0^{2\pi} \varphi(\vartheta) \cot\left(\frac{\vartheta-\vartheta_0}{2}\right) d\vartheta = \psi(\vartheta_0) - \frac{1}{2\pi}\int_0^{2\pi} \psi(\vartheta) d\vartheta, \quad (28.9)$$

$$-\frac{1}{2\pi}\int_0^{2\pi} \psi(\vartheta) \cot\left(\frac{\vartheta-\vartheta_0}{2}\right) d\vartheta = \varphi(\vartheta_0) - \frac{1}{2\pi}\int_0^{2\pi} \varphi(\vartheta) d\vartheta, \quad (28.10)$$

which must be understood in the sense that, *if* $\varphi(\vartheta)$ *and* $\psi(\vartheta)$ *satisfy the H condition, then each of the equations* (28.9), (18.10) *implies the other*. Leaving the proof to the reader (cf. the deduction of (28.11), (28.12)), some other inversion formulae will be proved which were given by Hellinger and Toeplitz [1] p. 1454 (with an error in sign, the authors quoting from Kellogg [1]). From Kellogg's formulae (28.9), (28.10) they deduced

$$\frac{1}{2\pi}\int_0^{2\pi} \varphi(\vartheta) \cot\left(\frac{\vartheta-\vartheta_0}{2}\right) d\vartheta + \frac{1}{2\pi}\int_0^{2\pi} \varphi(\vartheta) d\vartheta = \psi(\vartheta_0), \quad (28.11)$$

$$-\frac{1}{2\pi}\int_0^{2\pi} \psi(\vartheta) \cot\left(\frac{\vartheta-\vartheta_0}{2}\right) d\vartheta + \frac{1}{2\pi}\int_0^{2\pi} \psi(\vartheta) d\vartheta = \varphi(\vartheta_0), \quad (28.12)$$

which must be understood in the same sense as above.

Assume, for example, that (28.11) holds. Multiplying both sides of (28.11) by $d\vartheta_0$ and integrating from 0 to 2π, one obtains, using (*),

$$\int_0^{2\pi} \varphi(\vartheta) d\vartheta = \int_0^{2\pi} \psi(\vartheta) d\vartheta,$$

and hence, by (28.11),

$$\frac{1}{2\pi}\int_0^{2\pi} \varphi(\vartheta) \cot\left(\frac{\vartheta-\vartheta_0}{2}\right) d\vartheta = \psi(\vartheta_0) - \frac{1}{2\pi}\int_0^{2\pi} \psi(\vartheta) d\vartheta = \psi_0(\vartheta_0).$$

Since $\psi_0(\vartheta)$ satisfies the condition (28.6), one obtains from the above

$$\varphi(\vartheta_0) = -\frac{1}{2\pi}\int_0^{2\pi} \psi_0(\vartheta) \cot\left(\frac{\vartheta-\vartheta_0}{2}\right) d\vartheta + \text{const} =$$

$$= -\frac{1}{2\pi}\int_0^{2\pi} \psi(\vartheta) \cot\left(\frac{\vartheta-\vartheta_0}{2}\right) d\vartheta + C,$$

where C is some constant. Multiplying both sides of the above equation by $d\vartheta_0$ and integrating from 0 to 2π gives

$$2\pi C = \int_0^{2\pi} \varphi(\vartheta)d\vartheta = \int_0^{2\pi} \psi(\vartheta)d\vartheta.$$

Substitution of this value of C leads to the required formula (28.12). In exactly the same way (28.11) can be deduced from (28.12).

CHAPTER 4

CAUCHY INTEGRALS NEAR ENDS OF THE LINE OF INTEGRATION[1])

In this chapter some formulae and theorems on the behaviour of Cauchy integrals near the ends of the line of integration will be given. They are fundamental in the later work, whenever boundary problems or singular integral equations are considered for which the boundary or line of integration contains arcs. The results obtained below also find application in the more general case in which the density has discontinuities at several points on the line of integration. In this case the path of integration may be split up into separate arcs with the points of discontinuity as ends.

The results stated in the present chapter were obtained by the Author and were published (mainly without proof) in papers by N. I. Muskhelishvili [2] and also N. I. Muskhelishvili and D. A. Kveselava [1].

§ 29. Statement of the principal results.

Let $L = ab$ be a smooth arc and let $\varphi(t)$ be a function, given on L, satisfying the following conditions:

a) On any closed interval $a'b'$ of the arc ab, not including the ends a, b, the function $\varphi(t)$ satisfies the $H(\mu)$ condition

$$| \varphi(t_2) - \varphi(t_1) | \leqq A \, | t_2 - t_1 |^{\mu}, \tag{29.1}$$

where A does not depend on the position of t_1 and t_2 in the interval $a'b'$, but it may depend on the choice of a', b' (in fact, it may increase without limit when $a' \to a$ or $b' \to b$).

b) Near the ends a, b the function $\varphi(t)$ is of the form

$$\varphi(t) = \frac{\varphi^*(t)}{(t-c)^{\gamma}}, \; \gamma = \alpha + i\beta, \; 0 \leqq \alpha < 1, \tag{29.2}$$

where c is either of the ends a, b, α and β are real constants and $\varphi^*(t)$ satisfies the H condition near and at c. In (29.2), $(t-c)^{\gamma}$ is any definite branch which varies continuously on L. Further, note that for $\alpha > 0$ the condition (29.2) gives

---

1) The results of this chapter (§§ 29—33) will only be used in Parts IV, V.

$$\varphi(t) = \frac{\varphi^{**}(t)}{|t - c|^{\alpha}}, \tag{29.2a}$$

where $\varphi^{**}(t)$ is a bounded function such that $|t - c|^{\varepsilon}\varphi^{**}(t)$ satisfies the H condition for any $\varepsilon > 0$ and vanishes for $t = c$. In fact,

$$\varphi^{**}(t) = \varphi^{*}(t)e^{-i\alpha\vartheta}(t - c)^{-i\beta},$$

where $\vartheta = \arg(t - c)$; the proposition then follows by (7.3a) and (7.6).

Next examine

$$\Phi(z) = \frac{1}{2\pi i}\int_{L} \frac{\varphi(t)dt}{t - z}. \tag{29.3}$$

Under the assumed conditions for the point z, which is near c but not on L, $\Phi(z)$ is of the form:

1°. If $\gamma = 0$ [i.e, $\varphi^{*}(t) = \varphi(t)$],

$$\Phi(z) = \pm \frac{\varphi(c)}{2\pi i}\log\frac{1}{z - c} + \Phi_0(z), \tag{29.4}$$

where the upper sign is taken for $c = a$, the lower for $c = b$. By $\log\frac{1}{z - c} = -\log(z - c)$ is to be understood any branch, one-valued near c in the plane cut along L; $\Phi_0(z)$ is a bounded function tending to a definite limit when $z \to c$ along any path.

2° If $\gamma = \alpha + i\beta \neq 0$,

$$\Phi(z) = \pm \frac{e^{\pm\gamma\pi i}}{2i\sin\gamma\pi} \cdot \frac{\varphi^{*}(c)}{(z - c)^{\gamma}} + \Phi_0(z), \tag{29.5}$$

where the signs are chosen as in 1°, $(z - c)^{\gamma}$ is any branch, one-valued near c in the plane cut along L and taking the value $(t - c)^{\gamma}$ on the left side of L, and $\Phi_0(z)$ has the following properties: if $\alpha = 0$, it is bounded and tends to a definite limit when $z \to c$; if $\alpha > 0$,

$$|\Phi_0(z)| < \frac{C}{|z - c|^{\alpha_0}}, \tag{29.6}$$

where C and α_0 are real constants such that $\alpha_0 < \alpha$.

For the point $z = t_0$, lying on L, the following results hold:

3°. If $\gamma = 0$,

$$\Phi(t_0) = \pm \frac{\varphi(c)}{2\pi i}\log\frac{1}{t_0 - c} + \Phi^{*}(t_0), \tag{29.7}$$

where $\Phi^{*}(t_0)$ satisfies the H condition near and at c, the signs again being chosen as in 1°.

4°. If $\gamma = \alpha + i\beta \neq 0$,

$$\Phi(t_0) = \pm \frac{\cot \gamma\pi}{2i} \frac{\varphi^*(c)}{(t_0 - c)^\gamma} + \Phi^*(t_0), \tag{29.8}$$

and if $\alpha = 0$, $\Phi^*(t_0)$ satisfies the H condition near (and at) c; if $\alpha > 0$,

$$\Phi^*(t_0) = \frac{\Phi^{**}(t_0)}{|t_0 - c|^{\alpha_0}}, \tag{29.9}$$

where $\Phi^{**}(t_0)$ satisfies the H condition near and at c and $\alpha_0 < \alpha$; the signs are selected as in 1°.

The result 1° has already been proved in § 16 (cf. formulae (16.7), (16.8)), and 3° can be proved similarly, using the results of § 19 and the formula

$$\Phi(t_0) = \frac{1}{2\pi i} \int_L \frac{\varphi(t) - \varphi(c)}{t - t_0} dt + \frac{\varphi(c)}{2\pi i} \int_L \frac{dt}{t - t_0}.$$

The statements 2° and 4° will be proved in the following sections (§§ 31—32).

§ 30. An auxiliary estimate.

Retaining the notation of the preceding section, the integral

$$\Omega(z) = \frac{1}{2\pi i} \int_{ab} \frac{dt}{(t-a)^\gamma (t-z)}, \quad \gamma = \alpha + i\beta, \ 0 \leqq \alpha < 1, \ \gamma \neq 0, \tag{30.1}$$

will be examined.

By the Plemelj formula (§ 17)

$$\Omega^+(t_0) - \Omega^-(t_0) = (t_0 - a)^{-\gamma}.$$

For the function $(z - a)^{-\gamma}$, which is one-valued near a in the plane cut along $L = ab$ and takes the value $(t_0 - a)^{-\gamma}$ on the left side of L, one obviously has

$$[(t_0 - a)^{-\gamma}]^+ - [(t_0 - a)^{-\gamma}]^- = (1 - e^{-2\pi i\gamma})(t_0 - a)^{-\gamma}.$$

Consequently, putting

$$\omega(z) = \frac{(z-a)^{-\gamma}}{1 - e^{-2\pi i\gamma}} = \frac{e^{i\pi\gamma}}{2i \sin \gamma\pi} (z - a)^{\gamma-},$$

$[\Omega - \omega]^+ = [\Omega - \omega]^-$, and hence the function $\Omega(z) - \omega(z)$ is holomorphic in the neighbourhood of the point a (cf. § 15), i.e., in this neighbourhood

$$\Omega(z) = \frac{e^{i\gamma\pi}}{2i \sin \gamma\pi} (z - a)^{-\gamma} + A_0 + A_1(z - a) + A_2(z - a)^2 + \dots \tag{30.2}$$

Further, for $z = t_0$, $2\Omega(t_0) = \Omega^+(t_0) + \Omega^-(t_0)$, and it is easily seen that

$$\Omega(t_0) = \frac{1}{2i}(t_0 - a)^{-\gamma}\cot\gamma\pi + A_0 + A_1(t_0 - a) + \dots \qquad (30.3)$$

Similarly, in the neighbourhood of b,

$$\Omega(z) = -\frac{e^{-i\pi\gamma}}{2i\sin\gamma\pi}(z-b)^{-\gamma} + B_0 + B_1(z-b) + B_2(z-b)^2 + \dots, \qquad (30.2a)$$

$$\Omega(t_0) = -\frac{1}{2i}(t_0 - b)^{-\gamma}\cot\gamma\pi + B_0 + B_1(t_0 - b) + \dots \qquad (30.3a)$$

§ 31. Deduction of formula (29.5).

In the notation of § 29

$$\Phi(z) = \frac{1}{2\pi i}\int_{ab}\frac{\varphi^*(t)dt}{(t-c)^\gamma(t-z)}. \qquad (31.1)$$

In the case $\alpha = 0$, $\gamma = i\beta \neq 0$, rewrite (31.1) as

$$\Phi(z) = \frac{\varphi^*(c)}{2\pi i}\int_{ab}\frac{dt}{(t-c)^\gamma(t-z)} + \frac{1}{2\pi i}\int_{ab}\frac{[\varphi^*(t)-\varphi^*(c)](t-c)^{-i\beta}}{(t-z)}dt. \qquad (31.2)$$

Since $[\varphi^*(t) - \varphi^*(c)](t-c)^{-i\beta}$ satisfies the H condition near and at c (cf. § 7) and is 0 for $t = c$, the second integral on the right hand side is a function, bounded near c, tending to a definite limit as $z \to c$. The behaviour of the first integral is determined by (30.2) and (30.2a) of the preceding section and (29.5) follows for the case $\alpha = 0$.

Now consider the case $\alpha > 0$ (recalling that $\alpha < 1$). First some cruder estimate than (29.5) will be deduced, namely, that near c

$$\Phi(z) < \frac{\text{const}}{|z-c|^\nu}, \qquad (31.3)$$

where ν is some real number such that $\alpha < \nu < 1$, and therefore

$$|z-c|^\nu\Phi(z) = \frac{1}{2\pi i}\int_{ab}\frac{|z-c|^\nu - |t-c|^\nu}{|t-c|^\alpha(t-z)}\varphi^{**}(t)dt + $$
$$+ \frac{1}{2\pi i}\int_{ab}\frac{|t-c|^{\nu-\alpha}\varphi^{**}(t)dt}{t-z}.$$

As $|t-c|^{\nu-\alpha}\varphi^{**}(t)$ satisfies the H condition near c and vanishes for $t = c$, the second integral on the right side is bounded near c.

Now consider the first integral on the right side, denoting it by I. Since

$$\big|\, |z-c|^\nu - |t-c|^\nu \,\big| \leqq |z-t|^\nu,$$

$$|I| \leqq \frac{1}{2\pi}\int_{ab} \frac{|\varphi^{**}(t)|\, ds}{|t-c|^\alpha\, |t-z|^{1-\nu}}.$$

Writing $|t-c| = r$, $|z-c| = \delta$, noting that $|z-t| \geqq |r-\delta|$ and assuming that the arc ab is a standard arc (which, of course, does not affect the generality), one obtains

$$|I| \leqq B\int_{ab} \frac{|dr|}{r^\alpha\, |r-\delta|^{1-\nu}} \leqq C\int_0^R \frac{dr}{r^\alpha\, |r-\delta|^{1-\nu}},$$

where $R = |b-a|$ and B, C are certain constants. Since $0 < 1+\alpha-\nu < 1$, it is easily seen that the last integral is bounded. In fact,

$$\int_0^R \frac{dr}{r^\alpha\, |\delta-r|^{1-\nu}} = \int_0^\delta \frac{dr}{r^\alpha(\delta-r)^{1-\nu}} + \int_0^R \frac{dr}{r^\alpha(r-\delta)^{1-\nu}} \leqq$$

$$\leqq \int_0^{\delta/2} \frac{dr}{r^{1-\alpha-\nu}} + \int_{\delta/2}^{\delta} \frac{dr}{(\delta-r)^{1+\alpha-\nu}} + \int_\delta^R \frac{dr}{(r-\delta)^{1+\alpha-\nu}};$$

the last integral can be evaluated directly and the above statement verified. Thus the estimate (31.3) is obtained.

Now rewrite $\Phi(z)$ in the following manner:

$$\Phi(z) = \frac{1}{2\pi i}\int_{ab} \frac{\varphi^*(t)dt}{(t-c)^\gamma(t-z)} =$$

$$= \frac{\varphi^*(c)}{2\pi i}\int_{ab} \frac{dt}{(t-z)^\gamma(t-z)} + \frac{1}{2\pi i}\int_{ab} \frac{\varphi^*(t)-\varphi^*(c)}{(t-c)^\gamma(t-z)}dt.$$

The behaviour of the first term near c follows from (30.2) and (30.2a). The integrand in the second term may be rewritten

$$\frac{[\varphi^*(t)-\varphi^*(c)](t-c)^{-\varepsilon}}{(t-c)^{\gamma-\varepsilon}(t-z)},$$

where ε is a positive number sufficiently small for the numerator to satisfy the H condition near c. Now, using (31.3) to examine the second term and taking for $\nu = \alpha_0$ any number such that $\alpha - \varepsilon < \nu < \alpha$, the truth of (29.5) is clear in the case $\alpha > 0$.

§ 32. Deduction of formula (29.8).

The result of § 29, 4° remains to be proved. When $\alpha = 0$, this follows directly from

$$\Phi(t_0) = \frac{1}{2\pi i}\int_{ab} \frac{\varphi^*(t)(t-c)^{-i\beta}\,dt}{t-t_0} =$$

$$= \frac{\varphi^*(c)}{2\pi i}\int_{ab}\frac{(t-c)^{-i\beta}dt}{t-t_0} + \frac{1}{2\pi i}\int_{ab}\frac{[\varphi^*(t)-\varphi^*(c)](t-c)^{-i\beta}dt}{t-t_0}, \quad (32.1)$$

since the behaviour of the first term near c is determined by (30.3) or (30.3a) and the numerator of the integrand in the second term satisfies the H condition near c and vanishes for $t = c$.

Next consider the case $\alpha > 0$. For the sake of definiteness assume $c = a$ (the case $c = b$ is similar) and write

$$\Psi(t_0) = (t_0-a)^\gamma\,\Phi(t_0) = \frac{(t_0-a)^\gamma}{2\pi i}\int_{ab}\frac{\varphi^*(t)dt}{(t-a)^\gamma\,(t-t_0)}. \quad (32.2)$$

It is easily verified from (29.5), (29.6) and the formula

$$2\Phi(t_0) = \Phi^+(t_0) + \Phi^-(t_0)$$

that

$$\Psi(a) = \lim_{t_0\to\alpha}\Psi(t_0) = \frac{\cot\gamma\pi}{2i}\varphi^*(a).$$

The result will be proved, if it is shown that $\Psi(t_0)$ satisfies the H condition near a, since then

$$\Phi(t_0) = \frac{\Psi(t_0)}{(t_0-a)^\gamma} = \frac{\Psi(a)}{(t_0-a)^\gamma} + \frac{\Psi(t_0)-\Psi(a)}{(t_0-a)^\gamma} =$$

$$= \frac{\cot\gamma\pi}{2i}\,\frac{\varphi^*(a)}{(t_0-a)^\gamma} + \Phi^*(t_0),$$

where $\Phi^*(t_0)$ may obviously be written in the form (29.9). In the remainder of this section $\Psi(t_0)$ will be shown to satisfy the H condition near and at a.

One has

$$\Psi(t_0) = \frac{(t_0-a)^\gamma}{2\pi i}\int_{ab}\frac{\varphi^*(t)-\varphi^*(t_0)}{(t-a)^\gamma(t-t_0)}dt +$$

$$+ \frac{\varphi^*(t_0)(t_0-a)^\gamma}{2\pi i}\int_{ab}\frac{dt}{(t-a)^\gamma\,(t-t_0)}.$$

By (30.3) the last term satisfies the H condition near a.

Consider the first term

$$\frac{(t_0-a)^\gamma}{2\pi i}\int_{ab}\frac{\varphi^*(t)-\varphi^*(t_0)}{(t-a)^\gamma\,(t-t_0)}\,dt=$$

$$=\frac{1}{2\pi i}\int_{ab}\frac{[\varphi^*(t)-\varphi^*(t_0)][(t_0-a)^\gamma-(t-a)^\gamma]}{(t-a)^\gamma\,(t-t_0)}dt+\frac{1}{2\pi i}\int_{ab}\frac{\varphi^*(t)-\varphi^*(t_0)}{t-t_0}dt.$$

But

$$\frac{1}{2\pi i}\int_{ab}\frac{\varphi^*(t)-\varphi^*(t_0)}{t-t_0}\,dt=$$

$$=\frac{1}{2\pi i}\int_{ab}\frac{\varphi^*(t)-\varphi^*(a)}{t-t_0}\,dt+\frac{\varphi^*(a)-\varphi^*(t_0)}{2\pi i}\int_{ab}\frac{dt}{t-t_0}$$

obviously satisfies the H condition near and at a.

The integral

$$\Omega(t_0)=\frac{1}{2\pi i}\int_{ab}\frac{[\varphi^*(t)-\varphi^*(t_0)][(t_0-a)^\gamma-(t-a)^\gamma]dt}{(t-a)^\gamma(t-t_0)}$$

has still to be investigated.

One has

$$\Omega(t+h)-\Omega(t_0)=\frac{1}{2\pi i}\int_{ab}\left\{\frac{[(t_0+h-a)^\gamma-(t-a)\;][\varphi^*(t)-\varphi^*(t_0+h)]}{(t-a)^\gamma\,(t-t_0-h)}-\right.$$

$$\left.-\frac{[(t_0-a)^\gamma-(t-a)^\gamma][\varphi^*(t)-\varphi^*(t_0)]}{(t-a)^\gamma\,(t-t_0)}\right\}dt. \tag{32.3}$$

The position of t on ab will be defined by the arc coordinate s, measured from a. Let the point b correspond to $s=s_b$ and the points t_0 and t_0+h to s_0 and $s_0+\sigma$ respectively; obviously, without loss of generality, it may be assumed that $\sigma>0$ and, further, (since only the neighbourhood of the point a is of interest here) that $2\sigma<s_b-s_0$. Denote by t_2 the point, corresponding to $s_2=s_0+2\sigma$, and by t_1 the point $s_1=s_0-2\sigma$, if $s_0-2\sigma>0$, or the point a, if $s_0-2\sigma\leq 0$.

Take (32.3) as the sum of two integrals: I_0 referring to the portion $l=t_1t_2$, and I taken over the remaining part $l'=ab-t_1t_2$, so that

$$\Omega(t_0+h)-\Omega(t_0)=I_0+I. \tag{32.3a}$$

Since $(t-a)^\gamma=(t-a)^{\alpha+i\beta}$ satisfies the $H(\alpha)$ condition (§ 7, 2°),

$$\frac{|\,(t_0-a)^\gamma-(t-a)^\gamma\,|}{|\,t-t_0\,|^\alpha}\leqq M, \tag{32.4}$$

where M is a constant. Therefore, if λ is the Hölder index for $\varphi^*(t)$,

$$|I_0| \leqq A_0 \left\{ \int_l \frac{|t - t_0 - h|^\lambda \, ds}{|t - a|^\alpha \, |t - t_0 - h|^{1-\alpha}} + \int_l \frac{|t - t_0|^\lambda \, ds}{|t - a|^\alpha \, |t - t_0|^{1-\alpha}} \right\} \leqq$$

$$\leqq B_0 \left\{ \int_{s_1}^{s_2} \frac{ds}{s^\alpha \, |s - s_0 - \sigma|^{1-\alpha-\lambda}} + \int_{s_1}^{s_2} \frac{ds}{s^\alpha \, |s - s_0|^{1-\alpha-\lambda}} \right\},$$

where A_0, B_0 are constants, noting that $s_1 = s_0 - 2\sigma$ or 0, $s_2 = s_0 + 2\sigma$.

Substituting $s = s_0 + \sigma\varrho$, where ϱ is a new integration variable, it is easily seen that

$$|I_0| \leqq C_0 \sigma^\lambda, \tag{32.5}$$

where C_0 is some constant.

The integral I, taken over $l' = ab - t_1 t_2$, will be rewritten as

$$I = \frac{1}{2\pi i} \int_{l'} \left\{ \frac{(t_0 + h - a)^\gamma - (t - a)^\gamma}{(t - a)^\gamma \, (t - t_0 - h)} - \frac{(t_0 - a)^\gamma - (t - a)^\gamma}{(t - a)^\gamma \, (t - t_0)} \right\} [\varphi^*(t) -$$

$$- \varphi^*(t_0 + h)] dt + \frac{\varphi^*(t_0) - \varphi^*(t_0 + h)}{2\pi i} \int_{l'} \frac{(t_0 - a)^\gamma - (t - a)^\gamma}{(t - a)^\gamma \, (t - t_0)} dt. \tag{32.6}$$

The last integral is easily evaluated. In fact, by (32.4),

$$\left| \int_{l'} \frac{(t_0 - a)^\gamma - (t - a)^\gamma}{(t - a)^\gamma \, (t - t_0)} dt \right| \leqq A \int_{l'} \frac{ds}{s^\alpha \, |s - s_0|^{1-\alpha}} =$$

$$= A \left\{ \int_0^{s_1} \frac{ds}{s^\alpha (s_0 - s)^{1-\alpha}} + \int_{s_2}^{s_b} \frac{ds}{s^\alpha (s - s_0)^{1-\alpha}} \right\}, \tag{32.7}$$

where A is a constant; assume that in the first integral on the right side of (32.7) $s_1 = s_0 - 2\sigma > 0$ (since otherwise this integral does not occur). Using the substitution $s = s_0 \cdot \varrho$, one obtains

$$\int_0^{s_1} \frac{ds}{s^\alpha (s_0 - s)^{1-\alpha}} = \int_0^{1 - \frac{2\sigma}{s_0}} \frac{d\varrho}{\varrho^\alpha (1 - \varrho)^{1-\alpha}} \leqq \int_0^1 \frac{d\varrho}{\varrho^\alpha (1 - \varrho)^{1-\alpha}} = \text{const};$$

here and below const. is some positive constant.

Substitution of $s = s_0 + 2\sigma \cdot \varrho$ in the second integral on the right of (32.7) gives

$$\int_{s_2}^{s_b} \frac{ds}{s^\alpha (s - s_0)^{1-\alpha}} = \int_1^{\frac{s_b - s_0}{2\sigma}} \frac{d\varrho}{\left(\varrho + \frac{s_0}{2\sigma}\right)^\alpha \varrho^{1-\alpha}} \leqq \int_1^{\frac{s_b - s_0}{2\sigma}} \frac{d\varrho}{\varrho} =$$

$$= \log \frac{s_b - s_0}{2\sigma} < \text{const.} \ |\log \sigma|.$$

Thus the second term on the right side of (32.6) does not exceed const. $\sigma^\lambda \mid \log \sigma \mid$ and, a fortiori, const. $\sigma^{\lambda-\varepsilon}$, where ε is an arbitrarily fixed positive number.

Now evaluate the first term on the right side of (32.6) which will be written as the sum $I_1 + I_2$, where

$$I_1 = \frac{(t_0 + h - a)^\gamma - (t_0 - a)^\gamma}{2\pi i} \int_{l'} \frac{\varphi^*(t) - \varphi^*(t_0 + h)}{(t - a)^\gamma (t - t_0 - h)} dt,$$

$$I_2 = \frac{h}{2\pi i} \int_{l'} \frac{[\varphi^*(t) - \varphi^*(t_0 + h)][(t_0 - a)^\gamma - (t - a)^\gamma] dt}{(t - a)^\gamma (t - t_0 - h)(t - t_0)}.$$

One has

$$|I_1| \leqq A_1 \sigma^\alpha \left\{ \int_0^{s_1} \frac{ds}{s^\alpha (s_0 + \sigma - s)^{1-\lambda}} + \int_{s_2}^{s_b} \frac{ds}{s^\alpha (s - s_0 - \sigma)^{1-\lambda}} \right\},$$

where A_1 is a constant. If $\lambda \geqq \alpha$, one establishes as easily as for (32.7) that the expression in brackets is bounded for $\lambda > \alpha$, and that it does not exceed const. $|\log \sigma|$, if $\lambda = \alpha$; therefore, when $\lambda \geqq \alpha$,

$$|I_1| \leqq \text{const.} \ \sigma^{\alpha-\varepsilon},$$

where ε is an arbitrarily fixed positive constant (for $\lambda > \alpha$ one may take $\varepsilon = 0$).

If $\lambda < \alpha$, using the same substitution as for the estimate of (32.7), one obtains (for the first integral in the case $s_0 > 2\sigma$, because for $s_0 \leqq 2\sigma$ it does not occur)

$$\int_0^{s_1} \frac{ds}{s^\alpha (s_0 + \sigma - s)^{1-\lambda}} = \frac{1}{s_0^{\alpha-\lambda}} \int_0^{1 - \frac{2\sigma}{s_0}} \frac{d\varrho}{\varrho^\alpha \left(1 + \frac{\sigma}{s_0} - \varrho\right)^{1-\lambda}} \leqq$$

$$\leqq \frac{1}{s_0^{\alpha-\lambda}} \int_0^1 \frac{d\varrho}{\varrho^\alpha (1 - \varrho)^{1-\lambda}} < \text{const.} \ \sigma^{\lambda-\alpha},$$

$$\int_{s_2}^{s_b} \frac{ds}{s^\alpha (s-s_0-\sigma)^{1-\lambda}} = \frac{1}{(2\sigma)^{\alpha-\lambda}} \int_1^{\frac{s_b - s_0}{2\sigma}} \frac{d\varrho}{\left(\varrho + \frac{s_0}{2\sigma}\right)^\alpha \left(\varrho - \frac{1}{2}\right)^{1-\lambda}} <$$

$$< \frac{1}{(2\sigma)^{\alpha-\lambda}} \int_1^\infty \frac{d\varrho}{(\varrho - \frac{1}{2})^{1+\alpha-\lambda}} = \text{const. } \sigma^{\lambda-\alpha}.$$

Thus, for $\lambda < \alpha$,

$$|I_1| \leqq \text{const. } \sigma^\lambda.$$

Finally an estimate of I_2 will be given. One has

$$|I_2| \leqq A_2 \sigma \left\{ \int_0^{s_1} \frac{ds}{s^\alpha (s_0 + \sigma - s)^{1-\lambda} (s_0 - s)^{1-\alpha}} + \right.$$

$$\left. + \int_{s_2}^{s_b} \frac{ds}{s^\alpha (s - s_0 - \sigma)^{1-\lambda} (s - s_0)^{1-\alpha}} \right\},$$

where A_2 is a constant. With the same substitution as in the preceding case

$$\int_0^{s_1} \frac{ds}{s^\alpha (s_0 + \sigma - s)^{1-\lambda} (s_0 - s)^{1-\alpha}} =$$

$$= \frac{1}{s_0^{1-\lambda}} \int_0^{1-\frac{2\sigma}{s_0}} \frac{d\varrho}{\varrho^\alpha \left(1 + \frac{\sigma}{s_0} - \varrho\right)^{1-\lambda} (1-\varrho)^{1-\alpha}}.$$

Noting that $\frac{\sigma}{s_0} < \frac{1}{2}$, consider the two cases $\frac{\sigma}{s_0} > \frac{1}{4}$ and $\frac{\sigma}{s_0} \leqq \frac{1}{4}$.

In the first case, the last integral does not exceed

$$\int_0^{\frac{1}{2}} \frac{d\varrho}{\varrho^\alpha (1-\varrho)^{2-\lambda-\alpha}} = \text{const.};$$

in the second case, it does not exceed

$$\int_0^{1-\frac{2\sigma}{s_0}} \frac{d\varrho}{\varrho^\alpha (1-\varrho)^{2-\lambda-\alpha}} \leqq \int_0^{\frac{1}{2}} \frac{d\varrho}{\varrho^\alpha (1-\varrho)^{2-\lambda-\alpha}} + 2^\alpha \int_{\frac{1}{2}}^{1-\frac{2\sigma}{s_0}} \frac{d\varrho}{(1-\varrho)^{2-\lambda-\alpha}};$$

the first integral on the right side is bounded and the second can be evaluated directly and, as is easily seen, does not exceed const. $\left(\frac{\sigma}{s_0}\right)^{\lambda+\alpha-1}$, if $\lambda+\alpha \neq 1$, and const. $\left|\log\frac{\sigma}{s_0}\right|$, if $\lambda+\alpha=1$. Comparing the above inequalities it is concluded that in both cases

$$\sigma\int_0^{s_1}\frac{ds}{s^\alpha(s_0+\sigma-s)^{1-\lambda}(s_0-s)^{1-\alpha}} \leqq \text{const. } \sigma^{\lambda-\varepsilon},$$

where ε is an arbitrarily fixed positive constant.

Further,

$$\sigma\int_{s_2}^{s_b}\frac{ds}{s^\alpha(s-s_0-\sigma)^{1-\lambda}(s-s_0)^{1-\alpha}} = \frac{\sigma^\lambda}{2^{1-\lambda}}\int_1^{\frac{s_b-s_0}{2\sigma}}\frac{d\varrho}{\left(\varrho+\frac{s_0}{2\sigma}\right)^\alpha\left(\varrho-\frac{1}{2}\right)^{1-\lambda}\varrho^{1-\alpha}} \leqq$$

$$\leqq \frac{\sigma^\lambda}{2^{1-\lambda}}\int_1^{\frac{s_b-s_0}{2\sigma}}\frac{d\varrho}{(\varrho-\frac{1}{2})^{2-\lambda}};$$

evaluation of the last integral shows that

$$\sigma\int_{s_2}^{s_b}\frac{d\sigma}{s^\alpha(s_0+\sigma-s)^{1-\lambda}(s-s_0)^{1-\alpha}} \leqq \text{const. } \sigma^{\lambda-\varepsilon},$$

where ε is an arbitrarily fixed positive constant (for $\lambda < 1$ one may assume $\varepsilon = 0$).

Comparing the above estimates, one deduces

$$|I| \leqq C\sigma^{\mu-\varepsilon}, \tag{32.8}$$

where C is some positive constant, μ is the smaller of the numbers α, λ and ε is an arbitrary positive number.

The inequalities (32.5) and (32.8) have thus been proved.

§ 33. On the behaviour of a Cauchy integral near points of discontinuity.

The results stated in the preceding sections are also easily applied in the case in which the density $\varphi(t)$ of the Cauchy integral

$$\Phi(z) = \frac{1}{2\pi i}\int_L \frac{\varphi(t)dt}{t-z} \tag{33.1}$$

has discontinuities of a certain nature at a finite number of points of L, not necessarily coinciding with the ends. Obviously it is sufficient to study the case when $L = ab$ is a simple (as always, smooth) arc and $\varphi(t)$ has on ab only one point of discontinuity at c.

The following assumptions will be made. Let $\psi(t)$ be some function given everywhere on the arc ab, except possibly at the point c, and let its limits $\psi(c-0)$ and $\psi(c+0)$ exist, when the point t approaches the point c from a and b respectively. Then the function $\psi(t)$ will be ascribed the values $\psi(c-0)$ and $\psi(c+0)$ respectively at c. After this it is clear how, for example, the statement that the function $\psi(t)$ satisfies the H condition on each of the closed arcs ac and cb separately is to be understood.

Now assume that $\varphi(t)$ in (33.1) has the form

$$\varphi(t) = \frac{\varphi^*(t)}{(t-c)^\lambda}, \quad \gamma = \alpha + i\beta, \quad 0 \leqq \alpha < 1, \qquad (33.2)$$

where $\varphi^*(t)$ satisfies the H condition on each of the closed arcs ac and cb.

The expression $(t-c)^\gamma$ must be understood as follows. Let

$$ac = L', \quad cb = L''$$

and suppose the plane cut along L'' (Fig. 9). By the function $(t-c)^\gamma$ will be understood any branch, holomorphic near c in the plane cut along L''. Thus the function $(t-c)^\gamma$ takes on L' the same values from the left and from the right, and on L''

$$[(t-c)^\gamma]^- = e^{2\pi i\gamma}[(t-c)^\gamma]^+.$$

By $(t-c)^\gamma$ on $L = L' + L''$ will be understood the value of $(z-c)^\gamma$ from the left.

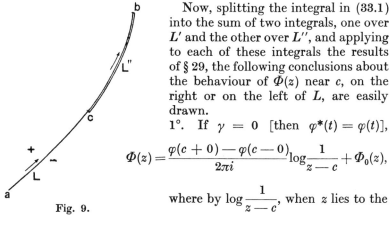

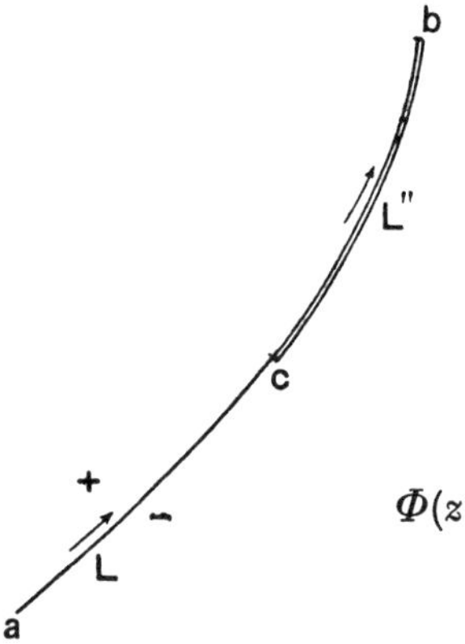

Fig. 9.

Now, splitting the integral in (33.1) into the sum of two integrals, one over L' and the other over L'', and applying to each of these integrals the results of § 29, the following conclusions about the behaviour of $\Phi(z)$ near c, on the right or on the left of L, are easily drawn.

1°. If $\gamma = 0$ [then $\varphi^*(t) = \varphi(t)$],

$$\Phi(z) = \frac{\varphi(c+0) - \varphi(c-0)}{2\pi i} \log \frac{1}{z-c} + \Phi_0(z),$$

where by $\log \frac{1}{z-c}$, when z lies to the right or to the left of L, is understood any branch holomorphic near c on the left or right of L; $\Phi_0(z)$ denotes a bounded function which tends to definite limits when z tends to c along any path remaining on the right or left of L.

2°. If $\gamma = \alpha + i\beta \neq 0$,

$$\Phi(z) = \frac{e^{\gamma\pi i}\varphi^*(c+0) - e^{-\gamma\pi i}\varphi^*(c-0)}{2i \sin \gamma\pi} \cdot \frac{1}{(z-c)^\gamma} + \Phi_0(z)$$

on the left of L,

$$\Phi(z) = \frac{e^{\gamma\pi i}\varphi^*(c+0) - e^{\gamma\pi i}\varphi^*(c-0)}{2i \sin \gamma\pi} \cdot \frac{1}{(z-c)^\gamma} + \Phi_0(z)$$

on the right of L,

where the function $\Phi_0(z)$ has the following properties. If $\alpha = 0$, it is bounded and tends to definite limits when z approaches c from the left or right of L; if $\alpha > 0$,

$$|\Phi_0(z)| < \frac{C}{|z-c|^{\alpha_0}},$$

where C and α_0 are positive constants and $\alpha_0 < \alpha$.

For the point $z = t_0$ on L one will have:

3°. If $\gamma = 0$,

$$\Phi(t_0) = \frac{\varphi(c+0) - \varphi(c-0)}{2\pi i} \log \frac{1}{t_0 - c} + \Phi^*(t_0),$$

where $\log \dfrac{1}{t_0 - c}$ is any branch, varying continuously on ac and on cb (except at the point c), and $\Phi^*(t_0)$ a function satisfying the H condition on each of the closed arcs ac and cb.

4°. If $\gamma = \alpha + i\beta \neq 0$,

$$\Phi(t_0) = \left\{ \frac{e^{\gamma\pi i}}{2i \sin \gamma\pi} \varphi^*(c+0) - \frac{\cot \gamma\pi}{2i} \varphi^*(c-0) \right\} \frac{1}{(t_0 - c)^\gamma} + \Phi^*(t_0)$$

for t_0 on L',

$$\Phi(t_0) = \left\{ \frac{\cot \gamma\pi}{2i} \varphi^*(c+0) - \frac{e^{-\gamma\pi i}}{2i \sin \gamma\pi} \varphi^*(c-0) \right\} \frac{1}{(t_0 - c)^\gamma} + \Phi^*(t_0)$$

for t_0 on L'',

where $\Phi^*(t_0)$ satisfies the H condition on each of the closed arcs ac and cb, if $\alpha = 0$, and if $\alpha > 0$,

$$\Phi^*(t_0) = \frac{\Phi^{**}(t_0)}{|t_0 - c|^{\alpha_0}}, \quad \alpha_0 < \alpha,$$

where Φ^{**} satisfies the H condition on each of the closed arcs L' and L'' [obviously it may be assumed that $\Phi^{**}(t_0)$ satisfies the H condition on the entire arc L, including c; for this, if necessary, α_0 may be replaced by a larger value].

PART II

THE HILBERT AND THE RIEMANN-HILBERT PROBLEMS AND SINGULAR INTEGRAL EQUATIONS (CASE OF CONTOURS)

In this Part the solution of a boundary problem of the theory of analytic functions, called the Hilbert problem, is given. Connected with this is the problem referred to as the Riemann-Hilbert problem (Chapter 5).

Further, an account is given of the theory of an important class of singular integral equations, defined in § 44. In future, unless stated otherwise, any singular equation will be of this class.

The solution of the Hilbert problem and the theory of singular equations are closely connected. The theory of singular equations may be, to a certain degree, developed without reference to the Hilbert problem. However, it is just the relation to this problem which makes the theory particularly simple and clear.

CHAPTER 5

THE HILBERT AND RIEMANN-HILBERT BOUNDARY PROBLEMS

§ 34. The homogeneous Hilbert problem.

Let S^+ be a connected region, bounded by smooth contours L_0, $L_1 \ldots L_p$, not intersecting one another, the first of which encloses all the others (cf. Fig. 8, § 24). The contour L_0 may be absent in which case S^+ is an infinite region (the plane with holes). By L will be denoted the union of L_0, $L_1 \ldots L_p$ (as before the positive direction of L will be such that S^+ lies to the left when L is described in that direction), by S^- that part of the plane which is the complement of $S^+ + L$, by S_0^-, $S_1^- \ldots S_p^-$ the components of S^-, bounded respectively by L_0, $L_1 \ldots L_p$ (the first of those regions is absent, if there is no L_0; S_0^- is infinite, if L_0 exists).

The following problem will be called the homogeneous Hilbert problem:

To find a sectionally holomorphic function $\Phi(z)$ of finite degree at infinity, under the boundary condition

$$\Phi^+(t) = G(t)\Phi^-(t) \text{ on } L, \tag{34.1}$$

where $G(t)$ is a non-vanishing function of the point t on L, satisfying the H condition.

The problem formulated above is often called the Riemann problem, but the Author considers this name to be incorrect (although he made use of it himself in his earlier works following an almost established tradition), because it was first considered by D. Hilbert[1]) essentially in the form in which it is stated. However, Hilbert considered the problem for less general conditions than it is done here; he assumed L to consist of a single analytic contour and $G(t)$ to possess a continuous second derivative on the arc. In a rather complicated way he transformed the problem into a Fredholm integral equation, the setting up of which requires a preliminary determination of Green's function (of the Neumann problem) for the regions S^+ and S^-; Hilbert did not give a complete investigation of the equation which he obtained.

However, the complete solution of the problem can be obtained in an elementary manner with the aid of Cauchy integrals. This solution, as a matter of fact, is contained in the final section of Plemelj's paper [1]. Plemelj considers only the particular case in which the so-called index (cf. the following section) is zero, but from this particular case the solution of the general case may be very simply inferred.

The same problem was also investigated by É. Picard [1]. For its solution he first constructed two integral equations, one of which is a Fredholm equation of the second kind, the other a singular equation; he did not investigate these equations and later quoted an elementary solution not connected with them, which is the same as Plemelj's (in the same particular case), without, however, referring to the latter.

The first general solution was given by F. D. Gakhov [1], [2]. This solution, with some simplifications and generalizations, is given below (Gakhov only considered the case in which L consists of a single contour; the solution of the general case was due to B. V. Khvedelidze [2]).

---

[1]) D. Hilbert: Götting. Nachrichten 1905; reproduced in Hilbert's book [2] pp. 83—94. The name "Riemann problem" apparently arose by Hilbert (cf. [2] p. 83) attributing it to the class of problems "im Sinne der Riemannschen Fragestellung". Thus this problem is treated in a chapter under the title "Riemanns Probleme in der Theorie der Funktionen einer komplexen Veränderlichen".

§ 35. General solution of the homogeneous Hilbert problem. The index.

Taking logarithms on both sides of (34.1) gives

$$[\log \Phi(t)]^+ - [\log \Phi(t)]^- = \log G(t). \qquad (*)$$

This relation immediately provides the means of finding $\log \Phi(z)$ by the methods stated in § 26, if it is one-valued in S^+ and S^-; consequently, the function $\log G(t)$ given on L must also be one-valued. (In exactly this way Plemelj gave the solution of the problem for a special case; see last section). But as generally speaking this is not true, a further investigation is required.

The function $\log G(t)$ will be studied first. When t moves along the contour L_k in the positive direction (i.e., counter-clockwise for $k = 0$ and clockwise for $k = 1, 2, \ldots p$), $\log G(t)$ increases by integral multiples of $2\pi i$. Thus

$$\frac{1}{2\pi i}[\log G(t)]_{L_k} = \frac{1}{2\pi}[\arg G(t)]_{L_k} = \lambda_k, \quad k = 0, 1, \ldots p, \qquad (35.1)$$

where the λ_k are integers, positive, negative or zero, and $[\ ]_{L_k}$ denotes the increment of the expression in the brackets as the result of a circuit around L_k.

The sum

$$\varkappa = \lambda_0 + \lambda_1 + \ldots + \lambda_p = \frac{1}{2\pi i}[\log G(t)]_L = \frac{1}{2\pi}[\arg G(t)]_L \quad (35.2)$$

will be called the index of the function $G(t)$ given on L, and also the index of the Hilbert problem (34.1). The index $\varkappa$ is a positive or negative integer or zero.

Let $a_1, a_2, \ldots, a_p$ be arbitrarily fixed points in the regions $S_1^-, \ldots, S_p^-$ and let the origin of the coordinates lie in S^+. Introduce

$$\Pi(z) = (z - a_1)^{\lambda_1}(z - a_2)^{\lambda_2} \ldots (z - a_p)^{\lambda_p} \qquad (35.3)$$

(if L consists only of a single contour L_0, $\Pi(z) \equiv 1$).
Then the argument of

$$G_0(t) = t^{-\varkappa}\Pi(t)G(t) \qquad (35.4)$$

will obviously return to its initial value after any circuit of the contours $L_0, L_1 \ldots L_p$, and hence

$$\log G_0(t)$$

is a definite function, one-valued and continuous (and satisfying the H condition) on L. On each of the contours L_k a branch of the function may be fixed arbitrarily.

Now, if $\Phi(z)$ is the unknown solution, $\Pi(z) \, . \, \Phi(z)$ is holomorphic in S^+ and $z^\varkappa \Phi(z)$ is holomorphic in S^- (except possibly at the point at infinity).

Multiplying both sides of (34.1) by $\Pi(z)$ and introducing the new, unknown, sectionally holomorphic function (of finite degree at infinity)

$$\Psi(z) = \begin{cases} \Pi(z)\Phi(z) & \text{in } S^+ \\ z^\varkappa \Phi(z) & \text{in } S^- \end{cases}, \tag{35.5}$$

the condition (34.1) may be written

$$\Psi^+(t) = G_0(t)\Psi^-(t). \tag{35.6}$$

It will be seen below that it is sufficient to obtain only some particular solution of the preceding problem. It is found in the following way. Purely formally, by taking logarithms, one obtains

$$\log \Psi^+(t) - \log \Psi^-(t) = \log G_0(t)$$

from which, assuming $\log \Psi(z)$ to be one-valued, sectionally holomorphic and vanishing at infinity, it follows, by § 26, that

$$\log \Psi(z) = \frac{1}{2\pi i} \int_L \frac{\log G_0(t)dt}{t - z}, \quad \Psi(z) = e^{\Gamma(z)},$$

where

$$\Gamma(z) = \frac{1}{2\pi i} \int_L \frac{\log G_0(t)dt}{t - z}. \tag{35.7}$$

Clearly $\Psi(z)$ is sectionally holomorphic, equal to unity at infinity and everywhere different from zero. By direct verification it is seen to be a (particular) solution of the problem (35.6). Indeed, by (35.7), $\Gamma^+(t_0) - \Gamma^-(t_0) = \log G_0(t_0)$, where t_0 is an arbitrary point on L, hence

$$\frac{\Psi^+(t_0)}{\Psi^-(t_0)} = e^{\log G_0(t_0)} = G_0(t_0),$$

i.e., (35.6). From the particular solution of the problem (35.6) a particular solution of the original problem (34.1) follows directly by (35.5):

$$X(z) = \begin{cases} \dfrac{1}{\Pi(z)} e^{\Gamma(z)} & \text{for } z \in S^+ \\ z^{-\varkappa} e^{\Gamma(z)} & \text{for } z \in S^- \end{cases}, \tag{35.8}$$

where $\Gamma(z)$ is given by (35.7).

Later on expressions will be required for $X^+(t_0)$ and $X^-(t_0)$. But by Plemelj's formulae

$$\Gamma^+(t_0) = \tfrac{1}{2}\log G_0(t_0) + \Gamma(t_0),$$
$$\Gamma^-(t_0) = -\tfrac{1}{2}\log G_0(t_0) + \Gamma(t_0),$$

and hence, by (35.8),

$$X^+(t_0) = \frac{1}{\Pi(t_0)} e^{\frac{1}{2}\log G_0(t_0)} e^{\Gamma(t_0)}, \qquad X^-(t_0) = t_0^{-\varkappa} e^{-\frac{1}{2}\log G_0(t_0)} e^{\Gamma(t_0)}, \tag{35.9}$$

or, using (35.4),

$$X^+(t_0) = \frac{e^{\Gamma(t_0)}\sqrt{G(t_0)}}{t_0^{\varkappa/2}\sqrt{\Pi(t_0)}}, \quad X^-(t_0) = \frac{e^{\Gamma(t_0)}}{t_0^{\varkappa/2}\sqrt{\Pi(t_0)}\sqrt{G(t_0)}}; \tag{35.10}$$

the last formulae define X^+ and X^- exactly, apart from sign, on each of the contours L_k of L. For an exact definition one has to remember that

$$\frac{\sqrt{G(t)}}{t^{\varkappa/2}\sqrt{\Pi(t)}} = \frac{e^{\frac{1}{2}\log G_0(t)}}{\Pi(t)}, \tag{35.10a}$$

where $\log G_0(t)$ has the meaning arbitrarily fixed at the outset, i.e., which appears in (35.7).

The particular solution $X(z)$ will be called a fundamental solution; its special feature is that it vanishes nowhere in the finite part of the plane (if $\varkappa > 0$, $X(\infty) = 0$), including the boundary values $X^+(t)$, $X^-(t)$. Further note that the degree of $X(z)$ at infinity is exactly $(-\varkappa)$.

Now it will be shown that *all solutions of the homogeneous Hilbert problem (always keeping in mind solutions, having finite degree at infinity) will be given by*

$$\Phi(z) = X(z)P(z), \tag{35.11}$$

where $P(z)$ is an arbitrary polynomial.

In fact, let $\Phi(z)$ be any solution. By hypothesis,

$$\Phi^+(t) = G(t)\Phi^-(t), \quad X^+(t) = G(t)X^-(t),$$

and hence, since $X^+(t) \neq 0$, $X^-(t) \neq 0$,

$$\frac{\Phi^+(t)}{X^+(t)} = \frac{\Phi^-(t)}{X^-(t)}.$$

Consequently the function $\Phi(z)/X(z)$ is holomorphic in the whole plane; but as it has finite degree at infinity it is a polynomial, and the statement is proved.

In particular, it follows from (35.11) that *the limiting values* $\Phi^+(t)$, $\Phi^-(t)$ *of any solution of the Hilbert problem satisfy the H condition*; naturally this is a consequence of the fact that the given function $G(t)$ was subjected to the H condition.

If k is the degree of $P(z)$, the degree of the solution (35.11) at infinity is $(-\varkappa + k)$. Hence the degree of (35.11) is not less than the degree $(-\varkappa)$ of the fundamental solution; the degrees of $\Phi(z)$ and $X(z)$ at infinity are equal only when $k = 0$, i.e., when $P(z) \equiv C$, where C is a constant, different from zero. Consequently the number $(-\varkappa)$, i.e., the index, is the lowest possible degree a solution of the homogeneous Hilbert problem, which is not identically zero, may have at infinity.

In future, any solution $CX(z)$, with C a constant different from zero, will be referred to as a fundamental solution. From the above it follows that a *fundamental solution is exactly defined by any of the following three properties*:

1°. It does not vanish anywhere in the finite part of the plane.

2°. It has the lowest possible degree $(-\varkappa)$ at infinity.

3°. Each solution of the homogeneous Hilbert problem is of the form (35.11).

From the point of view of application the solution of the homogeneous Hilbert problem, vanishing at infinity, is of particular interest. The following corollary of the above result will be mentioned:

If $\varkappa \leq 0$, *the homogeneous Hilbert problem has no solution, vanishing at infinity (apart from the trivial solution* $\Phi(z) \equiv 0$); *if* $\varkappa > 0$, *it has exactly the* $\varkappa$ *linearly independent solutions, vanishing at infinity,*

$$X(z), \quad zX(z), \quad \ldots \quad z^{\varkappa-1}X(z). \tag{35.12}$$

In fact, any solution, vanishing at infinity, is evidently given by by (35.11), where the degree of the polynomial $P(z)$ must not be greater than $\varkappa - 1$. Therefore all solutions, vanishing at infinity, are given by

$$\Phi(z) = X(z)P_{\varkappa-1}(z),$$

where

$$P_{\varkappa-1}(z) = C_0 z^{\varkappa-1} + C_1 z^{\varkappa-2} + \ldots + C_{\varkappa-1},$$

$C_0, C_1 \ldots C_{\varkappa-1}$ being arbitrary constants.

NOTE. Everything stated in this section also remains valid when L_0 is at infinity; in this case one must take in all the preceding formulae $\lambda_0 = 0$.

§ 36. Associate homogeneous Hilbert problems.

The homogeneous Hilbert problems, corresponding to the conditions

$$\Phi^+(t) = G(t)\Phi^-(t), \quad \Psi^+(t) = [G(t)]^{-1}\Psi^-(t),$$

will be called associate (or adjoint) to one another. (Note, however, that in the case of the Hilbert problem with several unknown functions the notions of the adjoint and associate problems do not coincide).

From the last section it follows immediately that, if $\varkappa$ is the index and $X(z)$ the fundamental solution of one of these problems, $(-\varkappa)$ and $[X(z)]^{-1}$ will be those of the other.

§ 37. The non-homogeneous Hilbert problem.

The following boundary problem will be called the non-homogeneous Hilbert problem:

To find the sectionally holomorphic function $\Phi(z)$, having finite degree at infinity and satisfying on L the boundary condition

$$\Phi^+(t) = G(t)\Phi^-(t) + g(t), \tag{37.1}$$

where $G(t)$ and $g(t)$ are functions given on L, satisfying the H condition, and $G(t) \neq 0$ everywhere on L (the remaining notation being as in § 34).

This problem, a natural generalization of the homogeneous Hilbert problem, was first considered (in a somewhat different form) by I. I. Privalov [4]. He studied the case in which L is a rectilinear contour, $G(t)$, $g(t)$ are functions, integrable in the Lebesgue sense and $0 < m < | G(t) | < M$; the only restriction on $\Phi(z)$ was that its limiting value is approached along a non-tangential path. However, he used the method by which Picard tried to solve the homogeneous problem (see § 34) and did not succeed in obtaining a complete solution. A solution of the problem was first given by F. D. Gakhov, [1], [2]; with some generalizations this is the solution reproduced here. Gakhov studied the case in which L is a single contour and this was generalized by B. V. Khvedelidze [2] to the solution given in the text. However, still earlier than I. I. Privalov and F. D. Gakhov, T. Carleman [1] solved incidentally a problem, analogous to the one studied here, for a special case (cf. also § 81); Gakhov's and Carleman's methods of solution are essentially alike.

The solution of the problem considered may be obtained immediately from the results of the preceding sections. In fact, let $X(z)$ be a fundamental solution of the homogeneous problem, obtained from (37.1) by putting $g(t) \equiv 0$. Then it follows from the equation $X^+(t) = G(t)X^-(t)$ that

$$G(t) = \frac{X^+(t)}{X^-(t)}.$$

Substitution of this value of $G(t)$ in (37.1) gives

$$\frac{\Phi^+(t)}{X^+(t)} - \frac{\Phi^-(t)}{X^-(t)} = \frac{g(t)}{X^+(t)}.$$

The function $\Phi(z)/X(z)$ has finite degree at infinity. Consequently, by § 26,

$$\frac{\Phi(z)}{X(z)} = \frac{1}{2\pi i}\int_L \frac{g(t)dt}{X^+(t)(t-z)} + P(z),$$

where $P(z)$ is an arbitrary polynomial; hence

$$\Phi(z) = \frac{X(z)}{2\pi i}\int_L \frac{g(t)dt}{X^+(t)(t-z)} + X(z)P(z). \tag{37.2}$$

This is the general solution of the non-homogeneous Hilbert problem. The function $X(z)$, being the fundamental solution of the homogeneous problem for $g(t) \equiv 0$, will be called the fundamental function corresponding to the non-homogeneous problem considered; the index of this problem is to be the index $\varkappa$ of the corresponding homogeneous problem.

Of special interest for applications is the solution of the non-homogeneous problem, vanishing at infinity.

The possibility of the existence of such solutions will be explained and the solutions found. Keeping in mind that $X(z)$ has exactly the degree $(-\varkappa)$ at infinity, the solution (37.2) is seen to vanish at infinity for $\varkappa \geqq 0$, if and only if the degree of the polynomial $P(z)$ is not greater than $(\varkappa - 1)$, while for $\varkappa = 0$ one has to take $P(z) \equiv 0$.

For $\varkappa < 0$, one has obviously $P(z) \equiv 0$ and, besides, the coefficients of $z^{-1}, z^{-2} \dots z^{-\varkappa}$ in the expansion

$$\frac{1}{2\pi i}\int_L \frac{g(t)dt}{X^+(t)(t-z)} = -\frac{z^{-1}}{2\pi i}\int_L \frac{g(t)dt}{X^+(t)} - \frac{z^{-2}}{2\pi i}\int_L \frac{tg(t)dt}{X^+(t)} - \cdots$$

must be zero. Thus the following results are obtained.

If $\varkappa \geqq 0$, the general solution of the non-homogeneous Hilbert problem (37.1), *vanishing at infinity, is given by*

$$\Phi(z) = \frac{X(z)}{2\pi i}\int_L \frac{g(t)dt}{X^+(t)(t-z)} + X(z)P_{\varkappa-1}(z), \tag{37.3}$$

where the $P_{\varkappa-1}(z)$ are arbitrary polynomials of degree not greater than $(\varkappa - 1)[P_{\varkappa-1}(z) \equiv 0$ for $\varkappa = 0]$.

If $\varkappa < 0$, provided the necessary and sufficient conditions for the existence of a solution vanishing at infinity

$$\int_L \frac{t^k g(t)dt}{X^+(t)} = 0, \; k = 0, 1, \ldots, (-\varkappa - 1) \tag{37.4}$$

are fulfilled, this solution is given by

$$\Phi(z) = \frac{X(z)}{2\pi i} \int_L \frac{g(t)dt}{X^+(t)(t-z)}. \tag{37.5}$$

Note that for $\varkappa = 0$ there is always one and only one solution vanishing at infinity; also, for $\varkappa < 0$, the solution vanishing at infinity, if it exists, is unique; for $\varkappa > 0$, there is always an unlimited number of solutions (in particular, a general solution contains $\varkappa$ arbitrary constants).

§ 38. On the extension to the whole plane of analytic functions given on a circle or half-plane.

In the study of the Hilbert problem the unknown function $\Phi(z)$ was sectionally holomorphic in the whole plane and the boundary conditions contained the values which this function assumes on either side of the boundary.

However, in many important problems, one has to deal with unknown functions, holomorphic in a part of the plane, while the boundary conditions involve the values of the unknown functions and their conjugates.

Very often it is possible to transform problems of the latter type into Hilbert problems by extending the unknown holomorphic function to a sectionally holomorphic function, defined on the whole plane (excepting the boundaries).

A function, holomorphic in a given region and continuous up to the boundaries, may be extended in an infinite number of ways to a sectionally holomorphic function, defined in the whole plane (e.g. it may be given the value 0 in those parts of the plane where it is not defined originally). But the methods of extension to the entire plane now to be given are especially useful (in Part IV, other special forms of extension suitable for specific problems will be indicated). These methods refer to the case of analytic functions, given on a circle or half-plane.

1°. Functions, given on a half-plane.

Let S^+ be the upper (lower) half-plane $y > 0$ $(y < 0)$, L its boundary (i.e., the Ox axis) and S^- the lower (upper) half-plane.

Let $\Phi(z)$ be a given function of the point z of S^+, and let it be connected with the function $\Phi_*(z)$ defined in S^- by the relation

$$\Phi_*(z) = \overline{\Phi(\bar{z})}; \tag{38.1}$$

thus, by definition, $\Phi(z)$ and $\Phi_*(z)$ take conjugate complex values at points lying symmetrically with respect to the Ox axis (i.e., at points $z = x + iy$, $z = x - iy$ which are the reflections of each other in the Ox axis). The formula (38.1) will also be written

$$\Phi_*(z) = \bar{\Phi}(z), \tag{38.2}$$

where by $\bar{\Phi}(z)$ must be understood the function defined by

$$\bar{\Phi}(z) = \overline{\Phi(\bar{z})}; \tag{38.3}$$

in other words, if $\Phi(z) = u(x, y) + iv(x, y)$, then, by definition,

$$\bar{\Phi}(z) = u(x, -y) - iv(x, -y). \tag{38.3a}$$

It is easily seen that, if $\Phi(z)$ is holomorphic or meromorphic in S^+, $\Phi_*(z) = \bar{\Phi}(z)$ is holomorphic or meromorphic[1]) in S^- and that (38.1) is symmetrical, as regards Φ_* and Φ, i.e., that

$$\Phi(z) = \overline{\Phi_*(\bar{z})}, \quad [\Phi_*(z)]_* = \Phi(z). \tag{38.4}$$

It is useful to note that, if $\Phi(z)$ is a rational function

$$\Phi(z) = \frac{a_0 z^m + a_1 z^{m-1} + \ldots + a_m}{b_0 z^n + b_1 z^{n-1} + \ldots + b_n},$$

then $\Phi_*(z) = \bar{\Phi}(z)$ is simply obtained from $\Phi(z)$ by replacing the coefficients by their conjugate complex values.

Now $\Phi(z)$ will be assumed to take a definite limiting value $\Phi^+(t)$ as $z \to t$ from S^+, where t is a point of the Ox axis, i.e. a real number. Then also $\Phi_*^-(t)$ exists and

$$\Phi_*^-(t) = \bar{\Phi}^-(t) = \overline{\Phi^+(t)}, \tag{38.5}$$

because for $z \to t$ from S^-, $\bar{z} \to t$ from S^+, and hence $\Phi_*(z) = \overline{\Phi(\bar{z})}$ tends to $\overline{\Phi^+(t)}$.

For simplicity it will now be assumed that $\Phi(z)$ is holomorphic in S^+, with the possible exception of the point at infinity, and continuous on L from the left.

---

1) In general, if $\Phi(z)$ is holomorphic in some region σ^+, $\bar{\Phi}(z)$ is holomorphic in σ^-, lying symmetrically to σ^+ with respect to the Ox axis.

This follows from the fact that, if $u(x, y)$ and $v(x, y)$ satisfy the Cauchy-Riemann equations, then

$$\frac{\partial u}{\partial x} = \frac{\partial v}{\partial y}, \quad \frac{\partial u}{\partial y} = -\frac{\partial v}{\partial x};$$

these equations remain valid when $u(x, y)$, $v(x, y)$ are replaced by $u(x, -y)$, $-v(x, -y)$.

Denote by $\Omega(z)$ the sectionally holomorphic function, defined by

$$\Omega(z) = \begin{cases} \Phi(z) & \text{for } z \in S^+ \\ \Phi_*(z) & \text{for } z \in S^- \end{cases}. \tag{38.6}$$

Then obviously, by (38.5),

$$\Omega^-(t) = \overline{\Omega^+(t)}, \; \Omega^+(t) = \overline{\Omega^-(t)}. \tag{38.7}$$

These relations enable the transformation of the boundary conditions (of any problem) containing $\Phi^+(t)$ and $\overline{\Phi^+(t)}$ [or $\Phi^-(t)$ and $\overline{\Phi^-(t)}$] into conditions involving $\Omega^+(t)$ and $\Omega^-(t)$.

Another important property of this method of extending $\Phi(z)$ will be mentioned. *If the imaginary part of $\Phi^+(t)$ is zero in any interval of the Ox axis, the function $\Phi_*(z)$ is the analytic continuation of $\Phi(z)$ through this interval.* This property is none other than Schwarz's "principle of reflection". It follows directly from a result proved in § 15, since on such an interval, by (38.5), $\Phi_*^-(t) = \Phi^+(t)$.

The above notation may be applied to functions defined not only on the half-plane. Let $\Psi(z)$ be a sectionally holomorphic function with the line of discontinuity L (Ox axis). Then $\Psi_*(z)$, defined as follows (for all z not on L),

$$\Psi_*(z) = \overline{\Psi(\bar{z})} = \bar{\Psi}(z),$$

will be another sectionally holomorphic function; obviously also

$$\Psi(z) = \overline{\Psi_*(\bar{z})} = \bar{\Psi}_*(z).$$

In particular, the sectionally holomorphic function $\Omega(z)$, defined by (38.6), has obviously the property

$$\Omega_*(z) = \Omega(z); \tag{38.8}$$

generally speaking, this is not true for an arbitrary sectionally holomorphic function.

It is easily seen that, similarly to (38.5),

$$\Psi_*^-(t) = \overline{\Psi^+(t)}, \quad \Psi_*^+(t) = \overline{\Psi^-(t)}. \tag{38.5a}$$

Finally, the following important formula will be stated. Let the sectionally holomorphic function be represented by a Cauchy integral:

$$\Psi(z) = \frac{1}{2\pi i} \int_L \frac{\varphi(t)dt}{t - z} \tag{38.9}$$

(L being the real axis); assume that the integral exists in the ordinary, or in some extended sense to be stated in § 43.

An expression for $\Psi_*(z)$ will be found. By definition, $\Psi_*(z) = \overline{\Psi(\bar{z})}$. But

$$\Psi(\bar{z}) = \frac{1}{2\pi i} \int_L \frac{\varphi(t)dt}{t - \bar{z}},$$

and hence

$$\Psi_*(z) = -\frac{1}{2\pi i} \int_L \frac{\overline{\varphi(t)}\, dt}{t - z}. \tag{38.10}$$

2°. FUNCTIONS GIVEN ON A CIRCLE OR ON THE PLANE WITH A CIRCULAR INSIDE BOUNDARY.

Now let S^+ be the region $|z| < 1$ (or $|z| > 1$), S^- the region $|z| > 1$ (or $|z| < 1$) and L the common boundary of these regions, i.e., the circle $|z| = 1$.

Let $\Phi(z)$ be the function given in S^+. This function will be compared with the function $\Phi_*(z)$, defined in S^- in an analogous manner to that used in the case of the half-plane, the only difference being that conjugate points are replaced by points, inverse with respect to the circle L, i.e., the points z and $1/\bar{z}$. Accordingly $\Phi_*(z)$ will be defined as

$$\Phi_*(z) = \overline{\Phi(1/\bar{z})} \tag{38.11}$$

or, if use is made of the notation introduced above,

$$\bar{\Phi}(z) = \overline{\Phi(\bar{z})},$$

so that

$$\Phi_*(z) = \bar{\Phi}(1/z). \tag{38.12}$$

The relation (38.11) is symmetrical, i.e,

$$\Phi(z) = \overline{\Phi_*(1/\bar{z})}, \ [\Phi_*(z)]_* = \Phi(z). \tag{38.13}$$

If $\Phi(z)$ is holomorphic or meromorphic in S^+, $\Phi_*(z)$ is holomorphic or meromorphic in S^-. In particular, if

$$\Phi(z) = \frac{a_0 z^m + a_1 z^{m-1} + \ldots + a_m}{b_0 z^n + b_1 z^{n-1} + \ldots + b_n},$$

then

$$\Phi_*(z) = \bar{\Phi}(1/z) = \frac{\bar{a}_0 z^{-m} + \bar{a}_1 z^{-m+1} + \ldots + \bar{a}_m}{\bar{b}_0 z^{-n} + \bar{b}_1 z^{-n+1} + \ldots + \bar{b}_n}.$$

Further, if

$$\Phi(z) = \sum_{-\infty}^{+\infty} a_k z^k \text{ in } S^+,$$

then

$$\Phi_*(z) = \sum_{-\infty}^{+\infty} \bar{a}_k z^{-k} \text{ in } S^-.$$

It is easily seen that, if $\Phi(z)$ has a pole (zero) of order k for $z = 0$ ($z = \infty$), $\Phi_*(z)$ has also a pole (zero) of the same order for $z = \infty$ ($z = 0$).

Now $\Phi(z)$ will be assumed to tend to a definite value $\Phi^+(t)$ as $z \to t$ on L from S^+. Then $\Phi_*^-(t)$ exists and

$$\Phi_*^-(t) = \bar{\Phi}^-(1/t) = \overline{\Phi^+(t)}, \tag{38.14}$$

since, if $z \to t$ from S^-, $1/\bar{z} \to t$ from S^+, and hence

$$\Phi_*(z) = \bar{\Phi}(1/z) = \overline{\Phi(1/\bar{z})}$$

tends to $\overline{\Phi^+(t)}$.

Next assume for simplicity that $\Phi(z)$ is holomorphic in S^+ (with the possible exception of the point at infinity) and continuous on L from the left. As before, let $\Omega(z)$ be the sectionally holomorphic function

$$\Omega(z) = \begin{cases} \Phi(z) & \text{for } z \in S^+, \\ \Phi_*(z) & \text{for } z \in S^-. \end{cases} \tag{38.15}$$

Then obviously, just as in 1°,

$$\Omega^-(t) = \overline{\Omega^+(t)}, \; \Omega^+(t) = \overline{\Omega^-(t)}. \tag{38.16}$$

Further, as in the case of the half-plane, *if the imaginary part of* $\Phi^+(t)$ *is zero in some interval of the circle* L, $\Phi_*(z)$ *is the analytic continuation of* $\Phi(z)$ *through this interval* (Schwarz's principle of reflection).

Thus, as in 1°, the function $\Psi_*(z)$ may be extended to any sectionally holomorphic function $\Psi(z)$ by putting

$$\Psi_*(z) = \overline{\Psi(1/\bar{z})} = \bar{\Psi}(1/z);$$

then obviously also

$$\Psi(z) = \overline{\Psi_*(1/\bar{z})} = \bar{\Psi}_*(1/z).$$

The sectionally holomorphic function $\Omega(z)$, introduced above, has the property

$$\Omega_*(z) = \Omega(z). \tag{38.8a}$$

It is easily seen that, analogously to (38.14),

$$\Psi_*^-(t) = \overline{\Psi^+(t)}, \; \Psi_*^+(t) = \overline{\Psi^-(t)}. \tag{38.14a}$$

Finally, suppose $\Psi(z)$ is represented by the Cauchy integral

$$\Psi(z) = \frac{1}{2\pi i} \int_L \frac{\varphi(t)dt}{t - z} \tag{38.17}$$

(L being the circle $|z| = 1$). Then

$$\Psi(1/\bar{z}) = \frac{1}{2\pi i} \int_L \frac{\varphi(t)dt}{t - 1/\bar{z}},$$

$$\Psi_*(z) = \overline{\Psi(1/\bar{z})} = -\frac{1}{2\pi i} \int_L \frac{\overline{\varphi(t)}\, \overline{dt}}{\bar{t} - 1/z}.$$

Since on the circle L: $t = e^{i\vartheta}$, $\bar{t} = e^{-i\vartheta} = 1/t$, $\overline{dt} = -ie^{-i\vartheta}d\vartheta = -\frac{dt}{t^2}$, one obtains

$$\Psi_*(z) = -\frac{1}{2\pi i} \int_L \frac{\overline{\varphi(t)}\, dt}{t - z} + \frac{1}{2\pi i} \int_L \overline{\varphi(t)} \frac{dt}{t}. \tag{38.18}$$

§ 39. The Riemann-Hilbert problem.

As an application of the results obtained above an important boundary problem of the theory of analytic functions will be considered which is a particular case of a very general problem, due to Riemann. This is the problem of finding a function, analytic in some region, for a given relation between the limiting values of its real and imaginary parts. This problem was mentioned by Riemann in his famous dissertation [1] (1851). The problem, of interest here, was first studied by Hilbert [1], [2], and hence it will be called the Riemann-Hilbert problem.

This problem is as follows. Let S^+ be a finite or infinite region, bounded by a single simple contour L. It is required:

To find a function $\Phi(z) = u + iv$, holomorphic in S^+ and continuous in $S^+ + L$, satisfying the boundary condition

$$\Re(a + ib)\Phi^+ \equiv au - bv = c \text{ on } L, \tag{39.1}$$

where a, b, c are real, continuous functions given on L.

Originally Hilbert [1] reduced this problem to singular integral equations of a type, indicated in § 44, in order to give an example of the application of such equations. Then he observed (Hilbert [2]) that the problem considered could be very simply applied to the solution of the two Dirichlet problems. Later F. Noether [1] used this solution for exactly the opposite purpose, i.e., for the study of singular integral equations of the above type. A full investigation of the problem, using such methods, may be found in a paper by I. N. Vekua [6]. The solution given here follows directly from the solution of the Hilbert problem obtained above, and use will be made of the method in § 38 of extending the unknown function, holomorphic in S^+, to a sectionally holomorphic function*).

This method will be applied to the case, where S^+ is the half-plane or a circular region. But the general case (of a simply connected region) may be reduced to one of these cases by means of conformal transformation. Therefore the solution of the problem for the circle will be given first.

Before taking up this problem, one important remark will be made. Let

$$\Phi_1(z) = u_1 + iv_1, \ \Phi_2(z) = u_2 + iv_2, \ \ldots \ \Phi_k(z) = u_k + iv_k$$

be any particular solutions of the homogeneous problem

$$au - bv = 0 \text{ on } L. \tag{39.2}$$

Then obviously also any linear combination

$$\Phi(z) = C_1\Phi_1(z) + C_2\Phi_2(z) + \ldots + C_k\Phi_k(z) \tag{39.3}$$

will be a solution, where the coefficients $C_1, C_2 \ldots C_k$ are real constants.

Therefore in the following sections (up to § 43 inclusive) a linear combination will be always understood to have real (constant) coefficients. Correspondingly the functions $\Phi_1(z), \Phi_2(z) \ldots \Phi_k(z)$ will be said to be linearly independent, if none of their linear combinations with real coefficients are identically zero, except when all coefficients are zero.

§ 40. Solution of the Riemann-Hilbert problem for the circle.

Let S^+ be the circular region $|z| < 1$ and L the circle $|z| = 1$. The boundary condition (39.1) of the Riemann-Hilbert problem may obviously be written as

$$2\Re(a + ib)\Phi^+(t) = (a + ib)\Phi^+(t) + (a - ib)\overline{\Phi^+(t)} = 2c \quad \text{on } L, \tag{40.1}$$

where $a = a(t)$, $b = b(t)$, $c = c(t)$ are given continuous, real functions of the point t on L. Further, it will be assumed that these functions satisfy the H condition and that

$$a^2 + b^2 \neq 0 \text{ everywhere on } L.$$

*) This method may also be employed in many other problems (cf. the Author's works on the theory of elasticity, in particular [1], and also Part IV of this book).

Extend the unknown function $\Phi(z)$ in S^+ by the function $\Phi_*(z)$, putting, as in § 38,

$$\Phi_*(z) = \overline{\Phi}(1/z) \text{ in } S^-$$

and again let $\Phi(z)$ be the sectionally holomorphic function which equals $\Phi(z)$ in S^+ and $\Phi_*(z)$ in S^- [in § 38 this function was denoted by $\Omega(z)$]. The function $\Phi(z)$, defined thus, has the property [cf. (38.8a)]

$$\Phi_*(z) = \overline{\Phi}(1/z) = \Phi(z) \text{ for } |z| \neq 1. \tag{40.2}$$

Further, it is obviously bounded at infinity.

With this notation the boundary condition (40.1) may be written as

$$(a + ib)\Phi^+(t) + (a - ib)\Phi^-(t) = 2c \tag{40.3}$$

or else

$$\Phi^+(t) = G(t)\Phi^-(t) + g(t), \tag{40.4}$$

where

$$G(t) = -\frac{a - ib}{a + ib}, \quad g(t) = \frac{2c}{a + ib}. \tag{40.5}$$

In this way one comes to the Hilbert problem studied in §§ 34—37. Now a solution of this problem must be sought, bounded at infinity. Let $\Phi(z)$ be any solution of the Hilbert problem (40.3), satisfying the last condition. This function may not be a solution of the original problem (40.1), because it may not satisfy the supplementary condition (40.2). However, with the help of $\Phi(z)$, a solution of the original problem may always be constructed. In fact, passing in (40.3) to the conjugate value, it is seen from (38.14a) that, if $\Phi(z)$ satisfies (40.3), $\Phi_*(z)$ satisfies

$$(a - ib)\Phi_*^-(t) + (a + ib)\Phi_*^+(t) = 2c,$$

i.e., it will also be a solution of the Hilbert problem (40.3), and consequently

$$\Omega(z) = \tfrac{1}{2}[\Phi(z) + \Phi_*(z)] \tag{40.6}$$

will be a solution of the original problem (40.1). Conversely, it is clear that every solution of this problem may be obtained in this manner. In fact, each solution $\Phi(z)$ of the original problem, extended to a sectionally holomorphic function as above, is a solution of the Hilbert problem (40.3) and

$$\Phi(z) = \Phi_*(z) = \tfrac{1}{2}[\Phi(z) + \Phi_*(z)].$$

As it is required to find the general solution of the Hilbert problem, the complete set of solutions of the Riemann-Hilbert problem must also be determined.

In order to investigate this set of solutions more closely, the homogeneous Riemann-Hilbert problem, obtained from (40.1) for $c \equiv 0$, will be considered.

Denote by $\varkappa$ the index of the function $G(t)$, i.e.,

$$\varkappa = \frac{1}{2\pi i}[\log G(t)]_L = \frac{1}{2\pi i}\left[\log \frac{a - ib}{a + ib}\right]_L = \frac{1}{2\pi}\left[\arg \frac{a - ib}{a + ib}\right]_L. \quad (40.7)$$

This formula may obviously be written as

$$\varkappa = \frac{1}{\pi i}[\log (a - ib)]_L = \frac{1}{\pi}[\arg(a - ib)]_L. \quad (40.8)$$

Thus $\varkappa$ is seen to be an even number, positive, negative or zero. In the case of discontinuous coefficients in the boundary conditions (40.1) [to be considered in Part IV (§ 93)] the formula (40.8) is not always true and the index may also take odd values. The number $\varkappa$ will be referred to as the index of the Riemann-Hilbert problem.

Let $X(z)$ be the fundamental function of the Hilbert problem (40.4). This function *) is given by

$$X(z) = C e^{\Gamma(z)} \text{ for } |z| < 1, \quad X(z) = C z^{-\varkappa} e^{\Gamma(z)} \text{ for } |z| > 1, \quad (40.9)$$

where C is an arbitrary constant, not zero, and

$$\Gamma(z) = \frac{1}{2\pi i}\int_L \frac{\log\,[t^{-\varkappa}G(t)]dt}{t - z} = \frac{1}{2\pi}\int_L \frac{\Theta(t)dt}{t - z}, \quad (40.10)$$

where

$$\Theta(t) = \arg\left[- t^{-\varkappa}\,\frac{a - ib}{a + ib}\right] \quad (40.11)$$

is a real function, continuous on L. By (38.18),

$$\Gamma_*(z) = \frac{1}{2\pi}\int_L \frac{\Theta(t)dt}{t - z} - i\alpha = \Gamma(z) - i\alpha, \quad (40.12)$$

where

$$\alpha = \frac{1}{2\pi i}\int_L \frac{\Theta(t)dt}{t} = \frac{1}{2\pi}\int_0^{2\pi} \Theta(t)d\vartheta \qquad (t = e^{i\vartheta}) \quad (40.13)$$

---

*) The fundamental function $X(z)$ is the fundamental solution of the homogeneous Hilbert problem

$$\Phi^+(t) = G(t)\,\Phi^-(t).$$

This solution is given by (35.4), (35.7), (35.8). In the present case in which L is a single contour, denoted by L_0 in § 35, $\Pi(z) \equiv 1$.

is a real constant. From these expressions it may be concluded that $X_*(z) = \bar{C}e^{\Gamma_*(z)} = \bar{C}e^{-i\alpha}e^{\Gamma(z)}$ for $|z| > 1$, $X_*(z) = \bar{C}e^{-i\alpha}z^{\varkappa}e^{\Gamma(z)}$ for $|z| < 1$, i.e., that for all z not on L

$$X_*(z) = \frac{\bar{C}}{C}e^{-i\alpha}\,z^{\varkappa}\,X(z).$$

Putting

$$C = e^{-\frac{i\alpha}{2}}, \tag{40.14}$$

one obtains the fundamental function $X(z)$ with the property

$$X_*(z) = z^{\varkappa}\,X(z). \tag{40.15}$$

A distinction will now be made between the two possible cases: $\varkappa \geqq 0$ and $\varkappa \leqq -2$. For $\varkappa \geqq 0$, the homogeneous Hilbert problem, obtained from (40.3) for $c(t) \equiv 0$, has non-zero solutions bounded at infinity all of which are given by

$$\Phi(z) = P(z)X(z), \tag{40.16}$$

where

$$P(z) = C_0z^{\varkappa} + C_1z^{\varkappa-1} + \ldots + C_{\varkappa} \tag{40.17}$$

is an arbitrary polynomial of degree not greater than $\varkappa$. In order that (40.16) be also a solution of the initial homogeneous Riemann-Hilbert problem, it is necessary and sufficient that $\Phi_*(z) = \Phi(z)$, i.e., that $X_*(z)P_*(z) = X(z)P(z)$ or, using $P_*(z) = \bar{P}(1/z)$ and $X_*(z) = z^{\varkappa}\,X(z)$, that

$$\begin{aligned} z^{\varkappa}\,\bar{P}(1/z) &= \bar{C}_0 + \bar{C}_1z + \ldots + \bar{C}_{\varkappa}\,z^{\varkappa} = \\ &= C_0z^{\varkappa} + Cz^{\varkappa-1} + \ldots + C_{\varkappa} = P(z), \end{aligned} \tag{40.18}$$

i.e.,

$$C_{\varkappa} = \bar{C}_{\varkappa-k},\ k = 0,\ 1,\ \ldots,\ \varkappa. \tag{40.18a}$$

Thus, if one puts

$$C_k = A_k + iB_k,\ \ k = 0,\ \ 1,\ \ldots,\ \frac{\varkappa}{2},$$

where the A_k, B_k are real numbers (and $B_{\varkappa/2} = 0$), then

$$C_k = A_{\varkappa-k} - iB_{\varkappa-k},\quad k = \frac{\varkappa}{2} + 1,\ \ldots,\ \varkappa.$$

Altogether there will be $(\varkappa + 1)$ arbitrary real constants. Denoting these constants in any order by D_0, D_1, ..., $D_{\varkappa}$, it is seen that for $\varkappa \geqq 0$ the general solution of the homogeneous Riemann-Hilbert problem has the form

$$\Phi(z) = D_0\Phi_0(z) + D_1\Phi_1(z) + \ldots + D_{\varkappa}\Phi_{\varkappa}(z), \tag{40.19}$$

where the D_0, D_1, . . ., $D_\varkappa$ are real arbitrary constants and the $\Phi_0(z)$, $\Phi_1(z)$, . . ., $\Phi_\varkappa(z)$ are linearly independent particular solutions of the same problem (where linear independence is to be understood in the sense, indicated in the last section).

For $\varkappa \leqq -2$, the homogeneous Hilbert problem, corresponding to (40.3), has no solutions, different from zero and bounded at infinity; hence there are also no non-zero solutions of the homogeneous Riemann-Hilbert problem under consideration.

Thus there are the following results concerning the homogeneous Riemann-Hilbert problem:

1°. *For* $\varkappa \geqq 0$, *the homogeneous Riemann-Hilbert problem has exactly* $\varkappa + 1$ *linearly independent solutions; the general solution is given by*

$$\Phi(z) = X(z)\,(C_0 z^\varkappa + C_1 z^{\varkappa-1} + \ldots + C_\varkappa),$$

where the C_0, C_1, . . ., $C_\varkappa$ *are constants, subject to* (40.18a), *but otherwise arbitrary. Here* $X(z)$ *is the fundamental function of the Hilbert problem* (40.3), *subject to* (40.15). The function $X(z)$ is obviously defined uniquely apart from an arbitrary constant real factor which may be determined from (40.9) — (40.11), (40.13), (40.14).

2°. *For* $\varkappa \leqq -2$, *the homogeneous Riemann-Hilbert problem has no solutions, different from zero.*

The non-homogeneous problem (40.1) will now be considered. To construct its general solution, it is sufficient to find only one particular solution, as the general solution is found by addition to the latter of the general solution of the homogeneous problem. However, to find a particular solution of the non-homogeneous Riemann-Hilbert problem, it is sufficient to find any particular solution of the Hilbert problem (40.3), bounded at infinity, because a particular solution of the Riemann-Hilbert problem is obtained from the latter by (40.6). On the other hand, it is known that, if the original Riemann-Hilbert problem has a solution, then the corresponding Hilbert problem has also a solution, bounded at infinity. Therefore, to resolve the question of the solution of the non-homogeneous problem, direct use may be made of the result in § 37, with an obvious small alteration due to the fact that the solution sought there vanished at infinity, while here it is only required to be bounded at infinity. Thus one obtains the result.

3°. *For* $\varkappa \geqq 0$, *the non-homogeneous Riemann-Hilbert problem has always a solution; the general solution involves linearly* $(\varkappa + 1)$ *real arbitrary constants.*

4°. *For* $\varkappa \leqq -2$, *this problem has a solution, if and only if the condition* (40.21) *or* (40.23), *given below, is satisfied, and in this case the solution is unique.*

The condition, referred to in 4°, is essentially the condition (37.4) in which now k must be given the values $0, 1, \ldots -\varkappa - 2$, because the solutions of the Hilbert problem, admitted here, do not vanish (but are bounded) at infinity. Thus these conditions have the form

$$\int_L \frac{t^k g(t)\,dt}{X^+(t)} = 0$$

or

$$\int_L \frac{t^k c(t)dt}{[a(t)+ib(t)]X^+(t)} = 0,\ k = 0,\ 1,\ \ldots,\ -\varkappa - 2. \qquad (40.20)$$

These conditions will now be transformed. By (40.10)

$$\Gamma^+(t_0) = \frac{i}{2}\,\Theta(t_0) + \frac{1}{2\pi}\int_L \frac{\Theta(t)\,dt}{t - t_0}$$

or, putting $t = e^{i\vartheta}$, $t_0 = e^{i\vartheta_0}$,

$$\Gamma^+(t_0) = \frac{i}{2}\Theta(t_0) + \frac{1}{4\pi}\int_0^{2\pi} \Theta(t)\cot\frac{\vartheta - \vartheta_0}{2}d\vartheta + \frac{i}{4\pi}\int_0^{2\pi}\Theta(t)d\vartheta,$$

and hence, noting that the last term equals $\frac{i\alpha}{2}$ by (40.13), that $X(z) = e^{-\frac{i\alpha}{2}}e^{\Gamma(z)}$ for $|z| < 1$ and that

$$e^{i\Theta(t_0)} = -t_0^{\varkappa}\frac{a(t_0) - ib(t_0)}{a(t_0) + ib(t_0)},$$

one obtains

$$X^+(t_0) = \pm\, t_0^{-\frac{\varkappa}{2}}\sqrt{-\frac{a(t_0) - ib(t_0)}{a(t_0) + ib(t_0)}}\exp\left\{\frac{1}{4\pi}\int_0^{2\pi}\Theta(t)\cot\frac{\vartheta - \vartheta_0}{2}\,d\vartheta\right\}.$$

Substitution in (40.20) gives

$$\int_0^{2\pi} e^{i\left(\frac{\varkappa}{2}+k\right)\vartheta}\,\Omega(\vartheta)c(\vartheta)d\vartheta = 0,\ k = 1,\ 2,\ \ldots,\ -\varkappa - 1, \qquad (40.21)$$

where

$$\Omega(\vartheta) = \frac{1}{\sqrt{a^2(\vartheta) + b^2(\vartheta)}}\exp\left\{-\frac{1}{4\pi}\int_0^{2\pi}\Theta(\vartheta_1)\cot\frac{\vartheta_1 - \vartheta}{2}\,d\vartheta_1\right\} \qquad (40.22)$$

and $a(t)$, $b(t)$, $c(t)$, $\Theta(t)$ are denoted by $a(\vartheta)$, $b(\vartheta)$, $c(\vartheta)$, $\Theta(\vartheta)$.

These conditions are equivalent to the following $(-\varkappa - 1)$ real conditions, written in real form

$$\int_0^{2\pi} \Omega(\vartheta)c(\vartheta)\cos k\vartheta d\vartheta = 0 \quad (k = 0, 1, \ldots, -\frac{\varkappa}{2} - 1),$$

$$\int_0^{2\pi} \Omega(\vartheta)c(\vartheta)\sin k\vartheta d\vartheta = 0 \quad (k = 1, \ldots, -\frac{\varkappa}{2} - 1). \tag{40.23}$$

The formulae, giving the solution of the initial non-homogeneous Riemann-Hilbert problem (40.1) when it exists, have yet to be written down.

If $\varkappa \leqq -2$ and if (40.20) is satisfied, the Hilbert problem (40.3) has the unique solution $\Phi(z)$ which follows from (37.5), using (40.5),

$$\Phi(z) = \frac{X(z)}{\pi i}\int_L \frac{c\,dt}{(a+ib)X^+(t)(t-z)}. \tag{40.24}$$

From the uniqueness of the solution, $\Phi(z)$ will also be a solution of the original problem (40.1); thus

For $\varkappa \leqq -2$, if the necessary and sufficient conditions (40.21) *or their equivalent* (40.23) *for the existence of the solution are satisfied, the* (*unique*) *solution of the Riemann-Hilbert problem* (40.1) *is given by* (40.24).

For $\varkappa \geqq 0$, the formula

$$\Psi(z) = \frac{X(z)}{\pi i}\int_L \frac{c\,dt}{(a+ib)X^+(t)(t-z)} \tag{*}$$

gives a particular solution of the problem (40.3); a particular solution $\Phi(z)$ of the original problem (40.1) is obtained by taking, corresponding to (40.6),

$$\Phi(z) = \tfrac{1}{2}[\Psi(z) + \Psi_*(z)]. \tag{**}$$

The function $\Psi_*(z)$ will be evaluated. Using (38.18), since

$$X_*(z) = z^{\varkappa}X(z),\ \overline{X^+(t)} = X_*^-(t) = t^{\varkappa}X^-(t)$$

and, by the definition of $X(z)$, $(a-ib)X^-(t) = -(a+ib)X^+(t)$, one easily obtains

$$\Psi_*(z) = X_*(z)\left\{-\frac{1}{\pi i}\int_L \frac{c\,dt}{(a-ib)\overline{X^+(t)}(t-z)} + \frac{1}{\pi i}\int_L \frac{c\,dt}{(a-ib)\overline{X^+(t)}t}\right\} =$$

$$= z^{\varkappa}X(z)\left\{\frac{1}{\pi i}\int_L \frac{t^{-\varkappa}c\,dt}{(a+ib)X^+(t)(t-z)} - \frac{1}{\pi i}\int_L \frac{t^{-\varkappa}c\,dt}{(a+ib)X^+(t)t}\right\}.$$

Substituting this expression in (**), one obtains the formula, giving a particular solution of the Riemann-Hilbert problem (40.1) for $\varkappa \geqq 0$,

$$\Phi(z) = \frac{X(z)}{2\pi i}\left\{\int_L \frac{c\,dt}{(a+ib)X^+(t)(t-z)} + z^\varkappa \int_L \frac{t^{-\varkappa}\,c\,dt}{(a+ib)X^+(t)(t-z)}\right\} - \\ - \frac{z^\varkappa X(z)}{2\pi i}\int_L \frac{t^{-\varkappa}\,c}{(a+ib)X^+(t)}\frac{dt}{t}. \quad (40.25)$$

When $\varkappa = 0$, the formula may be somewhat simplified:

$$\Phi(z) = \frac{X(z)}{\pi i}\int_L \frac{c\,dt}{(a+ib)X^+(t)(t-z)} - \frac{X(z)}{2\pi i}\int_L \frac{c\,dt}{(a+ib)X^+(t)t}. \quad (40.26)$$

§ 41. Example: The Dirichlet problem for a circle.

As the simplest example the Dirichlet problem for the circle $S^+(|z| < 1)$ will be considered, i.e., the problem of finding a function harmonic in S^+, continuous in $S^+ + L$ and satisfying the boundary condition,

$$u = f(t) \text{ on } L, \quad (41.1)$$

where $f(t)$ is a real continuous function, given on L. In order to apply the above result directly, one has to assume that $f(t)$ satisfies the H condition; however, the final result is easily seen to hold without this condition.

The Dirichlet problem above is a special case of the Riemann-Hilbert problem which is obtained by putting in (39.1) $a = 1, b = 0$, $c = f(t)$.

The corresponding homogeneous Hilbert problem, i.e., the problem $(a+ib)\Phi^+(t) + (a-ib)\Phi^-(t) = 0$, obtained from (40.3) for $c \equiv 0$, has here the form

$$\Phi^+(t) + \Phi^-(t) = 0. \quad (41.2)$$

The fundamental solution of this last problem is obviously: $X(z) = A$ for $|z| < 1$, $X(z) = -A$ for $|z| > 1$, where A is an arbitrary constant. The index $\varkappa = 0$. In order that the condition (40.15) is satisfied, i.e., in this case $X_*(z) = X(z)$, it is obviously sufficient to take $A = i$, so that $X(z) = i$ for $|z| < 1$, $X(z) = -i$ for $|z| > 1$. Corresponding to these, the general solution of the homogeneous Riemann-Hilbert problem will be Ci, where C is an arbitrary real constant. The general solution of the non-homogeneous problem is obtained from (40.26) by adding to the right side the general solution Ci of the homogeneous problem. Thus

$$\Phi(z) = \frac{1}{\pi i} \int_L \frac{f(t)dt}{t - z} - \frac{1}{2\pi i} \int_L f(t) \frac{dt}{t} + iC =$$

$$= \frac{1}{2\pi i} \int_L f(t) \frac{(t + z)}{(t - z)} \frac{dt}{t} + iC \tag{41.3}$$

which is the well-known Schwarz formula.

§ 42. Reduction of the general case to that of a circular region.

The general case of a (finite or infinite) region S^+, bounded by a simple contour L, will now be considered. It will be assumed that the contour L is not only smooth, but that the angle between its tangents and some constant direction satisfies the H condition.

Let $z = \omega(\zeta)$, $\zeta = \chi(z)$ be the transformations (each of which is the inverse of the other), mapping S^+ in the z plane conformally on the circle $|\zeta| < 1$ in the ζ plane; the circle $|\zeta| = 1$ will be denoted by γ. It is known from the theory of conformal transformations that, under the assumed conditions, not only the functions $\omega(\zeta)$ and $\chi(z)$, but also their derivatives $\omega'(\zeta)$ and $\chi'(z)$ are continuous from the left on γ and on L respectively; if σ and s are the arc coordinates of the corresponding points on γ and L, then the continuous derivatives $\frac{d\sigma}{ds}$ and $\frac{ds}{d\sigma}$ exist *).

Therefore it is obvious that, if $\Phi(t)$ is any function of the point t on L, satisfying the H condition on L, then the same function, in terms of the corresponding point τ of γ, i.e., $\psi(\tau) = \varphi(\omega(\tau))$, will satisfy the H condition on γ (with the same Hölder index); the same is valid, if the roles of L and γ are interchanged.

If it is required to solve the Riemann-Hilbert problem for a region S^+, satisfying the boundary condition

$$\Re(a + ib)\Phi^+ \equiv au - bv = c \text{ on } L, \tag{42.1}$$

where $a = a(t)$, $b = b(t)$, $c = c(t)$ are functions, given on L and satisfying the H condition, then, mapping the region S^+ conformally on the circle $|\zeta| < 1$, one arrives at exactly the same problem, but for the region $|\zeta| < 1$; the boundary condition for this latter problem is given by the same formula (42.1), if a, b, c are understood

*) The result regarding the continuity of $\omega'(\zeta)$ and $\chi'(z)$ is due to O. D. Kellogg [2] who also proved that $\omega'(\zeta)$ and $\chi'(z)$ satisfy the H condition. cf. also S. Warschawski [1].

to be the functions $a(\omega(\tau))$, $b(\omega(\tau))$, $c(\omega(\tau))$ of points on the circle γ.

Hence conclusions obtained for the case of a circular region may be directly transferred to the case of regions of the above type. The solution for the region S^+ and the conditions for the solubility of the problem are obtained in terms of z from the corresponding formulae for the circle $|\zeta| < 1$ by putting $\zeta = \chi(z)$.

In particular, the index $\varkappa$ may obviously be defined directly in terms of the given functions $a(t)$, $b(t)$, using the same formula as for the circle:

$$\varkappa = \frac{1}{2\pi i}\left[\log\frac{a - ib}{a + ib}\right]_L = \frac{1}{2\pi}\left[\arg\frac{a - ib}{a + ib}\right]_L \tag{42.2}$$

or, alternatively,

$$\varkappa = \frac{1}{\pi i}[\log (a - ib)]_L = \frac{1}{\pi}[\arg (a - ib)]_L. \tag{42.3}$$

NOTE 1. Let $\Phi(z)$ by any solution of the Riemann-Hilbert problem (42.1). It is easily seen that under the conditions, assumed regarding the contour L and the functions $a(t)$, $b(t)$, $c(t)$, the boundary value $\Phi^+(t)$ will satisfy the H condition. This follows from the fact that the limiting value of any solution of the Riemann-Hilbert problem for the circle satisfies the H condition, provided the functions a, b,c satisfy this condition; as was stated above, functions satisfying the H condition on γ (or on L) become, on conformal transformation, functions satisfying the H condition on L (or on γ).

NOTE 2. The case in which the boundary L of the (simply-connected) region under consideration extends to infinity does not differ essentially from the above; of course, it can also be reduced to the case of the circle by means of conformal transformation.

§ 43. The Riemann-Hilbert problem for the half-plane.

In accordance with the results of the preceding section, the Riemann-Hilbert problem for the half-plane may be reduced to the same problem for the circle; in this case it is particularly simple, since the conformal transformation of a half-plane on a circle leads to an inversion. However, a direct solution will be given.

Because of the analogy with § 42, the proof will only be outlined.

First some remarks must be made regarding Cauchy integrals taken along lines extending to infinity; the remarks will be restricted to the case of an integral, taken over an infinite straight line, as the results are easily extended to the case of an arbitrary smooth line.

1°. Let L be an infinite straight line and $\varphi(t)$ a function of the point t of this line. It will be assumed that $\varphi(t)$ satisfies the H

condition on L and that it takes the definite value $\varphi(\infty)$ when $t \to \infty$ or $-\infty$. Also suppose that for sufficiently large $|t|$

$$|\varphi(t) - \varphi(\infty)| < \frac{\text{const}}{|t|^{\alpha}}, \quad \alpha > 0 \tag{43.1}$$

which plays the part of the H condition at the point $t = \infty$. Then the integral

$$\Phi(z) = \frac{1}{2\pi i} \int_L \frac{\varphi(t)\,dt}{t - z} \tag{43.2}$$

has a definite meaning, if it is understood to be the limit of the integral taken over the segment ab, where a and b are points tending to infinity on either side and at equal distances from some fixed point. In fact, it is easily seen that

$$\Phi(z) = \frac{1}{2\pi i} \int_L \frac{\varphi(t) - \varphi(\infty)}{t - z}\,dt + \frac{\varphi(\infty)}{2\pi i} \int_L \frac{dt}{t - z} =$$

$$= \frac{1}{2\pi i} \int_L \frac{\varphi(t) - \varphi(\infty)}{t - z}\,dt \pm \tfrac{1}{2}\varphi(\infty) \tag{43.3}$$

for z not on L, where the upper sign must be taken when z is in S^+ and the lower when z is in S^-; S^+ and S^- denote the half-planes to the left and right of L respectively; obviously the first integral on the right converges absolutely by (43.1). For $z = t_0$ on L, the second term on the right of (43.3) must be changed to zero.

The indicated value of the integral is also called the Cauchy principal value at the point at infinity.

The Plemelj formulae (§ 17) are easily seen to remain true for the integral (43.2).

In contrast to the case of a finite line L, the integral (43.2) may not tend to zero as $z \to \infty$. In fact, as is shown by (43.3), $\Phi(z)$ tends to $\pm \frac{1}{2}\varphi(\infty)$, when $z \to \infty$ remaining in S^+ or in S^-, the upper sign being taken for S^+, the lower for S^-.

2°. Now consider the Riemann-Hilbert problem for the half-plane. Let it be required to find the function $\Phi(z) = u + iv$, holomorphic in the half-plane $S^+ (y > 0)$ and bounded at infinity, satisfying on the real axis L the boundary condition

$$au - bv = c \tag{43.4}$$

or

$$(a + ib)\Phi^+(t) + (a - ib)\overline{\Phi^+(t)} = 2c \text{ on } L, \tag{43.5}$$

where a, b, c are real continuous functions, given on L. In addition, these functions will be assumed to satisfy the H condition and a supplementary condition of the form (43.1); further, suppose that $a^2 + b^2 \neq 0$ everywhere on L, including $t = \infty$.

A method of solution will be used which is analogous to that used in the case of the circle. Again the sectionally holomorphic function, equal to $\Phi(z)$ in S^+ and $\Phi_*(z) = \overline{\Phi}(z)$ in S^- (cf. § 38, 1°), will be denoted by $\Phi(z)$; this function, which is now defined in the whole plane, satisfies the condition

$$\Phi_*(z) = \Phi(z). \tag{43.6}$$

In this notation the boundary condition (43.5) takes the form

$$(a + ib)\Phi^+(t) + (a - ib)\Phi^-(t) = 2c \text{ on } L \tag{43.7}$$

or

$$\Phi^+(t) = G(t)\Phi^-(t) + g(t) \text{ on } L, \tag{43.8}$$

where

$$G(t) = -\frac{a - ib}{a + ib}, \quad g(t) = \frac{2c}{a + ib}. \tag{43.9}$$

This is the Hilbert problem for the line of discontinuity L and it is required to find a solution bounded at infinity.

For simplicity, the total change of the argument of the function $G(t)$ will be assumed to be zero, when t goes on L from $-\infty$ to $+\infty$; this means that one takes the index $\varkappa = 0$. (The case of an arbitrary index may be considered as for the circle. However, generally speaking, when $\varkappa \neq 0$, it is preferable to reduce the problem by inversion to that for the circle). Under these conditions the function

$$\log G(t) = \log\left[-\frac{a - ib}{a + ib}\right]$$

(assuming a definite branch to be chosen) will be easily seen to satisfy the H condition and a supplementary condition of the form (43.1).

The fundamental solution (it is easily seen what has to be understood by this) of the homogeneous Hilbert problem, corresponding to (43.8), is given by

$$X(z) = e^{\Gamma(z)} \tag{43.10}$$

where

$$\Gamma(z) = \frac{1}{2\pi i}\int_L \frac{\log G(t)dt}{t - z} = \frac{1}{2\pi}\int_{-\infty}^{+\infty} \frac{\Theta(t)dt}{t - z} \tag{43.11}$$

and

$$\Theta(t) = \arg\left[-\frac{a - ib}{a + ib}\right] \tag{43.12}$$

is a real function of the real variable t. Therefore

$$X_*(z) = \overline{X}(z) = X(z).$$

The general solution of the Hilbert problem (43.8), bounded at infinity, will be

$$\Phi(z) = \frac{X(z)}{\pi i} \int_L \frac{c\,dt}{(a + ib)X^+(t)(t - z)} + CX(z), \qquad (43.13)$$

where C is a constant.

This function will be the general solution of the initial Riemann-Hilbert problem, if C is a real constant.

In fact, for $\Phi(z)$ to be a solution of the initial problem, it is necessary and sufficient that $\Phi_*(z) = \Phi(z)$. But the first term on the right of (43.13) satisfies this condition, as is easily seen using (38.10) and noting that $(a - ib)\overline{X^+(t)} = -(a + ib)X^+(t)$. Consequently, the second term on the right of (43.13) must also satisfy the above condition, i.e., $\overline{C}X_*(z) = CX(z)$ or, since $X_*(z) = X(z)$, one obtains $\overline{C} = C$. Thus the problem is solved.

In the particular case of the Dirichlet problem $a = 1$, $b = 0$, $c = f(t)$ one may take $X(z) = i$ for $y > 0$, $X(z) = -i$ for $y < 0$, and (43.13) gives

$$\Phi(z) = \frac{1}{\pi i} \int_L \frac{f(t)\,dt}{t - z} + Ci \qquad (43.14)$$

for the half plane $y > 0$, where C is an arbitrary real constant.

CHAPTER 6

SINGULAR INTEGRAL EQUATIONS WITH CAUCHY TYPE KERNELS (CASE OF CONTOURS)

§ 44. Singular equations and singular operators.

First some terms and notation required in the theory of singular integral equations with Cauchy type kernels will be given.

In future, an operator defined by

$$\mathbf{K}\varphi = A(t_0)\varphi(t_0) + \frac{1}{\pi i}\int_L \frac{K(t_0, t)\varphi(t)\,dt}{t - t_0} \tag{44.1}$$

will be called a singular operator (with Cauchy type kernel), where L is a smooth line, t and t_0 are points on L, and $A(t_0)$, $K(t_0, t)$ are functions given on L and satisfying conditions to be stated below. The singular operator is $\mathbf{K}$ (in general, singular operators will be printed in Clarendon type). Sometimes (44.1) is conveniently written as

$$\mathbf{K}\varphi = A(t_0)\varphi(t_0) + \frac{B(t_0)}{\pi i}\int_L \frac{\varphi(t)\,dt}{t - t_0} + \frac{1}{\pi i}\int_L k(t_0, t)\varphi(t)\,dt, \tag{44.2}$$

where

$$B(t_0) = K(t_0, t_0),\ k(t_0, t) = \frac{K(t_0, t) - K(t_0, t_0)}{t - t_0}. \tag{44.3}$$

The operator $\mathbf{K}^0$, defined by

$$\mathbf{K}^0\varphi = A(t_0)\varphi(t_0) + \frac{B(t_0)}{\pi i}\int_L \frac{\varphi(t)dt}{t - t_0}, \tag{44.4}$$

will be called the dominant part of the operator $\mathbf{K}$, and $A(t)$, $B(t)$ the coefficients of the dominant part. The function $K(t_0, t)/(t - t_0)$ will be referred to as the kernel of the operator $\mathbf{K}$.

In this and the following Parts it will always be assumed that:

1°. *The line L consists of a finite number of (smooth) non-intersecting contours.*

2°. *The functions $A(t)$ and $K(t_0, t)$ [and consequently also $B(t)$] satisfy the H condition.*

3°. *The functions*

$$S(t) = A(t) + B(t),\ D(t) = A(t) - B(t) \tag{44.5}$$

do not vanish anywhere on L.

The importance of this last condition will appear later (S and D play a more important part than A and B, the symbols being easily remembered as sum and difference); the operators, satisfying 3°, will be called regular type operators.

It follows from 2° that

$$k(t_0, t) = \frac{k^*(t_0, t)}{|t - t_0|^\lambda},\ 0 \leqq \lambda < 1, \tag{44.6}$$

where $k^*(t_0, t)$ satisfies the H condition.

In some cases, in addition to 1°—3°, it will be assumed that:

1°a. The line L is the boundary of some connected region (as in § 34).

The condition 1°a will only be used when dealing with the Hilbert problem (since for its solution such a condition was assumed). As it will become clear in the sequel, this condition has no influence on the generality of the results (cf. Note 2° of this section).

Equations of the type

$$\mathbf{K}\varphi \equiv A(t_0)\varphi(t_0) + \frac{1}{\pi i}\int_L \frac{K(t_0, t)\varphi(t)\,dt}{t - t_0} = f(t_0) \tag{44.7}$$

will be called singular integral equations (implying with Cauchy type kernel), where $\mathbf{K}$ is a singular operator, subject to the conditions 1°—3°, $f(t)$ is a given and $\varphi(t)$, the unknown function. The fact that the operator $\mathbf{K}$ satisfies 3° above will be expressed by saying that the equation (44.7) is regular.

In future, $f(t)$ will be assumed to satisfy the H condition and only such solutions will be sought which also satisfy this condition.

By (44.2) the equation (44.7) may also be written

$$\mathbf{K}\varphi \equiv A(t_0)\varphi(t_0) + \frac{B(t_0)}{\pi i}\int_L \frac{\varphi(t)dt}{t - t_0} + \frac{1}{\pi i}\int_L k(t_0, t)\varphi(t)dt = f(t_0). \tag{44.8}$$

The equation

$$\mathbf{K}^0\varphi \equiv A(t_0)\varphi(t_0) + \frac{B(t_0)}{\pi i}\int_L \frac{\varphi(t)dt}{t - t_0} = f(t_0) \tag{44.9}$$

will be called the dominant equation, corresponding to (44.7).

The theory of singular equations of the form (44.7) was first elaborated almost immediately after the development of the theory of the Fredholm equations. The foundations were laid by H. Poincaré [1] and D. Hilbert [1], [2] who encountered these equations in connection with quite different problems: Hilbert when studying some boundary problems of the theory of analytic functions, Poincaré while investigating the general theory of tides.

After them the theory was considerably developed by the work of a number of authors, referred to later on.

NOTE 1. The form

$$\mathbf{K}\varphi \equiv A(t_0)\varphi(t_0) + \frac{1}{\pi i}\int_L \frac{B(t)\varphi(t)\,dt}{t-t_0} + \frac{1}{\pi i}\int_L k_1(t_0, t)\varphi(t)\,dt = f(t_0), \tag{44.10}$$

where $B(t)$ is the same as above and

$$k_1(t_0, t) = \frac{K(t_0, t) - K(t, t)}{t - t_0}, \tag{44.11}$$

is sometimes more convenient than (44.8).

NOTE 2. It is easily shown that the shape and arrangement of the contours $L_0, L_1, \ldots, L_p$, constituting L, are of no fundamental significance (provided these contours are smooth and do not intersect one another). Otherwise there would be a conflict with the practical point of view. In applications the path of integration is, as a rule, closely connected with the problems under consideration and usually the elements, entering into the equation $\mathbf{K}\varphi = f$, have more or less simple geometrical interpretations which facilitate the investigation of the problem. In passing from L to another line of integration this advantage is often lost.

In fact, together with L another line Λ will be considered, consisting of the same number of smooth, non-intersecting contours $\Lambda_0, \Lambda_1, \ldots, \Lambda_p$. Points (and also their coordinates) on Λ will be denoted by τ and a (1,1) bi-continuous relation

$$t = t(\tau) \tag{44.12}$$

between L and Λ will be established.

This may always be done in such a way that the function $t'(\tau) = dt/d\tau$ is continuous on Λ and does not vanish anywhere. (For example, it is sufficient for this purpose to relate to one another points of L and Λ with proportional arc coordinates). In addition, $t'(\tau)$ will be assumed to satisfy the H condition. This can always be achieved, if, for example, the angles between the tangents to L, Λ and a constant direction satisfy the H condition. For this the relation between t and τ, suggested above, is sufficient. However, such a relation between L and Λ may be possible, when L and Λ are merely smooth lines; the simplest obvious example is the case in which the L_k and Λ_k, constituting L and Λ, differ only in their position in the plane or are geometrically similar.

Introducing in (44.7) the variable τ in place of t, one obtains the equation

$$A(\tau_0)\varphi(\tau_0) + \frac{1}{\pi i}\int_{\Lambda} \frac{K^*(\tau_0, \tau)\varphi(\tau)\,d\tau}{\tau - \tau_0} = f(\tau_0), \qquad (44.13)$$

where $$A(\tau) = A(t(\tau)),\ \varphi(\tau) = \varphi(t(\tau)),\ f(\tau) = f(t(\tau)),$$

$$K^*(\tau_0, \tau) = \frac{(\tau - \tau_0)t'(\tau)}{t(\tau) - t(\tau_0)} K(t(\tau_0), t(\tau)). \qquad (44.14)$$

It is easily verified, using the results of § 8, that $K^*(\tau_0, \tau)$ satisfies the H condition on Λ. Thus one obtains a singular equation of the same form but with the line of integration Λ.

In particular, if L does not satisfy the condition 1°a, then, obviously, it may always be replaced by a line Λ satisfying this condition.

NOTE 3. Further, the equation (44.7) may, of course, be written in a form in which the variables are real. This can be achieved by different methods; one of the simplest of these will be indicated.

For simplicity, the angle between the tangent to L and a constant direction will be assumed to satisfy the H condition and L to consist of a single contour only. Without loss of generality the length of this contour may be assumed to be 2π; correspondingly, the parametric representation of L may be taken in the form

$$x = x(s),\ y = y(s),\ 0 \leqq s \leqq 2\pi,$$

where s is the arc coordinate; by the assumed conditions, $x'(s)$, $y'(s)$ satisfy the H condition and $x(2\pi) = x(0)$, $y(2\pi) = y(0)$, $x'(2\pi - 0) = x'(+0)$, $y'(2\pi - 0) = y'(+0)$. Further, the functions

$$\frac{e^{is} - e^{is_0}}{t - t_0} = \omega(t_0, t), \quad \frac{t - t_0}{e^{is} - e^{is_0}} = \frac{1}{\omega(t_0, t)}$$

are easily seen to satisfy the H condition on L. (cf. § 8). [If $s - s_0$ had been taken instead of $e^{is} - e^{is_0}$, then $\dfrac{s - s_0}{t - t_0}$ would have been unbounded for $s \to 2\pi$, $s_0 \to 0$. Hence $e^{is} - e^{is_0}$ was chosen, since it has the period 2π and vanishes only for $s = s_0 + 2k\pi$, where k is an integer.]

Noting that

$$\frac{dt}{t-t_0}=\frac{\omega(t_0,t)t'(s)\,ds}{e^{is}-e^{is_0}}=\frac{\omega(t_0,t)t'(s)e^{-is_0}e^{-i\frac{s-s_0}{2}}}{e^{i\frac{s-s_0}{2}}-e^{-i\frac{s-s_0}{2}}}=$$

$$=\frac{\omega(t_0,t)t'(s)e^{-is_0}}{2i}\left\{\cot\frac{s-s_0}{2}-i\right\}ds \tag{44.15}$$

and substituting this expression in (44.7), one arrives, after some simple transformations similar to those by which (44.8) was obtained from (44.7), at an equation of the form

$$a(s_0)\psi(s_0)+\frac{b(s_0)}{2\pi}\int_0^{2\pi}\cot\frac{s-s_0}{2}\psi(s)ds+\int_0^{2\pi}l(s_0,s)\psi(s)ds=g(s), \tag{44.16}$$

where $a(s)$, $b(s)$, $g(s)$ are given functions, with periods 2π and satisfying the H condition, and $l(s_0, s)$ is also a given function which may be written in the form

$$l(s_0,s)=l^*(s_0,s)\left|\cot\frac{s-s_0}{2}\right|^{\lambda},\quad 0\leqq\lambda<1, \tag{44.17}$$

where $l^*(s_0, s)$ has the period 2π in s_0 and s and satisfies the H condition. Here $\psi(s)$ is the function $\varphi(t)$, expressed in terms of s.

Conversely, every equation of the form (44.16) may be transformed into one of the form (44.7) and the contour L may be chosen arbitrarily, provided it satisfies the smoothness condition above.

The simplest interpretation of an equation of the form (44.16) is that the integration is taken over the circumference of the circle L of radius 1 with its centre at the origin, s being the argument of the point t on L. Then $t=e^{is}$, $dt=i\,e^{is}\,ds$,

$$\frac{dt}{t-t_0}=\frac{i\,e^{is}\,ds}{e^{is}-e^{is_0}}=\tfrac{1}{2}\cot\frac{s-s_0}{2}ds+\frac{i}{2}ds, \tag{44.18}$$

and hence, noting that $i\,ds=\dfrac{dt}{t}$,

$$\tfrac{1}{2}\cot\frac{s-s_0}{2}ds=\frac{dt}{t-t_0}-\tfrac{1}{2}\frac{dt}{t}.$$

Substituting this expression and also that for ds in (44.16), one obviously obtains an equation of the form (44.8) or, what is the same thing, of the form (44.7).

§ 45. Fundamental properties of singular operators.

The singular operators, introduced in the last section, will now be studied more closely. They will only be applied to functions $\varphi(t)$ which satisfy the H condition.

The singular operator $\mathbf{K}$ *transforms any function* $\varphi(t)$, *satisfying the* H *condition, into a new function* $\psi(t)$ *which also satisfies the* H *condition.* This is a direct consequence of the result, proved at the beginning of § 20.

The relation between φ and ψ will usually be written as

$$\psi = \mathbf{K}\varphi,$$

but sometimes it will be written, for example,

$$\psi(t_0) = \mathbf{K}\varphi(t_0),$$

thus indicating that in $\psi(t)$, obtained from $\varphi(t)$ by application of the operator $\mathbf{K}$, the variable t is given the value t_0. Hence the symbol $\mathbf{K}\varphi(\ )$ must be understood as the symbol for the function $\psi(\ )$.

The following notation, introduced in the last section, is again needed:

$$\mathbf{K}\varphi = A(t_0)\varphi(t_0) + \frac{1}{\pi i}\int_L \frac{K(t_0, t)\varphi(t)\,dt}{t - t_0}, \tag{45.1}$$

$$B(t) = K(t, t), \tag{45.2}$$

$$S(t) = A(t) + B(t),\ D(t) = A(t) - B(t); \tag{45.3}$$

further, it will be recalled that it was agreed to consider only regular operators, i.e., such that $S(t)$ and $D(t)$ do not vanish anywhere on L.

Yet another concept will be introduced which plays an important role in the sequel. This is the integer

$$\varkappa = \frac{1}{2\pi i}\left[\log\frac{A - B}{A + B}\right]_L = \frac{1}{2\pi i}\left[\log\frac{D}{S}\right]_L \tag{45.4}$$

which will be called the index of the operator $\mathbf{K}$ or of the equation $\mathbf{K}\varphi = f$, where, as usual, the symbol $[\ ]_L$ stands for the increment, suffered by the function in braces, on one circuit of L in the positive direction.

It follows from the definition *that the index depends only on the dominant part of the operator* $\mathbf{K}$. Further, it is easily seen that *the index* $\varkappa$ *does not change, if the line of integration* L *is replaced by another line* Λ, *as in Note 2 of the last section.* In fact, as shown by (44.13) and (44.14), the coefficients A and B remain unchanged, i.e., their values at corresponding points of L and Λ are equal.

If $B(t) = 0$ and hence $A(t) \neq 0$ everywhere on L (by the condition S, D, $\neq 0$ on L), then the equation $\mathbf{K}\varphi = f$ becomes an ordinary Fredholm equation (of the second kind). Therefore, when $B \equiv 0$ or, what is the same thing, when $S = D$, the operator $\mathbf{K}$ will be called a Fredholm operator (of the second kind). In future, when speaking simply of Fredholm operators, those of the second kind will be understood. *The index of a Fredholm operator is zero,* as shown by (45.4).

Next consider the product of two singular operators. Let $\mathbf{K}_1$ and $\mathbf{K}_2$ be singular operators, defined by

$$\mathbf{K}_1\varphi \equiv A_1(t_0)\varphi(t_0) + \frac{1}{\pi i}\int\limits_L \frac{K_1(t_0, t)\varphi(t)\,dt}{t - t_0}, \tag{45.5}$$

$$\mathbf{K}_2\psi \equiv A_2(t_0)\psi(t_0) + \frac{1}{\pi i}\int\limits_L \frac{K_2(t_0, t)\psi(t)\,dt}{t - t_0}. \tag{45.6}$$

The operator

$$\mathbf{K}^* = \mathbf{K}_1\mathbf{K}_2, \tag{45.7}$$

defined by

$$\mathbf{K}^*\psi = \mathbf{K}_1(\mathbf{K}_2\psi), \tag{45.8}$$

will be called the product of the operators $\mathbf{K}_1$ and $\mathbf{K}_2$ in the indicated order (the operators $\mathbf{K}_1\mathbf{K}_2$ and $\mathbf{K}_2\mathbf{K}_1$ being, in general, different).

An expression will be found for $\mathbf{K}_1\mathbf{K}_2$. Carrying out the operation, indicated in (45.8), and making use of the Poincaré-Bertrand inversion formula (§ 23), one obtains directly

$$\begin{aligned}\mathbf{K}_1\mathbf{K}_2\psi \equiv {} & [A_1(t_0)A_2(t_0) + B_1(t_0)B_2(t_0)]\psi(t_0) + \\ & + \frac{1}{\pi i}\int\limits_L \frac{A_1(t_0)K_2(t_0, t) + K_1(t_0, t)A_2(t)}{t - t_0}\psi(t)dt + \\ & + \frac{1}{(\pi i)^2}\int\limits_L \left[\int\limits_L \frac{K_1(t_0, t_1)K_2(t_1, t)}{(t_1 - t_0)(t - t_1)}dt_1\right]\psi(t)\,dt,\end{aligned} \tag{45.9}$$

where, as before, $B_1(t) = K_1(t, t)$, $B_2(t) = K_2(t, t)$.

It is easily shown that

$$\int\limits_L \frac{K_1(t_0, t_1)K_2(t_1, t)}{(t_1 - t_0)(t - t_1)}dt_1 = \frac{k_{12}(t_0, t)}{|t - t_0|^\lambda}, \tag{45.10}$$

where $0 \leqq \lambda < 1$ and k_{12} satisfies the H condition. In fact, the function $F(t_0, t, t_1) = K_1(t_0, t_1)K_2(t_1, t)$ satisfies the H condition.

Further,

$$\int_L \frac{F(t_0, t, t_1)\, dt_1}{(t_1 - t_0)(t - t_1)} = \frac{1}{t - t_0}\left\{\int_L \frac{F(t_0, t, t_1)\, dt_1}{t_1 - t_0} - \int_L \frac{F(t_0, t, t_1)\, dt_1}{t_1 - t}\right\} =$$

$$= \frac{\omega_0(t_0, t) - \omega(t_0, t)}{t - t_0}, \tag{45.11}$$

where

$$\omega_0(t_0, t) = \int_L \frac{F(t_0, t, t_1)\, dt_1}{t_1 - t_0}, \quad \omega(t_0, t) = \int_L \frac{F(t_0, t, t_1) dt_1}{t_1 - t}.$$

As $\omega(t_0, t)$ and $\omega_0(t_0, t)$ satisfy the H condition and, besides, $\omega(t_0, t_0) = \omega_0(t_0, t_0)$, the right side of (45.11) can be written in the form $k_{12}(t_0, t)/|\, t_0 - t\,|^\lambda$, and the statement is proved.

Thus the dominant part of the operator $\mathbf{K}^*$ consists only of the first two terms on the right of (45.9).

Denoting by $A^*(t)$ and $B^*(t)$ the coefficients of the dominant part of $\mathbf{K}^*$, one has from (45.9)

$$A^* = A_1A_2 + B_1B_2, \quad B^* = A_1B_2 + B_1A_2, \tag{45.12}$$

and hence, putting

$$S_j = A_j + B_j, \quad D_j = A_j - B_j \quad (j = 1, 2),$$
$$S^* = A^* + B^*, \quad D^* = A^* - B^*,$$

one deduces

$$S^* = S_1S_2, \quad D^* = D_1D_2. \tag{45.13}$$

It follows from the preceding formulae, by (45.4), that *the index* $\varkappa^*$ *of* $\mathbf{K}^*$ *is equal to the sum of the indices* $\varkappa_1$ *and* $\varkappa_2$ *of* $\mathbf{K}_1$ *and* $\mathbf{K}_2$:

$$\varkappa^* = \varkappa_1 + \varkappa_2.$$

If $\mathbf{K}_1$ is such that $\mathbf{K}_1\mathbf{K}_2$ is a Fredholm operator, then $\mathbf{K}_1$ will be called the reducing operator of $\mathbf{K}_2$. For $\mathbf{K}^*$ to be a Fredholm operator, it is necessary and sufficient that $B^* = 0$ or, what comes to the same thing, that $S^* = D^*$. Thus, if $\mathbf{K}_1$ is an operator reducing $\mathbf{K}_2$, then

$$A_1B_2 + B_1A_2 = 0 \tag{45.14}$$

or, alternatively,

$$S_1S_2 = D_1D_2, \tag{45.15}$$

and conversely. It follows from the above that, if $\mathbf{K}_2$ is given, then its reducing operator $\mathbf{K}_1$ may be selected in an infinite number of ways. For example, one may give an arbitrary function $S^* = D^* \neq 0$, and then S_1, D_1 will be defined by

$$S_1 = \frac{S^*}{S_2}, \quad D_1 = \frac{S^*}{D_2}; \tag{45.16}$$

in particular, one may take $S^* = D^* = 1$. As a rule it is most convenient to satisfy the condition (45.14) by taking,

$$A_1 = A_2, \; B_1 = - B_2. \tag{45.17}$$

It is seen that, *in selecting the reducing operator, only the dominant part of the operator is of importance.*

Besides, it is seen that, if $\mathbf{K}_1$ is an operator reducing $\mathbf{K}_2$, then $\mathbf{K}_2$ reduces $\mathbf{K}_1$. [Here it should be noted that in general this is not true for operators, connected with systems of equations (Part VI)].

Since the index of the Fredholm operator is zero, *the indices of* $\mathbf{K}_1$ *and* $\mathbf{K}_2$, *reducing one another, are equal in magnitude and opposite in sign.*

Now a few words will be added on the multiplication of an arbitrary number of operators. It is easily verified that, if $\mathbf{K}_1$, $\mathbf{K}_2$, $\mathbf{K}_3$ are singular operators, then

$$\mathbf{K}_1(\mathbf{K}_2\mathbf{K}_3) = (\mathbf{K}_1\mathbf{K}_2)\mathbf{K}_3, \tag{45.18}$$

i.e., that the multiplication of singular operators is associative; hence one can write simply $\mathbf{K}_1\mathbf{K}_2\mathbf{K}_3$ instead of $\mathbf{K}_1(\mathbf{K}_2\mathbf{K}_3)$ or $(\mathbf{K}_1\mathbf{K}_2)\mathbf{K}_3$. A similar law will hold for any number of operators.

NOTE. The operator $\mathbf{k}$, defined by

$$\mathbf{k}\varphi = \frac{1}{\pi i}\int_L k(t_0, t)\varphi(t)\,dt, \tag{45.19}$$

will be considered, where

$$k(t_0, t) = \frac{k_0(t_0, t)}{|t - t_0|^{\lambda}}, \quad 0 \leqq \lambda < 1, \tag{45.20}$$

and $k_0(t_0, t)$ satisfies the H condition. Such operators will be called Fredholm operators of the first kind.

Let $\mathbf{K}$ be any singular operator. Then $\mathbf{Kk}$ and $\mathbf{kK}$ are easily seen to be Fredholm operators of the first kind. On the one hand, this follows from (45.12), because the Fredholm operator of the first kind may be considered as a particular case of the singular operator (but not belonging to the regular type) in which the coefficients A and B of the dominant part are identically zero.

On the other hand, the statement is easily verified directly. If

$$\mathbf{K}\psi \equiv A(t_0)\psi(t_0) + \frac{1}{\pi i}\int_L \frac{K(t_0, t)\psi(t)\,dt}{t - t_0},$$

then, e.g. by direct substitution of $\psi = \mathbf{k}\varphi$, one obtains

$$\mathbf{Kk}\varphi \equiv \frac{1}{\pi i}\int_L n(t_0, t)\varphi(t)\,dt,$$

where

$$n(t_0, t) = A(t_0)k(t_0, t) + \frac{1}{\pi i}\int_L \frac{K(t_0, t_1)k(t_1, t)\,dt_1}{t_1 - t}$$

and $n(t_0, t)$ is easily seen to satisfy a condition of the same form as (45.20) for $k(t_0. t)$. (The inversion of the order of integration is legitimate by the Note in § 23). This is easily verified directly, and it also follows from (45.10), since by (45.20) $k(t_0, t)$ may be represented in the form

$$k(t_0, t) = \frac{k^*(t_0, t)}{t - t_0},$$

where $k^*(t_0, t)$ satisfies the H condition (cf. the end of the Note in § 23).

Similarly for **kK**.

§ 46. Adjoint operators and adjoint equations.

The operators, defined by

$$\mathbf{K}\varphi \equiv A(t_0)\varphi(t_0) + \frac{1}{\pi i}\int_L \frac{K(t_0, t)\varphi(t)\,dt}{t - t_0} \tag{46.1}$$

and

$$\mathbf{K}'\psi \equiv A(t_0)\psi(t_0) - \frac{1}{\pi i}\int_L \frac{K(t, t_0)\psi(t)\,dt}{t - t_0}, \tag{46.2}$$

i.e., the operators obtained from one another by interchanging t_0 and t in the kernel $K(t_0, t)/t - t_0)$, are called adjoint. In particular, the operator adjoint to the dominant operator $\mathbf{K}^0$, defined by

$$\mathbf{K}^0\varphi \equiv A(t_0)\varphi(t_0) + \frac{B(t_0)}{\pi i}\int_L \frac{\varphi(t)\,dt}{t - t_0}, \tag{46.3}$$

is the operator $\mathbf{K}^{0\prime}$, defined by

$$\mathbf{K}^{0\prime}\psi \equiv A(t_0)\psi(t_0) - \frac{1}{\pi i}\int_L \frac{B(t)\psi(t)\,dt}{t - t_0}. \tag{46.4}$$

It should be noted that the operator $\mathbf{K}^{0\prime}$, adjoint to the dominant operator $\mathbf{K}^0$, is, in general, not the dominant part $\mathbf{K}'^0$ of $\mathbf{K}'$, because $\mathbf{K}'^0$ is given by

$$\mathbf{K}'^0\psi \equiv A(t_0)\psi(t_0) - \frac{B(t_0)}{\pi i}\int_L \frac{\psi(t)\,dt}{t - t_0}. \tag{46.5}$$

Hence a distinction must be made between $\mathbf{K}^{0\prime}$ and $\mathbf{K}'^0$.

Reverting to the general case, the following fundamental properties of adjoint operators will be noted.

1°. The indices of adjoint operators are equal in magnitude and opposite in sign.

2°. For any two operators

$$(\mathbf{K}_1\mathbf{K}_2)' = \mathbf{K}_2'\mathbf{K}_1'; \tag{46.6}$$

in general

$$(\mathbf{K}_1\mathbf{K}_2 \ldots \mathbf{K}_n)' = \mathbf{K}_n'\mathbf{K}_{n-1}' \ldots \mathbf{K}_2'\mathbf{K}_1'. \tag{46.6a}$$

3°. For any two functions φ and ψ, satisfying the H condition,

$$\int_L \psi\mathbf{K}\varphi dt = \int_L \varphi\mathbf{K}'\psi dt. \tag{46.7}$$

These properties are easily verified directly. The property 3° is fundamental for the concept of adjoint operators, i.e., if two singular operators $\mathbf{K}$ and $\mathbf{K}'$ fulfill (46.7), whatever are the functions φ and ψ satisfying the H condition, then $\mathbf{K}$ and $\mathbf{K}'$ are adjoint in the sense of the original definition.

The integral equations

$$\mathbf{K}\varphi = f \text{ and } \mathbf{K}'\psi = g \tag{46.8}$$

will be called adjoint, whatever may be f and g on the right.

The following important result is a consequence of (46.7): *If the equation* $\mathbf{K}\varphi = f$ *has a solution, then necessarily*

$$\int_L f\psi dt = 0, \tag{46.9}$$

where ψ *is any solution of the adjoint homogeneous equation* $\mathbf{K}'\psi = 0$. (Note that solutions are always understood to satisfy the H condition).

In fact, if φ is a solution of $\mathbf{K}\varphi = f$, then

$$\int_L f\psi dt = \int_L \psi\mathbf{K}\varphi dt = \int_L \varphi\mathbf{K}'\psi dt = 0.$$

The converse result also holds and will be proved in § 53.

§ 47. Solution of the dominant equation.

In the general theory of singular integral equations of the type under consideration great importance attaches to the solution of the dominant equation

$$\mathbf{K}^0\varphi \equiv A(t_0)\varphi(t_0) + \frac{B(t_0)}{\pi i}\int_L \frac{\varphi(t)\,dt}{t - t_0} = f(t_0). \tag{47.1}$$

A general and moreover quite elementary solution, in closed form, was obtained by I. N. Vekua [1] and will be given here.

The idea of the method of solution is due to T. Carleman [1] who, however, considered a special case different from the above (cf. the beginning of § 109). This same idea was used by S. G. Mikhlin [3] who, however, only gives the solution of (47.1) for $\varkappa = 0$ (assuming besides that $A(t) = 1$ and L is a simple contour with its curvature satisfying the H condition). For $\varkappa \neq 0$, S. G. Mikhlin limits himself to the solution of the homogeneous equation $\mathbf{K}^0\varphi = 0$, assuming L to be a circle.

It will be recalled that $A(t)$, $B(t)$, $f(t)$ are functions, satisfying the H condition, and that the unknown function $\varphi(t)$ is also subject to the H condition. Further, let the line L satisfy the additional condition 1°a of § 44.

The sectionally holomorphic function

$$\Phi(z) = \frac{1}{2\pi i}\int_L \frac{\varphi(t)\,dt}{t-z} \tag{47.2}$$

will be considered.

Noting that by the Plemelj formulae

$$\varphi(t_0) = \Phi^+(t_0) - \Phi^-(t_0), \quad \frac{1}{\pi i}\int_L \frac{\varphi(t)\,dt}{t-t_0} = \Phi^+(t_0) + \Phi^-(t_0), \tag{47.3}$$

it is seen, by (47.1), that $\Phi(z)$ must be a solution of the Hilbert problem

$$(A+B)\Phi^+ - (A-B)\Phi^- = f \tag{47.4}$$

which vanishes at infinity. Conversely, let the sectionally holomorphic function $\Phi(z)$, vanishing at infinity, be a solution of the problem (47.4). Let $\varphi(t_0)$ be defined by the first of the formulae (47.3). Then (§ 26) $\Phi(z)$ may be written in the form (47.2), and hence the second formula of (47.3) will also be true; this proves that $\varphi(t)$ is a solution of the original equation (47.1).

Thus a solution of (47.1) is equivalent to the solution of the problem (47.4) with the supplementary condition $\Phi(\infty) = 0$.

The boundary problem (47.4) has now to be solved. In order to make direct use of the formulae, introduced earlier, the condition (47.4) will be written as

$$\Phi^+(t_0) = G(t_0)\Phi^-(t_0) + \frac{f(t_0)}{A(t_0)+B(t_0)}, \tag{47.5}$$

where

$$G(t_0) = \frac{A(t_0)-B(t_0)}{A(t_0)+B(t_0)}. \tag{47.6}$$

Now the number $\varkappa$, called (§ 45) the index of the integral equation (47.1), is seen to be the index of the corresponding Hilbert problem (47.5).

Let $X(z)$ be the fundamental function (§ 37) of the Hilbert problem (47.5). Then (§ 37) the general solution of the problem (47.5), vanishing at infinity, is given for $\varkappa \geqq 0$ by

$$\Phi(z) = \frac{X(z)}{2\pi i} \int_L \frac{f(t)\, dt}{[A(t) + B(t)]X^+(t)(t - z)} + X(z)Q_{\varkappa-1}(z), \qquad (47.7)$$

where $Q_{\varkappa-1}(z)$ is an arbitrary polynomial of degree not greater than $\varkappa - 1$, and $Q_{\varkappa-1}(z) \equiv 0$ for $\varkappa = 0$. For $\varkappa < 0$, a solution exists only under the conditions

$$\int_L \frac{t^k f(t)\, dt}{[A(t) + B(t)]X^+(t)} = 0, \quad k = 0, 1, \ldots, -\varkappa - 1; \qquad (47.8)$$

if these conditions are satisfied, then the (unique) solution is given by (47.7) for $Q_{\varkappa-1}(z) \equiv 0$.

The solution of the original integral equation is obtained from the equation $\varphi(t_0) = \Phi^+(t_0) - \Phi^-(t_0)$ from which, by (47.7) and the Plemelj formulae, it follows that

$$\begin{aligned}\varphi(t_0) = {} & \frac{X^+(t_0) + X^-(t_0)}{2[A(t_0) + B(t_0)]X^+(t_0)} f(t_0) + \\ & + \frac{X^+(t_0) - X^-(t_0)}{2\pi i} \int_L \frac{f(t)\, dt}{[A(t) + B(t)]X^+(t)(t - t_0)} + \\ & + [X^+(t_0) - X^-(t_0)]Q_{\varkappa-1}(t_0). \end{aligned} \qquad (47.9)$$

Now it will be recalled that, by (35.8),

$$X(z) = \begin{cases} \dfrac{1}{\Pi(z)} e^{\Gamma(z)} & \text{for } z \in S^+, \\ z^{-\varkappa} e^{\Gamma(z)} & \text{for } z \in S^-, \end{cases} \qquad (47.10)$$

$$\Gamma(z) = \frac{1}{2\pi i} \int_L \frac{\log G_0(t)\, dt}{t - z}, \qquad (47.11)$$

$$\Pi(t) = \prod_1^p (t - a_k)^{\lambda_k}, \quad \varkappa = \lambda_0 + \lambda_1 + \ldots + \lambda_p,$$

$$\lambda_k = \frac{1}{2\pi i}[\log G(t)]_{L_k} = \frac{1}{2\pi i}\left[\log \frac{A - B}{A + B}\right]_{L_k},$$

$$G_0(t) = t^{-\varkappa}\Pi(t)G(t). \qquad (47.12)$$

By these formulae and by (35.9) it is easily seen that the solution (47.9) may be written as

$$\varphi(t_0) = \mathbf{K}^* f + B^*(t_0)Z(t_0)P_{\varkappa-1}(t_0), \tag{47.13}$$

where

$$\mathbf{K}^* f = A^*(t_0)f(t) - \frac{B^*(t_0)Z(t_0)}{\pi i}\int_L \frac{f(t)\,dt}{Z(t)\,(t-t_0)}, \tag{47.14}$$

$$\begin{aligned} Z(t_0) &= [A(t_0) + B(t_0)]X^+(t_0) = [A(t_0) - B(t_0)]X^-(t_0) = \\ &= [A(t_0) + B(t_0)]\frac{e^{\frac{1}{2}\log G_0(t_0)}e^{\Gamma(t_0)}}{\Pi(t_0)} = \\ &= [A(t_0) - B(t_0)]t_0^{-\varkappa}\, e^{-\frac{1}{2}\log G_0(t_0)}\, e^{\Gamma(t_0)}, \end{aligned} \tag{47.15}$$

$$A^*(t_0) = \frac{A(t_0)}{A^2(t_0) - B^2(t_0)}, \quad B^*(t_0) = \frac{B(t_0)}{A^2(t_0) - B^2(t_0)}, \tag{47.16}$$

and $P_{\varkappa-1}(t_0)$ is an arbitrary polynomial of degree not greater than $\varkappa - 1$. The expression (47.15) may also be written as

$$Z(t_0) = \frac{\sqrt{A^2(t_0) - B^2(t_0)}}{\sqrt{\Pi(t_0)}}\, t_0^{-\varkappa/2}\, e^{\Gamma(t_0)}. \tag{47.15a}$$

The choice of sign of the roots in the last formula is of no importance, if L consists of a single contour; otherwise (47.15) must be used, where $\log G_0(t)$ is the same as in (47.11).

The formula (47.13) gives the general solution of the integral equation (47.1) for $\varkappa \geqq 0$. The same formula also gives the solution for $\varkappa < 0$, if one puts $P_{\varkappa-1}(t_0) \equiv 0$ and assumes that $f(t)$ satisfies the (necessary and sufficient) conditions (47.8) which, by (47.15), may be written as

$$\int_L \frac{t^k f(t)\,dt}{Z(t)} = 0, \ k = 0, 1, \ldots, -\varkappa - 1. \tag{47.17}$$

The formulae (47.9)—(47.17) were given by I. N. Vekua [1] who deduced them for the case of a single contour [in which case $\Pi(t) \equiv 1$]; the formulae were generalized to the case of several contours by B. V. Khvedelidze [2].

When $f(t) \equiv 0$, i.e., when dealing with the homogeneous equation, the above results show that for $\varkappa \leqq 0$ the homogeneous equation has no solution, different from zero, and for $\varkappa > 0$ it has exactly $\varkappa$ linearly independent solutions which can be combined to give the general solution

$$\varphi(t) = B^*(t)Z(t)P_{\varkappa-1}(t), \tag{47.18}$$

where $P_{\varkappa-1}(t)$ is an arbitrary polynomial of degree not greater than $\varkappa - 1$.

The function $Z(t)$ will be called the fundamental function, corresponding to the equation (47.1). The results obtained are:

1°. *If* $\varkappa > 0$, *the homogeneous equation* $\mathbf{K}^0\varphi = 0$ *has exactly* $\varkappa$ *linearly independent solutions.*

2°. *If* $\varkappa \leqq 0$, *this equation has no solution different from zero.*

3°. *If* $\varkappa \geqq 0$, *the non-homogeneous equation* $\mathbf{K}^0\varphi = f$ *is soluble for any right part* f.

4°. *If* $\varkappa < 0$, *this equation is soluble, if and only if the function* f *satisfies the* $(-\varkappa)$ *conditions*

$$\int_L f\psi_k \, dt = 0, \quad k = 0, 1, \ldots, -\varkappa - 1, \tag{47.19}$$

where the ψ_k *are certain linearly independent functions* [cf.(47.17)]. Provided these conditions are satisfied, the equation has one and only one solution.

When $\varkappa > 0$, the sum of the solutions of the homogeneous equation is given by (47.18); when $\varkappa \geqq 0$, the general solution of the non-homogeneous equation is given by (47.13); when $\varkappa < 0$, the condition for solubility of the non-homogeneous equation is given by (47.17), and the solution by (47.13) for $P_{\varkappa-1}(t_0) \equiv 0$.

NOTE 1. In practice, it is often preferable to make direct use of (47.9); cf. for example the following Note.

NOTE 2. If in the dominant equation (47.1) $A(t_0) \equiv 0$ and consequently, by supposition, $B(t_0) \neq 0$ everywhere on L, one can assume $B(t_0) = 1$, without loss of generality; then the equation takes the form

$$\frac{1}{\pi i} \int_L \frac{\varphi(t)\, dt}{t - t_0} = f(t_0). \tag{47.20}$$

In this case $G(t) = -1$, $\varkappa = 0$, $\Pi(t) = 1$, and one may assume $\log G_0(t) = \log G(t) = +\pi i$; therefore

$$\Gamma(z) = \frac{1}{2\pi i} \int_L \frac{\pi i \, dt}{t - z} = \begin{cases} \pi i & \text{in } S^+ \\ 0 & \text{in } S^- \end{cases}, \quad X(z) = e^{\Gamma(z)} = \begin{cases} -1 & \text{in } S^+ \\ 1 & \text{in } S^- \end{cases}.$$

Thus $X^+(t) = -1$, $X^-(t) = 1$ and, by (47.9),

$$\varphi(t_0) = \frac{1}{\pi i} \int_L \frac{f(t)\, dt}{t - t_0}. \tag{47.21}$$

As was to be expected, the inversion formula, found in § 27 by another method, is obtained.

§ 48. Solution of the equation adjoint to the dominant equation.

Now the equation

$$\mathbf{K}^{0\prime}\psi \equiv A(t_0)\psi(t_0) - \frac{1}{\pi i}\int_L \frac{B(t)\psi(t)\,dt}{t-t_0} = g(t) \tag{48.1}$$

will be considered which is adjoint to the equation $\mathbf{K}^0\varphi = f$; the index $\varkappa'$ of this equation is equal to $(-\varkappa)$, if, as before, $\varkappa$ is the index of $\mathbf{K}^0$.

The equation (48.1) may also be transformed into a Hilbert problem by the following simple method which was proved by the Author (for much more general cases, i.e., systems of singular equations) and was published in a paper by N. I. Muskhelishvili and N. P. Vekua [1]. The equation considered is usually reduced to the dominant equation by means of the substitution $B(t)\psi(t) = \varphi(t)$; but such a substitution is not permissible, at least not without further investigation, if $B(t)$ is not different from zero everywhere. Introduce the sectionally holomorphic function

$$\Psi(z) = \frac{1}{2\pi i}\int_L \frac{B(t)\psi(t)\,dt}{t-z} \tag{48.2}$$

which vanishes at infinity. Using the formulae

$$B(t_0)\psi(t_0) = \Psi^+(t_0) - \Psi^-(t_0), \quad \frac{1}{\pi i}\int_L \frac{B(t)\psi(t)\,dt}{t-t_0} = \Psi^+(t_0) + \Psi^-(t_0),$$

the conclusion can be drawn that the equation (48.1) is equivalent to the following problem.

Find a function $\psi(t)$, satisfying the H condition, and a sectionally holomorphic function $\Psi(z)$, vanishing at infinity, under the conditions

$$\begin{aligned} A(t_0)\psi(t_0) &= \Psi^+(t_0) + \Psi^-(t_0) + g(t_0), \\ B(t_0)\psi(t_0) &= \Psi^+(t_0) - \Psi^-(t_0). \end{aligned} \tag{48.3}$$

These are equivalent to

$$(A+B)\psi = 2\Psi^+ + g, \quad (A-B)\psi = 2\Psi^- + g,$$

or

$$\psi = \frac{2\Psi^+}{A+B} + \frac{g}{A+B}, \quad \psi = \frac{2\Psi^-}{A-B} + \frac{g}{A-B}. \tag{48.4}$$

Comparing the right sides, one is led to the Hilbert problem

$$\Psi^+(t_0) = [G(t_0)]^{-1}\Psi^-(t_0) + \frac{B(t_0)g(t_0)}{A(t_0) - B(t_0)}, \tag{48.5}$$

where $G(t_0)$ is the same as in the preceding section, i.e.,

$$G(t_0) = \frac{A(t_0) - B(t_0)}{A(t_0) + B(t_0)}, \tag{48.6}$$

and it is required to find a solution vanishing at infinity. Having solved this problem, the solution of the original problem is obtained from either of the formulae (48.4).

Note that the homogeneous Hilbert problem

$$\Psi^+(t_0) = [G(t_0)]^{-1}\Psi^-(t_0), \tag{48.7}$$

obtained from (48.5) for $g \equiv 0$, is associate (§ 36) to the Hilbert problem, obtained from (47.5) for $f \equiv 0$. Hence, if $X(z)$ and $\varkappa$ are the fundamental solution and the index of this last problem, then $[X(z)]^{-1}$ and $\varkappa' = -\varkappa$ will be those of the problem (48.7). Correspondingly the solution of the problem (48.5), vanishing at infinity, is of the form

$$\Psi(z) = \frac{[X(z)]^{-1}}{2\pi i}\int_L \frac{X^+(t)B(t)g(t)dt}{[A(t) - B(t)](t - z)} + [X(z)]^{-1}Q_{\varkappa'-1}(z) \tag{48.8}$$

for $\varkappa' \geqq 0$ (i.e., for $\varkappa \leqq 0$), where $Q_{\varkappa'-1}(z)$ is an arbitrary polynomial of degree not greater than $\varkappa' - 1$ ($Q_{\varkappa'-1}(z) \equiv 0$ for $\varkappa' = 0$).

For $\varkappa' < 0$ (i.e., $\varkappa > 0$), the solution is also given by (48.8) with $Q_{\varkappa'-1}(z) \equiv 0$, if the following necessary and sufficient conditions of solubility are satisfied:

$$\int_L \frac{X^+(t)B(t)g(t)t^k\,dt}{A(t) - B(t)} = 0, \quad k = 0, 1, \ldots, -\varkappa' - 1. \tag{48.9}$$

From (48.8) one obtains by either of the formulae (48.4), after a simple transformation,

$$\psi(t_0) = \mathbf{K}^{*\prime}g + \frac{P_{\varkappa'-1}(t_0)}{Z(t_0)}, \tag{48.10}$$

where $\mathbf{K}^{*\prime}$ is the operator, associate to the $\mathbf{K}^*$ of the last section, i.e.,

$$\mathbf{K}^{*\prime}g = A^*(t_0)g(t_0) + \frac{1}{\pi i Z(t_0)}\int_L \frac{Z(t)B^*(t)g(t)dt}{t - t_0}, \tag{48.11}$$

with $A^*(t)$, $B^*(t)$, $Z(t)$ the same as in § 47 [(47.15), (47.16)].

The conditions (48.9) may be rewritten as

$$\int_L t^k Z(t) B^*(t) g(t)\, dt = 0, \quad k = 0, 1, \ldots, -\varkappa' - 1. \qquad (48.12)$$

The general solution of the homogeneous equation is given for $\varkappa' > 0$ by

$$\psi(t) = \frac{P_{\varkappa'-1}(t)}{Z(t)}, \qquad (48.13)$$

where $P_{\varkappa'-1}(t)$ is an arbitrary polynomial of degree not greater than $\varkappa' - 1$. Consequently, for $\varkappa' > 0$, the homogeneous equation has exactly $\varkappa'$ linearly independent solutions; for $\varkappa' \leqq 0$, the homogeneous equation has no solution different from zero.

Thus the equation $\mathbf{K}^{0\prime}\psi = g$ has the same properties as the equation $\mathbf{K}^0\varphi = f$, i.e., for the equation $\mathbf{K}^{0\prime}\psi = g$, the results 1°—4° of the last section hold, provided $\mathbf{K}^0$ and $\varkappa$ are replaced by $\mathbf{K}^{0\prime}$ and $\varkappa' = -\varkappa$.

Now it is seen that the functions ψ_k, figuring in the conditions of solubility (47.19) of $\mathbf{K}^0\varphi = f$, are a complete system of linearly independent solutions of the homogeneous adjoint equation $\mathbf{K}^{0\prime}\psi=0$; just as the condition of solubility (48.12) of $\mathbf{K}^{0\prime}\psi = g$ may be written as

$$\int_L g\varphi_k\, dt = 0, \qquad (48.14)$$

where the φ_k $(k = 1, 2, \ldots, \varkappa)$ are a complete system of linearly independent solutions of the homogeneous equation $\mathbf{K}^0\varphi = 0$, adjoint to the above equation. These results represent a particular case of an important theorem, to be proved in § 53.

Further note the following result, also a particular case of an important general theorem to be proved in § 53; the difference of the number k of the linearly independent solutions of the homogeneous equation $\mathbf{K}^0\varphi = 0$ and the number k' of the linearly independent solutions of the adjoint homogeneous equation $\mathbf{K}^{0\prime}\psi = 0$ equals the index of $\mathbf{K}^0$:

$$k - k' = \varkappa. \qquad (48.15)$$

In fact, for $\varkappa \geqq 0$, $k = \varkappa$, $k' = 0$; for $\varkappa \leqq 0$, $k = 0$, $k' = -\varkappa$.

§ 49. Some general remarks *).

Before going any further, some general remarks will be made. As is known, the difference between the Fredholm equation of the

*) In this section use will be made of some concepts of functional analysis; the corresponding paragraphs may be omitted without effecting the general understanding of what follows.

second kind

$$\mathbf{N}\varphi \equiv A(t_0)\varphi(t_0) + \int_L n(t_0, t)\varphi(t)dt = f(t_0), \tag{49.1}$$

where $A(t_0) \neq 0$ everywhere on L [introducing $A(t_0)$ to simplify the comparison with singular equations; one can always assume $A(t_0) = 1$ in (49.1)], and the Fredholm equation of the first kind

$$\int_L n(t_0, t)\varphi(t)\,dt = f(t_0), \tag{49.2}$$

obtained from (49.1) for $A(t_0) \equiv 0$, is quite essential.

There is no such difference between the singular equations

$$\mathbf{K}\varphi \equiv A(t_0)\varphi(t_0) + \frac{B(t_0)}{\pi i}\int_L \frac{\varphi(t)\,dt}{t - t_0} + \frac{1}{\pi i}\int_L k(t_0, t)\varphi(t)\,dt = f(t_0) \tag{49.3}$$

and

$$\frac{B(t_0)}{\pi i}\int_L \frac{\varphi(t)\,dt}{t - t_0} + \frac{1}{\pi i}\int_L k(t_0, t)\varphi(t)\,dt = f(t_0). \tag{49.4}$$

The singular equations (49.3) and (49.4) are sometimes called equations of the second and first kind (and sometimes this will also be done), but this classification is less justified than in the case of the Fredholm equations; the equation (49.4), as will be seen in the sequel and as has been seen already in a simple example (§ 47, Note 2), is simply a particular case of (49.3). The reason for this is, in general, as follows.

For simplicity, the kernel $n(t_0, t)$, the coefficient $A(t_0)$ and the function $f(t_0)$ in (49.1) and (49.2) are assumed to be continuous functions; likewise the solutions $\varphi(t)$ are sought in the functional space C, where the norm is defined by max $|\varphi(t)|$ and the distance between $\varphi_1(t)$ and $\varphi_2(t)$ by max $|\varphi_2(t) - \varphi_1(t)|$ on L.

If the unit operator, i.e., the operator such that $\mathbf{E}\varphi = \varphi$, is denoted by $\mathbf{E}$ and the operator $\mathbf{n}$ is defined by

$$\mathbf{n}\varphi = \int_L n(t_0, t)\varphi(t)\,dt, \tag{49.5}$$

then (49.1) may be rewritten as

$$A\mathbf{E}\varphi + \mathbf{n}\varphi = f. \tag{49.1a}$$

The linear operator $\mathbf{n}$ is a completely continuous operator, i.e., it converts any (in modulus) bounded set of functions $\varphi(t)$ into a compact set, while $\mathbf{E}$ is simply a bounded linear operator. Thus, in the operator $A\mathbf{E} + \mathbf{n}$, the parts $A\mathbf{E}$ and $\mathbf{n}$ have essentially different properties.

Now consider the singular equation (49.3) and write it as

$$\mathbf{K}\varphi \equiv A\mathbf{E}\varphi + B\mathbf{I}\varphi + \mathbf{k}\varphi = f, \tag{49.3a}$$

where $\mathbf{I}$ and $\mathbf{k}$ are operators defined by

$$\mathbf{I}\varphi = \frac{1}{\pi i}\int_L \frac{\varphi(t)\,dt}{t - t_0}, \quad \mathbf{k}\varphi = \frac{1}{\pi i}\int_L k(t_0, t)\varphi(t)\,dt. \tag{49.6}$$

For simplicity, assume $A(t_0)$, $B(t_0)$, $k(t_0, t)$ to satisfy the $H(\nu)$ condition on L and let μ be some fixed positive number such that $\mu \leqq \nu$, $\mu < 1$.

Now consider the set of all functions $\varphi(t)$, satisfying the $H(\mu)$ condition on L, and define the norm $\|\varphi(t)\|$ in the following manner:

$$\|\varphi(t)\| = M + M_0,$$

where $M = \max|\varphi(t)|$, $M_0 = \max\dfrac{|\varphi(t_2) - \varphi(t_1)|}{|t_2 - t_1|^\mu}$ on L (in the latter equality „max" denotes the exact upper bound); correspondingly the distance between $\varphi_1(t)$ and $\varphi_2(t)$ is defined as $\|\varphi_2(t) - \varphi_1(t)\|$. It is easily seen that the triangular inequality (just as the other axioms necessary for norms and distances) is satisfied and that the set of functions $\varphi(t)$ forms a complete metric linear space to be denoted by H^μ.

To prove the completeness of the space, let $\varphi_1, \varphi_2, \ldots$ be any convergent sequence in this space such that for every integer $p > 0$

$$\|\varphi_{n+p} - \varphi_n\| \leqq \varepsilon_n, \tag{*}$$

where ϵ_n is independent of p and $\lim\limits_{n\to\infty} \varepsilon_n = 0$.

It follows from (*) that the sequence $\varphi_1, \varphi_2, \ldots$ is also convergent in the space C, so that $\lim\limits_{n\to\infty} \varphi_n = \varphi$ exists in the sense that $\lim\limits_{n\to\infty} \max|\varphi - \varphi_n| = 0$. Further, it follows from (*) that

$$\max\left|\frac{\varphi_{n+p}(t_2) - \varphi_{n+p}(t_1)}{|t_2 - t_1|^\mu} - \frac{\varphi_n(t_2) - \varphi_n(t_1)}{|t_2 - t_1|^\mu}\right| \leqq \varepsilon_n,$$

and hence, since ϵ_n does not depend on p,

$$\max\left|\frac{\varphi(t_2) - \varphi(t_1)}{|t_2 - t_1|^\mu} - \frac{\varphi_n(t_2) - \varphi_n(t_1)}{|t_2 - t_1|^\mu}\right| \leqq \epsilon_n.$$

Therefore it is obvious that $|\varphi(t_2) - \varphi(t_1)|/|t_2 - t_1|^\mu$ is bounded and that hence φ belongs to the space H^μ; further, obviously, $\lim\limits_{n\to\infty}\|\varphi - \varphi_n\| = 0$ which proves the completeness of H^μ.

It is easily seen, using the estimates in § 19, that **I** is a linear operator in H^μ. From the inversion formula, taken from § 27, and from the results of § 20 it also follows that **I** transforms H^μ into itself, i.e., it has in this respect the same property as the operator **E**. Thus there is no essential difference between the first two terms on the left of (49.3a). Therefore, generally speaking, it is advisable to consider the dominant part $A\mathbf{E} + B\mathbf{I}$ of **K** as a single term.

As regards the operator **k**, it differs essentially from **E** and **I** as it is a completely continuous operator in H^μ, i.e., it transforms each bounded set of functions φ into a compact set (in the sense of the metric of the space H^μ). This will be proved.

Let $\{\varphi\}$ be a set of functions of H^μ such that $\| \varphi \| \leqq R$, where R is a constant. From the definition of the norm $\| \varphi \|$ it follows that the function φ is uniformly bounded and equi-continuous (because $| \varphi(t_2) - \varphi(t_1) | \leqq M_0 \, | t_2 - t_1 |^\mu = R \, | t_2 - t_1 |^\mu$). Consequently, by Arzela's theorem, one may select from each infinite subset of these functions φ a uniformly convergent subsequence $\varphi_1, \varphi_2, \ldots$ Denote its limit (in the usual sense, i.e., in the sense of the metric in C) by φ Let $\psi_n = \mathbf{k}\varphi_n$, $\psi = \mathbf{k}\varphi$. The statement will be proved, if it is shown that $\psi_n \to \psi$ in the sense of the metric of H^μ, i.e., that $\| \omega_n \| \to 0$ where $\omega_n = \psi - \psi_n$. But this follows directly from the fact that firstly $\max | \omega_n | \to 0$ and secondly

$$\omega_n(t_2) - \omega_n(t_1) = \int_L [k(t_2, t) - k(t_1, t)]\omega_n(t)\, dt,$$

so that

$$\frac{| \omega_n(t_2) - \omega_n(t_1) |}{| t_2 - t_1 |^\mu} \leqq \int_L \frac{| k(t_2, t) - k(t_1, t) |}{| t_2 - t_1 |^\mu} \, | \omega_n(t) | \, ds,$$

and consequently

$$\max \frac{| \omega_n(t_2) - \omega_n(t_1) |}{| t_2 - t_1 |^\mu} \to 0.$$

Thus the proposition is proved.

In particular, it is clear from the above that in the singular operator **K** its dominant part $A\mathbf{E} + B\mathbf{I}$ must play a similar role to that played by the part $A\mathbf{E}$ in the Fredholm operator **N**. However, one has also to keep in mind this essential difference; the operator $A\mathbf{E}$ has always a unique inverse, i.e., the equation $A\mathbf{E}\varphi = f$ has always a unique solution, while the operator $A\mathbf{E} + B\mathbf{I}$ cannot always be inverted, and if it can be inverted, then its inverse is not always unique. For it has been seen that the equation

$$\mathbf{K}^0\varphi \equiv A\mathbf{E}\varphi + \mathbf{I}\varphi = f$$

is soluble for each f, only if $\varkappa \geqq 0$, and has a unique solution for each f, only if $\varkappa = 0$. It will be seen below that the singular equation with $\varkappa = 0$ is in many respects analogous to the Fredholm equation (of the second kind).

Note. The fact that the terms $A\mathbf{E}\varphi$ and $B\mathbf{I}\varphi$ on the left of (49.3a) are, in a certain sense, equivalent may still be seen from the following. Substitute $\psi = \mathbf{I}\varphi$, where ψ is a new unknown function. Then, by the inversion formulae (§ 27), $\varphi = \mathbf{I}\psi$ and (49.3a) takes the form

$$B\mathbf{E}\psi + A\mathbf{I}\psi + \mathbf{k}_1\psi = f,$$

where $\mathbf{k}_1$ is easily seen to have the same form as $\mathbf{k}$; thus A and B simply change places.

§ 50. On the reduction of a singular integral equation.

The usual method of investigation of a singular integral equation consists in its reduction to a Fredholm equation. One of the methods of reduction will now be indicated. This method is the one most frequently used up to the present; it was suggested (in different particular applications) by the founders of the theory of singular equations — Poincaré and Hilbert. In its general form it was due to F. Noether [1].

Consider the singular integral equation

$$\mathbf{K}\varphi = f \tag{50.1}$$

and let $\mathbf{M}$ be any singular operator reducing $\mathbf{K}$ (§ 45).

Applying $\mathbf{M}$ to both sides of (50.1) one obtains the Fredholm equation

$$\mathbf{N}\varphi \equiv \mathbf{MK}\varphi = \mathbf{M}f. \tag{50.2}$$

Clearly each solution of (50.1) will also be a solution of (50.2); however, the converse is, in general, not true (see also § 53); consequently (50.1) and (50.2) are not, in general, equivalent. But it is easily seen that, having found all the solutions of the Fredholm equation (50.2), all the solutions of the original equation (50.1) may also be found. This will be discussed in detail in § 53; here only one direct consequence of the above will be noted:

The number of linearly independent solutions of the homogeneous singular equation

$$\mathbf{K}\varphi = 0 \tag{50.3}$$

is finite.

In fact, all solutions of (50.3) are solutions of the homogeneous Fredholm equation $\mathbf{MK}\varphi = 0$; however, the number of linearly independent solutions of the latter is known to be finite.

Note the following important fact. The Fredholm equation (50.2) has the form *)

$$\mathbf{N}\varphi \equiv a(t_0)\varphi(t_0) + \int_L n(t_0, t)\varphi(t)\,dt = g(t_0), \tag{50.4}$$

where it is easily verified, using the results of § 45, that $a(t_0)$ does not vanish anywhere on L and satisfies the H condition there, that $g(t_0) = \mathbf{M}f(t_0)$ also satisfies the H condition and that the kernel $n(t_0, t)$ has the form

$$n(t_0, t) = \frac{n^*(t_0, t)}{|t - t_0|^\lambda}, \quad 0 \leqq \lambda < 1, \tag{50.5}$$

where $n^*(t_0, t)$ satisfies the H condition.

The solutions of the Fredholm equation are normally sought in the class of continuous functions; but for the present purpose it is important to find solutions satisfying the H condition. However, there is no necessity to impose beforehand the latter additional condition on the solution of the Fredholm equation (50.4), because it will be satisfied automatically. In fact, it is easily verified that, under the assumed conditions, each continuous (and even merely bounded and integrable) solution of (50.4) necessarily satisfies the H condition. This will be proved in the next section.

§ 51. On the reduction of a singular integral equation (continued).

Every bounded and integrable solution of the equation (50.4) will be shown to satisfy the H condition. For this purpose it is obviously sufficient to prove that the function

$$\omega(t_0) = \int_L n(t_0, t)\varphi(t)\,dt = \int_L \frac{n^*(t_0, t)\varphi(t)\,dt}{|t - t_0|^\lambda} \tag{51.1}$$

satisfies the H condition, whatever may be the bounded integrable function $\varphi(t)$. Consequently it is sufficient to show that

$$\omega(t_0, t_1) = \int_L \frac{n^*(t_1, t)\varphi(t)\,dt}{|t - t_0|^\lambda}$$

satisfies the H condition for t_0 and t_1 (both of which are arbitrary points on L).

---

*) As always, it is assumed that the singular operator under consideration satisfies the conditions 1°—3°, stated in § 44, also that $f(t_0)$ of (50.1) satisfies the H condition and that the unknown function $\varphi(t)$ is subject to the same condition.

As regards t_1, this proposition is almost obvious; denoting by μ the Hölder index of $n^*(t_1, t)$ with respect to t_1, one obtains

$$| \omega(t_0, t_1 + h) - \omega(t_0, t_1) | \leqq \int_L \frac{|n^*(t_1 + h, t) - n^*(t_1, t)| \cdot | \varphi(t) | ds}{| t - t_0 |^\lambda} \leqq$$

$$\leqq B | h |^\mu \int_L \frac{ds}{| t - t_0 |^\lambda} \leqq B_0 | h |^\mu, \tag{51.2}$$

where B and B_0 are constants.

Now the proposition will be proved for t_0. One has

$$\omega(t_0 + h, t_1) - \omega(t_0, t_1) = \int_L \frac{| t - t_0 |^\lambda - | t - t_0 - h |^\lambda}{| t - t_0 |^\lambda | t - t_0 - h |^\lambda} n^*(t_0, t) \varphi(t) \, dt,$$

and hence, since $| t - t_0 |^\lambda$ satisfies the $H(\lambda)$ condition,

$$| \omega(t_0 + h, t_1) - \omega(t_0, t_1) | \leqq C | h |^\lambda \int_L \frac{ds}{| t - t_0 |^\lambda | t - t_0 - h |^\lambda} =$$

$$= C | h |^\lambda . I,$$

where C is a constant. Divide the integral I into two parts: $I = I_1 + I_2$, where I_1 extends over the standard arc l, cut from L by the standard circle of radius R with centre t_0, and I_2 is taken over the line $L - l$. Without loss of generality, it may be assumed that $| h | < \frac{1}{2} R$. Then obviously $| I_2 | < C_2$, where C_2 is a constant. Further, introducing the variable $r = | t - t_0 |$, one obtains

$$| I_1 | < C_1 \int_0^R \frac{dr}{r^\lambda | r - \delta |^\lambda},$$

where C_1 is a constant and $| h | = \delta$. Introducing a new integration variable ϱ by the relation $r = \delta . \varrho$, one obtains

$$| I_1 | < C_1 \delta^{1-2\lambda} \int_0^{R/\delta} \frac{d\varrho}{\varrho^\lambda | 1 - \varrho |^\lambda}.$$

If now $\lambda > \frac{1}{2}$, then

$$\int_0^{R/\delta} \frac{d\varrho}{\varrho^\lambda | 1 - \varrho |^\lambda} < \int_0^\infty \frac{d\varrho}{\varrho^\lambda | 1 - \varrho |^\lambda} = \text{const.}$$

(the last integral being convergent), and the proposition is proved, since it is always possible to assume $\lambda > \frac{1}{2}$ in (50.5) (there is no difficulty in obtaining a direct estimate for $\lambda \leqq \frac{1}{2}$ in the same way).

§ 52. On the resolvent of the Fredholm equation.

Now some known results of the theory of the Fredholm equation

$$\mathbf{N}\varphi \equiv a(t_0)\varphi(t_0) + \int_L n(t_0, t)\varphi(t)\,dt = g(t) \tag{52.1}$$

will be stated.

As before, L will be assumed to be the union of a finite number of non-intersecting smooth contours; for the other quantities in (52.1), a somewhat more general assumption than above will be made for the time being, namely $a(t)$ will be assumed to be a continuous non-vanishing function (as usually one may put $a(t) = 1$ without loss of generality) and

$$n(t_0, t) = \frac{n^*(t_0, t)}{|t - t_0|^\lambda}, \quad 0 \leqq \lambda < 1, \tag{52.2}$$

where $n^*(t_0, t)$ is continuous on L. The function $g(t)$ will also be assumed continuous.

The following will be assumed known from the theory of the Fredholm equation. If the homogeneous equation

$$\mathbf{N}\varphi = 0 \tag{52.3}$$

has no non-zero solution, then (52.1) has a unique solution for any $g(t)$ and this solution is given by

$$\varphi(t_0) = \mathbf{\Gamma} g \equiv \alpha(t_0)g(t_0) + \int_L \gamma(t_0, t)g(t)\,dt, \tag{52.4}$$

where $\alpha(t_0) = \dfrac{1}{a(t_0)}$ and $\gamma(t_0, t)$ is a certain unique function of the form

$$\gamma(t_0, t) = \frac{\gamma^*(t_0, t)}{|t - t_0|^\lambda}, \quad 0 \leqq \lambda < 1,$$

where $\gamma^*(t_0, t)$ is a continuous function. The function $\gamma(t_0, t)$ is called the Fredholm resolvent.

The case in which the homogeneous equation (52.3) has a non-zero solution will now be considered. As already mentioned, this equation can only have a finite number of linearly independent solutions, as has the equation adjoint to (52.3)

$$\mathbf{N}'\psi = 0. \tag{52.5}$$

The complete systems of linearly independent solutions of the homogeneous equations (52.3) and (52.5) will be denoted by φ_1, $\varphi_2, \ldots, \varphi_n$ and $\psi_1, \psi_2, \ldots, \psi_n$ respectively.

The non-homogeneous equation (52.1) has a solution, if and only if g satisfies the conditions

$$\int_L g(t)\psi_j(t)\,dt = 0, \quad j = 1, 2, \ldots, n. \tag{52.6}$$

Now it will be shown that also in this case there exists an operator $\mathbf{\Gamma}$ of the same form as the operator in (52.4), having the following property: if the conditions (52.6) are satisfied,

$$\varphi(t_0) = \mathbf{\Gamma} g \equiv \alpha(t_0)g(t_0) + \int_L \gamma(t_0, t)g(t)\,dt$$

is a solution (more precisely, one of the particular solutions) of (52.1); in this case the function $\gamma(t_0, t)$ will be called the generalized Fredholm resolvent. It was obtained by Fredholm himself; the very simple theory of the generalized resolvent was given in a paper by W. A. Hurwitz [1] who called it the pseudo-resolvent. The method of construction of the generalized resolvent, given in the text, is due to Hurwitz.

This method consists in the reduction of the problem to the case in which the homogeneous equation has no non-zero solution.

For this purpose two systems of functions will be introduced: $\xi_1(t), \xi_2(t) \ldots, \xi_n(t)$ and $\eta_1(t), \eta_2(t), \ldots, \eta_n(t)$, satisfying the H condition*) and connected with the functions $\varphi_1(t), \varphi_2(t), \ldots, \varphi_n(t)$ and $\psi_1(t), \psi_2(t), \ldots, \psi_n(t)$ by the following relations:

$$\int_L \varphi_i\xi_j\,dt = \delta_{ij}, \quad \int_L \psi_i\eta_j\,dt = \delta_{ij}, \tag{52.7}$$

where

$$\delta_{ij} = \begin{cases} 1 \text{ for } i = j \\ 0 \text{ for } i \neq j \end{cases}.$$

Such functions ξ_i and η_i can always be constructed in an infinite number of ways, as is shown in Appendix 3.

Together with (52.1) another Fredholm integral equation

$$a(t_0)\varphi(t_0) + \int_L \{n(t_0, t) + \sum_{i=1}^{n} \eta_i(t_0)\xi_i(t)\}\varphi(t)\,dt = g(t_0) \tag{52.8}$$

will be considered and it will be shown that each of its solutions is also a solution of the original equation (52.1), provided the condition (52.6) is satisfied.

In fact, let $\varphi(t)$ be any solution (if such exist) of (52.8); then from this equation

$$\mathbf{N}\varphi(t_0) + \sum_{i=1}^{n} a_i\eta_i(t_0) = g(t_0), \tag{52.9}$$

---

*) This condition is introduced, having in mind the application to singular integral equations; Hurwitz did not introduce it, as he restricted himself to the study of continuous functions (in the real field).

where the a_i are the constants

$$a_i = \int_L \varphi(t)\xi_i(t)\,dt. \tag{52.10}$$

Multiplying both sides of (52.9) by $\psi_k(t_0)dt_0$, integrating with respect to t_0 along L using (52.6) and (52.7) and noting that (§ 46, 3°)

$$\int_L \psi_k \mathbf{N}\varphi dt_0 = \int_L \varphi \mathbf{N}'\psi_k dt_0 = 0 \tag{52.11}$$

(because $\mathbf{N}'\psi_k = 0$), one obtains $a_k = 0$. Consequently, in (52.9) all $a_i = 0$ and therefore $\mathbf{N}\varphi = g(t_0)$, as was to be proved.

Finally it will be shown that the homogeneous equation, obtained from (52.8) for $g(t_0) \equiv 0$, has no solutions, different from zero.

Each solution of this last equation, as has just been seen, will also be a solution of the equation $\mathbf{N}\varphi = 0$ (because (52.6) is obviously satisfied for $g(t_0) \equiv 0$), besides all the constants $a_i = 0$. Consequently, on the one hand, $\varphi = b_1\varphi_1 + b_2\varphi_2 + \ldots + b_n\varphi_n$, where the $b_1, b_2, \ldots, b_n$ are constants; on the other hand, since the $a_i = 0$, all the $b_j = 0$ by (52.10) and (52.7), and the proposition is proved.

It follows from the above that the non-homogeneous equation (52.8) has a unique solution for every $g(t_0)$ and that this solution is of the form $\varphi = \mathbf{\Gamma} g$, where $\mathbf{\Gamma}$ is an operator of the above type. If, in addition, the conditions (52.6) are satisfied, then, as has been seen, $\mathbf{\Gamma} g$ will be one of the solutions of the original equation (52.1). Hence the existence of the generalized resolvent has been proved.

The general solution of (52.1), when the conditions (52.6) are satisfied, is obviously of the form

$$\varphi = \mathbf{\Gamma} g + c_1\varphi_1 + c_2\varphi_2 + \ldots + c_n\varphi_n. \tag{52.12}$$

Further, it is easily seen that the operator $\mathbf{\Gamma}'$, adjoint to $\mathbf{\Gamma}$, plays the same role with respect to $\mathbf{N}'\omega = g$, adjoint to $\mathbf{N}\varphi = g$, as $\mathbf{\Gamma}$ does with respect to the latter.

Finally the following fact will be noted. Assume that in (52.1) the coefficient $a(t_0)$ satisfies the H condition and that the kernel $n(t_0, t)$ has the form (52.2), where $n^*(t_0, t)$ also satisfies the H condition. Let $g(t_0)$ be an arbitrary function of the point t_0 on L, satisfying the H condition. Since $\varphi(t_0) = \mathbf{\Gamma} g(t_0)$ is a solution of (52.8), it follows, using the results of the last section, that $\mathbf{\Gamma} g(t_0)$ satisfies the H condition. The same is obviously true for $\mathbf{\Gamma}' g(t_0)$.

Now consider the functional relation between the resolvent and the kernel. For simplicity, take $a(t_0) \equiv 1$ and denote the kernel of (52.8) by $m(t_0, t)$, so that

$$m(t_0, t) = n(t_0, t) + \sum_{i=1}^{n} \eta_i(t_0)\xi_i(t).$$

Then, as is known from the theory of the Fredholm equation, one has identically

$$\gamma(t_0, t) + m(t_0, t) = -\int_L m(t_0, t_1)\gamma(t_1, t)\, dt_1 \tag{52.13}$$

and

$$\gamma(t_0, t) + m(t_0, t) = -\int_L \gamma(t_0, t_1)m(t_1, t)\, dt_1. \tag{52.14}$$

§ 53. Fundamental theorems.

A comparison of the results, obtained in the preceding section, with the foundations of the theory of the Fredholm equations leads easily to a general theory of the singular integral equation (where, as always, it is assumed that the conditions 1°—3° of § 44 are satisfied and that $f(t)$ and the unknown function $\varphi(t)$ satisfy the H condition)

$$\mathbf{K}\varphi \equiv A(t_0)\varphi(t_0) + \frac{1}{\pi i}\int_L \frac{K(t_0, t)\varphi(t)\, dt}{t - t_0} = f(t_0). \tag{53.1}$$

In principle, the general theory of (53.1) was first given by F. Noether [1]. The theory attained an extremely simple form in the works of I. N. Vekua, especially in his paper [5]. The results, due to I. N. Vekua, will be described in the following sections.

In the present section the method due to F. Noether will be used with some definitions and simplifications.

Let $\mathbf{M}$ be any operator, reducing $\mathbf{K}$ (cf. § 45). As in § 50, the Fredholm equation

$$\mathbf{N}\varphi \equiv \mathbf{MK}\varphi = \mathbf{M}f \tag{53.2}$$

is deduced from (53.1). The solutions of (53.2) include all the solutions of (53.1).

It is required to find the general solution of (53.1), if the general solution of the Fredholm equation (53.2) is known.

First of all the conditions for the solubility of (53.2) which have the form

$$\int_L \omega_j \mathbf{M}f dt_0 = 0, \quad j = 1, 2, \ldots, n \tag{53.3}$$

will be given, where $\omega_j = \omega_j(t_0)$, $j = 1, 2, \ldots, n$ is the complete system of linearly independent solutions of the homogeneous Fredholm equation

$$\mathbf{N}'\psi \equiv \mathbf{K}'\mathbf{M}'\psi = 0, \tag{53.4}$$

adjoint to (53.2). The conditions (53.3) which can be written as (§ 46, 3°)

$$\int_L f\mathbf{M}'\omega_j\,dt_0 = 0,\ j = 1,\ 2,\ \ldots,\ n \tag{53.5}$$

will be assumed satisfied. Then the general solution of (53.2) is of the form (cf. the last section)

$$\varphi(t_0) = \mathbf{\Gamma}\mathbf{M}f(t_0) + \sum_{i=1}^{n} c_i\chi_i(t_0), \tag{53.6}$$

where $\mathbf{\Gamma}$ is an operator of the type discussed in the last section, the χ_i $(i = 1, 2, \ldots, n)$ form a complete system of linearly independent solutions of the homogeneous equation $\mathbf{MK}\varphi = 0$ and the c_i are arbitrary constants.

But this solution of (53.2) may not be a solution of (53.1). For, if φ is any solution of (53.2) which may be written as

$$\mathbf{M}(\mathbf{K}\varphi - f) = 0,$$

then obviously

$$\mathbf{K}\varphi - f = \sum_{i=1}^{m} a_i\xi_i, \tag{53.7}$$

where $\xi_1, \ldots, \xi_m$ is a complete system of linearly independent solutions of $\mathbf{M}\xi = 0$ and the $a_1, a_2, \ldots, a_m$ are some constants. These constants are obviously fully defined, if the function φ is given, i.e., if the constants $c_1, c_2, \ldots, c_n$ in (53.6) are given.

In order that the solution (53.6) of the equation (53.2) will also be a solution of (53.1), it is necessary and sufficient that all the $a_i = 0$. The last condition will be expressed in terms of the given function f and the constants c_j; for this purpose the constants a_i of (53.7) will be given in terms of known quantities only. This will be done as follows: Let $\xi_1^*, \xi_2^*, \ldots, \xi_n^*$ be any definite functions, satisfying the H condition on L, such that

$$\int_L \xi_i(t_0)\xi_j^*(t_0)\,dt_0 = \delta_{ij},\ \text{where } \delta_{ij} = \begin{cases} 1 \text{ for } i = j \\ 0 \text{ for } i \neq j \end{cases};$$

such functions can always be found (cf. Appendix 3). Next multiply both sides of (53.7) by $\xi_j^*(t_0)dt_0$ and integrate along L. Then one obtains

$$a_j = \int_L \xi_j^*(\mathbf{K}\varphi - f)\,dt_0.$$

Introducing for $\varphi(t_0)$ its expression (53.6) and performing a simple transformation, such as is given in § 46, i.e.,

$$\int_L \xi_j^* \mathbf{K}\mathbf{\Gamma}\mathbf{M}f\,dt_0 = \int_L f\mathbf{M}'\mathbf{\Gamma}'\mathbf{K}'\xi_j^*\,dt_0,$$

one obtains expressions of the form

$$a_j = f_j + \sum_{i=1}^{n} A_{ji}c_i, \tag{53.8}$$

where the f_j are constants, expressed in terms of f by the formulae

$$f_j = \int_L f_j^*(t_0)f(t_0)\,dt_0 \tag{53.9}$$

in which the $f_j^*(t_0)$ are certain functions, satisfying the H condition and not depending on either f or the constants c_i, and the A_{ji} are certain constants, also independent of f and c_i. Consequently, in order that the function φ, defined by (53.6), will be a solution of (53.1), it is necessary and sufficient that the c_i satisfy the system of linear algebraic equations

$$\sum_{i=1}^{n} A_{ji}c_i + f_j = 0, \quad j = 1, 2, \ldots, n. \tag{53.10}$$

If this system has a solution, then the original integral equation has a solution, and vice versa. However, the condition of solubility of (53.10) is expressible by a certain number (of no consequence here) of relations of the form

$$\sum_{j=1}^{n} B_{ij}f_j = 0,$$

where the B_{ij} are definite constants. Since the constants f_j have the form (53.9), the preceding conditions can be written as

$$\int_L f(t)\psi_j^*(t)dt = 0, \tag{53.11}$$

where the $\psi_j^*(t)$ are certain functions, satisfying the H condition and not depending on f. Further, since the conditions (53.5) have the form (53.11), the following important conclusion can be drawn: *The necessary and sufficient condition for the solubility of the integral equation* (53.1) *may be expressed by a finite number of conditions of the form* (53.11), *where the* $\psi_j^*(t)$ *are definite functions, satisfying the* H *condition.*

Now it is not difficult to prove the following three theorems, due to F. Noether, which play the same role in the theory of singular integral equations of the type (53.1) as the well known Fredholm theorems do in the theory of the Fredholm equation. These theorems are as follows.

THEOREM I: *The necessary and sufficient conditions for the solubility of the equation*

$$\mathbf{K}\varphi = f \tag{53.1}$$

may be expressed as

$$\int_L f(t)\psi_j(t)\,dt = 0, \quad j = 1, 2, \ldots, k', \tag{53.12}$$

where the $\psi_1(t), \ldots, \psi_{k'}(t)$ *form a complete system of linearly independent solutions of the adjoint homogeneous equation* $\mathbf{K}'\psi = 0$.

THEOREM II. *The difference between the number* k *of linearly independent solutions of the homogeneous equation* $\mathbf{K}\varphi = 0$ *and the number* k' *of linearly independent solutions of the adjoint homogeneous equation* $\mathbf{K}'\psi = 0$ *depends only on the dominant part of the operator* $\mathbf{K}$.

THEOREM III. *The difference, referred to in Theorem II, is equal to the index of the operator* $\mathbf{K}$, i.e.,

$$k - k' = \varkappa. \tag{53.13}$$

Theorem II is obviously a direct consequence of Theorem III, since the index $\varkappa$, by definition, depends only on the dominant part of the operator $\mathbf{K}$. However, it was considered worthwhile to state Theorem II separately, since its proof, like that of Theorem I, can be carried out independently of the solution of the Hilbert problem. This provides an easy opportunity of extending these theorems to a series of other classes of singular equations for which the Hilbert problem or analogous problems cannot be used directly, because their application is too complicated. For this reason it has not been considered advisable to depart from the presentation of the proofs of Theorems I and II as given by F. Noether, although proofs by I. N. Vekua's [5] method, based from the beginning on the Hilbert problem, are much simpler. The proofs of Theorems I—III will now be given.

PROOF OF THEOREM I. The necessity of the conditions (53.12) is a direct consequence of the result at the end of § 46. Their sufficiency will be demonstrated. For this, in its turn, it will be shown that the conditions (53.11), known to be sufficient for the solubility of (53.1), are a consequence of (53.12). The following simple method (F. Noether) leads quickly to the desired result.

Let $g(t)$ be an arbitrary function, satisfying the H condition. The equation

$$\mathbf{K}\varphi = \mathbf{K}g$$

can be solved, since one of its solutions is $\varphi = g$. Then, because (53.11) is necessary,

$$\int_L \psi_j^* \mathbf{K} g dt = \int_L g \mathbf{K}' \psi_j^* dt = 0.$$

Since this must be true for any function g (satisfying the H condition), it is obviously necessary that $\mathbf{K}'\psi_j^* = 0$, i.e., the ψ_j^* are solutions of the homogeneous equation $\mathbf{K}'\psi = 0$. Consequently the $\psi_j{}^*$ are linear combinations of the functions ψ_i. Therefore the conditions (53.11) follow from (53.12), as was to be proved.

Proof of theorem II and III. As before, let $\mathbf{M}$ be any operator, reducing $\mathbf{K}$. Consider the two homogeneous Fredholm equations, mutually adjoint

$$\mathbf{MK}\varphi = 0, \quad \mathbf{K}'\mathbf{M}'\psi = 0. \tag{53.14}$$

These equations have exactly the same number of linearly independent solutions. They will be evaluated by two distinct methods and comparison of the results will then lead to Theorem II. (The proof given here agrees in principle with that by F. Noether, but it is incomparably simpler; it was stated by I. N. Vekua and published in the Author's note [3a]).

The linearly independent solutions of the equations

$$\mathbf{K}\varphi = 0, \quad \mathbf{K}'\psi = 0, \quad \mathbf{M}\chi = 0, \quad \mathbf{M}'\omega = 0$$

will be denoted by φ_j ($j = 1, \ldots, k$), ψ_j ($j = 1, \ldots, k'$), χ_j ($j = 1, \ldots, m$), ω_j ($j = 1, \ldots, m'$) respectively.

All the solutions of $\mathbf{MK}\varphi = 0$ satisfy the equation

$$\mathbf{K}\varphi = \sum_{j=1}^{m} a_j \chi_j, \tag{53.15}$$

where the a_j are constants. The constants must be chosen in such a way that the preceding equations are soluble, i.e., from Theorem I,

$$\sum_{j=1}^{m} A_{ij} a_j = 0 \quad (i = 1, 2, \ldots, k'), \tag{53.16}$$

where

$$A_{ij} = \int_L \psi_i \chi_j dt \quad (i = 1, 2, \ldots, k', \ j = 1, 2, \ldots, m). \tag{53.17}$$

Let r be the rank of the matrix $\| A_{ij} \|$; then, as is known, $m—r$ of the constants $a_1, a_2, \ldots, a_m$, satisfying the system (53.16), are completely arbitrary, and the remaining r are linear combinations of these. Denote the arbitrary constants by $b_1, b_2, \ldots, b_{m-r}$. Introducing on the right of (53.15) instead of the a_j their expressions in terms of the b_j, (53.15) is reduced to the form

$$\mathbf{K}\varphi = \sum_{i=1}^{m-r} b_i \xi_i ,$$

where the ξ_i are some linear combinations of the functions χ_j; the functions ξ_i are easily seen to be linearly independent.

If for some values of the b_i one has $\sum_{i=1}^{m-r} b_i\xi_i \equiv 0$, then $\sum_{i=1}^{m} a_i\chi_i$ from which the preceding sum was obtained by replacing a_i in terms of the b_j also vanishes, and consequently, by the linear independence of the χ_i, all the $a_i = 0$. But the b_j are linear combinations of the a_i, and hence all the b_j are zero.

The above equation is soluble for every b_i; solving it, one obtains

$$\varphi = \sum_{i=1}^{m-r} b_i\eta_i + \sum_{j=1}^{k} c_j\varphi_j ,$$

where the c_j are arbitrary constants, as are the b_i, and the η_i are any particular solutions of the equation $\mathbf{K}\varphi = \xi_i$ $(i = 1, 2, \ldots, m-r)$. The functions η_i, φ_j are easily seen to be linearly independent.

If for some values of the b_i, c_j one has $\sum_{i=1}^{m-r} b_i\eta_i + \sum_{j=1}^{k} c_j\varphi_j \equiv 0$, then, applying $\mathbf{K}$ to both sides of the last equation, one obtains $\sum_{i=1}^{m-r} b_i\xi_i \equiv 0$; hence it follows that all the $b_i = 0$. But then $\sum_{j=1}^{k} c_j\varphi_j = 0$, and so all the $c_j = 0$.

Consequently, the first equation of (53.14) has exactly $m - r + k$ linearly independent solutions.

A quite analogous result may be obtained for the second equation of (53.14); in the discussion one has only to replace $\mathbf{M}$ and $\mathbf{K}$ by $\mathbf{K}'$ and $\mathbf{M}'$ respectively, and consequently χ_i and φ_i by ψ_i and ω_i. Thus one obtains, instead of the A_{ij}, the constants

$$A'_{ij} = \int_L \chi_i\psi_j\,dt = A_{ji} ,$$

so that the rank of the matrix $\| A'_{ij} \|$ is the same as the rank r of the former matrix. Therefore the number of linearly independent solutions of the second equation of (53.14) is $k' - r + m'$. Comparing this result with the earlier one, one obtains

$$m' - r + k' = m - r + k,$$

and hence, finally,

$$k - k' = m' - m. \tag{53.18}$$

This equation proves Theorem II. because for all equations of the form $\mathbf{K}\varphi = 0$, having one and the same dominant part, one can take the same reducing operator $\mathbf{M}$, so that $(m'—m)$ will be the same.

In order to prove Theorem III, it is sufficient, by the preceding theorem, to retain in the operator $\mathbf{K}$ only its dominant part for the calculation of $(k - k')$; but then $\mathbf{K}\varphi = 0$ becomes the homogeneous dominant equation, and the theorem follows directly from (48.15).

For the deduction of (48.15) it was assumed that the line L satisfies the supplementary condition § 44, 1°a. But it is easily seen from the Note 2 at the end of the same section and from the fact that the index does not alter on replacement of one line of integration by another (§ 45) that this does not influence the generality of the results.

F. Noether himself, when determining $k - k'$, did not use the Hilbert problem, but the Riemann-Hilbert problem of which he gives in the same paper a not quite strict solution.

Note 1. A direct consequence of Theorem III should be noted. *If the index* $\varkappa > 0$, *then the homogeneous equation* $\mathbf{K}\varphi = 0$ *has at least* $\varkappa$ *linearly independent solutions*, because the number of solutions of this equation $k = \varkappa + k'$ and $k' \geqq 0$.

Note 2. It follows from Note 1 that, if the index $\varkappa$ of the equation

$$\mathbf{K}\varphi = f \tag{*}$$

is negative, then it is impossible to find a reducing operator $\mathbf{M}$ such that (*) will be equivalent to the Fredholm equation

$$\mathbf{MK}\varphi = \mathbf{M}f \tag{**}$$

for every choice of f. (This result was proved by C. G. Mikhlin [3]).

In fact, it is known (§ 45) that the index of the reducing operator must be equal to $-\varkappa > 0$. Assume that the equations (*) and (**) are equivalent for any choice of f. Take any function φ, satisfying the H condition, and put $f = \mathbf{K}\varphi + \chi$, where χ is any non-zero solution of the homogeneous equation $\mathbf{M}\chi = 0$ (as such a solution always exists, because the index of $\mathbf{M}$ is a positive number). But then $\mathbf{MK}\varphi = \mathbf{M}f$ and consequently, by the assumed equivalence of (*) and (**), $\mathbf{K}\varphi = f$, wherefrom $\chi \equiv 0$ which contradicts the condition.

§ 54. Real equations.

It may happen that the equation $\mathbf{K}\varphi = f$ which may provisionally be written as

$$\mathbf{K}\varphi \equiv A(t_0)\varphi(t_0) + \int_L N(t_0, t)\varphi(t)\,dt = f(t_0), \tag{54.1}$$

where

$$N(t_0, t) = \frac{1}{\pi i} \frac{K(t_0, t)}{t - t_0}, \tag{54.2}$$

is actually a real equation, i.e., that it is turned into a real equation, if the variable t is replaced by a suitably chosen real variable s.

Let $t = t(s)$ be a function of s such that, if s covers certain intervals $l_0, l_1, \ldots, l_p$, t describes (once and only once) the contours $L_0, L_1, \ldots, L_p$, constituting L. It will be assumed that the derivative $dt/ds = t'(s)$ is everywhere different from zero and satisfies the H condition; the reservations, necessary with regard to the ends of the intervals l_j, are obvious. If, for simplicity, the function of t, expressed in terms of s, is denoted by the former symbol, then (54.1) takes the form

$$\mathbf{N}\varphi \equiv A(s_0)\varphi(s_0) + \int_L N(s_0, s)t'(s)\varphi(s)\,ds = f(s). \tag{54.3}$$

By supposition, $A(s)$, $N(s_0, s)t'(s)$ and $f(s)$ are now real functions. The operator **K**, expressed in terms of s, has been denoted by another symbol (**N**) for a reason which will become clear later.

Evidently in this case, if $\varphi(s) = \xi(s) + i\eta(s)$, where $\xi(s)$, $\eta(s)$ are real functions, is any solution of (54.1) or, what is the same thing, of (54.3), then each of the functions $\xi(s)$ and $\eta(s)$ will also be solutions.

Let $\varphi_1(s), \varphi_2(s), \ldots, \varphi_k(s)$ be a complete system of linearly independent real solutions of the homogeneous equation

$$\mathbf{K}\varphi \equiv A(t_0)\varphi(t_0) + \int_L N(t_0, t)\varphi(t)\,dt = 0 \tag{54.4}$$

or again of the equation

$$\mathbf{N}\varphi \equiv A(s_0)\varphi(s_0) + \int_L N(s_0, s)t'(s)\varphi(s)ds = 0. \tag{54.5}$$

Every real solution of this equation will always be in the form of a linear combination

$$\varphi(s) = c_1\varphi_1(s) + c_2\varphi_2(s) + \ldots + c_k\varphi_k(s), \tag{54.6}$$

where the constants $c_1, c_2, \ldots, c_k$ take arbitrary real values.

By the above it is clear that also every complex solution of (54.4) or (54.5) is given by (54.6) in which, however, the constants $c_1, c_2, \ldots c_k$ may also take arbitrary complex values.

Thus the number of linearly independent solutions of (54.5) is always the same, whether the solutions considered are restricted to the real field or not, the only qualification being that in the first case linear combinations have to be understood as combinations with real coefficients, while in the second case complex coefficients are involved.

Now consider the equation, adjoint to (54.4), i.e.,

$$\mathbf{K}'\psi \equiv A(t_0)\psi(t_0) + \int_L N(t, t_0)\psi(t)\,dt = 0. \tag{54.7}$$

If the equation, adjoint to (54.5), is formed, understanding by this the real equation, obtained from (54.5) by rearrangement of the variables s_0 and s in the kernel, one gets

$$\mathbf{N}'\omega \equiv A(s_0)\omega(s_0) + t'(s_0)\int_L N(s, s_0)\omega(s)\,ds = 0. \tag{54.8}$$

This equation is not the same as (54.7), because the latter, after introduction of the variable s, takes the form

$$A(s_0)\psi(s_0) + \int_L N(s, s_0)t'(s)\psi(s)ds = 0 \tag{54.9}$$

(it is for this reason that a different notation was chosen for the operator, appearing on the left of the equations $\mathbf{K}\varphi = f$ and $\mathbf{N}\varphi = f$).

However, the difference between the equations (54.8) and (54.9) is unimportant; if in (54.9) a new unknown $\omega(s)$ is introduced instead of $\psi(s)$ by the relation

$$\psi(s) = [t'(s)]^{-1}\omega(s),$$

then this equation, after multiplication by $t'(s_0)$, is transformed into (54.8).

Let $\omega_j(s)$ $(j = 1, 2, \ldots, k')$ be a complete system of linearly independent real solutions of $\mathbf{N}'\omega = 0$. By the above, it is easily seen that the system of functions $\psi_j(s) = [t'(s)]^{-1}\omega_j(s)$ form a complete system of linearly independent solutions of $\mathbf{K}'\varphi = 0$ (in the complex field). It is known that the conditions of solubility of (54.1) have the form

$$\int_L f(t)\psi_j(t)\,dt = 0, \quad (j = 1, 2, \ldots, k'). \tag{54.10}$$

Replacing here the ψ_j by their expressions $[t']^{-1}\omega_j$, one obtains these conditions in the real form:

$$\int_L f(s)\omega_j(s)\,ds = 0, \quad (j = 1, 2, \ldots, k'). \tag{54.11}$$

Comparing the above results, the following conclusion is reached: *the fundamental theorems, formulated in the last section, also hold for the real equation, if consideration is restricted to solutions in the real field, and if by the equation, adjoint to* $\mathbf{N}\varphi = 0$, *is understood the real equation* $\mathbf{N}'\omega = 0$.

§ 55. I. N. Vekua's theorem of equivalence. An alternative proof of the fundamental theorems.

The method of reduction, indicated in § 50 and used in the proofs of the fundamental theorems, has the essential disadvantage that the resulting Fredholm equation is not always equivalent to the original equation. It was even seen (§ 53, Note 2) that, for $\varkappa < 0$, it is in general impossible to establish equivalence by such means. However, this does not mean that it is impossible to reduce a given singular equation to an equivalent Fredholm equation by any other means. On the contrary, one has the following

Theorem of equivalence (I. N. Vekua [5]): *The singular integral equation*

$$\mathbf{K}\varphi = f \tag{55.1}$$

is always equivalent (in the sense stated below) *to some Fredholm equation, obtained by the sole aid of quadrature.*

It will be assumed that the line of integration $\mathbf{L}$ satisfies the supplementary condition § 44, 1°a (which is known not to affect generality); the index of (55.1) will be denoted by $\varkappa$.

The following cases will be considered separately.

1°. $\varkappa \geqq 0$. Then an operator $\mathbf{M}$, reducing $\mathbf{K}$, exists such that the homogeneous equation $\mathbf{M}\omega = 0$ has no solution, different from zero. For example, such an operator is $\mathbf{K}^{0\prime}$, i.e., the operator, adjoint to the dominant part of $\mathbf{K}$; or the operator $\mathbf{K}'^{0}$, i.e., the dominant part of $\mathbf{K}'$, adjoint to $\mathbf{K}$. Both these operators have the index $-\varkappa \leqq 0$, and hence, for that choice of $\mathbf{M}$, the equation $\mathbf{M}\omega = 0$ has only the trivial solution $\omega \equiv 0$ (cf. §§ 47, 48). The reducing operator for this $\mathbf{M}$ follows from (45.14), (46.4) and (46.5).

Let $\mathbf{M}$ be any operator with the stated properties. Then *the Fredholm equation*

$$\mathbf{MK}\varphi = \mathbf{M}f \tag{55.2}$$

is equivalent to the original equation, since all solutions of $\mathbf{K}\varphi = f$ are obviously solutions of $\mathbf{MK}\varphi = \mathbf{M}f$ and, conversely, every solution of the latter equation will be a solution of the original one, because it follows from the last equation that $\mathbf{M}(\mathbf{K}\varphi - f) = 0$, and consequently $\mathbf{K}\varphi - f = 0$.

The fact that, for $\varkappa \geqq 0$, the equation $\mathbf{K}\varphi = f$ can be reduced to the equivalent Fredholm equation $\mathbf{MK}\varphi = \mathbf{M}f$ was proved by S. G. Mikhlin [3]. However, he restricted himself to the case in which $A(t) = 1$ and L consists of a single contour; assuming L to be a circle, he proved that $\mathbf{K}'^0$ can be taken as the operator $\mathbf{M}$ (using the same notation).

2°. $\varkappa < 0$. Then an operator $\mathbf{M}$ exists such that $\mathbf{K}$ will reduce $\mathbf{M}$ and that, further,

$$\mathbf{M}\psi = g \tag{55.3}$$

is soluble for every g (satisfying the H condition). As $\mathbf{M}$ one can take, for example, one of the same operators $\mathbf{K}^{0\prime}$ or $\mathbf{K}'^0$ as above. Since here $-\varkappa > 0$, they have the required property; such a choice has still the advantage that (55.3) is then solved for ψ in closed form by simple quadrature (§§ 47, 48).

Now the substitution

$$\varphi = \mathbf{M}\psi \tag{55.4}$$

will be made, where ψ is a new unknown function. Then (55.1) is reduced to the Fredholm equation

$$\mathbf{KM}\psi = f \tag{55.5}$$

which is equivalent to the original equation in the following sense: the equations (55.1) and (55.5) are either both soluble or insolube and, in the case of solubility, the solution of one of them reduces directly to the solution of the other; if one of the operators $\mathbf{K}^{0\prime}$, $\mathbf{K}'^0$ is taken as $\mathbf{M}$, then the passage from the solution of the one equation to that of the other is carried out by means of simple quadrature.

In fact, to every solution ψ of (55.5) there corresponds (one) solution φ of (55.1), and to every solution φ of (55.1) there corresponds (at least one) solution ψ of (55.5) which one obtains by solving (55.4) with respect to ψ. Thus the equations (55.1) and (55.5) are either soluble or insoluble. Now let these equations be soluble. The general solution of the Fredholm equation (55.5) is given by

$$\psi = \psi_0 + \sum_{i=1}^{m} a_i\psi_i, \tag{55.6}$$

where ψ_0 is any of its particular solutions, the ψ_i $(i = 1, \ldots, m)$ are linearly independent solutions of $\mathbf{KM}\psi = 0$ and the a_i are arbitrary constants. It is clear from the above that the general solution of the original equation is given by

$$\varphi = \mathbf{M}\psi_0 + \sum_{i=1}^{m} a_i\mathbf{M}\psi_i; \tag{55.7}$$

one cannot draw from this the conclusion that the number of linearly independent solutions of the homogeneous equation $\mathbf{K}\varphi = 0$ is m, because the functions $\mathbf{M}\psi_i$ $(i = 1, 2, \ldots, m)$ may be linearly dependent in spite of the linear independence of the ψ_i.

Similarly for the converse problem (which, however, is of no interest here) the general solution of (55.5) is found by the solution of (55.1) .

The above theorem provides the means of proving the fundamental theorems of § 53 in an extremely simple manner; this will now be done following I. N. Vekua's [5] treatment.

Proof of theorem I. The necessity of the conditions (53.12) follows directly from the results at the end of § 46. Now their sufficiency will be proved. For this it will be assumed that the conditions (53.12) are satisfied, and it will be shown that then $\mathbf{K}\varphi = f$ is soluble.

(a) First let $\varkappa \geqq 0$. Then (55.1) is equivalent to the Fredholm equation (55.2) and consequently, by the Fredholm theorem, it is soluble, if

$$\int_L \omega_j \mathbf{M} f \, dt = 0, \quad j = 1, 2, \ldots, n, \tag{55.8}$$

where ω_j $(j = 1, 2, \ldots, n)$ is a complete system of linearly independent solutions of the homogeneous equation $\mathbf{K}'\mathbf{M}'\omega = 0$, adjoint to (55.2). The last conditions can be written in the form

$$\int_L f \mathbf{M}'\omega_j \, dt = 0. \tag{55.9}$$

But since $\mathbf{K}'\mathbf{M}'\omega_j = 0$, then $\mathbf{M}'\omega_j$ is also a solution of the homogeneous equation $\mathbf{K}'\psi = 0$; hence the conditions (55.9) are satisfied by (53.12), and $\mathbf{K}\varphi = f$ is soluble.

(b) Now let $\varkappa < 0$. Then (55.1) is soluble, if the Fredholm equation (55.5) is soluble. The conditions of solubility of the latter are of the form

$$\int_L f \chi_j \, dt = 0, \tag{55.10}$$

where χ_j is a complete system of linearly independent solutions of the homogeneous equation $\mathbf{M}'\mathbf{K}'\chi = 0$, adjoint to (55.5). But $\mathbf{M}'\omega = 0$ cannot have a solution, different from zero, because, by supposition, the equation (53.3) is soluble for any $g(t)$ which would be impossible, if $\mathbf{M}'\omega = 0$ had a non-zero solution. Consequently $\mathbf{K}'\chi_j = 0$ and (55.10) is satisfied on the strength of (53.12). Thus the theorem is proved.

Proof of theorems II and III. It is sufficient to consider only the case $\varkappa \geqq 0$, because in the case $\varkappa < 0$ one can apply the same reasoning to the adjoint equation $\mathbf{K}'\psi = 0$ with the index $(-\varkappa)$.

Take one of the operators $\mathbf{K}^{0\prime}$ or $\mathbf{K}'^{0}$ as the reducing operator $\mathbf{M}$. Then $\mathbf{M}\varphi = 0$ has no solution, different from zero, while $\mathbf{M}'\psi = 0$ has exactly $\varkappa$ linearly independent solutions. The Fredholm equation

$\mathbf{MK}\varphi = 0$ is equivalent to $\mathbf{K}\varphi = 0$, and therefore the number of its solutions is equal to the number k of solutions of this latter equation. Consequently also the number of solutions of $\mathbf{K}'\mathbf{M}'\psi = 0$ must be equal to k. But the solution of $\mathbf{K}'\mathbf{M}'\psi = 0$ is obtained by solving the non-homogeneous equation

$$\mathbf{M}'\psi = c_1\psi_1 + c_2\psi_2 + \ldots + c_{k'}\psi_{k'}, \tag{55.11}$$

where $\psi_1, \psi_2, \ldots \psi_{k'}$ is a complete system of solutions of $\mathbf{K}'\psi = 0$ and the $c_1, c_2, \ldots, c_{k'}$, are arbitrary constants. Now (55.11) is always soluble and the number of its linearly independent solutions consists of the number k' of the linearly independent functions ψ_j on the right of (55.11) and of the number $\varkappa$ of solutions of $\mathbf{M}'\psi = 0$. Hence $k = k' + \varkappa$ and Theorems II and III are proved. (This proof of the theorems II and III is the same as that first given by V.D. Kupradze [1]).

§ 56. Comparison of a singular integral equation with a Fredholm equation. The Quasi-Fredholm singular equation. Reduction to the canonical form.

In applications of Fredholm integral equations to different problems the following theorem is known to play a fundamental part.

Let there be given the Fredholm equation $\mathbf{F}\varphi = f$, where $\mathbf{F}$ is a Fredholm operator. Then:

1°. The homogeneous equation $\mathbf{F}\varphi = 0$ has only a finite number of linearly independent solutions.

2°. *The adjoint homogeneous equations* $\mathbf{F}\varphi = 0$ *and* $\mathbf{F}'\psi = 0$ *have the same number of linearly independent solutions.*

3°. The non-homogeneous equation $\mathbf{F}\varphi = f$ is soluble for any f, if and only if the adjoint homogeneous equation $\mathbf{F}'\psi = 0$ (*or, what is the same thing, the homogeneous equation* $\mathbf{F}\varphi = 0$) has no solution, different from zero.

4°. The necessary and sufficient conditions for the solubility of the non-homogeneous equation are

$$\int_L f(t)\psi_k(t)\,dt = 0,$$

where the $\psi_k(t)$ form a complete system of linearly independent solutions of the adjoint homogeneous equation $\mathbf{F}'\psi = 0$.

Comparing these with the results, proved in the last sections for the singular equation $\mathbf{K}\varphi = f$, it is seen that the statements, contained in 1°—4° above, can be transferred to singular integral equations, *with the exception of those which are written in italics.* The fundamental difference between the Fredholm and the singular equations consists in that the proposition 2° must be replaced by

$$k - k' = \varkappa$$

(in the notation of the last sections).

However, among the singular equations one class may be singled out for which *without exception all* the propositions 1°—4° hold. These are those singular equations *which have the index* $\varkappa$ *equal to zero*. In this case $k = k'$ and the non-homogeneous equation $\mathbf{K}\varphi = f$ is soluble for every f (satisfying the H condition), if the homogeneous equation $\mathbf{K}\varphi = 0$ has no solution, different from zero (because then also the adjoint homogeneous equation $\mathbf{K}'\psi = 0$ has these properties).

Such singular equations are called Quasi-Fredholm equations and correspondingly the operators (i.e., the operators with indices zero) are called Quasi-Fredholm operators. (I. N. Vekua [5] uses in these cases the term „pseudo-regular".)

Now consider again singular equations of the general form. Recall that in many cases the Fredholm integral equations are not applied for the purpose of numerical solution of different problems, but for the purpose of their qualitative study. Therefore it is clear that, after having established the general properties of singular integral equations, these equations can be used directly for a similar purpose with the same success as the Fredholm equations. Consequently it is not in all cases expedient to reduce singular equations to Fredholm equations, as often these intermediate stages may only complicate the investigation.

In this connection there naturally arises the question: What is the simplest possible (singular) equivalent equation to which a given singular equation can be reduced? Of course, this question is vague. Here one very simple type of singular integral equation will be mentioned which was pointed out by I. N. Vekua [5] and to which one can reduce any other singular equation, without loss of equivalence; it will be assumed here that the line of integration L satisfies the supplementary condition § 44, 1°a, i.e., that L is the boundary of some connected region S^+.

Let there be given the equation

$$\mathbf{K}\varphi \equiv A(t_0)\varphi(t_0) + \frac{B(t_0)}{\pi i}\int_L \frac{\varphi(t)\,dt}{t - t_0} + \frac{1}{\pi i}\int_L k(t_0, t)\varphi(t)dt = f(t_0). \quad (56.1)$$

Let $\varkappa$ be the index of this equation. For the present it will be assumed that the contour L_0 is present (i.e., that S^+ is a finite region); now consider the operator, defined by

$$\mathbf{M}\psi \equiv a(t_0)\psi(t_0) + \frac{b(t_0)}{\pi i}\int_L \frac{\varphi(t)\,dt}{t - t_0}, \quad (56.2)$$

where

$$2a(t) = \frac{1}{A(t) + B(t)} + \frac{t^{\varkappa}}{A(t) - B(t)},$$
$$2b(t) = \frac{1}{A(t) + B(t)} - \frac{t^{\varkappa}}{A(t) - B(t)}. \tag{56.3}$$

It is easily seen that **M** is a Quasi-Fredholm operator, i.e., its index is zero (assuming, as always, that the origin of coordinates lies in S^+). Consequently the equation $\mathbf{M}\psi = 0$ has no solution, different from zero, and therefore (56.1) is equivalent to

$$\mathbf{N}\varphi \equiv \mathbf{MK}\varphi = \mathbf{M}f. \tag{56.4}$$

Then the dominant part $\mathbf{N}^0$ of **N** has the very simple form [making use of (45.12) or (45.13)]

$$\mathbf{N}^0\varphi \equiv \tfrac{1}{2}(1 + t_0^{\varkappa})\varphi(t_0) + \tfrac{1}{2}(1 - t_0^{\varkappa})\frac{1}{\pi i}\int_L \frac{\varphi(t)\,dt}{t - t_0}. \tag{56.5}$$

If the contour L_0 is at infinity, then the same result is easily obtained, writing in the preceding equation $(t - c)^{-\varkappa}$ instead of $t^{\varkappa}$, where c is any arbitrary point taken inside one of the contours $L_1, \ldots, L_p$.

The singular integral equation

$$\mathbf{N}\varphi = f, \tag{56.6}$$

having the dominant part $\mathbf{N}^0$ of the above-mentioned form, can be called the canonical form of the singular equation.

It is important to note that the general solution of the dominant equation

$$\mathbf{N}^0\varphi = f, \tag{56.7}$$

under the assumption that in the case $\varkappa < 0$ the conditions of solubility, stated below [see (56.11)], are satisfied, has also a very simple form; in fact, assuming for definiteness that the contour L_0 is present,

$$\varphi(t_0) = \mathbf{N}^*f + (1 - t_0^{-\varkappa})P_{\varkappa-1}(t_0), \tag{56.8}$$

where the $P_{\varkappa-1}(t_0)$ are arbitrary polynomials of degree not greater than $(\varkappa - 1)$, zero for $\varkappa \leq 0$, and

$$\mathbf{N}^*f \equiv \tfrac{1}{2}(1 + t_0^{-\varkappa})f(t_0) + \tfrac{1}{2}(1 - t_0^{-\varkappa})\frac{1}{\pi i}\int_L \frac{f(t)\,dt}{t - t_0}. \tag{56.9}$$

This last expression is easily deduced directly from (47.9), if it is noted that the function $X(z)$, corresponding to the equation $\mathbf{N}^0\varphi = f$, is given by

$$X(z) = \begin{cases} 1 & \text{for } z \in S^+, \\ z^{-\varkappa} & \text{for } z \in S^-. \end{cases} \tag{56.10}$$

Thus the boundary condition of the homogeneous Hilbert problem, corresponding to the equation $\mathbf{N}^0\varphi = 0$, has the form

$$\Phi^+(t) = t^{\varkappa}\Phi^-(t),$$

and clearly the function $X(z)$, defined by (56.10), will be the fundamental solution of this problem.

When $\varkappa < 0$, the conditions of solubility of (56.7) have the following form:

$$\int_L t^k f(t)\,dt = 0, \quad k = 0, 1, \ldots, -\varkappa - 1; \tag{56.11}$$

these follow directly from (47.8) and (56.10).

§ 57. Method of reduction, due to T. Carleman and I. N. Vekua.

In the last sections two methods of reduction of singular integral equations were described; the first of these consisted in the replacement of the equation $\mathbf{K}\varphi = f$ by $\mathbf{MK}\varphi = \mathbf{M}f$, where $\mathbf{M}$ is the reducing operator, while the second made use of the substitution $\varphi = \mathbf{M}\psi$. One more method will be indicated which was stated by T. Carleman [1] in one particular case (cf. § 109 below) and which was further developed in the works of I. N. Vekua [1]—[5]. In a number of cases this method is more convenient than those mentioned above.

Let there be given the singular integral equation

$$\mathbf{K}\varphi \equiv A(t_0)\varphi(t_0) + \frac{B(t_0)}{\pi i}\int_L \frac{\varphi(t)\,dt}{t - t_0} + \frac{1}{\pi i}\int_L k(t_0, t)\varphi(t)\,dt = f(t_0) \tag{57.1}$$

or

$$\mathbf{K}\varphi \equiv \mathbf{K}^0\varphi + \mathbf{k}\varphi = f, \tag{57.2}$$

where $\mathbf{K}^0$ is the dominant part of $\mathbf{K}$, i.e., as before,

$$\mathbf{K}^0\varphi \equiv A(t_0)\varphi(t_0) + \frac{B(t_0)}{\pi i}\int_L \frac{\varphi(t)\,dt}{t - t_0} \tag{57.3}$$

and

$$\mathbf{k}\varphi \equiv \frac{1}{\pi i}\int_L k(t_0, t)\varphi(t)\,dt \tag{57.4}$$

with $k(t_0, t)$ of the form (44.6).

The line of integration L will be assumed to satisfy the supplementary condition § 44, 1°a.

Write (57.2) in the form

$$\mathbf{K}^0\varphi = f - \mathbf{k}\varphi \tag{57.5}$$

and solve (57.5), as if the right side was a given function. *Leaving for the time being out of consideration the conditions of solubility for the case* $\varkappa < 0$, *to be stated later* [*see*(57.11)], one obtains, using the results of § 47,

$$\varphi(t_0) = \mathbf{K}^*f - \mathbf{K}^*\mathbf{k}\varphi + B^*(t_0)Z(t_0)P_{\varkappa-1}(t_0)$$

or

$$\varphi(t_0) + \mathbf{K}^*\mathbf{k}\varphi = \mathbf{K}^*f + B^*(t_0)Z(t_0)P_{\varkappa-1}(t_0), \tag{57.6}$$

where

$$\mathbf{K}^*f = A^*(t_0)f(t_0) - \frac{B^*(t_0)Z(t_0)}{\pi i}\int_L \frac{f(t)dt}{Z(t)(t-t_0)}, \tag{57.7}$$

$Z(t)$, $A^*(t)$, $B^*(t)$ are defined by (47.15), (47.16) and the $P_{\varkappa-1}(t_0)$ are arbitrary polynomials of degree not greater than $(\varkappa - 1)$ [$P_{\varkappa-1} = 0$ for $\varkappa \leqq 0$].

The operator $\mathbf{K}^*\mathbf{k}$, by the Note in § 45, is a Fredholm operator of the first kind, namely

$$\mathbf{K}^*\mathbf{k}\varphi \equiv \frac{1}{\pi i}\int_L N(t_0, t)\varphi(t)\,dt, \tag{57.8}$$

where

$$N(t_0, t) = \mathbf{K}^*k(t_0, t), \tag{57.9}$$

and for the application of $\mathbf{K}^*$ to $k(t_0, t)$ the variable is assumed to be t_0, while t is considered as a parameter, so that

$$N(t_0, t) = A^*(t_0)k(t_0, t) - \frac{B^*(t_0)Z(t_0)}{\pi i}\int_L \frac{k(t_1, t)\,dt_1}{Z(t_1)(t_1 - t_0)}. \tag{57.10}$$

By the results of § 47, the equation (57.6) is for $\varkappa \geqq 0$ equivalent to the original equation (57.1); if $\varkappa < 0$, then the equations

$$\int_L \frac{t^k\mathbf{k}\varphi(t)\,dt}{Z(t)} = \int_L \frac{t^k f(t)\,dt}{Z(t)},\ k = 0, 1, \ldots, -\varkappa - 1, \tag{57.11}$$

arising from the conditions (47.17), have to be added to (57.6); then the original equation (57.1) is equivalent to (57.6) together with (57.11).

The equation (57.6) is a Fredholm equation (of the second kind). One sees that, by the Note in § 45,

$$N(t_0, t) = \frac{N^*(t_0, t)}{|t - t_0|^\lambda}, \quad 0 \leqq \lambda < 1, \tag{57.12}$$

where $N^*(t_0, t)$ satisfies the H condition.

Thus a reduction of the original equation is obtained which has the very important feature of preserving equivalence. In fact, for $\varkappa < 0$, one obtains, in addition to a Fredholm equation of the second kind, the supplementary equations (57.11), but this is not essential, since in principle the problem reduces to the solution of the Fredholm equation (57.6).

Instead of using the formulae of § 47 for the reduction of a singular equation, one can also start from those of § 48. Namely, let there be given the equation

$$\mathbf{K}'\psi \equiv \mathbf{K}^{0\prime}\psi + \mathbf{k}'\psi = f(t_0), \tag{57.13}$$

where

$$\mathbf{K}^{0\prime}\psi \equiv A(t_0)\psi(t_0) - \frac{1}{\pi i}\int_L \frac{B(t)\psi(t)\,dt}{t - t_0}, \tag{57.14}$$

$$\mathbf{k}'\psi \equiv \frac{1}{\pi i}\int_L k(t, t_0)\psi(t)\,dt. \tag{57.15}$$

Of course, the equation $\mathbf{K}'\psi = f(t_0)$ may be considered without using the fact that it is adjoint to $\mathbf{K}\varphi = f$, because every singular operator $\mathbf{K}'$ can be represented in the form $\mathbf{K}^{0\prime} + \mathbf{k}'$.

Using the results of § 48, one obtains, as above, that (57.13) is, for $\varkappa' \geqq 0$ ($\varkappa'$ being its index), equivalent to the Fredholm equation (of the second kind)

$$\psi + \mathbf{K}^{*\prime}\mathbf{k}'\psi = \mathbf{K}^{*\prime}f + \frac{Q_{\varkappa'-1}(t_0)}{Z(t_0)}, \tag{57.16}$$

where the $Q_{\varkappa'-1}(t_0)$ are arbitrary polynomials of degree not greater than $\varkappa' - 1$ ($Q_{\varkappa'-1} \equiv 0$ for $\varkappa' = 0$).

For $\varkappa' < 0$, the equation (57.13) is equivalent to (57.16) for $Q_{\varkappa'-1}(t_0) \equiv 0$ and with the supplementary equations, obtained from the conditions (48.12),

$$\int_L t^k Z(t) B^*(t)\mathbf{k}'\varphi(t)\,dt = \int_L t^k Z(t) B^*(t) f(t)\,dt. \tag{57.17}$$

Note that even in the case where $\mathbf{K}'$ is the operator, adjoint to $\mathbf{K}$ the Fredholm equations (57.6) and (57.16) will not, in general, be adjoint, because $(\mathbf{K}^*\mathbf{k})' = \mathbf{k}'\mathbf{K}^{*\prime}$, and not $\mathbf{K}^{*\prime}\mathbf{k}'$.

All the fundamental theorems, studied in the preceding sections, can be obtained from the above method of reduction (in § 112 this method will be used to prove the fundamental theorems for the more complicated case, where L consists of arcs); this was done by I. N. Vekua in his papers [2] and [4] (the author used the first of the two methods stated in the present section); in addition, one can obtain a number of new results in this way, for example those given in the following section.

§ 58. Introduction of the parameter λ.

As in the theory of the Fredholm equation, one can also introduce a parameter λ into singular integral equations; for this the most simple results, analogous to those of the classical Fredholm theory, are obtained, if the parameter is introduced as a factor of the operator $\mathbf{k}$ in an equation of the form (57.2) or of the operator $\mathbf{k}'$ in an equation of the form (57.13), as it was done by I. N. Vekua [5]. This corresponds to the fact that in the Fredholm equation the parameter is usually introduced as a factor of the completely continuous part (cf. § 49) of the Fredholm operator; for any other method of introduction of the parameter the theory becomes considerably more complicated.

In agreement with these remarks consider the singular equation

$$\mathbf{K}\varphi \equiv \mathbf{K}^0\varphi + \lambda\mathbf{k}\varphi = f(t_0), \tag{58.1}$$

where λ is an arbitrary (in general complex) parameter.

Reducing this equation by the first of the methods, stated in the preceding section, one obtains for $\varkappa \geqq 0$ the equivalent Fredholm equation

$$\varphi + \lambda\mathbf{K}^*\mathbf{k}\varphi = f_0(t_0), \tag{58.2}$$

where

$$f_0(t_0) = \mathbf{K}^*f + B^*(t_0)Z(t_0)P_{\varkappa-1}(t_0) \tag{58.3}$$

and $P_{\varkappa-1}(t_0)$ is an arbitrary polynomial of degree not greater than $\varkappa - 1$. The homogeneous equation, adjoint to the preceding one, has the form

$$\psi + \lambda(\mathbf{K}^*\mathbf{k})'\psi \equiv \psi + \lambda\mathbf{k}'\mathbf{K}^{*\prime}\psi = 0. \tag{58.4}$$

From the theory of the Fredholm equation it is known that the adjoint homogeneous equations

$$\varphi + \lambda\mathbf{K}^*\mathbf{k}\varphi = 0 \text{ and } \psi + \lambda\mathbf{k}'\mathbf{K}^{*\prime}\psi = 0$$

either both have or both have not a solution, different from zero (and the number of linearly independent solutions of these equations is always the same); such solutions exist only, when the parameter

λ takes one of the infinite or finite sequence of „characteristic" values

$$\lambda_1, \lambda_2, \ldots, \tag{58.5}$$

having no point of accumulation in the finite part of the plane. If λ is different from the values (58.5), then the non-homogeneous equation (58.2) has a single-valued solution for every right side; the solution is of the form

$$\varphi(t_0) = f_0(t_0) + \int_L \Gamma(t_0, t; \lambda) f_0(t)\, dt, \tag{58.6}$$

where the Fredholm resolvent $\Gamma(t_0, t; \lambda)$ is a meromorphic function of λ with poles at the points (58.5).

For $f \equiv 0$, i.e., when the original equation is homogeneous,

$$f_0(t_0) = B^*(t_0)Z(t_0)(C_0 t_0^{\varkappa-1} + C_1 t_0^{\varkappa-2} + \ldots + C_{\varkappa-1}) \tag{58.7}$$

(C_j being arbitrary constants) and for λ, different from the characteristic values (58.5), the equation (58.2) has exactly $\varkappa$ linearly independent solutions, obtained by solution of the Fredholm equations

$$\varphi + \lambda \mathbf{K}^*\mathbf{k}\varphi = B^*(t_0)Z(t_0)t_0^k, \quad k = 0, 1, \ldots, \varkappa - 1. \tag{58.8}$$

Thus one arrives at the following results:

For $\varkappa \geqq 0$, the general solution of the singular integral equation $\mathbf{K}\varphi = f$ *is a meromorphic function of λ, containing linearly $\varkappa$ arbitrary constants. With the possible exception of a series of discrete characteristic values $\lambda_1, \lambda_2, \ldots$ the homogeneous equation* $\mathbf{K}\varphi = 0$ *has exactly $\varkappa$ linearly independent solutions which are also meromorphic functions of λ* [in all cases it is known from Note 1 in § 53 that the homogeneous equation has not less than $\varkappa$ linearly independent solutions].

To these results, pointed out by I. N. Vekua [5], the following may be added. Let now $\varkappa < 0$. Then the original equation (58.1) is equivalent to (58.2) with $P_{\varkappa-1}(t_0) \equiv 0$ and the supplementary conditions (57.11) of the last section. Introducing into the latter the expression (58.6) for φ, one is easily seen to obtain conditions of the form

$$\int_L \omega_j(t, \lambda) f(t) dt = 0, \quad j = 1, 2, \ldots, -\varkappa, \tag{58.9}$$

where the $\omega_j(t, \lambda)$ are certain meromorphic functions of λ with poles at the points (58.5). For λ, different from the values (58.5) the conditions (58.9) are necessary and sufficient for the solubility of (85.1), because in this case the Fredholm equation (58.2) is always soluble. Assuming λ to be different from the values (58.5), the conditions (58.9) will be expressed in another manner. Namely, consider the homogeneous equation

$$\mathbf{K}'\psi = \mathbf{K}^{0\prime}\psi + \lambda\mathbf{k}'\psi = 0, \tag{58.10}$$

adjoint to $\mathbf{K}\varphi = 0$. Let the $\psi_1, \psi_2, \ldots \psi_{k'}$, $(k' \geqq \varkappa' = -\varkappa)$ be its linearly independent solutions. Then the conditions (58.9) must be equivalent to the conditions

$$\int\limits_L \psi_j(t)f(t)\,dt = 0, \; j = 1, 2, \ldots, k', \tag{58.11}$$

because the latter are also necessary and sufficient conditions for the solubility of (58.1).

Hence it is easily concluded (cf. the proposition introduced at the end of appendix 3) that the functions $\omega_j(t, \lambda)$ are linear combinations of the $\psi_j(t)$, and vice versa. Consequently, for λ different from the values (58.5), all the functions $\omega_j(t, \lambda)$ $(j = 1, 2, \ldots, \varkappa')$ are linearly independent and $k' = \varkappa' = -\varkappa$.

Thus the following result is obtained:

When $\varkappa = -\varkappa' < 0$ for all values of λ, different from the characteristic values (58.5), *the conditions of solubility of $\mathbf{K}\varphi = f$ are $\varkappa' = -\varkappa$ relations of the type* (58.9), *where the $\omega_j(t, \lambda)$ are linearly independent functions, meromorphic with respect to λ, with poles at the points* (58.5); *if f satisfies these conditions, then the solution is given by* (58.6).

For $\varkappa = 0$, i.e., in the case of a Quasi-Fredholm equation, the results of this section coincide with the known results for Fredholm equations.

§ 59. Brief remarks on some other results.

As it is not possible to state here fully all interesting results, refering to singular integral equations, the following short remarks will refer to those which will not be treated in the remaining Parts of this book.

1°. Let $\mathbf{K}$ be a singular operator of the type considered above. In § 55 a method of reduction of the singular equation $\mathbf{K}\varphi = f$ to the equivalent Fredholm equation was stated which is due to I. N. Vekua. Somewhat earlier he indicated yet another method which is also of interest. This method is based on the method of reduction stated in § 57. If $\varkappa \geqq 0$, then $\mathbf{K}\varphi = f$ was seen to be directly reducible by this method to an equivalent Fredholm equation. I. N. Vekua [3], [4] also proved that, when $\varkappa < 0$, one may, by means of a simple transformation, reduce a given singular equation to a Fredholm equation, accompanied by additional conditions of the form

$$\int_L \omega_j(t) f(t)\, dt = 0; \tag{*}$$

this Fredholm equation and the functions $\omega_j(t)$ may be constructed by simple quadrature.

2°. Let T be the operator defined by

$$\mathbf{T}\varphi(t) = \frac{dt(s)}{ds} \cdot \varphi(t),$$

where, as always, t is the coordinate of a point on the contour L and s is the corresponding arc coordinate; consider the following operators: $\overline{\mathbf{K}}'\mathbf{TK}$, $\mathbf{U} =$ dominant part of $\overline{\mathbf{K}}'\mathbf{TK}$, $\mathbf{M} = \mathbf{U}'\overline{\mathbf{K}}'\mathbf{T}$, where the bar above any letter indicates the conjugate complex operator (i.e., by definition, $\overline{\mathbf{K}}\,\overline{\varphi} = \overline{\mathbf{K}\varphi}$).

It is easily proved that the operators $\overline{\mathbf{K}}'\mathbf{TK}$ and $\mathbf{U}$ are Quasi-Fredholm operators, while $\mathbf{MK}$ is a Fredholm operator, so that $\mathbf{M}$ is a reducing operator with respect to $\mathbf{K}$. Further, the following theorem holds:

The Fredholm equation

$$\mathbf{MK}\varphi = \mathbf{M}f$$

is soluble for every function f and is equivalent to $\mathbf{K}\varphi = f$, if the latter is soluble.

The operator $\mathbf{M}$ which is seen to be constructed in quite an elementary manner and which possesses the above property was given by I. N. Vekua [5] (the proof will be found in the same paper).

Still earlier V. D. Kupradze [1] indicated that in the case of the real equation $\mathbf{L}\varphi = f$ of the form (44.16) the operator $\mathbf{L}'$, adjoint to $\mathbf{L}$, has the same property. Kupradze [2] also constructed an operator, having this property, for the equation $\mathbf{K}\varphi = f$; but the construction of this latter operator required the determination of all the solutions of the adjoint homogeneous equation $\mathbf{K}'\psi = 0$.

3°. Numerous works of G. Giraud are devoted to singular integral equations. Their list can be found in one of his later papers (Giraud [2]).

It is the Author's impression that Giraud's work is of considerable interest only in those parts, concerning singular equations of several variables (cf. G. Giraud [1] and also other papers, listed in G. Giraud [2]). In the case of equations with a single variable (results obtained by him in this field are stated in their most complete form in G.

Giraud [2]), i.e., equations of the type considered here, Giraud's investigations are unnecessarily complicated and besides insufficiently complete: for example, the concept of index, fundamental to the entire theory, does not appear in his work, although this concept was introduced considerably earlier. Giraud bases his theory on the investigation of the dependence of the solution of an integral equation on the parameter λ which he did not introduce, as it is done in § 58; in fact, he considered the equation (written in the notation used here)

$$A(t_0)\varphi(t_0) + \frac{\lambda B(t_0)}{\pi i} \int\limits_L \frac{\varphi(t)dt}{t - t_0} + \lambda \int\limits_L k(t_0, t)\varphi(t)dt = f(t_0), \quad (59.1)$$

where λ also enters into the dominant part. This disturbs the analogy with the equations, considered in the classical Fredholm theory, and considerably complicates the investigation. In any case the general theory of an equation of the form (59.1) is by no means simpler, at least not with regard to the results obtained by Giraud, than is the theory of the much more general equation

$$A(t_0)\varphi(t) + \frac{1}{\pi i} \int\limits_L \frac{K(t_0, t; \lambda)\varphi(t)\, dt}{t - t_0} = f(t_0), \quad (59.2)$$

where $K(t_0, t; \lambda)$ satisfies the H condition with respect to t and t_0 and is a function of λ, meromorphic in some region. This equation is an analogue to the Fredholm equation which, with its kernel a meromorphic function of the parameter λ, was studied by J. D. Tamarkin [1].

Applying to the last equation the methods stated above, one can obtain Giraud's results much more simply than it was done by him and, in addition, find a number of more general results. This was done by D. F. Kharazov [1].

4°. Some work by S. G. Mikhlin is also devoted to the theory of singular integral equations. Reference has already been made to his paper [3]; attention is also called to his recently published note [4]. In other papers S. G. Mikhlin considers mainly singular equations in several dimensions; in the Author's opinion this work is of the greatest interest; only a reference will be made to his papers [1] and [2].

5°. References to the literature, dealing with multi-dimensional singular equations, can be found in the works of G. Giraud [1], S. G. Mikhlin [1], [2], and F. Tricomi [1].

PART III

APPLICATIONS TO SOME BOUNDARY PROBLEMS

The results of Parts I and II are useful for the solution of many important problems in the theory of analytic functions and mathematical physics. In recent times a series of results has been obtained by these methods which are of considerable interest. Some of these will be given in this Part.

The Dirichlet problem, one of the simplest problems of the type under consideration, will be treated first. Prior to the solution of this problem by use of singular integral equations, the classical Fredholm solution will be given with certain alterations, thus bringing this solution within the range of the problems considered in the present treatise.

The Dirichlet problem is a particular case of the Riemann-Hilbert problem, when one of the coefficients a, b in the boundary condition (39.1) is zero, i.e., provided one does not consider certain features, arising in the case of multiply-connected regions.

However, for the solution of the Riemann-Hilbert problem, use was made of conformal transformation of the given region S^+ on to the circle which is equivalent to the Dirichlet problem. Therefore the Dirichlet problem will be treated here independently, particularly since it will here be considered for regions bounded by an arbitrary number of contours.

CHAPTER 7

THE DIRICHLET PROBLEM

§ 60. Statement of the Dirichlet and the modified Dirichlet problems. Uniqueness theorems.

Let S^+ be a connected region, bounded by simple smooth non-intersecting contours $L_0, L_1, \ldots, L_p$ the first of which contains all the others. By L will be understood the union of these contours; as usual, the positive direction on L will be taken such that S^+ remains on the left. The contour L_0 may be absent in which case S^+ is infinite. The union of the finite regions $S_1^-, \ldots, S_p^-$, contained in $L_1, \ldots, L_p$

respectively, and (in the presence of L_0) the infinite region S_0^-, consisting of the points outside L_0, will be denoted by S^-.

The following condition will also be imposed on the contours $L_0, \ldots, L_p$: the angle between the tangent to L_j and some constant direction satisfies the H condition.

The classical Dirichlet problem for the region S^+ will be formulated as follows:

A. The Dirichlet problem. *To find the (real) function $u(x, y)$, harmonic in S^+, continuous in $S^+ + L$ and satisfying the boundary condition*

$$u = f(t) \text{ on } L, \tag{60.1}$$

where $f(t)$ is a (real) continuous function given on L; in the case in which S^+ is infinite, $u(x, y)$ is also required to be bounded at infinity.

The last condition regarding the boundedness of u at infinity is equivalent to the condition that u tends to a certain limit as z tends to infinity.

It will be recalled that every function $u(x, y)$, harmonic outside the circle $|z| > R_0$ and bounded at infinity, may be developed for $|z| > R_0$ into a series

$$u(x, y) = a_0 + \sum_{n=1}^{\infty} r^{-n}(a_n \cos n\vartheta + b_n \sin n\vartheta), \quad z = re^{i\vartheta}$$

which is absolutely and uniformly convergent outside any circle of radius $R > R_0$.

Therefore $u \to a_0$ as $r \to \infty$.

B. The modified Dirichlet problem. *To find the function $u(x, y)$, harmonic in S^+, continuous in $S^+ + L$ and satisfying the following conditions:*

1°. *The function $u(x, y)$ is the real part of some function $\Phi(z)$, holomorphic in S^+.*

2°. *It satisfies the boundary condition*

$$u = f(t) + a(t) \text{ on } L, \tag{60.2}$$

where $f(t)$ is a (real) continuous function given on L and

$$a(t) = a_j \text{ on } L_j, \; j = 0, 1, \ldots, p, \tag{60.3}$$

where the a_j are real constants which have to be determined; in the case of the infinite region the condition $u = f + a_0$ on L_0 is replaced by the condition that $u(x, y)$ is bounded at infinity.

It will be shown later that the constants a_j are uniquely defined by the conditions of the problem, provided one of them is fixed (arbitrarily). In the sequel, unless stated otherwise, a_0 will be assumed equal to zero (in the absence of L_0 this will be replaced by $u = 0$ at infinity).

The case in which L is a single contour will be considered separately. Here two cases must be distinguished:

a) $p = 0$. Then S^+ is a finite part of the plane bounded by L_0.

b) $p = 1$, and the contour L_0 is absent. Then S^+ is an infinite part of the plane bounded by L_1.

Obviously the problems A and B coincide in the case a) [if one assumes $a_0 = 0$]. In the case b) these problems lead directly to each other; e.g. if $u(x, y)$ is the solution of B (vanishing at infinity), then $u(x, y) - a_1$ will be the solution of A.

Next, the general case of arbitrary p will be considered. The difference between the problems A and B, i.e., between the classical and modified Dirichlet problems is as follows:

If $u(x, y)$ is the solution of A, then the function v, conjugate complex to u, may be found to be (and in general will be) multi-valued. The condition 1° in the statement of the problem B excludes this possibility and hence restricts the class of unknown harmonic functions; but, on the other hand, the condition 2° of Problem B requires only that u takes on the L_j values, given apart from additive constant terms.

Each of the problems A and B cannot have more than one solution [if it is agreed to put $a_0 = 0$ in B; instead of $a_0 = 0$, one can also take $a_k = 0$, where a_k is one of the constants $a_1, a_2, \ldots, a_p$]. In the case of the problem A this proposition leads to the result that a function, harmonic in S^+, continuous in $S^+ + L$ and zero on L, is zero throughout the region. The last result follows from the well known theorem that a harmonic function which is not constant attains its maximum and minimum values only on the boundary of a region. [The case in which the region is infinite, i.e., in which L_0 is absent, is reduced to the finite case by means of a simple inversion].

Next, the proof will be given of the same proposition for Problem B. In this case the statement is as follows:

If the function $u(x, y)$, harmonic in S^+ and continuous in $S^+ + L$, is the real part of $\Phi(z) = u + iv$, holomorphic in S^+, and if $u(x, y)$ takes the constant values a_j on L_j and if $a_0 = 0$, then necessarily $u = 0$, $a_1 = a_2 = \ldots = a_p = 0$ (if S is infinite, the condition $a_0 = 0$ is replaced by the condition $u = 0$ at infinity).

The idea of the proof given here is taken from J. Plemelj's book [3]. Another simple proof will be outlined before proceeding to the above-mentioned proof. From a well known theorem

$$\int_L u \frac{\partial u}{\partial n} ds = \iint_{S^+} \left[\left(\frac{\partial u}{\partial x} \right)^2 + \left(\frac{\partial u}{\partial y} \right)^2 \right] dx dy$$

(still assuming S^+ to be a finite region), where n is the outward normal. But the left side of this equation is zero, since

$$\int_L u \frac{\partial u}{\partial n} ds = \int_L u \frac{\partial v}{\partial s} ds = \sum_{j=0}^{p} a_j \int_L dv = 0,$$

as, by supposition, v is single-valued. Consequently, from the first of the above equations, u = const. in S^+. But $u = 0$ on L_0, hence $u = 0$ in S^+. Similarly for the case of an infinite region. However, this proof is not complete, as the assumption of the existence of $\frac{\partial u}{\partial n}$, $\frac{\partial v}{\partial s}$ and the relation $\frac{\partial u}{\partial n} = \frac{\partial v}{\partial s}$ on L (which does really hold under the assumed conditions) require justification.

Assume S^+ to be finite and denote by a_m the least of the constants $a_0, a_1, \ldots, a_p$ (if there are several of these constants, choose any one of them). Assume that $u(x, y)$ is not constant in S^+. Since $u = a_m$ on L_m, a_m is the minimum value of u in $S^+ + L$. Consequently $u > a_m$ everywhere in S^+, for, by assumption, the function u is not constant. Therefore, as is easily seen, one can always construct in S^+ two smooth contours L_m', L_m'' near L_m on which $u(x, y)$ assumes the constant values $a_m + \varepsilon'$, $a_m + \varepsilon''$, $0 < \varepsilon' < \varepsilon''$.

For this purpose, take on the (with respect to S^+) inward normal to L_m at t a segment tM of a sufficiently small constant length δ such that it lies entirely in S^+ for every t on L. When t describes L_m, the point M describes a continuous closed curve (the curve may intersect itself, but this is not important). Clearly on all these curves $u \geqq a_m + \varepsilon_0$, where ε_0 is some positive constant. Therefore, if the numbers ε_1 and ε_2 are chosen such that $0 < \varepsilon_1 < \varepsilon_2 < \varepsilon_0$, points t_1 and t_2 can be found on tM at which $u(x, y)$ takes the values $a_m + \varepsilon_1$ and $a_m + \varepsilon_2$. When t moves along L_m, the points t_1 and t_2 describe continuous closed curves $L_m^{(1)}$ and $L_m^{(2)}$. These curves have no points in common, since u assumes different values along them. Further, neither of these curves intersects itself, because, if, for example, $L_m^{(1)}$ formed a loop, then the harmonic function u, having the constant value $a_m + \varepsilon_1$ on the loop, would be constant throughout the region bounded by it, and hence constant throughout S^+. The curves $L_m^{(1)}$ and $L_m^{(2)}$, satisfying conditions of the form $u(x, y)$ = const., can only have a finite number of singular points, i.e., points at which $\frac{\partial u}{\partial x} = \frac{\partial u}{\partial y} = 0$. In fact, at such points $\Phi'(z) = 0$, but the number of zeros of the functions $\Phi'(z)$, holomorphic in S^+, lying at finite distances from the boundaries of S^+, is finite, because $\Phi'(z)$ is not identically zero (otherwise one would have: $\Phi(z)$ = const. in S^+). Consequently, since $u(x, y)$ is analytic, $L_m^{(1)}$ and $L_m^{(2)}$ consist of a finite number of arcs of analytic curves. The contours $L_m^{(1)}$ and $L_m^{(2)}$ bound some annular region Σ entirely in S^+. The number of points z_j in Σ, where $\Phi'(z) = 0$, is finite. On the other hand, if ε is any number such that $\varepsilon_1 < \varepsilon < \varepsilon_2$, one can similarly construct a contour L_m^0, entirely in Σ, such that $u = a_m + \varepsilon$ on L_m^0. Since there is only a finite number of points z_j, there is an infinity of values ε such that L_m^0 does not pass through these points, and hence there is a smooth analytic contour without singular points. By giving ε two values ε' and ε'', satisfying these conditions, the required contours L_m' and L_m'' are obtained.

Denoting by Σ the annular region between L_m' and L_m'' and by $\Lambda = L_m' + L_m''$ its boundary and using the well-known formula

$$\int_\Lambda u \frac{\partial u}{\partial n} ds = \iint_\Sigma \left[\left(\frac{\partial u}{\partial x}\right)^2 + \left(\frac{\partial u}{\partial y}\right)^2\right] dx\, dy, \qquad (*)$$

where n is the normal to Λ directed outwards from Σ, it is seen, since

$$\int_{L_m'} u \frac{\partial u}{\partial n} ds = \int_{L_m'} u \frac{\partial v}{\partial s} ds = (a_m + \varepsilon') \int_{L_m'} dv = 0$$

and similarly for the integral taken over L_m'', that the left side of (*) is zero; hence $u =$ const. in Σ, and hence in the entire region S^+. But since $u = a_0 = 0$ on L_0, $u = 0$ in S^+.

Thus the theorem is proved for the finite region.

When S^+ is an infinite region, the proof is quite similar; in this case one must understand by a_m the smallest of the constants $a_1, a_2, \cdots, a_p$ (or, if there are several, anyone of them).

NOTE. As before, suppose that $u = a_j$ on L_j, but omit the condition $a_0 = 0$. Then obviously $u =$ const. $= a_0 = a_1 = \ldots = a_p$.

Therefore, when the condition $a_0 = 0$ does not hold, two solutions of the problem B can differ by a constant.

§ 61. Solution of the modified Dirichlet problem by means of the potential of a double layer.

Firstly, the solution of the modified Dirichlet problem will be considered. In addition to the conditions and notation of the last section, the given function $f(t)$ will be assumed to satisfy the H condition [it is not difficult to dispose of this assumption (cf. Note 1 at the end of this section)]. The function $\Phi(z) = u(x, y) + iv(x, y)$ will be sought in the form

$$u(x, y) + iv(x, y) = \Phi(z) = \frac{1}{\pi i} \int_L \frac{\mu(t)\, dt}{t - z}, \qquad (61.1)$$

where $\mu(t)$ is an unknown real function satisfying the H condition.

Putting $t - z = re^{i\vartheta}$ and separating the real part, one obtains (cf. § 11)

$$u(x, y) = \frac{1}{\pi}\int_L \mu(t)d\vartheta = \frac{1}{\pi}\int_L \mu(t)\,\frac{\cos(r, n)}{r}\,ds, \qquad (61.2)$$

where n is the normal at t, directed towards the left of L, and (r, n) is the angle between n and $\overrightarrow{tz}$.

Thus $u(x, y)$ is in the form of the potential of a double layer.

Passing in (61.1) to the limit $z \to t_0$ and separating the real part, one obtains the boundary value of $u(x, y)$ [using the first of the Plemelj formulae (17.2) (cf. also § 12, (12.6) et seq.)]

$$u = \Re\Phi^+(t_0) = \mu(t_0) + \Re\frac{1}{\pi i}\int_L \frac{\mu(t)\,dt}{t - t_0} =$$

$$= \mu(t_0) + \frac{1}{\pi}\int_L \mu d\vartheta = \mu(t_0) + \frac{1}{\pi}\int_L \mu(t)\,\frac{\cos(r, n)}{r}\,ds, \qquad (61.3)$$

where $\vartheta = \vartheta(t_0, t)$ is the angle between the vector $\overrightarrow{t_0 t}$ and the Ox axis, $r = |t - t_0|$ and (r, n) is the angle between $\overrightarrow{tt_0}$ and n (Fig. 3, § 12).

Hence the boundary condition (60.2) takes the form

$$\mu(t_0) + \frac{1}{\pi}\int_L \mu(t)d\vartheta = f(t_0) + a(t_0) \qquad (61.4)$$

or

$$\mu(t_0) + \frac{1}{\pi}\int_L \mu(t)\,\frac{\cos(r, n)}{r}\,ds = f(t_0) + a(t_0). \qquad (61.4a)$$

The last equation is a Fredholm equation (this is even so, if one does not consider the term $a(t)$ which, as a rule, is used for the solution of the Dirichlet problem) with the kernel

$$K(t_0, t) = \frac{1}{\pi}\,\frac{\cos(r, n)}{r}$$

which, by the assumed conditions, is of the form [cf. § 8, (8.4)]

$$K(t_0, t) = \frac{K_0(t_0, t)}{|t - t_0|^a} = \frac{K_0(t_0, t)}{r^a}, \tag{61.5}$$

where $0 \leqq \alpha < 1$ and $K_0(t_0, t)$ satisfies the H condition with respect to both variables.

Consequently (§ 51) every (continuous) solution of the equation (61.4) also satisfies the H condition, because the function $f(t)$ was assumed to satisfy this condition. It can also be shown that $\mu(t)$ satisfies the H condition with the same index as the function $f(t)$ (cf. J. Schauder [1], p. 633).

The homogeneous equation

$$\mu(t_0) + \frac{1}{\pi}\int_L \mu(t)d\vartheta = 0, \tag{61.6}$$

corresponding to the equation (61.4), will now be considered. By direct verification it is seen to have the following solution (where the last of the equalities must be omitted in the case of an infinite region)

$$\mu = C_k \text{ on } L_k \ (k = 1, 2, \ldots, p), \ \mu = 0 \text{ on } L_0, \tag{61.7}$$

where the C_k are arbitrary constants. The equation (61.6) has no other solutions. In fact, let μ be any of its solutions. Then the real part of the function

$$\Phi(z) = \frac{1}{\pi i}\int_L \frac{\mu(t)\,dt}{t - z},$$

holomorphic in S^+, is zero on L. Hence $\Re\Phi(z) = 0$ in S^+ and the statement follows (§ 25, Note 1).

The general solution (61.7) of the homogeneous equation is a linear combination of the p linearly independent solutions $\mu_1(t), \ldots, \mu_p(t)$, where

$$\mu_j(t) = \begin{cases} 1 \text{ for } t \in L_j \ (j = 1, 2, \ldots, p), \\ 0 \text{ on the remaining contours.} \end{cases} \tag{61.8}$$

By the general theory of the Fredholm integral equation the non-homogeneous equation (61.4) is soluble, if and only if its right side satisfies the known p integral conditions. For the solution of the problem the constants a_j must be selected such that these conditions are satisfied.

However, the establishment of the above-mentioned conditions is, at least in practice, rather complicated, since one has to find all linearly independent solutions of the homogeneous equation, adjoint to (61.6). Moreover, the presence of non-zero solutions of (61.6) considerably complicates the solution of the original equation (16.4), even if the constants a_j have already been properly selected.

All these difficulties can be avoided by means of a very simple method which replaces the equation (61.4) by an equivalent equation, no longer containing the indeterminate constants a_j and having the property that its corresponding homogeneous equation has no non-zero solutions.

In fact, consider instead of (61.4) the different Fredholm integral equation

$$\mu(t_0) + \frac{1}{\pi}\int_L \mu(t)d\vartheta - \int_L k(t_0, t)\mu(t)\,ds = f(t_0), \qquad (61.9)$$

where s is the arc coordinate of the point t and the (real) function $k(t_0, t)$ is defined on L in the following manner:

$$k(t_0, t) = \begin{cases} \varrho_j(t) \text{ for } t_0, t \in L_j, \; j = 1, \ldots, p, \\ 0 \text{ on all other contours;} \end{cases} \qquad (61.10)$$

the $\varrho_j(t)$ are real continuous functions, given on L_j ($j = 1, \ldots, p$), satisfying the conditions

$$\int_{L_j} \varrho_j(t)ds \neq 0 \qquad (61.11)$$

and otherwise quite arbitrary. For example, one can take

$$\varrho_j(t) = 1, \quad j = 1, \ldots, p.$$

The expression

$$\int_L k(t_0, t)\mu(t)\,ds$$

on the left of (61.9) has on each of the contours L_j the constant values

$$\int_L k(t_0, t)\mu(t)\,ds = c_j \text{ for } t_0 \text{ on } L_j, \; j = 0, 1, \ldots, p, \qquad (61.12)$$

where the c_j are constant, i.e.,

$$c_j = \int_{L_j} \varrho_j \mu\,ds,$$

and $c_0 = 0$ [which has to be omitted for an infinite region, so that one must take then $j = 1, \ldots, p$ in (61.12)].

The equation (61.9) will now be shown to have the required property, i.e., that the homogeneous equation

$$\mu(t_0) + \frac{1}{\pi} \int_L \mu(t)\,d\vartheta - \int_L k(t_0, t)\mu(t)\,ds = 0 \qquad (61.13)$$

has no solutions different from zero.

Let μ be any solution of (61.13). By (61.12) and (61.13) the real part of the function

$$\Phi(z) = \frac{1}{\pi i} \int_L \frac{\mu\,dt}{t - z}, \qquad (*)$$

holomorphic in S^+, assumes on the contours L_j the constant values c_j, while $c_0 = 0$. Hence, by the result in the last section, $\Re\Phi(z) = 0$ in S^+. Therefore (§ 25, Note 1) $\mu(t) = b_j$ on L_j, where the b_j are constants, while $b_0 = 0$. Substituting these values of $\mu(t)$ in (61.13), one obtains

$$b_j \int_{L_j} \varrho_j\,ds = 0, \quad j = 1, \ldots, p,$$

i.e., by (61.11), all $b_j = 0$, and the statement is proved.

It follows from the above that *the non-homogeneous equation* (61.9) *always has a (unique) solution.*

The solution $\mu(t)$ of (61.9) gives the solution of the original equation (61.4), since from (61.12)

$$\mu(t_0) + \frac{1}{\pi} \int_L \mu(t)\,d\vartheta = f(t_0) + c_j \text{ on } L_j.$$

For the constants a_j in (61.4) one obtains the definite values

$$a_j = c_j = \int_{L_j} \varrho_j \mu\,ds. \qquad (61.14)$$

Thus the problem is completely solved.

The method stated was given by the Author in note [4]; for some additional consequences see note [5]. A generalization to a mixed problem was given in note [6], while the extension to three dimensional space appears in the paper [7].

NOTE 1. The function $\mu(t)$ was subjected to the H condition so that use could be made of the Plemelj formula for the passage to the limit $z \to t_0$ in (61.1). However, if one separates in (61.1) the real part before passing to the limit and makes use of the known properties of the potential of a double layer, one obtains the result (61.3), provided the function $\mu(t)$ is continuous.

Therefore all the above conclusions remain true, if the given function $f(t)$ and the unknown function $\mu(t)$ are only assumed to be continuous.

NOTE 2. The choice of the auxiliary kernel $k(t_0, t)$, of course, influences the solution $\mu(t)$ of the integral equation (61.9). However, it is easily seen that a change of the kernel $k(t_0, t)$ entails only the replacement of $\mu(t)$ by $\mu(t) + \alpha(t)$, where $\alpha(t)$ has a constant value on the different contours L_j, while $\alpha(t) = 0$ on L_0. In fact, if $\mu_1(t)$ and $\mu_2(t)$ are solutions, corresponding to the two different auxiliary kernels $k_1(t_0, t)$ and $k_2(t_0, t)$, then the difference $\mu(t) = \mu_2(t) - \mu_1(t)$ will obviously satisfy an equation of the form (61.4) for $f(t_0) \equiv 0$, and, as previously, the conclusion follows that $\Re\Phi(z) = 0$ in S^+, where $\Phi(z)$ is a function connected with $\mu(t)$ by (*), and hence, as before, $\mu(t) = \alpha_j =$ const. on L_j, while $\alpha_0 = 0$.

From the uniqueness theorem of the modified Dirichlet problem proved in § 60 it is, in addition, clear that *the constants* a_j *in* (61.14) *do not depend on the choice of the kernel* $k(t_0, t)$.

§ 62. Some corollaries.

Let $\Psi(z)$ be a function, holomorphic in S^+, such that $\Re\Psi(z)$ is continuous from the left on L; then the function $[\Re\Psi(t)]^+$ is known to be continuous on L (§ 14). For the present, the contour L_0 will be assumed to exist. If the function $[\Re\Psi(t)]^+$ is given the value $f(t)$ in the boundary condition (60.2) of the modified Dirichlet problem, this problem clearly has the solution $u(x, y) = \Re\Psi(z)$, while $a_1 = a_2 = \ldots = a_p = 0$ (and also, by hypothesis, $a_0 = 0$). By the uniqueness theorem the problem has no other solutions.

On the other hand, the solution of the problem is given by

$$u(x, y) = \Re \frac{1}{\pi i} \int_L \frac{\mu(t)\, dt}{t - z},$$

where $\mu(t)$ is a real continuous function, determined from the integral equation (61.9) with $f(t) = [\Re\Psi(t)]^+$. Here the statement in Note 1 of the last section is used; if it is desired to limit consideration to what has been said in the main part of § 61, it is sufficient to assume that $[\Re\Psi(t)]^+$ satisfies the H condition.

Hence

$$\Psi(z) = \frac{1}{\pi} \int_L \frac{\mu(t)\, dt}{t - z} + Ci, \tag{62.1}$$

where C is a real constant, and *the result follows that:*

Every function $\Psi(z)$, *holomorphic in the finite region* S^+ *and such* *that its real part is continuous on L from the left, is of the form* (62.1), *where $\mu(t)$ is a real continuous function and C is a real constant.*

From this result (and also directly from the result of § 25) it is clear that for a given function $\Psi(z)$ the function $\mu(t)$ is defined, apart from arbitrary (real) constants, on the inner contours $L_1, \ldots, L_p$ and uniquely on L_0; the constant C is completely defined.

In the case of an infinite region S^+, i.e., when L_0 is at infinity, it is easily seen that

$$\Psi(z) = \frac{1}{\pi i} \int_L \frac{\mu(t)\, dt}{t - z} + \Psi(\infty) \tag{62.1a}$$

holds under the same conditions relating to $\Psi(z)$, where $\mu(t)$ is a real continuous function defined on the contours L_j apart from arbitrary constants.

Now assume $[\Re\Psi(t)]^+$ to satisfy the H condition on L. Then, by the results of the last section, the function $\mu(t)$, defined by the integral equation (61.9) for $f(t) = [\Re\Psi(t)]^+$, will also satisfy the H condition.

Consequently, by the results of § 16 and § 19, the limiting value $\Psi^+(t)$ exists (remembering that $\Psi^+(t)$ referred to the case when the limit is attained by any path remaining in S^+) and satisfies the H condition. Thus one obtains the following result (Theorem of I. I. Privalov [2]; the theorem is proved there for the circle, but the generalization to the case of an arbitrary region of the type considered here can be directly obtained by means of conformal transformation):

If the real part of a function $\Psi(z)$, holomorphic in S^+, assumes on L a given boundary value satisfying the H condition, then also its imaginary part will have this property.

It is easily seen (cf. § 61) that, if $[\Re\Psi(t)]^+$ satisfies the $H(\alpha)$ condition, then also $[\Im\Phi(t)]^+$ satisfies the $H(\alpha)$ condition, if $\alpha < 1$, and the $H(1-\varepsilon)$ condition, if $\alpha = 1$, where ε is any arbitrarily small positive number.

§ 63. Solution of the Dirichlet problem.

After the solution of the modified Dirichlet problem the solution of the classical Dirichlet problem (60.1) presents no difficulty; it can be reduced to the former by many methods (and in the case of a finite simply connected region both problems coincide).

One of the simplest methods of reduction will be given here. Let $z_1, z_2, \ldots, z_p$ be arbitrarily fixed points in the regions $S_1^-, S_2^-, \ldots, S_p^-$ respectively. Write the unknown harmonic function in the form

$$u(x, y) = U(x, y) + \sum_{k=1}^{p} A_k \log | z - z_k |, \tag{63.1}$$

where $U(x, y)$ is a new unknown harmonic function and the A_k are real constants, for the present not defined. When the contour L_0 is at infinity, assume

$$\sum_{k=1}^{p} A_k = 0; \tag{63.2}$$

this condition ensures that $\Sigma A_k \log | z - z_k |$ is zero at infinity, since for large $| z |$

$$\log | z - z_k | = \log | z | + \log \left| 1 - \frac{z_k}{z} \right| = \log | z | + O\left(\frac{1}{| z |}\right),$$

and hence

$$\sum_{k=1}^{p} A_k \log | z - z_k | = \log | z | \sum_{k=1}^{p} A_k + O\left(\frac{1}{| z |}\right).$$

By (60.1) the function U satisfies the boundary condition

$$U = f(t) - \sum_{k=1}^{p} A_k \log | t - z_k | \text{ on } L. \tag{63.3}$$

If the constants A_j are assumed arbitrarily fixed and if the function U is required to be the real part of a function holomorphic in S^+ and, in the absence of L_0, to vanish at infinity, then the problem (63.3) is, in general, insoluble, but there will always be a solution of the modified Dirichlet problem with the boundary condition

$$U = f(t) - \sum_{k=1}^{p} A_k \log | t - z_k | + a_j \text{ on } L_j, \tag{63.4}$$

$$j = 0, 1, \ldots, p, \ a_0 = 0$$

[where, in the absence of L_0, the condition $a_0 = 0$ must be omitted and $j = 1, 2, \ldots, p$].

The solution of this latter problem gives $U(x, y)$ and also determines the constants $a_1, \ldots, a_p$. These are easily obtained in the form

$$a_j = f_j + \sum_{k=1}^{p} \gamma_{jk} A_k, \tag{63.5}$$

where the γ_{jk} are certain constants, independent of the function $f(t)$, and the f_j $(j = 1, \ldots, p)$ are also constants which depend on $f(t)$ and are zero for $f(t) \equiv 0$.

In order that the function u, given by (63.1), is the solution of the Dirichlet problem, it is necessary and sufficient that the $a_j = 0$ $(j = 1, \ldots, p)$, i.e., that the A_k satisfy the system of linear equations

$$\sum_{k=1}^{p} \gamma_{jk} A_k + f_j = 0, \quad j = 1, \ldots, p. \tag{63.6}$$

Firstly consider the case of a finite region. The determinant formed by the γ_{jk} is different from zero. In fact, if this determinant were zero, the homogeneous system, obtained from (63.6) for $f(t) \equiv 0$ (i.e., when $f_j = 0$, $j = 1, \ldots, p$) would have the solutions $A_1, \ldots, A_p$ different from zero. But then one would obtain the function

$$u = U + \sum_{k=1}^{p} A_k \log |z - z_k|,$$

harmonic and not identically zero in S^+, but zero on L, which is impossible.

Furthermore, this expression cannot be identically zero in S^+, unless all the A_k are zero, since otherwise the function

$$U + iV + \sum_{k=1}^{p} A_k \log (z - z_k),$$

where V is the function conjugate to U, would be constant. But this is impossible, since, by hypothesis, $U + iV$ is a one-valued function.

Hence the system (63.6) is always soluble; having determined the A_k the solution of the original problem is obtained.

In the case of an infinite region the equation (63.2) must be added to (63.6). This new system will, in general, be incompatible; it is not difficult to understand this, since the function u, presented in the form (63.1), vanishes at infinity, if (63.2) holds, and, in general, there is no such solution of the Dirichlet problem (as in the statement of this problem the unknown function was only required to be bounded at infinity).

At first the boundary condition will be satisfied apart from a constant (which is the same on all L_j), i.e., replace the condition $a_1 = a_2 = \ldots = a_p = 0$ by $a_1 = A$, $A_2 = A, \ldots, A_p = A$, where A is some constant. Then one arrives at the system

$$\begin{aligned} &\sum_{k=1}^{p} \gamma_{jk} A_k - A + f_j = 0 \quad (j = 1, \ldots, p), \\ &\sum_{k=1}^{p} A_k = 0, \end{aligned} \tag{63.7}$$

instead of (63.6), i.e., one obtains $(p + 1)$ equations with the $(p + 1)$ unknowns $A, A_1, \ldots, A_p$. Similarly as above, it can be shown that the determinant of this system is different from zero and that therefore it is always soluble. Thus a harmonic function u has been found, vanishing at infinity and taking the value $f(t) + A$ on L. Hence, the function

$$u - A \tag{63.8}$$

will be the solution of the original problem.

NOTE. It can be shown by a different method that the above results (concerning the uniqueness of the solution, but not its representation as the potential of a double layer) also hold under much more general assumptions with regard to the boundary of the region (the results can be considerably generalized in these directions as it was done e.g. by J. Radon [1]. For example, it is sufficient to assume that the contours forming the boundary are Jordan curves (so that they need not even be rectifiable). Such a region can always be mapped conformally (and continuously up to the boundary) on a region bounded, for instance, by analytic curves (or even merely by circles; cf. any detailed treatment of complex function theory).

§ 64. Solution of the modified Dirichlet problem, using the modified potential of a simple layer.

The modified Dirichlet problem was solved in § 61 by use of the potential of a double layer. But in some cases, important from the point of view of applications, one has to introduce the solution in the form of the modified potential of a simple layer. This problem will be treated in full, partly with a view to giving one of the simplest and most direct applications of the theory of singular integral equations.

The modified Dirichlet problem was stated in § 60; now the additional condition $a_0 = 0$, used in the preceding sections, will be abandoned, and hence it will be assumed that none of the constants $a_0, a_1, \ldots, a_p$, appearing in the statement of the problem, are given beforehand.

The solution of the problem will be sought in the form

$$u(x, y) = \Re\Phi(z), \tag{64.1}$$

where this time

$$\Phi(z) = \frac{1}{\pi}\int_L \frac{\mu(t)dt}{t-z} = \frac{1}{\pi i}\int_L \frac{i\mu(t)dt}{t-z} \tag{64.2}$$

and $\mu(t)$ is again an unknown real function satisfying the H condition.

Thus $u(x, y)$ will have the form (cf. § 11)

$$u(x, y) = \frac{1}{\pi}\int_L \mu(t)\frac{dr}{r(z, t)}, \tag{64.3}$$

where $r(z, t) = |z - t|$. The integral on the right is the modified potential of a simple layer (§ 11).

It is easily seen that this problem cannot have two different solutions of the form (64.1), (64.2). In fact, as shown in the Note at the end of § 60, if $[\Re\Phi]^+ = a_j$ on L_j ($j = 0, 1, \ldots, p$), where the

a_j are constants, then $u = \text{const.} = a_0 = a_1 = \ldots = a_p$. But here, by the Plemelj formula, $[\Re\Phi]^+ = [\Re\Phi]^-$. Hence it may also be concluded that $\Re\Phi(z) = \text{const.} = a_0 = \ldots = a_p$ in S^-. But $\Phi(\infty) = 0$; hence $a_0 = a_1 = \ldots = a_p = 0$, $\Re\Phi(z) \equiv 0$, the reasoning also being true in the absence of L_0, when one has to consider $j = 1, 2, \ldots, p$.

Substituting (64.3) in the boundary condition (60.2) and noting that

$$\Re\Phi^+(t) = \frac{1}{\pi}\int_L \frac{\mu(t)\,dr}{r(t_0, t)},$$

one obtains the integral equation

$$\frac{1}{\pi}\int_L \mu(t)\frac{dr}{r(t_0, t)} = f(t_0) + a(t_0), \tag{64.4}$$

where, as in § 61, $f(t_0)$ is a given real function, satisfying the H condition, and $a(t_0) = a_j = \text{const.}$ on L_j, where $j = 0, 1, \ldots, p$ in the presence of L_0 and $j = 1, \ldots, p$ when L_0 is at infinity; in contrast to § 61, as has been mentioned already, a_0 will not be assumed to be zero.

The expression on the left of (64.4)

$$\frac{1}{\pi}\int_L \mu(t)\frac{dr}{r(t_0, t)} = \frac{1}{\pi}\int_L \mu(t)\frac{\cos\alpha(t_0, t)}{r(t_0, t)}\,ds, \tag{64.5}$$

where $\alpha(t_0, t)$ is the angle between the positive tangent at t and the vector $\overrightarrow{t_0 t}$ (Fig. 3, § 12), is a singular operator of the Cauchy type. Actually, if

$$\vartheta = \vartheta(t_0, t) = \arg(t - t_0), \text{ then } \log r = \log(t - t_0) - i\vartheta,$$

so that

$$\frac{dr}{r(t_0, t)} = \frac{dt}{t - t_0} - i\,d\vartheta(t_0, t),$$

and therefore

$$\frac{1}{\pi}\int_L \mu(t)\frac{dr}{r(t_0, t)} = \frac{1}{\pi}\int_L \frac{\mu(t)dt}{t - t_0} + \frac{1}{\pi i}\int_L \mu(t)\frac{d\vartheta}{ds}\,ds. \tag{64.6}$$

It is known that (cf. § 61)

$$\frac{d\vartheta}{ds} = \frac{\sin\alpha\,(t_0, t)}{r} = \frac{\cos\,(r, n)}{r} = \frac{K_0(t_0, t)}{r^\alpha},\ 0 \leq \alpha < 1,$$

where $K_0(t_0, t)$ satisfies the H condition. Therefore the first term on the right of (64.6) is the dominant part of the operator. Thus the index of the equation (64.4) is seen to be zero, i.e., it is a quasi-Fredholm equation (§ 56). This fact greatly facilitates the investigation (in particular, the equation (64.4) is easily reduced to an equivalent Fredholm equation by any of the methods given in Part II).

Since this singular equation is real, one need only consider the case of real functions $\mu(t)$ [cf. § 54].

The homogeneous equation, corresponding to (64.4), viz.

$$\frac{1}{\pi}\int \frac{\mu(t)\,dr}{r(t_0, t)} = 0, \tag{64.7}$$

has the solution

$$\mu(t) = C_j \text{ on } L_j, \tag{64.8}$$

$j = 0, 1, \ldots, p$ (in the presence of L_0),
$j = 1, \ldots, p$ (when L_0 is at infinity),

where the C_j are arbitrary (real) constants. There are no other solutions. For, let $\mu(t)$ be any solution of (64.7), then $0 = \Re\Phi^+(t) = \Im\Psi^+(t)$, where

$$\Psi(z) = \frac{1}{\pi i}\int_L \frac{\mu(t)\,dt}{t - z} = \frac{1}{i}\Phi(z),$$

and therefore (§ 25, Note 2) $\mu = C_j$ on L_j, thus proving the statement. (Also note that here $\Im\Psi^+ = \Im\Psi^-$ and therefore $\Im\Psi(z) \equiv 0$ in S^+ as well as in S^-.)

The singular equation will now be treated in a similar manner as the Fredholm equation in § 61, i.e., it will be replaced by the equation

$$\frac{1}{\pi}\int_L \mu(t)\frac{dr}{r(t_0, t)} - \int_L k(t_0, t)\mu(t)ds = f(t_0), \tag{64.9}$$

where $k(t_0, t)$ is defined in the following manner:

$$k(t_0, t) = \begin{cases} \varrho_j(t) \text{ for } t_0, t \in L_j, \\ 0 \text{ in all other cases,} \end{cases} \tag{64.10}$$

the $\varrho_j(t)$ being arbitrarily chosen (real) functions, subject only to the conditions

$$\int_{L_j} \varrho_j\,ds \neq 0, \tag{64.11}$$

with $j = 0, 1, \ldots, p$ and $j = 1, \ldots, p$ respectively, when L_0 is present or absent.

This differs from § 61 only in the fact that here $k(t_0, t)$ is not assumed zero when t_0 and t lie on L_0; but in § 61 something similar could have been done, thereby leading to some more general integral equation (cf. I. I. Muskhelishvili [5] and [7]).

Reasoning as in § 61, one arrives at the conclusion that the homogeneous equation corresponding to (64.9) has no solutions different from zero. Hence the adjoint of this homogeneous equation also has no solutions different from zero (since here the index is zero) and (64.9) is therefore always soluble.

Having solved (64.9), the definite function $\mu(t)$ can be found and the constants a_j will be given by

$$a_j = \int_{L_j} \varrho_j \mu \, ds, \tag{64.12}$$

$j = 0, 1, \ldots, p$ in the presence of L_0,
$j = 1, \ldots, p$ when L_0 is absent.

Thus the problem is solved. In the case of an infinite region the solution clearly vanishes at infinity. If for a finite region one requires $u = f$ on L_0, it is sufficient to replace u by $u - a_0$.

NOTE 1. Similarly, the result in Note 2 at the end of § 61 is easily established, namely, that, although the choice of the auxiliary kernel $k(t_0, t)$ influences the solution $\mu(t)$ of the integral equation, a change of the kernel can only alter $\mu(t)$ by a value, constant on each of the L_j.

The constants a_j do not depend on the choice of the auxiliary kernel; this follows directly from the uniqueness theorem proved at the beginning of this section.

NOTE 2. Reasoning similarly as in § 62, the conclusion is easily reached that, if $\Psi(z)$ is a function holomorphic in S^+ such that $[\Re\Psi(z)]^+$ exists and satisfies the H condition, then

$$\Psi(z) = \frac{1}{\pi}\int_L \frac{\nu(t)\,dt}{t-z} + C + iC',$$

where $\nu(t)$ is a real function satisfying the H condition and C and C' are real constants. The constant C is fully determined, while the function $\nu(t)$ is defined on every contour L_j apart from a constant term.

In the case of a finite region S^+, i.e., when L_0 is present, addition of a suitable real constant to $\nu(t)$ on L_0 will give $C' = 0$, and hence

$$\Psi(z) = \frac{1}{\pi}\int_L \frac{\nu(t)\,dt}{t-z} + C. \tag{64.13}$$

In this way $\nu(t)$ is completely determined on L_0 and, apart from constant terms, on L_j $(j = 1, \ldots, p)$.

When S^+ is infinite, i.e. when the contour L_0 is absent, one will have $C + iC' = \Psi(\infty)$ and therefore

$$\Psi(z) = \frac{1}{\pi} \int\limits_L \frac{\nu(t)\,dt}{t - z} + \Psi(\infty), \tag{64.14}$$

and $\nu(t)$ is again determined on the L_j apart from constant terms.

These results (even in a somewhat more general form) can also be obtained directly from the results of § 62 by the simple substitution of Ψ for $i\Psi$.

§ 65. Solution of the Dirichlet problem by the potential of a simple layer. Fundamental problem of electrostatics.

From the solution of the modified Dirichlet problem, obtained in the last section, it is not difficult to arrive by a method, entirely analogous to that used in § 63, at the solution of the classical Dirichlet problem. In order to avoid repetition this will not be done here, but instead the Dirichlet problem will be solved directly by two methods: with the help of the modified potential of a simple layer and by means of the ordinary potential of a simple layer; as a by-product the solution of the so-called fundamental problem of electrostatics will be obtained.

The object being to give a simple application of the method of singular equations, only the case will be considered in which L is a single contour L_0 bounding the finite region S^+. The assumption made in § 60 regarding the properties of the tangent to L will be retained.

Thus let it be required to find the function $u(x, y)$, harmonic in S^+ and continuous in $S^+ + L$, satisfying on L the boundary condition

$$u = f(t), \tag{65.1}$$

where $f(t)$ is a real function given on L; as before, $f(t)$ will be assumed to satisfy the H condition.

1°. Solution by means of the modified potential of a simple layer.

The solution of the problem will be written in the form

$$u(x, y) = \Re\Phi(z), \quad \Phi(z) = \frac{1}{\pi} \int\limits_L \frac{\mu(t)dt}{t - z}, \tag{65.2}$$

where $\mu(t)$ is an unknown real function satisfying the H condition.

As in the last section, substituting (65.2) in (65.1), one obtains the integral equation

$$\frac{1}{\pi}\int_L \frac{\mu(t)\,dr}{r(t_0, t)} \equiv \frac{1}{\pi}\int_L \mu(t)\frac{\cos\alpha(t_0, t)}{r(t_0, t)}\,ds = f(t_0). \qquad (65.3)$$

This is a real singular integral equation (cf. § 64) with index zero. The entire investigation will be carried through in the field of real functions (cf. § 54). The homogeneous equation, corresponding to (65.3),

$$\frac{1}{\pi}\int_L \mu(t)\frac{\cos\alpha(t_0, t)}{r(t_0, t)}\,ds = 0 \qquad (65.4)$$

has the obvious solution $\mu(t) = C$, where C is an arbitrary constant; it has no other solutions (cf. § 64). Hence also the adjoint equation to (65.4)

$$\frac{1}{\pi}\int_L \nu(t)\frac{\cos\alpha(t, t_0)}{r(t, t_0)}\,ds = 0 \qquad (65.5)$$

has one and only one linearly independent solution (because the index $\varkappa = 0$), so that, if $\nu_0(t)$ is any definite solution of (65.5) not identically zero, then all other solutions are given by $\nu(t) = C\nu_0(t)$, where C is an arbitrary constant (again restricting consideration to real solutions).

From the general theory of singular equations (cf. § 54) the original equation (65.3) has a solution, if and only if

$$\int_L \nu_0(t)f(t)\,ds = 0; \qquad (65.6)$$

provided this condition is satisfied, the general solution of (65.3) has the form

$$\mu(t) = \mu_0(t) + C, \qquad (65.7)$$

where $\mu_0(t)$ is any particular solution and C is an arbitrary constant.

The character of the condition (65.6) will be studied more closely. By the results of § 13, (65.5) is equivalent to the following equation:

$$\frac{d}{ds_0}\int_L \nu(t)\log r(t_0, t)ds = 0$$

or

$$\int_L \nu(t)\log r(t_0, t)\,ds = \text{const.} \qquad (65.8)$$

It will be shown that

$$m_0 = \int_L \nu_0(t)\,ds \neq 0. \tag{65.9}$$

In fact, if this were not so, the potential of a simple layer

$$U(x, y) = -\int_L \nu_0(t) \log |t - z|\, ds = \int_L \nu_0(t) \log \frac{1}{r(t, z)}\, ds \tag{65.10}$$

would be a function harmonic in S^- including the point at infinity; this means that $U(x, y)$ takes a definite finite value at infinity; in the present case this value is zero, since for large $|z|$

$$U(x, y) = -\int_L \nu_0(t) \log |t - z|\, ds = -\log |z| \int_L \nu_0(t) ds -$$
$$-\int_L \nu_0(t) \log \left|1 - \frac{t}{z}\right| ds = -\log |z| \int_L \nu_0(t) ds + O\left(\frac{1}{|z|}\right).$$

Furthermore, since by (65.8) $U^- = U^+ = U =$ const. on L, one would have that $U(x, y) =$ const. in S^+ as well as in S^-. Here use is made of a known property of the potential of a simple layer, namely, that it remains continuous for the passage through L, and for this to be so it is sufficient that the density $\nu_0(t)$ is bounded and integrable. But by a known formula of potential theory

$$-2\pi\nu_0(t_0) = \left(\frac{\partial U}{\partial n}\right)^+ - \left(\frac{\partial U}{\partial n}\right)^-, \tag{65.11}$$

where n is the normal to L at t_0, directed to the left; in order that (65.11) will hold true, it is sufficient that the density $\nu_0(t)$ be continuous. Hence, in the present case, one would obtain $\nu_0(t) = 0$ on L, and this contradicts the condition.

It follows from (65.9) that one can always choose the constant a_0 such that

$$\int_L (f + a_0)\nu_0 ds = \int_L f\nu_0 ds + a_0 \int_L \nu_0 ds = 0 \tag{65.12}$$

which is the condition of solubility of the equation

$$\frac{1}{\pi} \int_L \mu(t) \frac{\cos \alpha(t_0, t)}{r(t_0, t)}\, ds = f(t_0) + a_0. \tag{65.13}$$

If $\mu(t)$ is any of the solutions of (65.13) (all the other solutions being obtained from the formula $\mu +$ const.), then

$$u(x, y) = \Re \frac{1}{\pi} \int_L \frac{\mu(t)\, dt}{t - z} - a_0 = \frac{1}{\pi} \int_L \frac{\mu(t) dr}{r(z, t)} - a_0 \quad (65.14)$$

is the solution of the original Dirichlet problem.

2°. THE FUNDAMENTAL PROBLEM OF ELECTROSTATICS

The following problem of the theory of the logarithmic potential has been solved.

To find the density $\nu(t)$ of a mass distribution on the boundary L of the region S^+ such that the corresponding potential

$$U(x, y) = \int_L \nu(t) \log \frac{1}{r}\, ds \quad (65.15)$$

remains constant in S^+. Here $r = r(t, z) = |\, t - z \,|$.

This problem is the two-dimensional analogue of the fundamental problem of electrostatics dealing with the distribution of charge on the boundary L of a conductor S^+. It corresponds approximately to the problem of distribution of charge over a long cylindrical conductor with the cross-section S^+.

If $U = \text{const.}$ in S^+, then $U = \text{const.}$ on L, and vice versa. Hence the problem leads to the Fredholm integral equation of the first kind

$$\int_L \nu(t) \log r(t_0, t) ds = \text{const.}$$

and thus, by differentiation with respect to s_0, the homogeneous singular equation (65.5) is obtained.

This was seen to have non-zero solutions. If $\nu_0(t)$ was one solution, then all others were obtainable from the formula $\nu(t) = C\nu_0(t)$. The corresponding potential is given by

$$U(x, y) = CU_0(x, y), \quad (65.16)$$

where

$$U_0(x, y) = \int_L \nu_0(t) \log \frac{1}{r}\, ds. \quad (65.17)$$

The potential $U_0(x, y)$ has a constant value in $S^+ + L$; this value will be denoted by k_0, i.e.,

$$U_0(x, y) = k_0 \text{ in } S^+ + L. \quad (65.18)$$

It may happen that $k_0 = 0$; such a case will be called particular.

The general solution (65.16) of the present problem is seen to contain the arbitrary constant C. The problem becomes determinate, if one superimposes simultaneously a distribution of „masses" or

„charges" m on L such that

$$m = \int_L \nu(t)ds = C \int_L \nu_0 ds = Cm_0. \tag{65.19}$$

This relation determines C, since by (65.9) $m_0 \neq 0$.

Instead of giving m, one may give the value k of the potential $U(x, y)$ on $S^+ + L$. Then one has the equation $k = k_0 C$ for the determination of C which gives C in all cases, except when $k_0 = 0$ in which case $U(x, y) = 0$ on $S^+ + L$ for all C. This particular case has no analogue in three dimensional space, i.e., for the Newtonian potential.

3°. Solution of the Dirichlet problem by means of the potential of a simple layer.

A direct generalization of the above result is the determination of the potential of a simple layer (65.15) under the condition $U = f(t)$ on L; this is the Dirichlet problem, it being required to find the solution by means of the potential of a simple layer. The problem leads to the Fredholm integral equation of the first kind

$$-\int_L \nu(t) \log r(t_0, t)ds = f(t_0). \tag{65.20}$$

If the unknown solution $\nu(t)$ is to satisfy the H condition, then the given function $f(t_0)$ must be assumed to have a derivative with respect to s_0, satisfying the H condition, because the left side of (65.20) is easily seen to have that property.

Differentiating both sides of (65.20) with respect to s_0, one obtains the singular integral equation

$$-\int_L \nu(t) \frac{\cos \alpha(t, t_0)}{r(t, t_0)} ds = \frac{df}{ds_0}. \tag{65.21}$$

This equation is adjoint to (65.3) which has been obtained, when solving the Dirichlet problem by means of the modified potential of a simple layer.

The homogeneous equation, adjoint to (65.21), will be (65.4) which is known to have the unique linearly independent solution $\mu_0(t) = 1$.

The condition of solubility of (65.21) is always fulfilled, because

$$\int_L \mu_0(t) \frac{df}{ds} ds = \int_L \frac{df}{ds} ds = 0. \tag{65.22}$$

Hence (65.21) is always soluble. Its general solution has the form

$$\nu(t) = \nu^*(t) + C\nu_0(t), \tag{65.23}$$

where $\nu^*(t)$ is any particular solution, $\nu_0(t)$ a solution of the corresponding homogeneous equation (65.5) and C an arbitrary constant.

Let $\nu(t)$ be any solution of (65.21). Then the potential $U(x, y)$ defined by (65.15) will obviously satisfy the boundary condition

$$U = f(t_0) - k \text{ on } L, \tag{65.24}$$

where k is a certain constant. Thus, the function

$$u(x, y) = U(x, y) + k \tag{65.25}$$

is the solution of the Dirichlet problem (65.1), so that the solution is in the form of a sum of the potential of a simple layer and some constant k. If, however, one requires to present the solution by means of the potential of a simple layer, one has still to find the potential of a simple layer, having the constant value k on L and hence in $S^+ + L$. This was seen to be always possible with the exception of the particular case when $k_0 = 0$ in (65.18).

The integral equation (65.21) was obtained by G. Bertrand [2] who reduced it to a Fredholm integral equation by a method which is a special form of that stated in § 50. However, (not having the general theory of singular equations at his disposal) he did not succeed in giving any complete solution, in spite of rather complicated calculations and considerable restrictions imposed on the contour L and the given function $f(t)$.

NOTE. It follows directly from the result of 3° that, if the particular case $k_0 = 0$ is excluded, then each function $U(x, y)$, harmonic in S^+, bounded on L and having a derivative with respect to s satisfying the H condition, is of the form,

$$U(x, y) = \int_L \nu(t) \log \frac{1}{r(t, z)} ds, \tag{65.26}$$

where $\nu(t)$ is a real function satisfying the H condition. Such a representation is easily seen to be unique; this follows from the fact that, if $U(x, y) \equiv 0$ in S^+ and hence on L, then necessarily $\nu(t) \equiv 0$.

However, in the particular case, one is easily seen to have instead

$$U(x, y) = \int_L \nu(t) \log \frac{a}{r(t, z)} ds, \tag{65.27}$$

where a is an arbitrarily fixed positive constant, different from unity. For fixed a (65.27) defines $\nu(t)$ completely, when the function $U(x, y)$ is given.

Furthermore, in all cases, every harmonic function, satisfying the above-mentioned conditions, is seen to be expressible in the form (65.27), if a is any fixed positive constant such that

$$\int_L \nu_0(t) \log \frac{a}{r(t_0, t)} ds = m_0 \log a + k_0 \neq 0, \tag{65.28}$$

where m_0 and k_0 are determined by (65.9) and (65.18) respectively. Thus the choice of a depends only on the region S^+ and not on the represented harmonic function.

Hence it is seen that the particular case ($k_0 = 0$) is not essentially a special case; it can be avoided by a simple change in scale.

It also follows from the above that, if the real part $U(x, y)$ of the function $\Psi(z)$ holomorphic in S^+ satisfies the above conditions, then the function $\Psi(z)$ can be written in the form

$$\Psi(z) = \int_L \nu(t) \log \frac{a}{t-z} ds + iC, \tag{65.29}$$

where $\nu(t)$ is a real function satisfying the H condition, a is a certain positive constant which satisfies the condition (65.28) and can be fixed once and for all for a given region S^+, and C is a real constant (depending on the represented function). Of course, it has been assumed that the value of the logarithm has been fixed in a suitable manner; it is left to the reader to ascertain how this is done [cf. (69.2) for $m = 1$].

For a fixed and a definite choice of the logarithm the function $\Psi(z)$ is completely determined by the function $\nu(t)$ and the constant C.

CHAPTER 8

VARIOUS REPRESENTATIONS OF HOLOMORPHIC FUNCTIONS BY CAUCHY AND ANALOGOUS INTEGRALS

One of the most natural methods of solution of various boundary problems involving holomorphic functions consists of expressing the unknown holomorphic function in the form of a Cauchy or analogous integral; this expression, substituted into the boundary condition, reduces to an integral equation. In fact, such a method was used in the last section for the solution of the classical and modified Dirichlet problems.

It is of course very important to make a suitable choice of the integral representation of the unknown function, adapted to the solution of any given problem.

In the present chapter some of the simplest methods of representation of a function holomorphic in a region will be given using Cauchy and analogous integrals.

§ 66. General remarks.

Let S^+ be a connected region bounded by the union of smooth contours $L_0, L_1, \ldots, L_p$, of which L_0 contains all the others; L_0 can be absent in which case S^+ is an infinite region. As before, let L be the union of the contours $L_0, L_1, \ldots, L_p$ and assume that the positive direction on L leaves S^+ on the left. Again let S^- be the complement of $S^+ + L$; it consists of finite regions $S_1^-, \ldots, S_p^-$ bounded by $L_1, \ldots, L_p$ respectively and, in the presence of L_0, of the infinite region S_0^- bounded by L_0.

Let $\Phi(z)$ be a function, holomorphic in S^+ and continuous on L from the left, and let $\Phi(\infty) = 0$ when S^+ is infinite. This function can always be represented by the integral

$$\Phi(z) = \frac{1}{2\pi i} \int_L \frac{\Phi^+(t)\, dt}{t - z} \tag{66.1}$$

which is a particular form of the integral of the Cauchy type.

The question arises regarding the possibility of representing the same function $\Phi(z)$ by different Cauchy integrals in which some other function of the points on the boundary takes the place of $\Phi^+(t)$. This leads to the following: *what are the necessary and sufficient conditions for the integrals*

$$\Phi(z) = \frac{1}{2\pi i}\int_L \frac{\varphi(t)\,dt}{t-z} \quad \text{and} \quad \Psi(z) = \frac{1}{2\pi i}\int_L \frac{\psi(t)dt}{t-z} \tag{66.2}$$

to represent one and the same holomorphic function in S^+, where $\varphi(t)$ and $\psi(t)$ are continuous functions of the points on L?

This question is very easily answered. In fact, the condition

$$\int_L \frac{\varphi(t)-\psi(t)}{t-z}\,dt = 0$$

must hold for all z in S^+. But then, by a statement in § 24,

$$\varphi(t)-\psi(t) = \Omega^-(t), \tag{66.3}$$

where $\Omega^-(t)$ is the boundary value of some function $\Omega(z)$, holomorphic in S^-, continuous on L from the right and (when L_0 is present) vanishing at infinity. Alternatively, if (66.3) holds, then $\Phi(z) \equiv \Psi(z)$ in S^+.

It should be remembered that $\Omega(z)$ is the union of the functions $\Omega_1(z), \ldots, \Omega_p(z)$, holomorphic in $S_1^-, \ldots, S_p^-$ respectively, and (when L_0 is present) of the function $\Omega_0(z)$, holomorphic in S_0^- and vanishing at infinity.

In the case of an infinite region, if $\Phi(\infty) \neq 0$, one has instead of (66.1)

$$\Phi(z) = \frac{1}{2\pi i}\int_L \frac{\Phi^+(t)\,dt}{t-z} + \Phi(\infty) \tag{66.1a}$$

and hence such a function can also be represented in an infinite number of ways by Cauchy integrals, accurate apart from a constant term.

Similar results will be obtained after interchanging S^+ and S^-.

§ 67. Representation by a Cauchy integral with real or imaginary density.

Using the arbitrariness of the function $\Omega(z)$ in the last section, one can give the Cauchy integral, representing a given holomorphic function, various forms convenient from different points of view.

In § 62 the simplest representations of a function, holomorphic in a given region, by Cauchy integrals has been given. In this section reference will be made to these representations, in order to link them with the statements made here.

Let L, S^+ and S^- be the same as in § 66, but, in addition, assume that the tangent to L forms with some constant direction an angle satisfying the H condition.

1°. First assume that S^+ is a finite region, i.e., that L_0 is present, and let $\Phi(z)$ be a given function, holomorphic in S^+ and continuous on L from the left.

It has been seen in § 62 that $\Phi(z)$ can be written in the form

$$\Phi(z) = \frac{1}{2\pi i} \int_L \frac{\mu(t)\, dt}{t - z} + Ci = \frac{1}{2\pi i} \int_L \frac{\mu(t) + Ci}{t - z}\, dt, \qquad (67.1)$$

where $\mu(t)$ is a real continuous function and C a real constant. It has also been shown there that (67.1) does not only hold, when $\Phi(z)$ is continuous on L from the left, but also when this property refers only to $\Re\Phi(z)$.

On the other hand, $\Phi(z)$ can be written as the Cauchy integral

$$\Phi(z) = \frac{1}{2\pi i} \int_L \frac{\Phi^+(t)\, dt}{t - z}. \qquad (67.2)$$

Hence, by § 66, the function $\Omega(z)$, holomorphic in S^-, continuous on L from the right and vanishing at infinity, exists such that

$$\Phi^+(t) = \mu(t) + Ci + \Omega^-(t). \qquad (67.3)$$

Putting

$$\Phi(z) = U(x, y) + iV(x, y),\ \Omega(z) + Ci = u(x, y) + iv(x, y), \qquad (67.4)$$

one will have, by (67.3),

$$v^- = V^+ \text{ on } L. \qquad (67.5)$$

The value of V^+ on L can be assumed known, since $\Phi(z)$ is given in S^+. Hence the values of $v(x, y)$ in $S_0^-, S_1^-, \ldots, S_p^-$ will be found, thus solving the Dirichlet problem for these regions. After this the values of $\Omega_0(z), \Omega_1(z), \ldots, \Omega_p(z)$ of $\Omega(z)$ in these regions can be determined. The value of $\Omega_0(z)$ is obviously completely determined, if $\Omega(\infty) = 0$ is taken into consideration; similarly the constant C is determined as the value of $v(x, y)$ at infinity. The values of $\Omega_1(z), \ldots, \Omega_p(z)$ are fixed apart from arbitrary real constants.

Now $\mu(t)$ may be found from

$$\mu(t) = U^+ - u^-, \qquad (67.6)$$

using (67.3); once again $\mu(t)$ is seen to be fully determined on L_0 and, apart from arbitrary (real) constants, on $L_1, \ldots, L_p$.

In § 62, the determination of $\mu(t)$ was connected with the solution of an integral equation, linked, in its turn, with the solution of the

Dirichlet problem for the multiply-connected region $S^+(p > 0)$. Here, however, $\mu(t)$ is given by the solution of the Dirichlet problem for the simply-connected regions S_0^-, S_1^-, . . ., S_p^-. But in § 62 the function $\Phi(z)$ was subject to a more general condition than here: there, only the real part of $\Phi(z)$ had to be continuous on the contour from the left; here, the real as well as the imaginary part of $\Phi(z)$ have to be continuous on the contour from the left.

2°. If S^+ is infinite, i.e., if L_0 is absent, then, as shown in § 62, any function $\Phi(z)$, holomorphic in S^+ and continuous on L from the left, is representable in the form

$$\Phi(z) = \frac{1}{2\pi i}\int_L \frac{\mu(t)\,dt}{t-z} + \Phi(\infty), \tag{67.7}$$

where $\mu(t)$ is a real continuous function [(67.7) also being true when only $\Re\Phi(z)$ is continuous on L from the left (§ 62)]. As above, the determination of $\mu(t)$ can again be reduced to the solution of the Dirichlet problem for the simply-connected regions S_1^-, . . ., S_p^-. The function $\mu(t)$ is determined on L_1, . . ., L_p apart from arbitrary constants.

3°. Replacing $\Phi(z)$ by $i\Phi(z)$, as was indicated in Note 2 of § 64, one obtains for $\Phi(z)$, holomorphic in S^+ and continuous on L from the left, the representation

$$\Phi(z) = \frac{1}{2\pi}\int_L \frac{\nu(t)dt}{t-z} + C, \tag{67.8}$$

when S^+ is finite, and

$$\Phi(z) = \frac{1}{2\pi}\int_L \frac{\nu(t)\,dt}{t-z} + \Phi(\infty), \tag{67.9}$$

when S^+ is infinite; in these formulae $\nu(t)$ is a real continuous function and C a real constant. The function $\nu(t)$ is determined on L_1, . . ., L_p, apart from arbitrary constants, and completely on L_0; the constant C is fully determined. The representations (67.8) and (67.9) also hold, if $\Im\Phi(z)$ is only continuous on L from the left; this follows from the results of § 62.

The determination of $\nu(t)$ can be carried out in the same way as that of $\mu(t)$ in the preceding case.

§ 68. Representation by a Cauchy integral with density of the form $(a + ib)\mu$.

The representations derived in the preceding section are particular cases of

$$\Phi(z) = \frac{1}{2\pi i} \int_L \frac{(a + ib)\mu(t)}{t - z} dt + C, \tag{68.1}$$

where $a = a(t)$, $b = b(t)$ are given real continuous functions (independent of Φ) such that $a^2 + b^2 \neq 0$ everywhere on L and $\mu(t)$ is a real continuous function which must be chosen in correspondence with $\Phi(z)$, just as the constant C. For $a = 1$, $b = 0$ and a suitable C one obtains (67.1) or (67.7), and for $a = 0$, $b = 1$ and a suitable C, (67.8) or (67.9).

As in § 67, the angle between the tangent to L and a constant direction will be assumed to satisfy the H condition; the same condition is assumed to be satisfied by $a(t)$ and $b(t)$.

Further, it will be assumed that $\Phi(z)$ is continuous on L from the left, and that $\Phi^+(t)$ satisfies the H condition.

First let S^+ be a finite region, i.e., L_0 is present. Then (68.1) can also be written as

$$\Phi(z) = \frac{1}{2\pi i} \int_L \frac{(a + ib)\mu(t) + C}{t - z} dt. \tag{*}$$

In order to arrive at (*), one must, by the result in § 66, find the function $\Omega(z)$, holomorphic in S^-, vanishing at infinity and continuous on L from the right, under the condition

$$\Phi^+(t) = (a + ib)\mu(t) + C + \Omega^-(t) \text{ on } L. \tag{68.2}$$

Introducing

$$iC + i\Omega(z) = \omega(z), \tag{68.3}$$

one arrives at the relation

$$\Re \frac{\omega^-(t)}{a + ib} = \Re \frac{i\Phi^+(t)}{a + ib}$$

or

$$\Re(a - ib)\omega^-(t) = \Re i(a - ib)\Phi^+(t). \tag{68.4}$$

Thus the determination of $\omega(z)$ and hence of $\Omega(z)$ leads to the solution of the Riemann-Hilbert problem for the simply connected regions S_0^-, S_1^-, . . ., S_p^-, because the right side of (68.4) is a function given on L, if $\Phi(z)$ is given in S^+.

Take any of the regions S_j^-. The corresponding Riemann-Hilbert problem will always be soluble, if the corresponding index $\varkappa_j$ is non-negative. This index is given by

$$\varkappa_j = \frac{1}{\pi}[\arg (a - ib)]_{L_j}. \tag{68.5}$$

In order to see this, consult the end of § 42. Applying (42.3), one must firstly remember that b should be replaced by $-b$, as, since the positive direction on L_j leaves, in the present case, S_j^- on the right, one has to reverse the sign on the right side. Therefore

$$\varkappa_j = -\frac{1}{\pi}[\arg(a+ib)]_L = \frac{1}{\pi}[\arg(a-ib)]_{L_j},$$

i.e., (68.5) agrees formally with (42.3).

Thus the representation (68.1) is always possible, if $\varkappa_j \geqq 0$. When some of the $\varkappa_j$ are negative, such a representation is only possible, if $\Phi(z)$ satisfies some additional conditions which will not be given here.

When S^+ is an infinite region, similar results hold true. The constant C is then determined from the outset:

$$C = \Phi(\infty). \tag{68.6}$$

Applying to the function $\Phi(z) - \Phi(\infty)$ a reasoning, similar to the above, one arrives at the solution of the Riemann-Hilbert problem (68.4) for the finite simply connected regions $S_1^-, \ldots, S_p^-$.

These problems are always soluble, if all $\varkappa_j \geqq 0$. If this is not the case, (68.1) is only possible, when $\Phi(z)$ satisfies some additional conditions.

In the following section one of the representations of the form (68.1) will be considered in detail.

§ 69. Integral representation by I. N. Vekua.

In many important problems, not only the value of the unknown function, but also those of its derivatives up to a certain order enter into the boundary conditions. Therefore it is important to have formulae giving the integral representations of a series of successive derivatives of a given holomorphic function. One of these representations which is very suitable for many applications was given by I. N. Vekua [6], [7]. It will be reproduced here.

1°. Let S^+ be a finite simply connected region, bounded by a contour L the tangent to which forms with a given direction an angle satisfying the H condition. The region complementary to $S^+ + L$ will be denoted by S^-.

The following proposition will be proved (I. N. Vekua):

THEOREM. *Let the m-th derivative of the function $\Phi(z)$, holomorphic in S^+, take on L a limiting value satisfying the H condition. Then, assuming that the origin of coordinates lies in S^+, the function is representable in the following form:*

for $m = 0$,

$$\Phi(z) = \int_L \frac{\mu(t)ds}{1 - \frac{z}{t}} + iC, \tag{69.1}$$

for $m \geqq 1$,

$$\Phi(z) = \int_L \mu(t)\left(1 - \frac{z}{t}\right)^{m-1} \log\left(1 - \frac{z}{t}\right) ds + \int_L \mu(t)ds + iC, \tag{69.2}$$

where $\mu(t)$ *is a real function satisfying the H condition and C is a real constant;* $\mu(t)$ *and C are uniquely determined by* $\Phi(z)$.

By $\log\left(1 - \frac{z}{t}\right)$, for given t, that branch is to be understood which becomes zero for $z = 0$; s is the arc coordinate.

First the proof will be given when $m = 0$. Noting that $ds = t'^{-1}dt = \bar{t}'\,dt$, where

$$t' = \frac{dt}{ds} = \frac{dx}{ds} + i\frac{dy}{ds}, \quad \bar{t}' = \frac{dx}{ds} - i\frac{dy}{ds} = t'^{-1}, \tag{69.3}$$

(69.1) can be written

$$\Phi(z) = \frac{1}{2\pi i}\int_L \frac{2\pi i\mu(t)t\bar{t}'\,dt}{t - z} + iC = \frac{1}{2\pi i}\int_L \frac{2\pi i\mu(t)t\bar{t}' + iC}{t - z}\,dt,$$

and hence, by the results of the last section,

$$2\pi i\, t\bar{t}'\mu(t) = \Phi^+(t) - \Omega^-(t) - iC, \tag{69.4}$$

where $\Omega^-(t)$ is the limiting value of some function, holomorphic in S^-, continuous on L from the right and vanishing at infinity. Put $\Omega_0(z) = \Omega(z) + iC$. The function $\Omega_0(z)$ must assume a purely imaginary value at infinity and satisfy on L the following boundary condition:

$$\Re\frac{\Omega_0^-(t)}{t\bar{t}'} = \Re\frac{\Phi^+(t)}{t\bar{t}'}.$$

One thus obtains for S^- the Riemann-Hilbert problem in the form

$$\Re(a - ib)\Omega_0^-(t) = c, \tag{69.5}$$

where $a - ib = a(t) - ib(t) = \frac{1}{t\bar{t}'} = \frac{t'}{t}$, $c = c(t) = \Re\frac{\Phi^+(t)}{t\bar{t}'}$; under the assumed conditions $a(t)$, $b(t)$ and $c(t)$ clearly satisfy the H condition.

The corresponding index is easily seen to be zero and hence, by the results of §§ 40, 42, the problem (69.5) always has a solution which is of the form

$$\Omega_0(z) = \omega(z) + A\chi(z),$$

where $\omega(z)$ is a certain particular solution of (69.5), $\chi(z)$ is a particular solution of the homogeneous problem, i.e., of the problem,

$$\Re \frac{t'}{t} \chi^-(t) = 0, \tag{69.6}$$

and A is a real constant. It now remains to select A in such a way that $\Re\Omega_0(\infty) = 0$, i.e., that $\Re[\omega(\infty) + A\chi(\infty)] = 0$. This is always possible, since $\chi(\infty)$ is a real quantity different from zero. In fact, $\chi(\infty) \neq 0$, because the solution of the homogeneous Riemann-Hilbert problem for zero index is always different from zero. Actually, for a circular region, this follows directly from the formulae in § 40. For an arbitrary region this property is obviously retained, since conformal transformation has no effect upon it.

It remains to show that $\Im\chi(\infty) = 0$. But this follows from (69.6) which gives

$$0 = \Re \int_L \frac{\chi^-(t)t'\,ds}{t} = \Re \int_L \frac{\chi^-(t)\,dt}{t} = \Re[2\pi i\chi(\infty)].$$

Thus A is completely determined. Furthermore, since $\Omega_0^-(t)$ satisfies the H condition (cf. Note in § 42), $\mu(t)$, determined by (69.4), also satisfies this condition.

Thus the theorem is proved in the case $m = 0$. (This proof differs from I. N. Vekua's; in contrast, the following proof for $m \geqq 1$ is a reproduction of Vekua's [6] proof). By the same way of reasoning it is clear that $\mu(t)$ and C in (69.1) are uniquely determined by $\Phi(z)$; the uniqueness of the representation (69.1) is also easily shown directly and it is sufficient to assume that $\mu(t)$ is a continuous (real) function (see below).

Next the proof will be given for $m \geqq 1$. First of all note that, if $\mu(t)$ satisfies the H condition, the right side of (69.2) is a function holomorphic in S^+, such that the boundary value of its m-th derivative satisfies this condition. Actually, differentiating m times, one obtains

$$\Phi^{(m)}(z) = (-1)^m(m-1)!\int_L \frac{\mu(t)ds}{t^{m-1}(t-z)} =$$

$$= (-1)^m(m-1)!\int_L \frac{\mu(t)\bar{t}'dt}{t^{m-1}(t-z)},$$

and hence, by the theorem in § 19, the above statement follows.

Now it will be shown that, if the representation (69.2) holds, then the function $\mu(t)$ and the constant C are uniquely determined for a given $\Phi(z)$. This leads to the statement that, if

$$\int_L \mu(t)\left(1-\frac{z}{t}\right)^{m-1}\log\left(1-\frac{z}{t}\right)ds + \int_L \mu(t)ds + iC = 0 \quad (69.7)$$

for all z in S^+, then necessarily $\mu(t) \equiv 0$ (and then obviously also $C = 0$). The truth of the last statement will now be verified; for its proof it is sufficient to assume that $\mu(t)$ is a continuous (real) function.

Expanding the left side of (69.7) near the point $z = 0$ in powers of z, one easily obtains

$$0 = \int_L \mu(t)t^{-k}\,ds = \int_L \mu(t)\bar{t}'t^{-k}dt,\ k = 0,\ 1,\ 2,\ \ldots \quad (69.8)$$

But it follows from this that the Cauchy integral

$$\omega(z) = \frac{1}{2\pi i}\int_L \frac{\mu(t)\bar{t}'\,dt}{t-z}$$

is zero for all z in S^+; to verify this it is sufficient to expand $\omega(z)$ in powers of z near the point $z = 0$ and to take (69.8) into consideration. Hence $\omega^+(t) = 0$ and therefore

$$\mu(t)t' = \omega^-(t). \quad (69.9)$$

It follows from (69.8), for $k = 0$, that near the point at infinity an expansion of the form

$$\omega(z) = \frac{a_{-2}}{z^2} + \frac{a_{-3}}{z^3} + \ldots$$

holds true and hence the function

$$\omega_0(z) = \int_{z_0}^{z} \omega(z)dz,$$

where z_0 is an arbitrarily fixed point in $S^- + L$ and the integral is taken over any path in this region, is holomorphic in S^- including $z = \infty$. Assuming z_0 to lie on L, one deduces from (69.9)

$$\omega_0^-(t) = \int_{z_0}^{t} \omega^-(t)\,dt = \int_0^s \mu(t)\,ds$$

(assuming the arc-coordinate s to be measured from z_0); and hence it follows that $\Im\omega_0^-(t) = 0$ on L, so that $\omega_0(z) = \text{const}$ in S^-,

$\omega(z) = \omega_0'(z) = 0$ in S^-, and thus $\mu(t) = 0$; this proves the proposition regarding the uniqueness of a representation of the form (69.2). Exactly the same proof is also applicable to the representation (69.1).

Before proceeding further, the following convention will be introduced. Throughout this section linear combinations of several (real or complex) functions will be linear combinations with real constant coefficients; linear dependence or independence of such functions will be defined accordingly.

With this convention, the following proposition is immediately inferred based only upon the property (already proved) of uniqueness of the representations of the function $\Phi(z)$ in the form (69.1) or (69.2):

Let $\mu_j(t)$ be any continuous real function. Assuming $z \in S^+$, put

$$\Phi_j(z) = \int_L \frac{\mu_j(t)\,ds}{1 - z/t} \quad (m = 0) \tag{69.1a}$$

or

$$\Phi_j(z) = \int_L \mu_j(t)(1 - z/t)^{m-1}\log(1 - z/t)\,ds + \int_L \mu_j(t)\,ds \quad (m \geqq 1). \tag{69.2a}$$

From the linear dependence or independence of the functions $\mu_j(t)$ $(j = 1, 2, \ldots, k)$ *follows then the linear dependence or independence of the functions* $\Phi_j(z)$ *(and vice versa).*

After these preliminary remarks the proof of the possibility of the representation (69.2) will be given. By assumption, let $\Phi(z)$ be a given function, holomorphic in S^+, such that $[\Phi^{(m)}(t)]^+$ exists and satisfies the H condition. If the representation (69.2) holds true, then, differentiating m times, passing to the limit $z \to t_0$ (from S^+) and denoting $[\Phi^{(m)}(t)]^+$ simply by $\Phi^{(m)}(t)$, one obtains

$$\Phi^{(m)}(t_0) = (-1)^m(m-1)!\,\pi i\, t_0^{1-m}\bar{t}_0'\mu(t_0) + \\ + (-1)^m(m-1)! \int_L \frac{\mu(t)ds}{t^{m-1}(t - t_0)}.$$

Dividing both sides by $(-1)^m(m-1)!\,\pi i\,\bar{t}_0' t_0^{1-m}$ and using $t'\bar{t}' = 1$, one deduces

$$\mu(t_0) + \frac{1}{\pi i}\int_L \frac{t_0^{m-1} t_0'\mu(t)\,ds}{t^{m-1}(t - t_0)} = \frac{t_0^{m-1} t_0'\Phi^{(m)}(t_0)}{(-1)^m(m-1)!\,\pi i}, \tag{69.10}$$

and hence, comparing the real parts,

$$\mu(t_0) + \int_L \Re\left[\frac{t_0^{m-1} t_0'}{\pi i\, t^{m-1}(t - t_0)}\right]\mu(t)ds = \Re\,\frac{t_0^{m-1} t_0'\Phi^{(m)}(t_0)}{(-1)^m(m-1)!\,\pi i}. \tag{69.11}$$

This is easily seen to be a real Fredholm integral equation with a kernel of the form

$$\frac{K^*(t_0, t)}{|t - t_0|^{\alpha}}, \quad 0 \leqq \alpha < 1, \tag{*}$$

where $K^*(t_0, t)$ satisfies the H condition; in fact,

$$\frac{1}{\pi i} \frac{t_0^{m-1} t_0'}{t^{m-1}(t - t_0)} = \frac{1}{\pi i} \frac{t_0^{m-1} - t^{m-1}}{t - t_0} \cdot \frac{t_0'}{t^{m-1}} + \frac{1}{\pi i} \frac{t_0'}{t - t_0}.$$

The first term obviously satisfies the H condition, while the second term can be written

$$\frac{1}{\pi i} \frac{t_0'}{t - t_0} = -\frac{1}{\pi i} \frac{d}{ds_0} \log(t - t_0) = -\frac{1}{\pi i} \frac{d \log r}{ds_0} - \frac{1}{\pi} \frac{d\vartheta(s_0, s)}{ds_0},$$

where $r = |t - t_0|$, $\vartheta(s_0, s) = \arg(t - t_0)$. Hence

$$\Re \frac{1}{\pi i} \frac{t_0'}{t - t_0} = -\frac{1}{\pi} \frac{d\vartheta(s_0, s)}{ds_0}.$$

The last expression, however, has the form (*) [cf. § 8].

The right side of (69.11) also satisfies the H condition. Therefore any of the continuous solutions of (69.11) will also satisfy the H condition (§ 51).

Next, the solubility of the Fredholm equation (69.11) will be proved. For this consider its adjoint homogeneous equation

$$\nu(t_0) + \int_L \Re \left[\frac{t^{m-1} t'}{\pi i t_0^{m-1}(t_0 - t)} \right] \nu(t)\, ds = 0 \tag{69.12}$$

which, assuming here and in the sequel that $\nu(t)$ is a real function, can also be written as

$$\Re \left\{ \nu(t_0) - \frac{t_0^{1-m}}{\pi i} \int_L \frac{t^{m-1} \nu(t)\, dt}{t - t_0} \right\} = 0. \tag{69.13}$$

Let $\nu(t)$ be any (continuous) solution of (69.12); then $\nu(t)$ satisfies the H condition. Now consider the function

$$\Omega(z) = \frac{1}{z^{m-1}} \cdot \frac{1}{\pi i} \int_L \frac{t^{m-1} \nu(t)\, dt}{t - z};$$

it is holomorphic in S^-, vanishes at infinity like z^{-m} and, by (69.13),

$$\Re \Omega^-(t) = 0 \text{ on } L.$$

Consequently, $\Omega(z) \equiv 0$ in S^-. Therefore (§ 24)

$$t^{m-1}\nu(t) = \omega^+(t) \text{ on } L, \tag{69.14}$$

where $\omega(z)$ is some function, holomorphic in S^+. It follows from (69.14) that

$$\Re[it^{1-m}\omega^+(t)] = 0 \text{ on } L.$$

Hence $\omega(z)$ is a solution of the homogeneous Riemann-Hilbert problem $\Re(a+ib)\omega^+ = 0$ for S^+ with $(a+ib) = it^{1-m}$. The index is $2m - 2 \geqq 0$. Therefore (§§ 40 and 42) it has $(2m-1)$ linearly independent solutions. Consequently the homogeneous Fredholm equation (69.12) has $(2m-1)$ linearly independent solutions and the homogeneous equation, corresponding to the equation (69.11), has also $(2m-1)$ solutions.

In spite of this *the non-homogeneous equation* (69.11) *is always soluble.* In fact, by the known Fredholm theorem on the solubility of the equation (69.11), it is necessary and sufficient that the right side of this equation satisfies the condition

$$\int_L \nu(t)\,\Re\left[\frac{t^{m-1}t'\Phi^{(m)}(t)}{(-1)^m(m-1)!\,\pi i}\right]ds = \Re\int_L \frac{\nu(t)\,t^{m-1}t'\Phi^{(m)}(t)\,ds}{(-1)^m(m-1)!\,\pi i} = 0,$$

where $\nu(t)$ is any (real) solution of the homogeneous equation (69.12). But, by (69.14), $\nu(t) = t^{1-m}\omega^+(t)$, and the above condition can be written as

$$\Re\frac{1}{\pi i}\int_0 \omega^+(t)\Phi^{(m)}(t)\,dt = 0.$$

This condition is always fulfilled, because the integrand is the boundary value of the function $\omega(z)\Phi^{(m)}(z)$, holomorphic in S^+.

Thus the solubility of the equation (69.11) is proved. The general solution of this equation has the form

$$\mu(t) = \mu_0(t) + c_1\mu_1(t) + \ldots + c_{2m-1}\mu_{2m-1}(t), \tag{69.15}$$

where $\mu_1(t), \ldots, \mu_{2m-1}(t)$ is a complete system of linearly independent solutions of the homogeneous equation corresponding to (69.11) and $\mu_0(t)$ is a particular solution of (69.11), chosen so that $\mu(t) = 0$, if the right side of this equation is zero; $c_1, c_2, \ldots, c_{2m-1}$ are real constants.

Now the general solution of (69.11), obtained from the equation (69.10) by separating the real part, will also be shown to satisfy the latter equation. In fact, put (for $z \in S^+$)

$$\Psi(z) = \int_L \mu(t)(1 - z/t)^{m-1}\log(1 - z/t)ds + \int_L \mu(t)\,ds + iC, \tag{69.16}$$

where $\mu(t)$ is defined by (69.15) and C is a real constant. Now, noting that (69.11) only expresses the condition

$$\Re\left[\frac{t_0^{m-1}t_0'\Psi^{(m)}(t_0)}{(-1)^m(m-1)!\pi i}\right]^+ = \Re\left[\frac{t_0^{m-1}t_0'\Phi^{(m)}(t_0)}{(-1)^m(m-1)!\pi i}\right]^+ \tag{69.11a}$$

and putting $\Psi(z) - \Phi(z) = X(z)$, the function $X(z)$ is seen to satisfy the boundary condition

$$\Re[it_0^{m-1}t_0' X^{(m)}(t_0)] = 0 \text{ on } L,$$

where, for brevity, $X^{(m)}(t_0)$ has been written instead of $[X^{(m)}(t_0)]^+$. But the last condition is the homogeneous Riemann-Hilbert problem for $X^{(m)}(z)$, and the index of this problem is $-2m$, i.e., it is negative. Hence $X^m(z) = 0$, i.e., $\Psi^{(m)}(z) = \Phi^{(m)}(z)$, and this proves the proposition, because the equation (69.10) simply expresses the condition

$$[\Psi^{(m)}(t_0)]^+ = [\Phi^{(m)}(t_0)]^+. \tag{69.10a}$$

It should also be noted that

$$\Psi(z) = \Phi(z) + Q(z), \tag{69.17}$$

where $Q(z)$ is a polynomial of degree not greater than $m-1$.

In particular, it follows from the above that the functions $\mu_1(t), \mu_2(t), \ldots, \mu_{2m-1}(t)$, being solutions of the homogeneous equation corresponding to (69.11), are at the same time solutions of the homogeneous equation corresponding to (69.10). Let the $\Psi_j(z)$ be functions connected with the $\mu_j(t)$ $(j = 1, \ldots, 2m-1)$ by the formulae (69.2a) with $\Psi_j(z)$ replacing $\Phi_j(z)$. Then, since the $\mu_j(t)$ are now the solutions of the homogeneous equation corresponding to (69.10), i.e., of the equation obtained from (69.10) for $\Phi(z) = 0$, it is seen from (69.17) that the $\Psi_j(z)$ are polynomials of degree not greater than $m-1$.

Substituting in (69.16) the value of $\mu(t)$ given by (69.15), one therefore obtains

$$\Psi(z) = \Psi_0(z) + iC + c_1\Psi_1(z) + \ldots + c_{2m-1}\Psi_{2m-1}(z), \tag{69.18}$$

where the $\Psi_j(z)$ $(j = 0, 1, \ldots, 2m-1)$ are defined by (69.2a), again $\Psi_j(z)$ replacing $\Phi_j(z)$. It has just been seen that the $\Psi_j(z)$, for $j \geqq 1$, are polynomials of degree not greater than $m-1$.

It will be noted from (69.2a) that the quantities $\Psi_j(0)$ are real numbers.

The functions $\mu_1(t), \ldots, \mu_{2m-1}(t)$ are, by definition, linearly independent. Consequently, from what has been said above [as a result of (69.2a)], the polynomials $\Psi_1(z), \ldots, \Psi_{2m-1}(z)$ are also linearly independent and so are the polynomials

$$i,\ \Psi_1(z),\ \Psi_2(z),\ \ldots,\ \Psi_{2m-1}(z); \tag{69.19}$$

in fact, if $Ci + c_1\Psi_1(z) + \ldots + c_{2m-1}\Psi_{2m-1}(z) \equiv 0$, where C, c_1, ..., c_{2m-1} are real constants, then, putting $z = 0$ and taking into consideration that the $\Psi_j(0)$ are real numbers, one obtains $C = 0$; but, by the linear independence of the functions $\Psi_1(z), \ldots, \Psi_{2m-1}(z)$, the constants $c_1 = c_2 = \ldots = c_{2m-1}$ are also zero, which proves the proposition.

Hence it is easily concluded that any polynomial of degree not greater than $(m - 1)$ is a completely defined combination of the polynomials (69.19). For let $P(z)$ be such a polynomial. It will be shown that this polynomial may be expressed (uniquely) in the form

$$P(z) = iC + c_1\Psi_1(z) + \ldots + c_{2m-1}\Psi_{2m-1}(z),$$

where the $C, c_1, \ldots, c_{2m-1}$ are real constants. Comparing the coefficients of equal powers of z and separating real and imaginary parts, one obtains $2m$ (real) linear equations in $C, c_1, \ldots, c_{2m-1}$. The determinant of this system of equations is different from zero; for, if this were not the case, one could choose the constants $C, c_1, \ldots, c_{2m-1}$, not all zero, so that $P(z) \equiv 0$, which is impossible, since the functions (69.19) are linearly independent.

Applying now (69.17) to the function $\Psi_0(z)$, one sees that

$$\Phi(z) - \Psi_0(z) = Q_0(z),$$

where $Q_0(z)$ is a certain polynomial of degree not greater than $(m - 1)$. Hence, if the given function $\Phi(z)$ is to be equal to $\Psi(z)$, real constants $C, c_1, \ldots, c_{2m-1}$ must be selected such that

$$Q_0(z) = iC + c_1\Psi_1(z) + \ldots + c_{2m-1}\Psi_{2m-1}(z).$$

By the above this will always be possible in a unique manner.

Hence the theorem is proved for $m \geqq 1$ (for $m = 0$ it has already been proved).

2°. Under the same conditions as in 1°, the only difference being that now S^+ is an infinite region bounded by L, *the function* $\Phi(z)$ *can be represented by formulae which differ from* (69.1) *and* (69.2) *only in as far as* z/t *is replaced by* t/z; *this time it must be assumed that the origin of coordinates lies outside* S^+ (before it was assumed to lie inside L) *and that by* $\log(1 - t/z)$ *must be understood the branch which becomes zero for* $z = \infty$.

This result can be obtained either by repetition of the preceding reasoning, with the necessary minor alterations, or else by reduction of the case of the infinite region to the case of a finite region by the substitution $z = 1/\zeta$, $t = 1/\tau$.

If $\Phi(\infty) = 0$, the term iC in the corresponding formulae vanishes.

3°. Using the notation of § 66, let S^+ be a multiply-connected region of the same form as in that section. Further, let $\Phi(z)$ be a function, holomorphic in S^+ and continuous on L from the left. Then it can be represented by the Cauchy integral

$$\Phi(z) = \frac{1}{2\pi i} \int_L \frac{\Phi^+(t)\,dt}{t-z} = \sum_{j=0}^{p} \Phi_j(z), \tag{69.20}$$

where

$$\Phi_j(z) = \frac{1}{2\pi i} \int_{L_j} \frac{\Phi^+(t)\,dt}{t-z}, \quad j = 0, 1, \ldots, p;$$

if the contour L_0 is at infinity, one has to assume $\Phi_0(z) \equiv 0$ and to add on the right of (69.20) the term $\Phi(\infty)$.

Hence $\Phi(z)$ is the sum of $\Phi_0(z)$, holomorphic in the simply-connected region bounded by L_0, and of the functions $\Phi_j(z)$ $(j = 1, 2, \ldots, p)$, holomorphic, respectively, in the infinite simply connected regions bounded by the contours L_j $(j = 1, 2, \ldots, p)$ and vanishing at infinity.

If, moreover, the boundary value of $[\Phi^{(m)}(t)]^+$ satisfies the H condition on L, one arrives at the general representation of $\Phi(z)$ by applying to $\Phi_0(z)$ the representation of 1° and to the $\Phi_j(z)$ $(j \geqq 1)$ the representation of 2°; the origin of coordinates must each time be chosen in a suitable manner.

Regarding the generalization of (69.1) and (69.2) to the case of multiply-connected regions see I. N. Vekua [7].

CHAPTER 9

SOLUTION OF THE GENERALIZED RIEMANN-HILBERT-POINCARÉ PROBLEM

Of the many boundary problems which can be solved by means of the results stated above one particular problem will be considered in this chapter *) which is of considerable interest, both in itself and from the point of view of application. This problem is a generalization of the Riemann-Hilbert problem above and of the Poincaré problem (cf. § 70).

§ 70. Preliminary remarks.

The Riemann-Hilbert problem was considered in detail in § 40—42. Closely related to it is the problem at which Poincaré arrived when investigating the mathematical theory of tides (H. Poincaré [1]). This problem, to be called the Poincaré problem, is as follows: *To find a function, harmonic in some region S^+, which satisfies on the boundary L of this region the condition*

$$A(s)\frac{\partial u}{\partial n} + B(s)\frac{\partial u}{\partial s} + c(s)u = f(s), \tag{70.1}$$

where $A(s)$, $B(s)$, $c(s)$, $f(s)$ are real functions given on L, s is the arc coordinate and n the normal to L.

It was in connection with this problem that Poincaré was led to the study of singular integral equations, that he deduced the inversion formula and stated one of the methods of reduction of singular equations. Poincaré himself gave a (incomplete) solution of the problem for the case in which $A(s) = 1$, $c(s) = 0$, assuming L, $B(s)$ and $f(s)$ to be analytic.

A complicated (and also not quite complete) solution was recently given by W. Pogorzelski [1] who assumed that $A(s) = 1$, $B(s)$, $c(s)$ and $f(s)$ are analytic functions and L an analytic contour.

The incompleteness of the solutions of the problem by the abovementioned authors (apart from the restrictions imposed on the

---

*) The essential parts of this chapter (§ 71—75) have been taken by the Author (sometimes without any changes) from the work of I. N. Vekua [6].

boundary and the given functions) consists of the fact that the question of equivalence of the Fredholm integral equations obtained by them (which, in their turn, resulted from the reduction of singular integral equations) with the original problem was not elucidated; therefore even the question of the existence of a solution remained open.

The first complete solution of the problem (70.1) was given by B. V. Khvedelidze [1], [2] under the following assumptions: the region S^+ is bounded by a finite number of contours the tangent to which forms with a constant direction an angle satisfying the H condition; the functions $A(s)$, $B(s)$, $c(s)$ and $f(s)$ satisfy the H condition.

In § 74 the solution of the Poincaré problem will be obtained as a particular case of the solution of a much more general problem to be stated in the next section.

§ 71. The generalized Riemann-Hilbert-Poincaré problem (Problem V). Reduction to an integral equation.

Let S^+ be a finite region bounded by a simple contour L. The angle between the tangent to L and a constant direction will be assumed to satisfy the H condition.

The following problem has been referred to in the title of this section (Problem V):

To find a function $\Phi(z)$, holomorphic in S^+, for the boundary condition

$$\Re(\mathbf{L}\Phi) = f(t_0) \text{ on } L, \tag{V}$$

where **L** *is the integro-differential operator, defined by*

$$\mathbf{L}\Phi \equiv \sum_{j=0}^{m} \{a_j(t_0)\Phi^{(j)}(t_0) + \int_L h_j(t_0, t)\Phi^{(j)}(t)\, ds\} \tag{71.1}$$

with $a_0(t), \ldots, a_m(t)$ being functions, generally complex, given on **L** *and satisfying the H condition; $f(t)$ is a given real function satisfying the H condition and the $h_j(t_0, t)$ are functions, given on* **L** *and in general complex, of the form*

$$h_j(t_0, t) = \frac{h_j^0(t_0, t)}{|t - t_0|^\alpha}, \quad 0 \leq \alpha < 1,$$

where $h_j^0(t_0, t)$ satisfies the H condition.

By $\Phi^{(j)}(t_0)$ will be understood the boundary value $[\Phi(^{(j)}t_0)]^+$ of the j-th derivative of the function $\Phi(z)$; thus the existence of these boundary values up to and including $j = m$ is assumed. Moreover, it will be assumed that $\Phi^{(m)}(t)$ satisfies the H condition on L [and hence the same condition must be satisfied by all the $\Phi^{(j)}(t)$ for $j \leqq m$].

In particular, if one takes $m = 0$, $h_j(t_0, t) \equiv 0$, one obtains the Riemann-Hilbert problem; the Poincaré problem is also a special case of the stated problem (cf. § 74).

Many important boundary problems can be reduced to Problem V, particularly those connected with partial differential equations of the elliptic type (cf. § 76, 3° and 4°).

The problem, in the general form stated above, was given and ably solved (including in this also the general representation of holomorphic functions, stated in § 69) by I. N. Vekua in the paper [6], quoted earlier; his treatment will be followed here. However, it should be mentioned that one particular case of the problem V, namely that when the $h_j(t_0, t)$ are zero, was considered earlier by F. D. Gakhov [2]. Following a method, used by Hilbert in one particular case, Gakhov reduced the problem to a complicated singular integral equation with a kernel containing the Green's function. He also imposed some additional restrictions on the unknown function which do not follow from the character of the problem.

First of all the following property of the homogeneous problem, corresponding to the problem V, i.e., to the problem $\Re(\mathbf{L}\Phi) = 0$, should be noted: if the $\Phi_1(z), \ldots, \Phi_k(z)$ are particular solutions of this problem, then any of their linear combinations $C_1\Phi_1 + \ldots + C_k\Phi_k$ with the real coefficients $C_1, \ldots, C_k$ are also solutions.

In the whole of this chapter a linear combination will be understood to be a linear combination with real (constant) coefficients and the notion of linear dependence and independence will be similar to that in § 40—42, when considering the Riemann-Hilbert problem, and also to that in § 69.

The method of solution to be used here consists in the transformation of the problem into a singular integral equation (which will also be found to be of the Fredholm type).

The case $m \geqq 1$ will be considered first. Assuming the origin of coordinates to lie inside S^+, the unknown function will be represented in the form (§ 69)

$$\Phi(z) = \int_L \mu(t)(1 - z/t)^{m-1}\log(1 - z/t)ds + \int_L \mu(t)ds + iC, \quad (71.2)$$

where $\mu(t)$ is an unknown real function, satisfying the H condition, and C is an unknown real constant; the following elementary functions will be introduced:

$$N_0(z, t) = (1 - z/t)^{m-1}\log(1 - z/t) + 1, \quad (71.3)$$

$$N_l(z, t) = \frac{d^l}{dz^l}[(1 - z/t)^{m-1}\log(1 - z/t)] =$$

$$= (-1)^l \frac{(m-1)(m-2)\dots(m-l)}{t^l}(1 - z/t)^{m-l-1}\Big\{\log(1 - z/t) +$$

$$+ \frac{1}{m-1} + \frac{1}{m-2} + \dots + \frac{1}{m-l}\Big\}, \qquad (71.4)$$

$$(l = 1, 2, \dots, m-1),$$

$$N_m(z, t) = \frac{d^m}{dz^m}\Big[(1-z/t)^{m-1}\log(1-z/t)\Big] = \frac{(-1)^m(m-1)!}{t^{m-1}(t-z)}, \quad (71.5)$$

where t is an arbitrary point on L and z a point of S^+. As in § 69, by $\log(1 - z/t)$ is understood the branch vanishing (for fixed t) at $z = 0$. The functions $N_l(z, t)$ are holomorphic (with respect to z) in S^+. Keeping t arbitrarily fixed and passing to the limit $z \to t_0$, where t_0 and t are also points on L, one obtains the functions $N_l(t_0, t)$, uniquely defined on L, which, with the exception of $N_{m-1}(t_0, t)$ and $N_m(t_0, t)$, satisfy the H condition on L with respect to both variables. The function $N_{m-1}(t_0, t)$ has for $t = t_0$ a singularity of the logarithmic type, while $N_m(t_0, t)$ has one of the type $(t - t_0)^{-1}$.

The boundary values of $\Phi(z)$ and its derivatives up to the $(m-1)$-th order are now easily seen to occur in the form

$$\Phi(t_0) = \int_L N_0(t_0, t)\mu(t)ds + iC,$$

$$\Phi^{(l)}(t_0) = \int_L N_l(t_0, t)\mu(t)ds \quad (l = 1, 2, \dots, m-1). \quad (71.6)$$

By the Plemelj formula, the boundary value of the m-th derivative is given by

$$\Phi^{(m)}(t_0) = (-1)^m(m-1)!\,\pi i t_0^{1-m}\bar{t}_0'\mu(t_0) + \int_L N_m(t_0, t)\mu(t)ds. \quad (71.7)$$

Therefore the boundary condition (V) acquires the form

$$\mathbf{N}\mu \equiv A(t_0)\mu(t_0) + \int_L N(t_0, t)\mu(t)\,ds = f(t_0) - C\sigma(t_0), \qquad (71.8)$$

where

$$A(t_0) = \Re\{(-1)^m(m-1)!\,\pi i t_0^{1-m}\bar{t}_0' a_m(t_0)\}, \qquad (71.9)$$

$$\sigma(t_0) = \Re\{ia_0(t_0) + i\int_L h_0(t_0, t)\,ds\}, \qquad (71.10)$$

$$N(t_0, t) = \sum_{l=0}^{m} \Re\{a_l(t_0)N_l(t_0, t) + \int_L h_l(t_0, t_1)N_l(t_1, t)ds_1\} +$$
$$+ \Re\{(-1)^m(m-1)!\pi i h_m(t_0, t)t^{1-m}\bar{t}'\}. \tag{71.11}$$

Thus one has obtained for the determination of $\mu(t)$ a real singular integral equation with a kernel which is easily seen to have the form

$$N(t_0, t) = \frac{K(t_0, t)}{t - t_0}, \tag{71.12}$$

where $K(t_0, t)$ satisfies the H condition. The functions $A(t_0)$ and $\sigma(t_0)$ likewise satisfy the H condition.

It follows from this result that the obtained integral equation is equivalent to the original boundary problem.

The case $m \geqq 1$ has been considered. In the case $m = 0$ the condition (V) has the form

$$\Re\{a_0(t_0)\Phi(t_0) + \int_L h_0(t_0, t)\Phi(t)\,ds\} = f(t_0); \tag{71.13}$$

the representation (69.1) will now be used, viz.

$$\Phi(z) = \int_L \frac{\mu(t)\,ds}{1 - z/t} + iC. \tag{71.14}$$

Then, similarly as before, a singular integral equation equivalent to the original boundary problem is obtained. A simple calculation shows that this integral equation is given by the same formula (71.8), while $A(t)$, $\sigma(t)$, $N(t_0, t)$ are given by (71.9), (71.10) and (71.11) respectively in which one has to take $m = 0$, $(-1)^m(m-1)! = 1$; by $N_0(t_0, t)$ must here be understood the expression (71.5) for $m = 0$, i.e.,

$$N_0(t_0, t) = \frac{t}{t - t_0}. \tag{71.15}$$

NOTE. With a view to further applications, the following modification of the problem will be noted. Instead of the operator **L** consider the operator **L***, defined by

$$\mathbf{L}^*\Phi \equiv \sum_{j=0}^{m} \{a_j(t_0)\,\Phi^{(j)}(t_0) + \int_0^{t_0} h_j^*(t_0, t)\Phi^{(j)}(t)dt\}, \tag{71.16}$$

where now the $h_i^*(t_0, t)$ satisfy the conditions $h_m^*(t_0, t) \equiv 0$ [this assumption simplifies the formulae; it may be removed by subjecting $h_m^*(t_0, t)$ to the same conditions as the other $h_j^*(t_0, t)$]; the other functions are defined for all t_0 on L and for all t in $S^+ + L$. These functions are holomorphic in S^+ and continuous in $S^+ + L$ with respect to t; when both t_0 and t lie on L, these functions satisfy the H condition. The integral in (71.16) is taken over any path joining the point $t = 0$ in S^+ with the point t_0 on L without leaving S^+ (owing to the assumed conditions the integrals are independent of the path); instead of $t = 0$ one can, of course, take any other point in S^+.

All the above formulae and also those of the next section will hold true, if, wherever it occurs, $h_j(t_0, t)ds$ is replaced by $h_j^*(t_0, t)dt = h_j^*(t_0, t)t'ds$ and the integration is taken from 0 to t_0 instead of over L. One obtains in this case an integral equation of the form (71.8), where this time

$$\sigma(t_0) = \Re\{ia_0(t_0) + i\int_0^{t_0} h_j^*(t_0, t)\,dt\}, \tag{71.17}$$

$$N(t_0, t) = \sum_{l=0}^{m} \Re\{a_l(t_0)N_l(t_0, t) + \int_0^{t_0} h_l^*(t_0, t_1)N_l(t_1, t)dt_1\}. \tag{71.18}$$

§ 72. Investigation of the solubility of Problem V.

Next the singular integral equation

$$\mathbf{N}\mu \equiv A(t_0)\mu(t_0) + \int_L N(t_0, t)\mu(t)\,ds = f(t_0) - C\sigma(t_0) \tag{72.1}$$

will be considered (i.e., the equation (71.8) of § 71). This is a real equation. In the sequel, solutions of this equation and also solutions of its adjoint equation will be understood to be real solutions satisfying the H condition.

The index of (72.1) will be determined first and for this purpose the dominant part

$$\mathbf{N}^0\mu \equiv A(t_0)\mu(t_0) + \frac{B(t_0)}{\pi i}\int_L \frac{\mu(t)\,dt}{t - t_0}$$

of $\mathbf{N}\mu$ will be separated. Here $A(t_0)$ is given by (71.9) which can be written

$$A(t_0) = \tfrac{1}{2}(-1)^m(m-1)!\,\pi i\,[t_0^{1-m}\bar{t}_0' a_m(t_0) - \bar{t}_0^{1-m} t_0' \overline{a_m(t_0)}].$$

The coefficient $B(t_0)$, however, is clearly determined by the following term entering into $N(t_0, t)ds$ of (71.8):

$$\Re\{a_m(t_0)N_m(t_0, t)ds\} = \tfrac{1}{2}(-1)^m(m-1)!\left\{\frac{t^{1-m}a_m(t_0)}{t - t_0} + \frac{\bar{t}^{1-m}\overline{a_m(t_0)}}{\bar{t} - \bar{t}_0}\right\}ds.$$

Noting that

$$\frac{ds}{t-t_0}=\frac{\bar{t}'\,dt}{t-t_0},$$

$$\frac{ds}{\bar{t}-\bar{t}_0}=\frac{\bar{t}'\,d\bar{t}}{\bar{t}-\bar{t}_0}=t'd\log(\bar{t}-\bar{t}_0)=t'd\log(t-t_0)+t'd\log\frac{\bar{t}-\bar{t}_0}{t-t_0}=$$

$$=\frac{t'dt}{\bar{t}-t_0}+t'd\log\frac{\bar{t}-\bar{t}_0}{t-t_0}$$

and that the last term does not influence the dominant part of the integral equation, it is seen that $B(t_0)$ is determined by the following terms in $N(t_0, t)ds$:

$$\tfrac{1}{2}(-1)^m(m-1)!\left\{\frac{t^{1-m}\bar{t}'a_m(t_0)}{t-t_0}+\frac{\bar{t}^{1-m}t'\overline{a_m(t_0)}}{t-t_0}\right\}dt$$

from which one readily obtains

$$B(t_0)=\tfrac{1}{2}(-1)^m(m-1)!\,\pi i[t_0^{1-m}\bar{t}_0'a_m(t_0)+\bar{t}_0^{1-m}t_0'\,\overline{a_m(t_0)}].$$

Thus

$$A(t_0)+B(t_0)=(-1)^m(m-1)!\,\pi i t_0^{1-m}\bar{t}_0'a_m(t_0),$$

$$A(t_0)-B(t_0)=(-1)^m(m-1)!\,\pi i\bar{t}_0^{1-m}t_0'\overline{a_m(t_0)}.$$

In order that the singular equation will be regular (§ 44), it is necessary and sufficient that

$$a_m(t_0)\neq 0 \text{ everywhere on } L.$$

In the sequel, this condition will be assumed satisfied and correspondingly Problem V under consideration will be said to be of the regular type.

The index $\varkappa$ of (72.1) which will also be called the index of Problem V is given by

$$\varkappa=\frac{1}{2\pi}\left[\arg\frac{t^{m-1}t'\overline{a_m(t)}}{\bar{t}^{m-1}\bar{t}'a_m(t)}\right]_L=2(m+n), \tag{72.2}$$

$$\text{where}\quad n=\frac{1}{2\pi}[\arg\overline{a_m(t)}]_L. \tag{72.3}$$

For the investigation of the solubility of the integral equation (72.1) the homogeneous equation

$$N'\nu=A(t_0)\nu(t_0)+\int_L N(t,t_0)\nu(t)\,ds=0, \tag{72.4}$$

adjoint to (72.1), must be considered. It will be remembered that

(§ 53, Theorem III)

$$k - k' = \varkappa, \tag{72.5}$$

where k and k' respectively denote the number of linearly independent solutions of the adjoint homogeneous equations $\mathbf{N}\mu = 0$ and $\mathbf{N}'\nu = 0$.

The case in which Problem V is soluble for every $f(t)$ on the right of (72.1) is of particular interest. This problem is solved by the following fundamental theorem:

THEOREM: *For Problem V to be soluble for any function $f(t_0)$, it is necessary and sufficient that $k' = 0$ or $k' = 1$, where in the second case the solution* (determined apart from a constant factor, because $k' = 1$)$\nu(t)$ *of the equation* $\mathbf{N}'\nu = 0$ *must satisfy the condition*

$$\int_L \nu(t)\sigma(t)ds \neq 0; \tag{72.6}$$

in both cases $\varkappa \geqq 0$ and the homogeneous problem $\Re\{\mathbf{L}\Phi\} = 0$ has exactly $\varkappa + 1$ linearly independent solutions.

Firstly, the proof will be given for the sufficiency of the stated condition and of that part of the theorem dealing with the number of solutions of the homogeneous problem $\Re\{\mathbf{L}\Phi\} = 0$.

1°. If $k' = 0$, then by Theorem I of § 53 the equation (72.1) is soluble for any function $f(t)$ and any constant C. The homogeneous equation $\mathbf{N}\mu = 0$ has by (72.5) $k = \varkappa$ linearly independent solutions; since $k \geqq 0$, then also $\varkappa \geqq 0$. Consequently the general solution of (72.1) for arbitrary C has the form

$$\mu(t) = \mu^*(t) + C\mu_0(t) + C_1\mu_1(t) + \ldots + C_\varkappa\mu_\varkappa(t), \tag{72.7}$$

where the C, C_1, . . ., $C_\varkappa$ are arbitrary real constants, the $\mu_1(t)$, . . ., $\mu_\varkappa(t)$ are linearly independent solutions of the homogeneous equation $\mathbf{N}\mu = 0$ and $\mu^*(t)$ and $\mu_0(t)$ are any particular solutions of the equations $\mathbf{N}\mu = f(t_0)$ and $\mathbf{N}\mu = -\sigma(t_0)$ respectively.

Substituting (72.7) in (71.2) or, when $m = 0$, in (71.14), one obtains the general solution of the original problem V in the form

$$\Phi(z) = \Phi^*(z) + C(i + \Phi_0(z)) + C_1\Phi_1(z) + \ldots + C_\varkappa\Phi_\varkappa(z), \tag{72.8}$$

where the C, C_1, . . ., $C_\varkappa$ are arbitrary real constants, $\Phi_0(z)$, $\Phi_1(z)$, . . ., $\Phi_\varkappa(z)$ are functions, holomorphic in S^+ and related to $\mu_0(t)$, $\mu_1(t)$, . . ., $\mu^k(t)$ by the formulae (69.2a) or (69.1a), while $\Phi^*(z)$ is also holomorphic in S^+ and related to $\mu^*(t)$ in a similar manner.

Due to the linear independence of the $\mu_j(t)$ $(j = 1, \ldots, \varkappa)$ also the corresponding $\Phi_j(z)$ are linearly independent. Furthermore, since the $\Phi_j(0)$ $(j = 0, 1, \ldots, \varkappa)$ are real quantities, the linear independence of the functions $i + \Phi_0(z)$, $\Phi_1(z)$, . . ., $\Phi_\varkappa(z)$ easily follows (cf. end of § 69).

Correspondingly the homogeneous equation $\Re(\mathbf{L}\Phi) = 0$ has exactly $\varkappa + 1$ linearly independent solutions.

2°. Now let $k' = 1$ and the condition (72.6) be satisfied. Then the equation (72.1) will be soluble, provided

$$\int_L \nu(t)[f(t) - C\sigma(t)]\, ds = 0;$$

this determines uniquely the value of C, because, by supposition, (72.6) is fulfilled.

With this value of C one arrives at the solution of (71.2) in the form

$$\mu(t) = \mu^*(t) + C_1\mu_1(t) + \ldots + C_{\varkappa+1}\mu_{\varkappa+1}(t), \tag{72.9}$$

where the $C_1, C_2, \ldots, C_{\varkappa+1}$ are arbitrary real constants and the $\mu_1(t), \mu_2(t), \ldots, \mu_{\varkappa+1}(t)$ are linearly independent solutions of the homogeneous equation $\mathbf{N}\mu = 0$; by (72.5), there will be $\varkappa + 1$ solutions of this type. Since $\varkappa + 1 \geqq 0$ and $\varkappa$ is an even number, one has necessarily $\varkappa \geqq 0$.

Accordingly the general solution of the original problem is of the form

$$\Phi(z) = \Phi^*(z) + C_1\Phi_1(z) + \ldots + C_{\varkappa+1}\Phi_{\varkappa+1}(z), \tag{72.10}$$

where the $C_1, C_2, \ldots, C_{\varkappa+1}$ are arbitrary real constants and the $\Phi_1(z)$, $\Phi_2(z), \ldots, \Phi_{\varkappa+1}(z)$ are linearly independent functions, holomorphic in S^+. Moreover, it is clear from the above that the homogeneous problem $\Re(\mathbf{L}\Phi) = 0$ has exactly $\varkappa + 1$ linearly independent solutions.

Next the necessity of the conditions stated in the theorem will be proved. Let $\nu_j(t)$ $(j = 1, 2, \ldots, k')$ be the complete system of linearly independent solutions of the equation $\mathbf{N}'\nu = 0$ which will be assumed orthogonalised and normalised, so that

$$\int_L \nu_i\nu_j\, ds = \delta_{ij} \left(\delta_{ij} = \begin{cases} 1 \text{ for } i = j \\ 0 \text{ for } i \neq j \end{cases}\right). \tag{72.11}$$

Problem V will be assumed soluble for any $f(t_0)$. Then the integral equation (72.1) must also be soluble for any $f(t_0)$ and proper choice of C. Hence for any $f(t)$

$$\int_L \nu_j(t)[f(t) - C\sigma(t)]\, ds = 0, \; j = 1, 2, \ldots, k', \tag{72.12}$$

provided the constant C be suitably chosen. It has to be shown that necessarily $k' = 0$ or $k' = 1$ and that in the second case (72.6) must be satisfied.

Consider the two possibilities:

(a) all the numbers

$$\int_L \sigma(t)\nu_j(t)\,ds = 0, \quad j = 1, 2, \ldots, k'; \tag{72.13}$$

(b) at least one of them is different from zero, say

$$\int_L \sigma(t)\nu_1(t)\,ds \neq 0. \tag{72.14}$$

For $k' \geqq 1$, the first assumption is impossible, because then the condition (72.12) cannot be satisfied for an arbitrary choice of $f(t)$. Under the second assumption, for $k' \geqq 2$ [giving $f(t)$ the particular value $\nu_2(t)$], one concludes from (72.12) for $j = 1,2$ that

$$C = 0, \quad \int_L [\nu_2(t)]^2 ds = 0,$$

but the last condition is impossible. Hence necessarily $k' \leqq 1$, while for $k' = 1$ the inequality (72.6) must be satisfied.

Thus the theorem is proved.

Corollary: *If $\sigma(t) \equiv 0$, then Problem V is soluble for any function $f(t_0)$, if and only if $k' = 0$; in this case the homogeneous problem $\Re(\mathbf{L}\Phi) = 0$ has exactly $\varkappa + 1$ linearly independent solutions.*

Now consider the case, when the conditions in the theorem proved above are not satisfied (i.e., for example, when $\varkappa < 0$). Then Problem V is not soluble for every $f(t_0)$: a solution exists, if and only if (72.12) is satisfied for a suitably chosen C.

Here again the two possible assumptions (a) and (b), stated above, will be considered.

For (a), i.e., when (72.13) holds, the non-homogeneous problem V is soluble, if and only if

$$\int_L \nu_j(t)f(t)\,dt = 0, \quad j = 1, 2, \ldots, k' \tag{72.15}$$

and the integral equation $\mathbf{N}\mu = -C\sigma(t_0)$, corresponding to the homogeneous problem $\Re(\mathbf{L}\Phi) = 0$, is soluble for arbitrary C as a consequence of (72.13). The general solution of the last integral equation has the form $\mu(t) = C\mu_0(t) + C_1\mu_1(t) + \ldots + C_k\mu_k(t)$, where, as before, $\mu_0(t)$ is a particular solution of $\mathbf{N}\mu = -\sigma(t_0)$ and the $\mu_1(t), \ldots, \mu_k(t)$ are linearly independent solutions of the homogeneous equation $\mathbf{N}\mu = 0$. Correspondingly, the homogeneous problem has exactly $k + 1 = \varkappa + k' + 1$ linearly independent solutions.

For (b) one has to assume $k' > 1$, because for $k' \leqq 1$ the conditions of the theorem proved above would be satisfied. The first of the relations (72.12) uniquely determines C:

$$C = \frac{\int_L \nu_1(t) f(t)\, ds}{\int_L \nu_1(t)\sigma(t) ds}.$$

Substituting this value of C in the remaining equations (72.12), one obtains the conditions of solubility of the non-homogeneous problem V in the form

$$\int_L \nu_j^*(t) f(t)\, ds = 0, \tag{72.16}$$

where

$$\nu_j^*(t) = \nu_{j+1}^*(t) - \nu_1(t) \frac{\int_L \sigma \nu_{j+1}\, ds}{\int_L \sigma \nu_1\, ds}, \quad j = 1, 2, \ldots, k' - 1. \tag{72.17}$$

In the case of the homogeneous problem $\Re(\mathbf{L}\Phi) = 0$ one must, in correspondence with the above, take $C = 0$. Therefore the homogeneous problem $\Re(\mathbf{L}\Phi) = 0$ in the case considered has exactly $k = \varkappa + k'$ linearly independent solutions.

In particular, it follows from the above that the homogeneous problem $\Re(\mathbf{L}\Phi) = 0$ has no solutions different from zero, except in the case (b), and then only when $k = \varkappa + k' = 0$.

§ 73. Criteria of solubility of Problem V.

As was seen in the last section, the number k' of linearly independent solutions of the homogeneous equation $\mathbf{N}'\nu = 0$ plays a fundamental role for the solubility of Problem V. Therefore it is very important to be able to determine the number k' without actually having to solve this equation. In this connection I. N. Vekua obtained again very interesting results (loc. cit.).

The function

$$\Omega^*(t_0, z) = \sum_{j=0}^{m} \{a_j(t_0) N_j(t_0, z) + \int_L h_j(t_0, t) N_j(t, z)\, ds\} \tag{73.1}$$

will be considered, where the $N_j(t_0, z)$ are the elementary functions defined by (71.3)—(71.5), except that here the first argument t_0 is a point on L, while the second argument z is an arbitrary point of the plane. In the sequel the function $\Omega^*(t_0, z)$ will only have to be considered for $z \in S^-$. Then the term $\log(1 - t_0/z)$, entering into $N_j(t_0, z)$, will be understood as referring to that branch which is holomorphic with respect to z in S^- and vanishes for $z = \infty$. Under these conditions $\Omega^*(t_0, z)$ will be holomorphic with respect to z in S^- and it is easily seen that

$$\Omega^*(t_0, \infty) = a_0(t_0) + \int_L h_0(t_0, t)\, ds. \tag{73.2}$$

Using $\Omega^*(t_0, z)$, the equation $\mathbf{N}'\nu = 0$ is easily expressed in the form

$$\Re \Psi^-(t_0) = 0, \tag{73.3}$$

where

$$\Psi(z) = \int_L \nu(t)\Omega^*(t, z)\, ds. \tag{73.4}$$

It is inferred from (73.3) that for $z \in S^-$

$$\Psi(z) = iC, \tag{73.5}$$

where C is a real constant, defined by

$$C = \int_L \nu(t)\Im\Omega^*(t, \infty)\, ds. \tag{73.6}$$

Now introduce the function

$$\Omega(t_0, t) = \Omega^*(t_0, z) - i\Im\Omega^*(t_0, \infty). \tag{73.7}$$

It follows directly from (73.4), (73.6) and (73.5) that *the equation* $\mathbf{N}'\nu = 0$ *is equivalent to the functional equation*

$$\int_L \nu(t)\Omega(t, z)\, ds = 0 \quad \text{for all } z \in S^-. \tag{A}$$

In other words, the function $\nu(t)$ is required to be orthogonal to the given function $\Omega(t, z)$ for any $z \in S^-$; the function $\Omega(t, z)$ will sometimes be called the „kernel".

Using the well-known theorem on the uniqueness of analytic functions, one can in a variety of ways replace the equation (A) by equivalent relations, convenient for application to a definite case. Following I. N. Vekua, the following relations, equivalent to (A), will be studied in detail:

$$\int_L \nu(t)\omega_j(t)\, ds = 0, \tag{B}$$

where one may understand by the $\omega_j(t)$ ($j = 0, 1, 2, \ldots,$) any of the systems of functions below:

1°. $\omega_j(t) = \Omega(t, z_j),\ j = 0, 1, 2, \ldots,$ $\quad$ (B_1)

where $z_0, z_1, \ldots$ is any sequence of points of S^-, having at least one limit point in S^-; in particular, this point can be $z = \infty$.

2°. $$\omega_j(t) \equiv \Omega_j(t, z_0) = \left[\frac{d^j\Omega(t, z)}{dz^j}\right]_{z=z_0}, \quad j = 0, 1, 2, \ldots, \tag{B$_2$}$$

where z_0 is an arbitrarily fixed point of S^-.

3°. $$\omega_j(t) = \chi_j(t), \quad j = 0, 1, 2, \ldots,$$

where

$$\begin{aligned} \chi_0(t_0) &= \Re\{a_0(t_0) + \int_L h_0(t_0, t)\, ds\}, \\ \chi_k(t_0) &= \mathbf{L}\psi, \quad \psi = t^k, \quad k = 1, 2, \ldots, \end{aligned} \tag{B$_3$}$$

and $\mathbf{L}$ is the same operator as before, defined by

$$\mathbf{L}\psi \equiv \sum_{j=0}^{m} \{a_j(t_0)\psi^{(j)}(t_0) + \int_L h_j(t_0, t)\psi^{(j)}(t)\, ds\}.$$

The functions $\chi_j(t)$ differ only by constant non-zero multipliers from the coefficients in the series expansion of $\Omega(t, z)$ in decreasing powers of z near the point $z = \infty$. This is easily seen from (73.7), (73.1) and (71.3)—(71.5).

The relation (B) will be expressed by saying that $\nu(t)$ is orthogonal to all the elements of the sequence $\{\omega_j(t)\}$.

If the functions $\nu(t)$ (which are always understood to be real functions, satisfying the H condition) are different from zero and do not form an orthogonal system with $\Omega(t, z)$ for any $z \in S^-$, then the kernel $\Omega(t, z)$ will be called complete; if only a finite number of linearly independent functions $\nu(t)$, orthogonal to $\Omega(t, z)$, exist, then the kernel will be called almost complete.

Similarly, if the functions $\nu(t)$, different from zero and orthogonal to all the elements of the sequence $\{\omega_j(t)\}$, do not exist, then this sequence will be called complete; it will be called almost complete, if there is only a finite number of linearly independent functions $\nu(t)$, orthogonal to all the elements of this sequence.

Let k' be the maximum number of linearly independent (real) functions $\nu_j(t)$ $(j = 1, 2, \ldots, k')$, orthogonal to the kernel $\Omega(t, z)$ or to all the elements of the sequence $\{\omega_j(t)\}$; this number k' will be called the deficiency of the kernel $\Omega(t, z)$ or of the sequence $\{\omega_j(t)\}$. Clearly, if the functions $\nu_1(t), \nu_2(t), \ldots, \nu_{k'}(t)$ are added to the sequence $\{\omega_j(t)\}$, then a complete sequence is obtained. Therefore the deficiency of the sequence $\{\omega_j(t)\}$ can be defined as the number of linearly independent functions $\nu_j(t)$, orthogonal to the $\omega_j(t)$, which must be added to the sequence $\{\omega_j(t)\}$ to form a complete sequence.

In the present case the number of linearly independent functions $\nu(t)$, orthogonal to the kernel $\Omega(t, z)$ or to the elements of the sequence $\{\omega_j(t)\}$, where the $\omega_j(t)$ are defined by (B_1), (B_2) or (B_3), is finite; it is equal to the number k' of linearly independent solutions of the homogeneous integral-equation $N'\nu = 0$. Hence *the kernel* $\Omega(t, z)$ *and the sequence* $\{\omega_j(t)\}$ *are here almost complete.*

In the above terminology the proposition proved in the last section can be written:

For Problem V to have solutions for any function $f(t)$ on the right, it is necessary and sufficient that the deficiency of the sequence $\{\omega_j(t)\}$ [or the kernel $\Omega(t, z)$] is zero or unity, where in the latter case the function $\nu(t)$, orthogonal to all the elements of the sequence $\{\omega_j(t)\}$ [or the kernel $\Omega(t, z)$], must not be orthogonal to the function $\sigma(t)$.

§ 74. The Poincaré Problem (Problem P).

Consider again the Poincaré problem (§ 70) and let S^+ and L be the same as in § 71.

The boundary condition

$$A(s)\frac{\partial u}{\partial n} + B(s)\frac{\partial u}{\partial s} + c(s)u = f(s) \tag{74.1}$$

(where the $A(s)$, $B(s)$, $c(s)$ and $f(s)$ are real functions, given on L, u is the unknown function, harmonic in S^+, n the normal to L at the point with arc coordinate s, directed to the left of L) can be written in a somewhat different form, using the relations

$$\frac{\partial u}{\partial s} = \frac{\partial u}{\partial x}\cos\theta + \frac{\partial u}{\partial y}\sin\theta, \quad \frac{\partial u}{\partial n} = -\frac{\partial u}{\partial x}\sin\theta + \frac{\partial u}{\partial y}\cos\theta, \tag{74.2}$$

where θ is the angle between the positive tangent to L and the Ox axis. Then (74.1) assumes the form

$$a(t)\frac{\partial u}{\partial x} + b(t)\frac{\partial u}{\partial y} + c(t)u = f(t) \text{ on } L, \tag{74.3}$$

where the $a(t)$, $b(t)$, $c(t)$ and $f(t)$ are real functions, given on L. In the sequel these functions will be assumed to satisfy the H condition; in addition, the functions $a(t)$, $b(t)$ will be assumed not to vanish simultaneously, so that

$$a(t) + ib(t) \neq 0 \text{ anywhere on } L. \tag{74.4}$$

The partial derivatives $\dfrac{\partial u}{\partial x}, \dfrac{\partial u}{\partial y}$ of the unknown function will be required to have on L boundary values satisfying the H condition (which then will be also true for u).

The problem (74.3) will be called Problem P. The condition (74.3) can also be written (where now t_0 is written for t)

$$\Re\{(a+ib)\Phi'(t_0)+c\Phi(t_0)\}=f(t_0) \text{ on } L, \tag{P}$$

where, as before, $\Phi(t_0)$, $\Phi'(t_0)$ is written instead of $\Phi^+(t_0)$, $\Phi'^+(t_0)$ and where

$$\Phi(z)=u(x,y)+iv(x,y) \tag{74.5}$$

is a function, holomorphic in S^+, such that

$$\Re\Phi(z)=u(x,y).$$

The condition (P) is a particular case of the condition (V), when in the latter

$$m=1,\ h_0=h_1=0,\ a_1(t)=a(t)+ib(t),\ a_0(t)=c(t).$$

Applying to Problem P the method stated in the last section, the unknown function $\Phi(z)$ is represented by (71.2) for $m=1$:

$$\Phi(z)=\int_L \mu(t)\log(1-z/t)\,ds+\int_L \mu(t)ds+iC, \tag{74.6}$$

where $\mu(t)$ is a real unknown function satisfying the H condition and C is a real constant.

In the case considered the function $\sigma(t)$, defined by (71.10), is identically zero and the integral equation (71.8) has the form

$$\mathbf{N}\mu\equiv\Re\{-\pi i\bar{t}_0'[a(t_0)+ib(t_0)]\}\mu(t_0)+$$
$$+\int_L \mu(t)\Re\left\{c(t_0)\log(1-t_0/t)-\frac{a(t_0)+ib(t_0)}{t-t_0}\right\}ds=f(t_0). \tag{74.7}$$

The index of this equation is, by (72.2),

$$\varkappa=2(n+1),\quad n=\frac{1}{2\pi}[\arg(a-ib)]_L. \tag{74.8}$$

The homogeneous equation, adjoint to (74.7), has the form

$$\mathbf{N}'\nu\equiv\Re\{-\pi i\bar{t}_0'[a(t_0)+ib(t_0)]\}\nu(t_0)+$$
$$+\int_L \nu(t)\,\Re\left\{c(t_0)\log(1-t/t_0)+\frac{a(t)+ib(t)}{t-t_0}\right\}ds=0. \tag{74.9}$$

It is equivalent to the functional equation (cf. § 73)

$$\int_L \nu(t)\Omega(t,z)\,ds=0 \text{ for all } z\in S^-, \tag{A}$$

where in the present case

$$\Omega(t, z) = \frac{a(t) + ib(t)}{t - z} + c(t) \log (1 - t/z). \tag{74.10}$$

Before formulating the fundamental results which may be deduced directly from this, the following will be noted: if $\Phi(z)$ is any solution of Problem P, then obviously $\Phi(z) + iC$, where C is a real constant, is also a solution; accordingly, the homogeneous problem, obtained from (P) for $f(t) \equiv 0$, has the obvious solution iC. Since the addition of iC to $\Phi(z)$ does not alter the function $u(x, y)$, but only changes the imaginary part $v(x, y)$ of $\Phi(x, y)$ which does not enter into the original statement of the problem, it will be agreed not to consider as different those solutions of Problem P which differ by terms of the form iC; in other words, two solutions $\Phi_1(z)$, $\Phi_2(z)$ of Problem P will be assumed different only when their real parts $u_1(x, y)$ and $u_2(x, y)$ are different.

Remembering now that in the present case $\sigma(t) \equiv 0$ and using the deductions from the theorem of § 72, stated towards the end of that section, one arrives at the following fundamental result (I. N. Vekua, loc. cit.):

THEOREM:

For Problem P to be soluble for any function $f(t)$, it is necessary and sufficient that the equation (74.9) *has no solutions, different from zero, or, what amounts to the same thing, that the kernel $\Omega(t, z)$ is complete. When this condition is satisfied, the homogeneous problem, corresponding to Problem P, has exactly $\varkappa$ linearly independent solutions* (not considering, as indicated above, trivial solutions of the form iC).

Next assume that (74.9) or the functional equation (A) have the k' linearly independent (real) solutions $\nu_j(t)$ $(j = 1, \ldots, k')$. Then Problem P is soluble, if the following necessary and sufficient conditions are satisfied:

$$\int_L \nu_j(t) f(t)\, ds = 0, \quad j = 1, 2, \ldots, k'; \tag{74.11}$$

however, the homogeneous problem corresponding to P has exactly $k = \varkappa + k'$ linearly independent solutions (neglecting solutions of the form iC). If, in particular, $\varkappa + k' = 0$, then Problem P has one and only one solution, provided the conditions (74.11) are satisfied.

As stated in the last section, the equation (A) is equivalent to the conditions

$$\int_L \nu(t) \omega_j(t)\, ds = 0, \quad j = 0, 1, \ldots, \tag{B}$$

if the system of functions $\omega_j(t)$ is understood to be any of the following:

$$1°.\ \omega_j(t) = \Omega(t, z_j) = \frac{a(t) + ib(t)}{t - z_j} + c(t) \log (1 - t/z_j),$$
$$j = 0, 1, \ldots, \qquad (B_1)$$

where $z_0, z_1, \ldots$ is any sequence of points of S^-, having at least one limit point in S^-.

$2°.\ \omega_j(t) = \Omega_j(t, z_0),\ j = 0, 1, \ldots,$

where

$$\Omega_0(t, z_0) = \Omega(t, z_0) = \frac{a(t) + ib(t)}{t - z_0} + c(t) \log (1 - t/z_0),$$

$$\Omega_k(t, z_0) = \left[\frac{d^k\Omega(t, z)}{dz^k}\right]_{z=z_0} = \qquad (B_2)$$

$$= k!\, \frac{a(t) + ib(t)}{(t - z_0)^{k+1}} - (k-1)!c(t)\left\{\frac{1}{(t - z_0)^k} + \frac{(-1)^{k-1}}{z_0^k}\right\},$$

$k = 1, 2, \ldots,$ and z_0 any fixed point in S^-.

$3°.\ \omega_j(t) = \chi_j(t) = j[a(t) + ib(t)]t^{j-1} + c(t)t^j,\ j = 0, 1, \ldots \qquad (B_3)$

It will be remembered that *the completeness of one of the systems 1°—3° is the necessary and sufficient condition for the solubility of Problem P for any arbitrary function $f(t)$.*

Finally, one direct consequence of the results obtained will be noted. *If the homogeneous problem corresponding to Problem P has no solutions different from zero and if the index $\varkappa = 0$, then Problem P is uniquely soluble for any function $f(t)$.* In fact, the number of solutions of the homogeneous problem was seen to be $k = \varkappa + k'$; therefore, if $k = \varkappa = 0$, then also $k' = 0$, and hence the proposition follows.

Often the homogeneous problem is easily seen directly to have no solutions different from zero. One such case was indicated by B. V. Khvedelidze [1], [2].

Let the boundary condition be given in the form (74.1) and let $A(s) = 1$. Further, it will be assumed that $B(s)$ has an integrable derivative, that $c(s) \not\equiv 0$ and that

$$\frac{1}{2}\frac{dB(s)}{ds} - c(s) \geqq 0. \qquad (74.12)$$

Then the homogeneous problem

$$\frac{\partial u}{\partial n} + B(s)\frac{\partial u}{\partial s} + c(s)u = 0 \qquad (*)$$

has no solutions different from zero.

In fact, substituting in the well-known formula

$$\iint\limits_S \left[\left(\frac{\partial u}{\partial x}\right)^2 + \left(\frac{\partial u}{\partial y}\right)^2\right] dxdy = -\int\limits_L u\frac{\partial u}{\partial n}\, ds$$

the value of $\dfrac{\partial u}{\partial n}$ from (*), one obtains after a simple transformation, using integration by parts,

$$\iint\limits_S \left[\left(\frac{\partial u}{\partial x}\right)^2 + \left(\frac{\partial u}{\partial y}\right)^2\right] dxdy = -\int\limits_L \left[\frac{1}{2}\,\frac{dB(s)}{ds} - c(s)\right] u^2\, ds \leqq 0,$$

and hence it is concluded that $\dfrac{\partial u}{\partial x} = \dfrac{\partial u}{\partial y} = 0$, i.e., $u = \text{const.}$ But then it follows from (*) and from $c(s) \not\equiv 0$ that $u = 0$.

Obviously the criterion stated above for the absence of non-zero solutions of the homogeneous problem also remains true in the case of a finite multiply-connected region.

NOTE. If the boundary condition of Problem P is given in the form (74.1), then one will have the following formula for the determination of the index:

$$\varkappa = \frac{1}{\pi}\,[\arg A + iB]_L; \tag{74.13}$$

this follows from (74.8), because, as is easily seen from (74.2),

$$A + iB = i(a - ib)e^{i\theta}.$$

§ 75. Examples.

Some simple examples will now be given which illustrate the results of the preceding sections (§ 71—§ 74); it must be remembered that in these sections S^+ is a finite simply-connected region and that the point $z = 0$ lies inside S^+. It will be noted that the connection between Problem V (or, in particular, P) and the sequence of functions $\{\omega_j(t)\}$ can be used in two ways: if the question of completeness of these sequences can be resolved directly, then the answer to the problem of solubility of the problem will be obtained; conversely, if the question of solubility of the problem V is decided by any method whatsoever, then there arises the possibility of establishing the completeness or incompleteness of the sequence $\{\omega_j(t)\}$.

1°. THE DIRICHLET PROBLEM

$$u = f \text{ on } L \tag{75.1}$$

is a particular case of Problem V for $m = 0$, $a_0(t) = 1$, $h(t_0, t) = 0$.

The index of the corresponding integral equation (72.1) is zero by (72.2); the function $\sigma(t) \equiv 0$. The system of functions $\{\chi_j(t)\}$ is as follows [(B_3) of § 73, where now $L\psi \equiv \psi$]:

$$1,\ t,\ t^2,\ \ldots. \tag{75.2}$$

From the above the Dirichlet problem is known to be soluble for any function f. Hence it is concluded that the system (75.2) is complete, i.e., that, if

$$\int_L t^k \nu(t) ds = 0, \quad k = 0,\ 1,\ 2,\ \ldots, \tag{75.3}$$

then $\nu(t) = 0$; as always, $\nu(t)$ is understood to be a real function satisfying the H condition. (Actually this conclusion is easily seen to remain true, if $\nu(t)$ is only assumed continuous).

Putting $t = re^{i\varphi}$ and separating real and imaginary parts, a conclusion is reached regarding the completeness of the system of real functions given on L:

$$1,\ r\cos\varphi,\ r\sin\varphi,\ r^2\cos 2\varphi,\ r^2\sin 2\varphi,\ \ldots \text{ on } L, \tag{75.4}$$

where $r = |\,t\,|$, $\varphi = \arg t$ on L.

If L is a circle with centre O, then one obtains the well known theorem on the completeness of the system of functions

$$1,\ \cos\varphi,\ \sin\varphi,\ \cos 2\varphi,\ \sin 2\varphi,\ \ldots. \tag{75.5}$$

2°. THE NEUMANN PROBLEM

$$\frac{\partial u}{\partial n} = f \text{ on } L, \tag{75.6}$$

where n is the normal, assumed to be directed to the left of L [i.e., it is the inside normal], is a particular case of Problem P of the last section for

$$a(t) = \cos(n, x) = -\sin\theta,\ b(t) = \sin(n, x) = \cos\theta,\ c(t) = 0, \tag{75.7}$$

where θ is the angle between the tangent to L and the Ox axis; thus

$$a + ib = ie^{i\theta} = it',\ a - ib = -ie^{-i\theta} = -i\bar{t}'; \tag{75.8}$$

by (74.8), the index of the equation $N\mu = f(t_0)$ corresponding to the problem is zero (and, in addition, this equation is easily seen in the present case to be a Fredholm equation). It follows from (74.10) that

$$\Omega(t, z) = \frac{it'}{t - z}. \tag{75.9}$$

This kernel is incomplete, since the functional equation (A) of § 74

$$\int_L \nu(t)\Omega(t, z)ds \equiv i \int_L \frac{\nu(t)t'\,ds}{t-z} \equiv i \int_L \frac{\nu(t)\,dt}{t-z} = 0, \; z \in S^-,$$

has the real non-zero solution $\nu(t) = 1$; there are no other solutions (linearly independent of it) (§ 25); therefore the deficiency is 1.

Hence the problem (75.6) is soluble, if and only if

$$\int_L \nu(t)f(t)\,ds = 0, \; \text{i.e.,} \int_L f(t)\,ds = 0,$$

and this is a well-known result.

The system of functions $\{\chi_k(t)\}$ is in the present case from (B_3), § 73 (with $\mathbf{L}\psi = (a + ib)\psi' = ie^{i\theta}\psi' = it_0\psi'(t_0)$) or § 74 as follows:

$$\chi_k(t) = kit't^{k-1}, \; k = 1, 2, \ldots \tag{75.10}$$

By the above this system is incomplete with the deficiency 1. But if the function $\nu(t) = 1$ is added to this system, one arrives at the complete system (omitting the factor ik)

$$1, \; t', \; t't, \; t't^2, \; \ldots$$

3°. Next, the somewhat more general problem

$$\frac{\partial u}{\partial n} + c(t)u = f(t) \tag{75.11}$$

will be considered which is obtained from Problem **P**, assuming a and b given by (75.8). Here again the index of the corresponding equation $\mathbf{N}\mu = f$ is zero. In the present case $\mathbf{L}\psi = it_0'\psi'(t_0) + c(t_0)\psi(t_0)$ and, by (B_3), § 73 or § 74,

$$\chi_k(t) = ikt't^{k-1} + c(t)t^k, \; k = 0, 1, \ldots \tag{75.12}$$

If the system of functions (χ_k) is complete, then the problem (75.11) has a solution for each function $f(t)$, and conversely.

The case when L is the circle of radius 1 with centre at O will now be considered; let $c(t) = c$ be a constant. Then $t' = it$ and

$$\chi_k(t) = (c - k)t^k, \; k = 0, 1, \ldots \tag{75.13}$$

or, putting $t = e^{i\varphi}$,

$$\chi_k(t) = (c - k)(\cos k\varphi + i \sin k\varphi). \tag{75.13a}$$

If c is neither a positive integer nor zero, then the above system is obviously complete and the problem has one and only one solution for every function $f(t)$.

Now let $c = m$, where m is a non-negative integer. Then the system (75.13a) is incomplete. For $m = 0$, it is made complete by addition of the single function $\nu(t) = 1$; for $m \geqq 1$, it is completed by addition of the two linearly independent functions: $\nu_1(t) = \cos mt$, $\nu_2(t) = \sin mt$. Hence, for $m = 0$, the deficiency is 1, and it is 2 for $m \geqq 1$. The homogeneous problem corresponding to (75.11) has therefore one solution for $m = 0$ and two solutions for $m \geqq 1$; the non-homogeneous problem, however, is soluble, if and only if

$$\int_0^{2\pi} f(t)d\varphi = 0, \text{ for } m = 0, \tag{75.14}$$

$$\int_0^{2\pi} f(t)\cos m\varphi d\varphi = 0, \quad \int_0^{2\pi} f(t)\sin m\varphi d\varphi = 0, \text{ for } m \geqq 1. \tag{75.15}$$

4°. PROBLEM OF THE DIRECTIONAL DERIVATIVE. Finally, the particular case of Problem $\dot{\text{P}}$ will be considered when $c(t) = 0$, so that the boundary condition has the form

$$a(t)\frac{\partial u}{\partial x} + b(t)\frac{\partial u}{\partial y} = f(t) \text{ on } L. \tag{75.16}$$

The problem of determining the harmonic function for this boundary condition is sometimes called "the problem of the directional derivative" (problème de la dérivée oblique), since the boundary condition (75.16) can be written

$$\sqrt{a^2 + b^2}\frac{\partial u}{\partial l} = f(t) \text{ on } L, \tag{75.17}$$

where l is the direction of the vector (a, b) which is generally inclined to the normal or tangent. In particular, this problem has been treated in an extensive paper recently published by A. Liénard [1]. It can be solved by the method, studied in § 74; however, in the notation adopted here, it is more natural to reduce it directly to the Riemann-Hilbert problem which can be done very simply.

In fact, as in the last section, assume $u(x, y) = \Re\Phi(z)$ and, in addition, introduce the function

$$U + iV = \Psi(z) = \Phi'(z) = \frac{\partial u}{\partial x} - i\frac{\partial u}{\partial y},$$

holomorphic in S^+.

Then (75.16) can be written as $a(t)U - b(t)V = f(t)$ or

$$\Re(a + ib)\Psi^+(t) = f(t),$$

so that the problem leads directly to the Riemann-Hilbert problem, already solved in § 40—42 for simply-connected regions (regarding its solution for multiply-connected regions see 2° of the next section).

§ 76. Some generalizations and applications.

The methods and results studied in the preceding sections can be effectively applied to the solution of a series of important problems. Only a few brief remarks will be made here concerning some generalizations and applications *).

1°. The solution of Problem P for multiply-connected regions was given by B. V. Khvedelidze in his paper [2] in which, as in his paper [1] which deals with simply-connected regions, he represents, following Poincaré, the unknown function in the form of the potential of a simple layer. In his paper [3] he gives the solution of the more general problem using this time I. N. Vekua's representation (cf. 3° below).

2°. THE SOLUTION OF THE HILBERT PROBLEM FOR MULTIPLY-CONNECTED REGIONS (in § 40—42 it was solved for simply-connected regions) can be obtained, if it is considered as a particular case of Problem V and if the method of § 71—73 is applied in a form suitably extended to the case of multiply-connected regions.

3°. THE POINCARÉ PROBLEM FOR AN EQUATION OF THE ELLIPTIC TYPE. The problem, called above Problem P (Poincaré problem), which arises in connection with the mathematical theory of tides is not a problem, restricted to this theory.

The theory of tides leads directly to the problem of finding a solution, regular in some region S, of the differential equation of the elliptic type

$$\Delta u + X(x, y)\frac{\partial u}{\partial x} + Y(x, y)\frac{\partial u}{\partial y} + Z(x, y)u = F(x, y), \qquad (76.1)$$

or even, under more general assumptions, to the solution of the integro-differential equation of the form

$$\Delta u + X(x, y)\frac{\partial u}{\partial x} + Y(x, y)\frac{\partial u}{\partial y} + Z(x, y)u + \\ + \iint_S K(x, y; \xi, \eta)u(\xi, \eta)d\xi\, d\eta = F(x, y) \qquad (76.2)$$

*) Generalizations to the cases of problems with discontinuous functional coefficients and with several unknown functions will be given in Parts V and VI respectively.

with a boundary condition of the form (n and s being the normal and arc coordinate respectively)

$$A(s)\frac{\partial u}{\partial n} + B(s)\frac{\partial u}{\partial s} + c(s)u = f(s) \text{ on } L; \qquad (76.3)$$

here Δ is the Laplace operator, $X(x, y)$, $Y(x, y)$, $Z(x, y)$, $F(x, y)$ and $K(x, y; \xi, \eta)$ are functions given in S, and $A(s)$, $B(s)$, $c(s)$ and $f(s)$ are functions, given on the boundary L of the region S, satisfying the known conditions of regularity which will not be discussed further.

H. Poincaré [1] considered the case of the equation (76.1), assuming that in the boundary condition $A(s) = 1$, $c(s) = 0$; the case of the equation (76.2) was considered in the extensive work of G. Bertrand [3] under the same assumptions, i.e., $A(s) = 1$, $c(s) = 0$; the case of the same equation, assuming $A(s) = 1$, was dealt with by W. Pogorzelski [1]. All these authors assumed the boundary L of the region S to be an analytic contour and the given functions in the boundary condition likewise to be analytic.

Problem P, i.e., the same boundary problem as here, but in the case where the equation (76.1) or (76.2) is reduced to its simplest form $\Delta u = 0$, was used by the authors for the purpose of finding the Green function corresponding to a boundary condition of the form (76.3); using this Green function they then formed, by the well-known method, the Fredholm integral equation for the solution of the original problem.

The intermediate stage, i.e., the solution of Problem P, as has been partly explained earlier, was not studied with sufficient completeness by any of these authors, since the question of equivalence of the Fredholm equations, obtained by them (as a result of the reduction of singular equations), remained open (as a matter of fact, generally speaking, equivalence did not enter into their consideration). Nothing was said about the fact that the structure of the Fredholm integral equation (with the region of integration S) finally obtained (by means of the Green function) was very complicated.

For one rather wide and important class of equations of the form (76.1) the problem considered can be directly reduced to the problem solved in §§ 71, 72.

In fact, consider the homogeneous equation of the elliptic type

$$\Delta u + X(x, y)\frac{\partial u}{\partial x} + Y(x, y)\frac{\partial u}{\partial y} + Z(x, y)u = 0, \qquad (76.4)$$

where the $X(x, y)$, $Y(x, y)$, $Z(x, y)$ are integral holomorphic functions with their arguments real for real x, y (the fact that consideration is restricted to a homogeneous equation does not reduce the generality; cf. below).

The (real) solutions of the equation (76.4) will be sought, regular (i.e., continuous together with their derivatives up to the second order) in some region S which is, for simplicity, assumed simply-connected.

I. N. Vekua [8], [9], [10] has shown that each such solution can be written in the form (assuming that the point $z = 0$ belongs to S)

$$u(x, y) = \Re\{\alpha(z, \bar{z})\varphi(z) + \int_0^z \beta(z, \bar{z}; \zeta)\varphi(\zeta)d\zeta\}, \qquad (76.5)$$

where $\alpha(z, \bar{z})$ and $\beta(z, \bar{z}; \zeta)$ are completely defined integral holomorphic functions, depending only on the coefficients of the equation (76.4); the first of these is determined in an elementary manner (quadrature), while the second is obtained by means of the always converging algorithm of successive approximation. By $\varphi(z)$ is denoted some function, holomorphic in S; whatever may be the holomorphic function $\varphi(z)$, the function $u(x, y)$, corresponding to it by (76.5), will be a regular solution of (76.4), and, conversely, to every regular solution corresponds a completely defined holomorphic function $\varphi(z)$, provided the condition $\Im\varphi(0) = 0$ is satisfied.

If, instead of the homogeneous equation (76.4), one has a non-homogeneous equation (i.e., an equation of the form (76.1)), then one must add yet another term to the right side of (76.5) which depends only on the function $F(x, y)$ and on the coefficients in the equation; this term is not found to have any great influence upon the method of solution.

As has been indicated, the function $\alpha(z, \bar{z})$ can be determined by elementary means. In some important cases the function $\beta(z, \bar{z}; \zeta)$ also has a very simple form. For example, in the case of the equation

$$\Delta u + \lambda^2 u = 0,$$

where λ^2 is a real constant (for $\lambda^2 > 0$ it is the equation of vibrations of a membrane), $\alpha(z, \bar{z}) = 1$, and

$$\beta(z, \bar{z}; \zeta) = -\frac{\partial}{\partial\zeta} J_0(\lambda\sqrt{\bar{z}(z-\zeta)}),$$

where $J_0(z)$ is the Bessel function of zero order.

Returning to the boundary problem for the equation (76.4), if (76.5) is substituted in the condition (76.3), then this condition assumes the form

$$\Re\{a_1(t_0)\varphi'(t_0) + a_0(t_0)\varphi(t_0) + \int_0^{t_0} h_0^*(t_0, t)\varphi(t)\,dt\} = f(t_0)$$

on L, where $a_1(t_0)$, $a_0(t_0)$, $h_0^*(t_0, t)$, $f(t_0)$ are given functions having the properties stated in § 71 and in the Note at the end of that section (for $m = 1$).

Therefore one can apply the method of § 72. The problem was solved in this way by B. V. Khvedelidze [3] who also extended the method to the case of multiply-connected regions, using a generalization of the representation (76.5), likewise given by I. N. Vekua [12] (cf. the original paper).

4°. BOUNDARY PROBLEMS CONNECTED WITH EQUATIONS OF THE ELLIPTIC TYPE. The method of 3° can be extended to some other classes of differential equations of the elliptic type (including also equations of order greater than 2). In the presence of a general representation of the solution similar to (76.5), the (linear) boundary problems considered can in most cases be converted to Problem V, as above. The general representations suitable for this purpose were given for a number of important cases in the quoted papers by I. N. Vekua to which the reader is again referred.

PART IV

THE HILBERT PROBLEM IN THE CASE OF ARCS OR DISCONTINUOUS BOUNDARY CONDITIONS AND SOME OF ITS APPLICATIONS

From the point of view of application great interest lies in the solution of the Hilbert boundary problem for the case in which the boundary consists of arcs or where the given functions entering into the boundary condition have discontinuities of a certain form. The first case may be reduced to the second (in the same way as the case where the boundary consists of arcs and the given functions are discontinuous).

However, a start will be made with the study of the Hilbert problem for continuous boundary conditions in the case of boundaries consisting of arcs, since, firstly, this problem is of independent interest and, secondly, the corresponding theory is somewhat simpler and easily transferred to the case of discontinuous boundary problems.

It will also be noted that all the results stated below can be easily generalized to the case where the boundary is the union of arcs and contours.

The solution of the Hilbert problem will be given in Chapter 10. The other chapters of Part IV will be devoted to various applications and can be read independently of one another.

CHAPTER 10

THE HILBERT PROBLEM IN THE CASE OF ARCS OR DISCONTINUOUS BOUNDARY CONDITIONS *)

§ 77. Definitions.

1°. Up to and including § 85, L will be understood to be the union of smooth, non-intersecting arcs $L_1, L_2, \ldots, L_p$ with definite positive directions. The ends of the arcs L_j ($j = 1, 2, \ldots, p$)

*) The results of this chapter were obtained by the Author in his paper [2] and extended in the paper [1] by N. I. Muskhelishvili and D. A. Kveselava.

will be denoted by a_j, b_j in such a way that the positive direction of L_j is from a_j to b_j; "the ends a" and "the ends b" will sometimes be distinguished accordingly. When it is of no importance which end is referred to, it will be denoted by c or c_j. The plane cut along $L = L_1 + L_2 + \ldots + L_p$ will be denoted by S; the boundary L does not belong to S.

2°. In correspondence with the terminology of § 15 the function $\Phi(z)$, holomorphic in S, except possibly at the point at infinity, and continuous on L from the left and from the right with the possible exception of the ends c, near which the inequality

$$|\Phi(z)| < \frac{\text{const.}}{|z - c|^{\alpha}}, \quad 0 \leqq \alpha < 1 \tag{77.1}$$

is to hold, will be called sectionally holomorphic with the line of discontinuity or boundary L.

It will be remembered that, if the expansion of $\Phi(z)$ near the point at infinity in decreasing powers of z has only a finite number of terms with positive exponents, i.e., if for large $|z|$

$$\Phi(z) = A_k z^k + A_{k-1} z^{k-1} + \ldots, \tag{77.2}$$

then, by definition, $\Phi(z)$ has the finite degree k at infinity; for $k > 0$, the point $z = \infty$ is a pole of order k, while for $k < 0$ it is a zero of order $(-k)$.

3°. Let $\varphi(t)$ be some function of the point t on L. The function $\varphi(t)$ will be said to belong to the class H on L, if it satisfies, for some $\mu > 0$, the $H(\mu)$ condition on each of the closed arcs L_j of L (i.e., including the ends); if $\varphi(t)$ satisfies the $H(\mu)$ condition only in the neighbourhood of some end c of L, including c, then $\varphi(t)$ will be said to belong to the class H in the neighbourhood of c.

4°. If the function $\varphi(t)$, given on L, satisfies the $H(\mu)$ condition on every closed part of L not containing ends, and if near any end c it is of the form

$$\varphi(t) = \frac{\varphi^*(t)}{(t - c)^{\alpha}}, \quad 0 \leqq \alpha < 1, \tag{77.3}$$

where $\varphi^*(t)$ belongs to the class H, then $\varphi(t)$ will be said to belong to the class H^* on L. If (77.3) holds only in the neighbourhood of a given end c, then $\varphi(t)$ will be said to belong to the class H^* near c.

5°. Finally, if $\varphi(t)$ belongs to H^* near c for every arbitrarily small α in (77.3), i.e., if $(t-c)^{\varepsilon}\varphi(t)$ belongs to the class H near c for arbitrarily small $\varepsilon > 0$, then $\varphi(t)$ will be said to belong to the class H^*_{ε} near c; it will be said to belong to this class on L, if the above condition is satisfied for all the ends and if $\varphi(t)$ satisfies the H condition everywhere, except possibly near the ends.

For example (§ 7, 2°) the function $(t-c)^{i\beta}$, where β is any real number, belongs to the class H_ε^* near c; this is an example of a bounded function of the class H_ε^*; the function $\log(z-c)$ is an example of an unbounded function of the same class.

6°. In the sequel, one will often come across functions $\Phi(z)$ of the point z in S, having the property that $|z-c|^\varepsilon \Phi(z) \to 0$ for $z \to c$ whatever the positive constant ε may be. Such functions will be called almost bounded near c. For example, $\log(z-c)$ is a function almost bounded near c.

7°. In the sequel, when stating that some boundary condition is satisfied on L, this will be understood to exclude the ends of L.

§ 78. Definition of a sectionally holomorphic function for a a given discontinuity.

Let $\varphi(t)$ be a function of the class H^* given on L. By the results of § 29 it is clear that *the function*

$$\Phi(z) = \frac{1}{2\pi i}\int_L \frac{\varphi(t)\,dt}{t-z} \tag{78.1}$$

is a sectionally holomorphic function, vanishing at infinity (*i.e.*, of degree not greater than -1).

The following problem will be considered (compare § 26): *To find the sectionally holomorphic function $\Phi(z)$ for the given discontinuity*

$$\Phi^+(t_0) - \Phi^-(t_0) = \varphi(t_0) \text{ on } L \tag{78.2}$$

(cf. § 77, 7°), *assuming that $\varphi(t)$ belongs to the class H^* and that at infinity $\Phi(z)$ is of degree not greater than $k \geqq -1$.*

The function

$$\Phi(z) = \frac{1}{2\pi i}\int_L \frac{\varphi(t)\,dt}{t-z} + P_k(z) \tag{78.3}$$

is obviously the solution, where $P_k(z)$ is an arbitrary polynomial of degree not greater than k, while $P_{-1}(z) \equiv 0$. The problem is easily seen to have no other solutions.

In fact, if $\Phi_1(z)$ and $\Phi_2(z)$ are two possible solutions, then the function $P(z) = \Phi_2(z) - \Phi_1(z)$, if it is given the value $P^+(t) = P^-(t)$ on L, will be holomorphic everywhere in the finite part of the plane, except possibly at the ends a_j and b_j; but these points can only be isolated singularities, because near them the degree of infinity of the function $P(z)$ is necessarily less than unity, and hence $P(z)$ is bounded near these points. Thus $P(z)$ can be assumed holomorphic in the entire plane and, since at infinity $P(z) = O(z^k)$, $P(z)$ is a polynomial of degree not greater than k.

Note, in particular, that *for* $k = -1$ (corresponding to the requirement $\Phi(\infty) = 0$) *the problem has the unique solution*

$$\Phi(z) = \frac{1}{2\pi i} \int_L \frac{\varphi(t)\,dt}{t - z}. \tag{78.4}$$

79. The homogeneous Hilbert problem for open contours.

This problem may be formulated similarly as in § 34.

To find the sectionally holomorphic function, having finite degree at infinity, for the boundary condition

$$\Phi^+(t) = G(t)\Phi^-(t) \text{ on } L, \tag{79.1}$$

where $G(t)$ *is a given function of the class* H *which does not vanish on* L.

Let $\log G(t)$ be any value of this multi-valued function which varies continuously over each of the arcs L_j, and put

$$\Gamma(z) = \frac{1}{2\pi i} \int_L \frac{\log G(t)\,dt}{t - z.}. \tag{79.2}$$

Using the Plemelj formula, the function $e^{\Gamma(z)}$ is seen directly to satisfy the boundary condition (79.1). However, one may not have satisfied all the conditions of the problem; in fact, the condition (77.1) may be found to be violated.

Now the ends a_j, b_j of the arcs L_j ($j = 1, 2, \ldots, p$) will be denoted in any order by c_k ($k = 1, 2, \ldots, 2p$). By (29.4) one will have near c_k

$$\Gamma(z) = \mp \frac{\log G(c_k)}{2\pi i} \log (z - c_k) + \Gamma_0(z),$$

where $\Gamma_0(z)$ remains bounded near c_k and takes a certain value there; the upper sign has to be taken for $c_k = a_j$, the lower for $c_k = b_j$. Therefore

$$e^{\Gamma(z)} = (z - c_k)^{\alpha_k + i\beta_k}\Omega(z)$$

near c_k, where α_k, β_k are real constants given by

$$\alpha_k + i\beta_k = \mp \frac{\log G(c_k)}{2\pi i} \tag{79.3}$$

(upper sign for $c_k = a_j$, lower sign for $c_k = b_j$), and $\Omega(z)$ is a non-vanishing bounded function which assumes a definite value at the point c.

Now select integers λ_k, satisfying the conditions

$$-1 < \alpha_k + \lambda_k < 1, \tag{79.4}$$

and put

$$\Pi(z) = \prod_{k=1}^{2p} (z - c_k)^{\lambda_k}. \tag{79.5}$$

Then the function

$$X(z) = \Pi(z)e^{\Gamma(z)} \tag{79.6}$$

obviously satisfies all the conditions of the problem and will thus be one of its particular solutions, because the conditions $-1 < \alpha_k + \lambda_k$ then ensure that $X(z)$ becomes infinite at the ends with degree less than 1. The reason for the condition $\alpha_k + \lambda_k < 1$ will become clear below.

The solution $X(z)$, generally speaking, is not completely defined by the conditions (79.4); in fact, it is only completely defined in the case where α_k is an integer; correspondingly the integer λ_k is uniquely determined: $\lambda_k = -\alpha_k$.

The ends c_k for which α_k is an integer, i.e., for which $G(c_k)$ is a real positive quantity, will be called special ends.

For non-special ends the numbers λ_k are determined apart from the terms ± 1, and, in fact, λ_k can be chosen such that $\alpha_k + \lambda_k > 0$ or $\alpha_k + \lambda_k < 0$.

The choice of λ_k will be completely determined, if one more condition is introduced. This condition will now be stated.

Corresponding to the above, those solutions of the problem (79.1) will be admitted which become infinite at the ends (with degree less than 1). Sometimes it is expedient to require that the unknown solution be bounded at non-special ends $c_1, c_2, \ldots, c_q$, given beforehand.

The solutions $\Phi(z)$, satisfying this condition, will be called solutions of the class $h(c_1, c_2, \ldots, c_q)$; the class, corresponding to $q = 0$, will be denoted by $h(0)$ or by h_0. If m is the number of non-special ends, then the class $h(c_1, c_2, \ldots, c_m)$ will sometimes be denoted by h_m. The class h_0 contains all other classes, while the class h_m is a subclass of all other classes. For the definition of the classes of solutions attention is not turned to the special ends, because, as will be seen later, each solution of the homogeneous Hilbert problem (79.1) is necessarily bounded near all special ends.

Now it will be agreed to produce a completely determined solution of the form (79.6) for each possible class of solutions, and, in fact, to select the numbers λ_k in such a way that at the non-special ends, where the solutions of a given class are bounded, $\alpha_k + \lambda_k > 0$, while at the remaining non-special ends $\alpha_k + \lambda_k < 0$ (at special ends, as has already been mentioned, $\alpha_k + \lambda_k = 0$).

The particular solution $X(z)$ thus determined and likewise each solution differing from $X(z)$ by a constant non-zero multiplier will be called the fundamental solutions of a given class.

Next, the boundary values $X^+(t)$ and $X^-(t)$ will be found. Applying to (79.6) the Plemelj formula, one obtains

$$X^+(t_0) = e^{\frac{1}{2}\log G(t_0)}\,\Pi(t_0)e^{\Gamma(t_0)}, \quad X^-(t_0) = e^{-\frac{1}{2}\log G(t_0)}\,\Pi(t_0)e^{\Gamma(t_0)}$$

or

$$X^+(t_0) = \sqrt{G(t_0)}\,X(t_0), \quad X^-(t_0) = \frac{X(t_0)}{\sqrt{G(t_0)}}, \tag{79.7}$$

where

$$X(t_0) = \Pi(t_0)e^{\Gamma(t_0)}; \tag{79.8}$$

the branch of the root in (79.7) is fixed by

$$\sqrt{\;_0)G(t} = e^{\frac{1}{2}\log G(t_0)}.$$

It is easily seen from (29.7) that

$$X(t_0) = \omega(t_0) \prod_{k=1}^{2p} (t_0 - c_k)^{\gamma_k}, \tag{79.9}$$

where $\omega(t_0)$ is a function of the class H, not vanishing on L, and

$$\gamma_k = \alpha_k + i\beta_k + \lambda_k. \tag{79.10}$$

Thus the function $X(t)$ belongs to the class H^* on L and to the class H near the ends $c_1, c_2, \ldots, c_q$ at which it becomes zero; near special ends it belongs to the class H^*_ε, remaining bounded there. The same obviously applies to $X^+(t)$, $X^-(t)$.

From the above it follows that $X(z)$ *is nowhere zero in the finite part of the plane including* L *with the exception of the ends* $c_1, c_2, \ldots, c_q$; when stating that $X(z)$ is not zero on L (with the exception of the ends, referred to) this will be understood to mean that $X^+(t)$ and $X^-(t)$ are not zero on L.

The notation

$$\varkappa = -\sum_{j=1}^{2p} \lambda_j \tag{79.11}$$

will be introduced and $\varkappa$ will be called the index of the given class of solutions $h(c_1, c_2, \ldots, c_q)$. From (79.6) it follows that $X(z)$ has the degree $(-\varkappa)$ at infinity, i.e., a zero of order $\varkappa$, if $\varkappa > 0$, a pole of order $(-\varkappa)$, if $\varkappa < 0$. For $\varkappa = 0$, $X(\infty) = 1$. In all cases

$$\lim_{z\to\infty} z^\varkappa X(z) = 1. \tag{79.12}$$

Obviously

$$\Phi(z) = X(z)P(z), \tag{79.13}$$

where $P(z)$ is an arbitrary polynomial, is also a solution of the given class.

Now it will be shown that, conversely, *each solution* $\Phi(z)$ *of a given class is determined by* (79.13) *for a suitable choice of the polynomial* $P(z)$.

In fact, it follows from $\Phi^+(t) = G(t)\Phi^-(t)$, $X^+(t) = G(t)X^-(t)$ that

$$\frac{\Phi^+(t)}{X^+(t)} = \frac{\Phi^-(t)}{X^-(t)} \quad \text{on } L.$$

This equation indicates that the function $\Phi(z)/X(z)$ is holomorphic in the entire plane, with the possible exception of end points and the point at infinity. Near the end points it may become infinite with degree less than unity. This follows from the way in which the numbers λ_k were chosen above (in fact, they were chosen so for this reason); hence the ends are isolated singular points. Further, since $\Phi(z)$ is, by assumption, of finite degree at infinity $\Phi(z)/X(z)$ is a polynomial, and the proposition is proved.

A number of direct consequences of (79.13) will now be indicated. First of all it is clear that *all solutions* $\Phi(z)$ *of the homogeneous problem* (79.1) *remain bounded near all special ends, since* $X(z)$ *has this property.*

Furthermore, *if a solution remains bounded near a given non-special end* c, *then it necessarily becomes zero at this end.* In fact, if c is one of the (non-special) ends determining the class of $X(z)$, then, due to its general construction, $X(z)$ becomes zero for $z = c$. If, however, c is one of the other non-special ends, then it is necessary for the boundedness of $\Phi(z)$ near c that the polynomial $P(z)$ should have a factor $(z - c)$ in which case $\Phi(z) = 0$ for $z = c$.

From (79.13) it also follows that the degree of $\Phi(z)$ at infinity is $-\varkappa + k$, where $\varkappa$ is the index of the class under consideration and k is the degree of the polynomial $P(z)$. Hence the solution of a given class having the lowest possible degree at infinity is obtained by putting $P(z) = C = \text{const} \neq 0$. Thus *the lowest possible degree at infinity of the solution of a given class is* $(-\varkappa)$; all solutions of a class having this degree are contained in

$$\Phi(z) = CX(z),$$

i.e., they are the fundamental solutions of the class.

Hence *a fundamental solution of a given class can be defined as the solution of the lowest possible degree at infinity* (*in that class*).

In addition, it follows directly from the above that *the funda-*

mental solution of a given class $h(c_1, c_2, \ldots, c_q)$ can be defined as that solution which is nowhere zero except at the ends $c_1, c_2, \ldots, c_q$, where it is zero with degree less than unity, and possibly at $z = \infty$.

As before, let m be the number of all nonspecial ends $c_1, c_2, \ldots, c_m$. The most general class of solutions, i.e., the class of solutions on which no additional conditions are imposed at the ends, was denoted above by h_0; the corresponding fundamental solution will be denoted by $X_0(z)$. The function $X_0(z)$ will be obtained by taking $\alpha_k + \lambda_k < 0$ for all non-special ends. This function is then zero nowhere in the finite part of the plane (including ends).

The index $\varkappa_0$ of the class h_0 is then larger than the indices of all other classes.

Likewise it is obvious that the fundamental solution $X(z)$ and the index $\varkappa$ of every class $h(c_1, c_2, \ldots, c_q)$ is connected with $X_0(z)$ and $\varkappa_0$ by the relations

$$X(z) = (z - c_1)(z - c_2) \ldots (z - c_q)X_0(z), \qquad \varkappa = \varkappa_0 - q. \tag{79.14}$$

The class $h(c_1, c_2, \ldots, c_m)$, being a subclass of all other classes, will be denoted as above by h_m; the corresponding fundamental solution and index will be denoted by $X_m(z)$ and $\varkappa_m$. From this it is clear that

$$X_m(z) = (z - c_1) \ldots (z - c_m)X_0(z), \qquad \varkappa_m = \varkappa_0 - m. \tag{79.15}$$

NOTE. The number of all classes is easily determined; there is one class h_0, m classes $h(c_j)$ $(j = 1, 2, \ldots, m)$, $\dfrac{m(m-1)}{2!}$ classes $h(c_j, c_k)$, etc., and altogether there are

$$1 + \frac{m}{1} + \frac{m(m-1)}{2!} + \ldots + \frac{m}{1} + 1 = 2^m$$

classes.

§ 80. The associate homogeneous Hilbert problem. Associate classes.

Together with the problem (79.1) the problem

$$\Psi^+(t) = [G(t)]^{-1}\Psi^-(t) \text{ on } L \tag{80.1}$$

will be considered which will be called the problem associate to (79.1). Between the solutions of associate problems there exists a close connection which has a particularly simple form, if the concept of associate classes of solutions is introduced. This concept is the basis of the theory of singular integral equations given in Part V. It was introduced in the paper [1] by N. I. Muskhelishvili and D. A. Kveselava.

It is easily seen from the definition of special and non-special ends that the ends which are special (or non-special) for a given problem will likewise be special (or non-special) for the associate problem.

As before, let $c_1, c_2, \ldots, c_m$ be the non-special ends (for the two associate problems). The classes

$$h = h(c_1, c_2, \ldots, c_q) \text{ and } h' = h(c_{q+1}, \ldots, c_m)$$

will be called associate classes, where $c_1, c_2, \ldots, c_q$ are any non-special ends, while $c_{q+1}, \ldots, c_m$ are all the remaining non-special ends.

It is easily seen from the definition of indices and fundamental solutions that the fundamental solutions $X(z)$ and $X'(z)$ of the associate problems in associate classes are connected by the relation

$$X'(z) = C[X(z)]^{-1}, \tag{80.2}$$

while the indices $\varkappa$ and $\varkappa'$ of the classes h and h' respectively satisfy the relation

$$\varkappa' = -\varkappa; \tag{80.3}$$

C is an arbitrary non-zero constant which can always be assumed to equal unity.

§ 81. Solution of the non-homogeneous Hilbert problem for arcs.

Now the following problem will be considered:

To find the sectionally holomorphic function $\Phi(z)$, having finite degree at infinity, for the boundary condition

$$\Phi^+(t) = G(t)\Phi^-(t) + g(t) \text{ on } L, \tag{81.1}$$

where $G(t)$, $g(t)$ are functions of the class H, given on L, and $G(t) \neq 0$ everywhere on L.

Again ends will be called special, if they are special ends of the homogeneous problem (79.1). Similarly as in § 79, all possible solutions of the problem (81.1) will be divided into classes $h(c_1, c_2, \ldots, c_q)$ by means of the non-special ends $c_1, c_2, \ldots, c_q$ near which the solutions of a given class must be bounded, without imposing conditions on the behaviour of the solutions near the other non-special and special ends (excepting, of course, the conditions imposed by definition on every sectionally holomorphic function).

A solution of the non-homogeneous problem (81.1), belonging to a given class $h(c_1, c_2, \ldots, c_q)$, will now be sought. Let $X(z)$ be the fundamental solution of this class of the homogeneous problem (79.1). This function will be called the fundamental function of the given class of the problem (81.1). Noting that

$$G(t) = \frac{X^+(t)}{X^-(t)},$$

the condition (81.1) will be written

$$\frac{\Phi^+(t)}{X^+(t)} - \frac{\Phi^-(t)}{X^-(t)} = \frac{g(t)}{X^+(t)}. \tag{81.2}$$

Since, by assumption, the function $\Phi(z)$ remains bounded near those ends $c_1, c_2, \ldots, c_q$ at which $X(z)$ becomes zero, one has near these ends

$$\left| \frac{\Phi(z)}{X(z)} \right| < \frac{\text{const.}}{|z - c_j|^\alpha}, \quad \alpha < 1;$$

the function $\Phi(z)/X(z)$ is thus sectionally holomorphic and has, in addition, finite degree at infinity. Therefore one concludes from § 78 that

$$\Phi(z) = \frac{X(z)}{2\pi i} \int_L \frac{g(t)dt}{X^+(t)(t-z)} + X(z)P(z), \tag{81.3}$$

where $P(z)$ is an arbitrary polynomial. This is the general solution of the given class of the non-homogeneous problem (81.1). The second term of (81.3) is the general solution of the given class of the corresponding homogeneous problem, while the first term is a particular solution of the original non-homogeneous problem.

It follows from the results of § 29 that the solution (81.3) is actually bounded near the ends $c_1, c_2, \ldots, c_q$.

Near the special ends, as is again easily seen from § 29, the solution remains also bounded, with the possible exclusion of those ends where the numbers β_k of (79.3) are zero, i.e., in other words, at those ends c_k where $G(c_k) = 1$. Near the latter the solutions may be unbounded, but they will certainly be almost bounded (of the logarithmic type).

Note also that in the present case, in contrast to that of the homogeneous problem, the solutions remaining bounded near any non-special ends are not, generally speaking, zero at these ends.

From the point of view of application, special interest is attached to the determination of those solutions of the non-homogeneous problem (81.1) which vanish at infinity.

Noting that the degree of the function $X(z)$ at infinity is exactly

$(-\varkappa)$, i.e., the index of the class $h(c_1, c_2, \ldots, c_q)$ with opposite sign, the following conclusions are easily drawn from (81.3):

1°. For $\varkappa \geqq 0$, the solutions of a given class, vanishing at infinity, are given by

$$\Phi(z) = \frac{X(z)}{2\pi i} \int_L \frac{g(t)dt}{X^+(t)(t-z)} + X(z)P_{\varkappa-1}(z), \tag{81.4}$$

where $P_{\varkappa-1}(z)$ is an arbitrary polynomial of degree not greater than $\varkappa - 1$ ($P_{\varkappa-1}(z) = 0$ for $\varkappa = 0$).

2°. For $\varkappa < 0$, the solutions of a given class vanishing at infinity exist, if and only if the conditions

$$\int_L \frac{t^j g(t)\,dt}{X^+(t)} = 0, \quad j = 0, 1, \ldots, -\varkappa - 1, \tag{81.5}$$

expressing that $\Phi(\infty) = 0$, are satisfied. Provided these conditions are fulfilled, the (unique) solution is given again by (81.4) with $P_{\varkappa-1}(z) = 0$.

The solutions of the homogeneous and non-homogeneous Hilbert *) problems given here were obtained by a method which is a generalization of one developed by T. Carleman [1] for one particular case. This author regarded the line L as a single segment of the real axis and restricted himself to the case, corresponding (in the above notation) to the value $\varkappa_0 = 1$; Carleman, of course, introduced neither the notion of index nor that of classes of solutions.

NOTE. The general solution of a given class $h(c_1, c_2, \ldots, c_q)$ can be represented by an expression somewhat different from (81.3). In fact, let $X_1(z)$ be the fundamental function of any class $h(c_1, c_2, \ldots, c_q, c_{q+1}, \ldots, c_r)$ which is a subclass of $h(c_1, c_2, \ldots, c_q)$. Then the function

$$\Phi_1(z) = \frac{X_1(z)}{2\pi i} \int_L \frac{g(t)\,dt}{X_1^+(t)(t-z)},$$

being a particular solution of the class $h(c_1, c_2, \ldots, c_q, c_{q+1}, \ldots, c_r)$, will be at the same time a particular solution of the class $h(c_1, c_2, \ldots, c_q)$. On the other hand, it is clear that the general solution of the class $h(c_1, c_2, \ldots, c_q)$ of the non-homogeneous problem is of the form

$$\Phi(z) = \Phi_1(z) + \Psi(z),$$

*) These solutions were given in the Author's paper [2] and extended by the introduction of classes of functions in the paper [1] by N. I. Muskhelishvili and D. A. Kveselava. A somewhat different method of solution, independent of the one here, was given by F. D. Gakhov, see end of §§ 85, 87.

where $\Phi_1(z)$ is any particular solution of this class of the non-homogeneous problem and $\Psi(z)$ is the general solution of this class of the homogeneous problem. Consequently, the general solution of the class $h(c_1, c_2, \ldots, c_q)$ of the non-homogeneous problem can be written

$$\Phi(z) = \frac{X_1(z)}{2\pi i} \int_L \frac{g(t)dt}{X_1^+(t)(t-z)} + X(z)P(z), \tag{81.3a}$$

where $P(z)$ is an arbitrary polynomial, $X(z)$ is the fundamental solution of the class $h(c_1, \ldots, c_q)$ of the homogeneous problem and $X_1(z)$ is the fundamental solution of the same homogeneous problem of any class $h(c_1, \ldots, c_q, c_{q+1}, \ldots, c_r)$, i.e., of a class which is a subclass of $h(c_1, \ldots, c_q)$. In particular, one can always take $X_1(z)$ in place of the fundamental solution $X_m(z)$ of the class $h(c_1, c_2, \ldots, c_m)$.

§ 82. The concept of the class *h* of functions given on L.

In the preceding sections the concept of the class $h(c_1, c_2, \ldots, c_q)$ for the solutions $\Phi(z)$ of the Hilbert problem was introduced. For the sequel it is very important to establish an analogous idea for the functions $\varphi(t)$ given on L.

As before, let $c_1, c_2, \ldots, c_m$ $(m \leqq 2p)$ be all non-special ends, corresponding to a given Hilbert problem, i.e., (79.1) or (81.1); these ends are completely defined by the function $G(t)$ on L.

Let $\varphi(t)$ be a function of the class H^*, given on L. The function $\varphi(t)$ will be said to belong to the class $h(c_1, c_2, \ldots, c_q)$, $q \leqq m$, if it belongs to the class H near the ends $c_1, c_2, \ldots, c_q$, while no supplementary conditions are imposed near the other (special or non-special) ends.

The classes $h(c_1, c_2, \ldots, c_q)$ and $h(c_{q+1}, \ldots, c_m)$ will be called associate classes. The class corresponding to $q = 0$ will be denoted by h_0, and the class $h(c_1, \ldots, c_m)$ by h_m.

Thus the same designation will be applied to functions of a different form, i.e., to the sectionally holomorphic function $\Phi(z)$ which is a solution of the Hilbert problem and to the function $\varphi(t)$ of the class H^* on L. However, no confusion can be caused by this.

§ 83. Some generalizations.

1°. It has been seen that the general solution of the homogeneous Hilbert problem, including solutions of all classes, can be written in the form

$$\Phi(z) = X_0(z)P(z), \tag{83.1}$$

where $X_0(z)$ is the fundamental solution of the class h_0 and $P(z)$ is a polynomial. It follows directly from this that each solution of the homogeneous problem, almost bounded near any end c (§ 77, 6°), is necessarily bounded at c; if this end is non-special, then the solution, almost bounded at c, is necessarily zero at c.

Furthermore, by a remark at the end of § 81, each solution of the non-homogeneous Hilbert problem can be presented in the form

$$\Phi(z) = \frac{X_m(z)}{2\pi i} \int_L \frac{g(t)dt}{X_m^+(t)(t-z)} + X_0(z)P(z), \qquad (83.2)$$

where, as always, $X_m(z)$ is the fundamental function of the class $h(c_1, \ldots, c_m)$ and $X_0(z)$ is the fundamental function of the class h_0.

The first term is bounded at all non-special ends, as follows from § 29. From this the conclusion is easily drawn that every solution of the non-homogeneous problem, almost bounded at any non-special end c, is necessarily bounded there; because, if c is such an end, then necessarily $P(c) = 0$ and hence $P(z)$ has the factor $(z - c)$.

Thus, *for the definition of the class $h(c_1, \ldots, c_q)$ of the solutions of the Hilbert problem, one can, without altering the results, replace the condition of boundedness of the solution at the ends $c_1, c_2, \ldots, c_q$ by the condition of almost boundedness.* This remark will be useful, since it permits the establishment of uniformity in a number of statements.

2°. So far it has been assumed that the term $g(t)$ in the boundary condition (81.1) of the non-homogeneous problem belongs to the class H. The function $g(t)$ will now be assumed to belong to the class $h(c_1, c_2, \ldots, c_q)$, where $c_1, c_2, \ldots, c_q$ are given non-special ends (cf. last section). It is easily seen that the solutions $\Phi(z)$ of the problem (81.1), belonging to the same class $h(c_1, c_2, \ldots, c_q)$, are given by the same formulae as in the case, when $g(t)$ belongs to the class H, i.e., by (81.3), for the general solution, and by (81.4), for solutions vanishing at infinity; it is also easily seen that the necessary and sufficient conditions for the existence of solutions of the class $h(c_1, \ldots, c_q)$, vanishing at infinity, are given by the formulae (81.5).

It will be noted that the solutions may now be unbounded to any degree less than 1 near the special ends, depending on the behaviour of $g(t)$ there; but if, in particular, $g(t)$ belongs to the class H_ε^* near a given special end, then the solutions will be almost bounded at that end.

§ 84. Examination of the problem $\Phi^+ + \Phi^- = g$.

In the sequel, one particular case of the Hilbert problem will often have to be considered, consisting of the determination of a sectionally holomorphic function $\Phi(z)$ which satisfies the boundary condition

$$\Phi^+(t) + \Phi^-(t) = g(t) \text{ on } L, \tag{84.1}$$

where $g(t)$ is a function of the class H, given on L. This condition is the particular case of (81.1), when $G(t) = -1$. Therefore an expression for the fundamental solution can be written down immediately, i.e., for the fundamental solution of a given class of the corresponding homogeneous problem

$$\Phi^+(t) + \Phi^-(t) = 0, \tag{84.2}$$

as well as for the general solution of the non-homogeneous problem (84.1).

The fundamental solution of the class $h(c_1, c_2, \ldots, c_q)$ of the homogeneous problem (84.2) will obviously be

$$X(z) = C\frac{\sqrt{R_1(z)}}{\sqrt{R_2(z)}} = C\sqrt{\frac{R_1(z)}{R_2(z)}}, \tag{84.3}$$

where C is an arbitrarily fixed non-zero constant and

$$R_1(z) = \prod_{k=1}^{q} (z - c_k), \quad R_2(z) = \prod_{k=q+1}^{2p} (z - c_k). \tag{84.4}$$

The quantity

$$\sqrt{\frac{R_1(z)}{R_2(z)}} \tag{*}$$

is understood to refer to that branch which is holomorphic in S, i.e., in the plane cut along L.

The expression (84.3) is inferred immediately from the general results of § 79; however, it is still simpler to verify directly that it satisfies all the conditions to be fulfilled by the fundamental solution.

Since one has often to deal with roots of the type (*), the following convention will be arranged and adhered to throughout the later work. This root will always indicate that branch of the function, holomorphic in S, the expansion of which in decreasing powers of z has near the point at infinity the form

$$\sqrt{\frac{R_1(z)}{R_2(z)}} = z^{q-p} + A_1 z^{q-p-1} + A_2 z^{q-p-2} + \ldots. \tag{84.5}$$

The roots $\sqrt{R_1(z)}$ and $\sqrt{R_2(z)}$ will only be encountered in the ratios

$$\frac{\sqrt{R_1(z)}}{\sqrt{R_2(z)}}, \quad \frac{\sqrt{R_2(z)}}{\sqrt{R_1(z)}}. \tag{**}$$

The first of the ratios (**) will always be understood to be the same as (*), and the second the inverse of (*). Finally, the boundary value taken by the root (*) on L from the left will be simply denoted by

$$\sqrt{\frac{R_1(t)}{R_2(t)}} \quad \text{or} \quad \frac{\sqrt{R_1(t)}}{\sqrt{R_2(t)}},$$

so that, by definition,

$$\left[\sqrt{\frac{R_1(t)}{R_2(t)}}\right]^+ = \left[\frac{\sqrt{R_1(t)}}{\sqrt{R_2(t)}}\right]^+ = \sqrt{\frac{R_1(t)}{R_2(t)}} = \frac{\sqrt{R_1(t)}}{\sqrt{R_2(t)}}; \tag{84.6}$$

further, it is obvious that

$$\left[\sqrt{\frac{R_1(t)}{R_2(t)}}\right]^- = -\left[\sqrt{\frac{R_1(t)}{R_2(t)}}\right]^+ = -\sqrt{\frac{R_1(t)}{R_2(t)}}. \tag{84.7}$$

It follows from the expression (84.3) for the fundamental solution of the class $h(c_1, c_2, \ldots, c_q)$ that the index of this class

$$\varkappa = p - q, \tag{84.8}$$

because the degree of $X(z)$ at infinity is $q - p$. Likewise it is clear that all the ends are non-special.

The fundamental function $X_0(z)$ of the largest class h_0 will obviously be ($q = 0$)

$$X_0(z) = \frac{C}{\sqrt{R(z)}}, \tag{84.9}$$

where C is a constant,

$$R(z) = \prod_{k=1}^{2p} (z - c_k) = \prod_{j=1}^{p} (z - a_j)(z - b_j); \tag{84.10}$$

the index of the class h_0 is

$$\varkappa_0 = p. \tag{84.11}$$

The fundamental solution $X_{2p}(z)$ of the smallest class $h_{2p} = h(c_1, \ldots, c_{2p})$ is

$$X_{2p} = C\sqrt{R(z)} \tag{84.12}$$

with the corresponding index

$$\varkappa_{2p} = -p. \tag{84.13}$$

If $q = p$, then the index of the corresponding class will be zero. For example, a fundamental solution with index zero is the solution of the class $h(a_1, a_2, \ldots, a_p)$

$$X_a(z) = C\frac{\sqrt{R_a(z)}}{\sqrt{R_b(z)}}, \tag{84.14}$$

where C is a constant and

$$R_a(z) = \prod_{k=1}^{p} (z - a_k), \quad R_b(z) = \prod_{k=1}^{p} (z - b_k). \qquad (84.15)$$

In correspondence with the above, the general solution of the class $h(c_1, \ldots, c_q)$ of the non-homogeneous problem (84.1), having finite degree at infinity, is given by

$$\Phi(z) = \frac{\sqrt{R_1(z)}}{2\pi i \sqrt{R_2(z)}} \int_L \frac{\sqrt{R_2(t)} g(t) dt}{\sqrt{R_1(t)}(t - z)} + \frac{\sqrt{R_1(z)}}{\sqrt{R_2(z)}} Q(z), \qquad (84.16)$$

where $Q(z)$ is an arbitrary polynomial.

The following results are obtained for the solutions of the class $h(c_1, c_2, \ldots, c_q)$, vanishing at infinity:

For $\varkappa = p - q \geqq 0$, *the general solution of the class* $h(c_1, \ldots, c_q)$, *vanishing at infinity, always exists and is given by*

$$\Phi(z) = \frac{\sqrt{R_1(z)}}{2\pi i \sqrt{R_2(z)}} \int_L \frac{\sqrt{R_2(t)} g(t)\, dt}{\sqrt{R_1(t)}\, (t - z)} + \frac{\sqrt{R_1(z)}}{\sqrt{R_2(z)}} Q_{p-q-1}(z), \qquad (84.17)$$

where $Q_{p-q-1}(z)$ *is an arbitrary polynomial of degree not greater than* $p - q - 1$ $[Q_{p-q-1}(z) \equiv 0$ for $p = q]$.

For $\varkappa = p - q < 0$, *a solution of the class* $h(c_1, c_2, \ldots, c_q)$ *exists, if and only if* $g(t)$ *satisfies the conditions*

$$\int_L \frac{\sqrt{R_2(t)} t^k g(t) dt}{\sqrt{R_1(t)}} = 0, \quad k = 0, 1, \ldots, q - p - 1; \qquad (84.18)$$

it is then unique and given by (84.17) *for* $Q_{p-q-1}(z) = 0$.

Obviously these results also hold, if $g(t)$ belongs to the class $h(c_1, \ldots, c_q)$ [cf. § 83].

Note. In correspondence with the Note at the end of § 81, the general solution of a given class can also be expressed by formulae, somewhat different from (84.16). For example, the general solution of the class h_0, vanishing at infinity, can obviously be written in the form

$$\Phi(z) = \frac{\sqrt{R_1(z)}}{2\pi i \sqrt{R_2(z)}} \int_L \frac{\sqrt{R_2(t)} g(t) dt}{\sqrt{R_1(t)}(t - z)} + \frac{Q_{p-1}(z)}{\sqrt{R(z)}}, \qquad (84.19)$$

where $R_1(z)$, $R_2(z)$ and $R(z)$ are the same as in (84.4), (84.10), while it is now necessary to assume $q \leqq p$, in order to ensure that $\Phi(\infty) = 0$.

§ 85. The Hilbert problem in the case of discontinuous coefficients.

The case in which L is a smooth contour will now be considered, but the given functions $G(t)$, $g(t)$, entering into the boundary conditions of the Hilbert problem (81.1), will have first order discontinuities at a finite number of points of the contour L. For simplicity, the case where L is a single contour will be considered first, the case in which L consists of several contours being treated quite similarly (§ 86). After the results of the last two sections, the solution of the Hilbert problem in this case offers no difficulties. On the other hand, the problem considered in §§ 83, 84 can, in its turn, be converted into the case which is of interest here, as will be shown in § 87.

Firstly, however, the following convention will be adopted. In fact, that direction of L will be assumed positive which leaves the finite region S^+, bounded by L, on the left; as always, S^- will be the complement of $S^+ + L$.

The points of discontinuity of the functions $G(t)$ and $g(t)$, given on L, will be denoted by $d_1, d_2, \ldots, d_r$ in the order in which they are encountered, when traversing L in the positive direction; the points d_{j+1} and d_{j-1} for $j = r$ and $j = 1$ respectively will be understood to be the points d_1 and d_r. The same points $d_1, \ldots, d_r$, but taken in any order, will be denoted by $c_1, c_2, \ldots, c_r$. The arcs $d_j\, d_{j+1}$ $(j = 1, \ldots, r)$ will be denoted by L_j, and the points d_j or c_j will again be called ends.

The idea of the classes H, H^*, H_ε^* for the functions $\varphi(t)$, given on the arcs L_j, will be retained. A function $\varphi(t)$ will be assumed given on L, if it is given on each of the closed arcs L_j; the values of $\varphi(t)$ at the points d_j are immaterial. The limits of $\varphi(t)$ (if they exist) when $t \to d_j$, moving respectively in the positive or negative direction, will be denoted by $\varphi(d_j - 0)$, $\varphi(d_j + 0)$.

The function $\varphi(t)$, given on L, will be said to belong to the class H_d, if the limits $\varphi(d_j - 0)$, $\varphi(d_j + 0)$ exist for all points d_j and if on each of the open arcs $L_j = d_j d_{j+1}$ this function separately satisfies the H condition, provided it is ascribed the values $\varphi(d_j + 0)$ and $\varphi(d_{j+1} - 0)$ at the ends d_j, d_{j+1}.

If the function $\varphi(t)$, given on L, satisfies on everyone of the arcs L_j the H^* (or H_ε^*) condition, then $\varphi(t)$ will be said to satisfy the H^* (or H_ε^*) condition on L.

The notion of a function $\varphi(t)$, satisfying the H_d, H^* or H_ε^* conditions near a given end d_j, requires no explanation.

A sectionally holomorphic function $\Phi(z)$ will be understood to be a function, holomorphic in each of the regions S^+, S^-, except perhaps at the point $z = \infty$, and continuous everywhere on L from the left and from the right, with the possible exception of the points c_k near which the inequalities

$$|\Phi(z)| < \frac{\text{const.}}{|z - c_k|^\alpha}, \quad \alpha < 1$$

hold.

Finally, the condition adopted in § 77, 7° and the notions of almost bounded functions $\Phi(z)$ (§ 77, 6°) and of classes $h(c_1, c_2, \ldots, c_q)$ (§ 82) will be retained, no new definitions being required.

The following problem may now be identified with the homogeneous Hilbert problem:

To find a sectionally holomorphic function $\Phi(z)$, having finite degree at infinity, under the boundary condition

$$\Phi^+(t) = G(t)\Phi^-(t) \text{ on } L, \tag{85.1}$$

where $G(t)$ is a function of the class H_d, given on L, which does not vanish anywhere including the values $G(d_j \pm 0)$.

After having solved the Hilbert problem for the cases considered in § 35 and § 79, the solution in the present case is clear.

Firstly, the notion of the index must be introduced. This will be defined as in § 35, viz.

$$\varkappa = \frac{1}{2\pi i}[\log G(t)]_L = \frac{1}{2\pi}[\arg G(t)]_L, \tag{85.2}$$

with additional reservations. In that section $[\arg G(t)]_L$ represented the increase in the argument of $G(t)$, when t traversed L in the positive direction, as long as the argument changed continuously. This stipulation uniquely defined $\varkappa$. In the present case the requirement of continuity of the argument is not fulfilled, when t passes through the points c_j. Therefore, retaining the condition of continuity on the individual arcs L_j, one may replace the continuity condition for the passage of t through the point c_j by the condition

$$-1 < \frac{1}{2\pi}\Delta_j \arg G(t) < 1, \tag{85.3}$$

where $\Delta_j \arg G(t)$ is the jump of the argument of $G(t)$ for the passage of t through c_j in the positive direction.

Introduce the following notation:

$$\alpha_j + i\beta_j = -\frac{1}{2\pi i}\log\frac{G(c_j+0)}{G(c_j-0)} =$$

$$= -\frac{1}{2\pi i}[\log G_j(c+0) - \log G(c_j-0)], \tag{85.4}$$

where, as above, $G(c_j - 0)$, $G(c_j + 0)$ are the limits of $G(t)$ for $t \to c_j$ along L in the positive and negative directions respectively.

Then (85.3) can be written as

$$-1 < \alpha_j < 1. \tag{85.5}$$

The conditions (85.3) or (85.5) will always be satisfied. If the ratio $G(c_j + 0)/G(c_j - 0)$ is a real positive number, then α_j is uniquely defined, i.e., $\alpha_j = 0$. The ends c_j for which $\alpha_j = 0$ will be called special, and the others non-special. For the non-special ends c_j one may subject the α_j at will to one of the two conditions

$$0 < \alpha_j < 1 \text{ or } -1 < \alpha_j < 0. \tag{85.6}$$

Commencing from some internal point c_0 of one of the arcs L_j and selecting (arbitrarily) one of the values of $\log G(t)$ at c_0, $\log G(t)$ will be continued on L by varying t in the positive direction along L and adopting a definite choice of one of the inequalities (85.6) for passage through any non-special end. On return to the starting point, a definite value for the increase in argument of $G(t)$ will be obtained which is also to be used in (85.2) above.

Now put

$$G_0(t) = t^{-\varkappa}G(t), \tag{85.7}$$

assuming the origin of coordinates to lie inside S^+.

Then the function

$$\log G_0(t) = \log [t^{-\varkappa}G(t)] = -\varkappa \log t + \log G(t)$$

will return to the original value after one cycle of L, if the argument of $G(t)$ alters as described above, while the argument of $t^{-\varkappa}$ changes continuously. Therefore $\log G_0(t)$ can be regarded as a single-valued function on L, having the first order points of discontinuity $c_1, c_2, \ldots, c_r$ (the point c_0, i.e., the starting point of the cycle, not being a point of discontinuity); the values of $\log G_0(t)$ at the points $c_1, c_2, \ldots, c_r$ may be given at will.

Finally, the sectionally holomorphic function

$$\Gamma(z) = \frac{1}{2\pi i}\int_L \frac{\log G_0(t)\,dt}{t - z} \tag{85.8}$$

will be introduced. Putting

$$X(z) = \begin{cases} e^{\Gamma(z)} & \text{for } z \in S^+, \\ z^{-\varkappa}e^{\Gamma(z)} & \text{for } z \in S^-, \end{cases} \tag{85.9}$$

it is easily seen (compare § 79) that $X(z)$ is a particular solution of the problem (85.1), having the following properties: it becomes zero or infinite like $(z - c_j)^{\alpha_j}$ at those ends, where the α_j satisfy the first or the second of the inequalities (85.6) respectively, and it remains bounded near the special ends. With the exception of the ends mentioned and possibly also the point at infinity, $X(z)$ is nowhere zero. The degree of the solution at infinity is $(-\varkappa)$.

Divide, as in § 79, all possible solutions of the homogeneous problem (85.1) into classes, putting into the class $h(c_1, c_2, \ldots, c_q)$ those solutions which remain bounded at the given non-special ends $c_1, c_2, \ldots, c_q$. Accordingly, a certain particular solution of the form (85.9) will be established in each class and this solution will be chosen, so that for the ends $c_1, c_2, \ldots, c_q$ the first, while for the remaining non-special ends the second of the inequalities (85.6) holds. This particular solution $X(z)$, or any other solution differing from it by a constant non-zero factor, will be called the fundamental solution of the given class and the corresponding index $\varkappa$ the index of that class.

Similarly as in § 79, one can easily establish *that the general solution of a given class of the homogeneous problem* (85.1) *is given by*

$$\Phi(z) = X(z)P(z), \tag{85.10}$$

where $X(z)$ is the fundamental solution of the given class and $P(z)$ is an arbitrary polynomial.

The implications of (85.10) are quite similar to those of (79.13) and therefore they will not be repeated. It will only be noted that, as in § 79, *the fundamental solution of a class $h(c_1, c_2, \ldots, c_q)$ may be defined as a solution which is nowhere zero, with the exception of the points $c_1, c_2, \ldots, c_q$, where it becomes zero with degree less than* 1, *and perhaps the point $z = \infty$.*

In the present instance let again $c_1, c_2, \ldots, c_m$ be the non-special ends $(m \leqq r)$, h_m the class $h(c_1, c_2, \ldots, c_m)$, h_0 the class containing all the solutions and $X_m(z)$, $X_0(z)$, $\varkappa_m$, $\varkappa_0$ the corresponding fundamental solutions and indices.

Next, the non-homogeneous Hilbert problem will be considered:

To find a sectionally holomorphic function $\Phi(z)$, having finite degree at infinity, subject to the boundary condition

$$\Phi^+(t) = G(t)\Phi^-(t) + g(t), \tag{85.11}$$

where $G(t)$ and $g(t)$ are functions of the class H_d, given on L, and $G(t)$ does not become zero in the same way as $G(c_j \pm 0)$.

Here again all possible solutions will be divided into classes, just as in § 81. Then the general solution of a given class of the non-homogeneous problem is easily seen to be given by

$$\Phi(z) = \frac{X(z)}{2\pi i} \int_L \frac{g(t)dt}{X^+(t)(t-z)} + X(z)P(z), \tag{85.12}$$

where $X(z)$ is the fundamental solution of the given class of the homogeneous problem and $P(z)$ is an arbitrary polynomial. The function $X(z)$ will be called the fundamental function of the given class of the problem (85.11).

The solutions of a given class, vanishing at infinity, are given by the following formulae: for $\varkappa \geqq 0$,

$$\Phi(z) = \frac{X(z)}{2\pi i} \int_L \frac{g(t)dt}{X^+(t)(t-z)} + X(z)P_{\varkappa-1}(z), \qquad (85.13)$$

where $P_{\varkappa-1}(z)$ is an arbitrary polynomial of degree not greater than $\varkappa - 1$ [$P_{\varkappa-1}(z) \equiv 0$ for $\varkappa = 0$]; for $\varkappa < 0$, a solution vanishing at infinity exists, if and only if

$$\int_L \frac{t^j g(t)\,dt}{X^+(t)} = 0, \quad j = 0, 1, \ldots, -\varkappa - 1. \qquad (85.14)$$

Provided these conditions are satisfied, the (unique) solution is given by (85.13) for $P_{\varkappa-1}(z) \equiv 0$.

The results of § 83 are easily seen to remain true in the present case.

The method of solution of the Hilbert problem studied here was given by D. A. Kveselava [1]. F. D. Gakhov [3] solved the problem by a somewhat different method, namely by means of reduction to the case of continuous coefficients (similarly to the analogous case treated by Plemelj [2]); he obtained his original solution without division into classes, independently of N. I. Muskhelishvili [2] and N. I. Muskhelishvili and D. A. Kveselava [1]; he extended it later using the second of the references above.

§ 86. The Hilbert problem in the case of discontinuous coefficients (continued).

The case when L consists of several contours $L_0, L_1, \ldots, L_p$, distributed as in § 35, is treated quite similarly.

In this instance, as in § 35, the equation (85.7) must be replaced by

$$G_0(t) = t^{-\varkappa}\Pi(t)G(t), \qquad (86.1)$$

where

$$\Pi(t) = (t-a_1)^{\lambda_1}(t-a_2)^{\lambda_2} \ldots (t-a_p)^{\lambda_p}, \qquad (86.2)$$

$$\varkappa = \lambda_0 + \lambda_1 + \ldots + \lambda_p, \qquad (86.3)$$

$$\lambda_k = \frac{1}{2\pi i}\,[\log G(t)]_{L_k} = \frac{1}{2\pi}\,[\arg G(t)]_{L_k}, \qquad (86.4)$$

and where (86.4) must be interpreted as in the last section [cf. the remarks following (85.2)]. The remaining notation follows from § 35.

The fundamental solution of a given class is defined, apart from a constant non-zero multiplier, by

$$X(z) = \begin{cases} \dfrac{1}{\Pi(z)} e^{\Gamma(z)} & \text{for } z \in S^+, \\ z^{-\varkappa} e^{\Gamma(z)} & \text{for } z \in S^-, \end{cases} \tag{86.5}$$

where

$$\Gamma(z) = \frac{1}{2\pi i} \int_L \frac{\log G_0(t)\, dt}{t - z} \tag{86.6}$$

(as in § 35).

§ 87. Connection with the case of arcs.

Consider again the case in which L consists of the arcs $L_1 = a_1b_1$ $L_2 = a_2b_2, \ldots, L_p = a_pb_p$ (§ 77). The points b_1 and a_2, b_2 and $a_3, \ldots,$ b_p and a_1 will be connected by smooth arcs, so that a smooth simple contour will be obtained which will be denoted by L_0.

The homogeneous Hilbert problem, considered in § 79, viz.

$$\Phi^+(t) = G(t)\Phi^-(t) \text{ on } L, \tag{87.1}$$

may obviously be converted into the case where the boundary is the contour L_0, if one sets $G(t) = 1$ on the additional parts b_ja_{j+1}; for then the function $\Phi(z)$ will be analytically continued through these additional portions and will therefore be a sectionally holomorphic function in the plane cut along L.

Analogously, the non-homogeneous Hilbert problem

$$\Phi^+(t) = G(t)\Phi^-(t) + g(t) \text{ on } L \tag{87.2}$$

is transformed into the same problem for the contour L_0, provided one sets $G(t) = 1$, $g(t) = 0$ on the additional parts.

F. D. Gakhov [3] solved the Hilbert problem in this way for the case of arcs; firstly, he transformed the problem to that for a contour, but with discontinuous coefficients, and afterwards converted it into the problem for a contour involving continuous coefficients (cf. the end of § 85).

CHAPTER 11

INVERSION FORMULAE FOR ARCS *)

Passing on to a number of important and simple applications of the results of the last chapter, a start will be made with the deduction of the inversion formulae of the Cauchy integral taken over a union of arcs.

The results inferred below are a generalization of known formulae giving the solution of the singular integral equation

$$\int_{-a}^{+a} \frac{\varphi(x)\,dx}{x - x_0} = f(x)_0,$$

where $f(x)$ is a given and $\varphi(x)$ is the unknown function. This equation plays an important role in the theory of thin aerofoils for which reason it has been considered by many authors (e.g. K. Schröder [1], H. Söhngen [1], J. Weissinger [1]). The solution given below for the most general case (where the integral is taken over the union of smooth arcs of arbitrary shape instead of a single segment of the real axis; H. Söhngen considered by a very complicated method also the case of two segments of the real axis) is considerably simpler than the solutions known to the Author, which are always distinguished by their extraordinary complexity. In this connection the Author does not refer to these investigations the complicated nature of which is caused by the fact that the unknown function $\varphi(t)$ and the given function $f(x)$ are subject to less restrictive conditions than those imposed here.

§ 88. The inversion of a Cauchy integral.

Let it be required to solve the singular integral equation

$$\frac{1}{\pi i} \int_L \frac{\varphi(t)\,dt}{t - t_0} = f(t_0) \text{ on } L, \tag{88.1}$$

---

*) The contents of this chapter are taken from the Author's paper [2].

where $L = L_1 + L_2 + \ldots + L_p$ is the union of smooth arcs as in § 77, $f(t)$ is a given and $\varphi(t)$ an unknown function of the point t on L. The functions $f(t)$ and $\varphi(t)$ will be assumed to belong to the classes H and H^* respectively (§ 77).

It will be assumed that (88.1) must be satisfied for all t_0 on L, with the possible exception of the ends.

The equation (88.1) is a particular case of the singular integral equation which will be studied in detail in Part V. It will be considered independently here, since it is of considerable interest in itself (cf. the introduction to this chapter) and its theory is simpler than that of the general one.

The sectionally holomorphic function

$$\Phi(z) = \frac{1}{2\pi i}\int_L \frac{\varphi(t)\,dt}{t - z}, \tag{88.2}$$

vanishing at infinity, will be considered. One has

$$\Phi^+(t_0) + \Phi^-(t_0) = \frac{1}{\pi i}\int_L \frac{\varphi(t)\,dt}{t - t_0}, \tag{88.3}$$

and hence the equation (88.1) is equivalent to the problem

$$\Phi^+(t) + \Phi^-(t) = f(t) \text{ on } L \tag{88.4}$$

with the additional condition $\Phi(\infty) = 0$.

Having found $\Phi(z)$, $\varphi(t)$ will be determined from the formula

$$\varphi(t) = \Phi^+(t) - \Phi^-(t). \tag{88.5}$$

The problem (88.4) has already been solved in § 84; its most general solution (i.e., the solution of the class h_0), vanishing at infinity, may be written in the form (cf. Note in § 84)

$$\Phi(z) = \frac{\sqrt{R_1(z)}}{2\pi i\sqrt{R_2(z)}}\int_L \frac{\sqrt{R_2(t)}f(t)\,dt}{\sqrt{R_1(t)}(t - z)} + \frac{Q_{p-1}(z)}{\sqrt{R(z)}}, \tag{88.6}$$

where

$$R_1(z) = \prod_{k=1}^{q}(z - c_k),\quad R_2(z) = \prod_{k=q+1}^{2p}(z - c_k),$$

$$R(z) = \prod_{k=1}^{p}(z - a_k)(z - b_k) = R_1(z)R_2(z), \tag{88.7}$$

and $q \leqq p$.

The general solution $\varphi(t)$ of the original integral equation will be obtained by use of (88.5). Remembering (84.7) and using the Plemelj formula (17.4), one gets

$$\varphi(t_0) = \frac{\sqrt{R_1(t_0)}}{\pi i \sqrt{R_2(t_0)}} \int_L \frac{\sqrt{R_2(t)} f(t) dt}{\sqrt{R_1(t)}(t - t_0)} + \frac{P_{p-1}(t_0)}{\sqrt{R(t_0)}} \tag{88.8}$$

From (88.8) results the following property of the solutions of (88.1): *if any solution (of the class H^*) of the integral equation* (88.1) *remains bounded near any end c_j, then it necessarily becomes zero there and belongs to the class H near this end.*

In fact, it may always be assumed that $(z - c_j)$ is a factor of $R_1(z)$. But then, by what has been said in § 29, the first term on the right side belongs to the class H near c_j and it becomes zero for $t_0 = c_j$ (as follows from (29.8) for $\gamma = \frac{1}{2}$). Further, since by supposition $\varphi(t_0)$ is bounded near c_j, the polynomial $P_{p-1}(t_0)$ is necessarily divisible by $(t_0 - c_j)$, and hence the proposition becomes obvious.

In the same manner as the solutions of the Hilbert problem were divided into classes, the solutions of the equation (88.1) may be separated into classes by putting into the class $h(c_1, c_2, \ldots, c_q)$ all the solutions remaining bounded near the ends $c_1, c_2, \ldots, c_q$; a solution of the class $h(c_1, c_2, \ldots, c_q)$ was seen to belong necessarily to the class H near these ends (and to become zero there); therefore the present definition of the class $h(c_1, c_2, \ldots, c_q)$ agrees with the definition of § 82.

Next, let it be required to find all the solutions of a given class $h(c_1, c_2, \ldots, c_q)$ of the equation (88.1). It is clear from the above that to each solution $\varphi(t)$ of this class there corresponds, by use of (88.2), a solution $\Phi(z)$ of an identical class of the Hilbert problem (88.4); conversely, every solution $\Phi(z)$ of the class $h(c_1, c_2, \ldots, c_q)$ of the Hilbert problem corresponds, by (88.5), to a solution of an identical class of the equation (88.1).

Now, using the results of § 84, the following conclusion is easily reached:

For $\varkappa = p - q \geqq 0$, the solutions of the class $h(c_1, c_2, \ldots, c_q)$ of the equation (88.1) *always exist and are given by*

$$\varphi(t_0) = \frac{\sqrt{R_1(t_0)}}{\pi i \sqrt{R_2(t_0)}} \int_L \frac{\sqrt{R_2(t)} f(t) dt}{\sqrt{R_1(t)}(t - t_0)} + \frac{\sqrt{R_1(t_0)}}{\sqrt{R_2(t_0)}} P_{p-q-1}(t_0), \tag{88.9}$$

where $P_{p-q-1}(t_0)$ is an arbitrary polynomial of degree not greater than $p - q - 1$ (it is identically zero for $p = q$);

for $\varkappa = p - q < 0$, the (unique) solution of the class $h(c_1, c_2, \ldots, c_q)$ exists, if and only if $f(t)$ satisfies the conditions

$$\frac{\sqrt{R_2(t)}\, t^k f(t) dt}{\sqrt{R_1(t)}} = 0, \quad k = 0, 1, \ldots, q - p - 1; \tag{88.10}$$

provided these conditions are satisfied, the solution is given again by (88.9) *for* $P_{p-q-1}(t_0) \equiv 0$.

The result (88.9) can be called the inversion formula of the integral on the left of (88.1).

In particular, the inversion formula corresponding to the class $h(a_1, a_2, \ldots, a_p)$ will be noted; in this case, in the notation of § 84 [formulae (84.15)],

$$\varphi(t_0) = \frac{\sqrt{R_a(t_0)}}{\pi i \sqrt{R_b(t_0)}} \int\limits_L \frac{\sqrt{R_b(t)}f(t)\,dt}{\sqrt{R_a(t)}(t - t_0)}. \tag{88.11}$$

It is easily seen *that all the above results remain true, if the function* $f(t)$ *is assumed to belong to the class* $h(c_1, c_2, \ldots, c_q)$, *instead of to the class* H; *moreover, the solutions belong to the same class* (*cf.* § 83, 2°).

§ 89. Some variations of the inversion problem.

It has been seen that, generally speaking, a solution of the equation (88.1), remaining bounded near all ends, does not exist. However, it is easy to show that *the equation*

$$\frac{1}{\pi i} \int\limits_L \frac{\varphi(t)\,dt}{t - t_0} = f(t_0) + P(t_0) \text{ on } L \tag{89.1}$$

(*where* $f(t_0)$ *is a given function of the class* H *and* $P(t_0)$ *is a polynomial of degree not greater than* $p - 1$, *not given beforehand*) *always has one and only one solution bounded everywhere*; *for this the polynomial* $P(t_0)$ *is also completely determined.*

From what has been said in the last section it will be recalled that a solution bounded everywhere necessarily belongs to the class H.

Firstly, the proof will be given of the fact that, if the solution exists, then it is unique. This leads to the statement that, if

$$\frac{1}{\pi i} \int\limits_L \frac{\varphi(t)dt}{t - t_0} = P(t_0), \tag{89.2}$$

where $\varphi(t)$ is a function of the class H and $P(t)$ a polynomial of degree not greater than $p - 1$, then necessarily $\varphi(t) \equiv 0$, $P(t) \equiv 0$.

This fact will now be proved. If a solution of the class H of the equation (89.2) exists, then it is necessarily given, for example, by the formula (88.11)

$$\varphi(t_0) = \frac{\sqrt{R_a(t_0)}}{\pi i \sqrt{R_b(t_0)}} \int\limits_L \frac{\sqrt{R_b(t)}P(t)dt}{\sqrt{R_a(t)}(t - t_0)}, \tag{89.3}$$

because this formula gives not only all the solutions of the class H, but also all the solutions of the class H^*, bounded at the ends a_j.

The integral

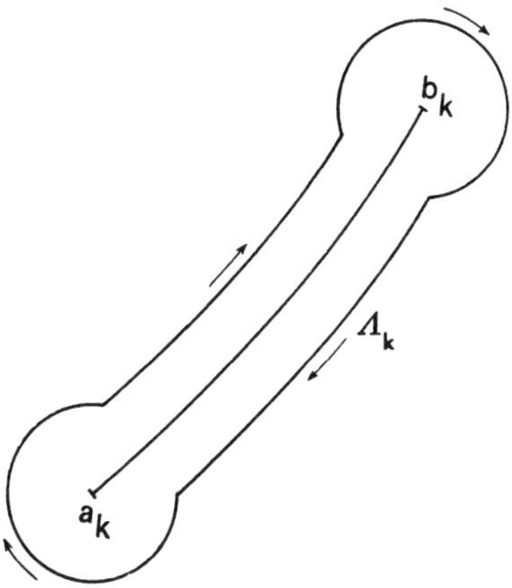

Fig. 10.

$$J(t_0) = \frac{1}{\pi i} \int_L \frac{\sqrt{R_b(t)}P(t)dt}{\sqrt{R_a(t)}(t - t_0)}$$

is easily evaluated in finite form.

In fact, put

$$\Omega(z) = \frac{1}{2\pi i} \int_L \frac{\sqrt{R_b(t)}P(t)dt}{\sqrt{R_a(t)}(t - z)},$$

where z is a point not on L. Then obviously $J(t_0) = \Omega^+(t_0) + \Omega^-(t_0)$. On the other hand, it is clear that

$$\Omega(z) = \frac{1}{4\pi i} \int_\Lambda \frac{\sqrt{R_b(t)}P(t)dt}{\sqrt{R_a(t)}(t - z)},$$

where Λ is the union of p simple contours Λ_k, surrounding the arcs L_k, described in the clockwise direction and situated so close to them that the point z lies outside each of the contours Λ_k (Fig. 10).

By the Cauchy residue theorem one obtains immediately

$$\Omega(z) = \frac{1}{2} \frac{\sqrt{R_b(z)}}{\sqrt{R_a(z)}} P(z) + \tfrac{1}{2}P^*(z),$$

where $P^*(z)$ is some polynomial of degree $p - 1$ which is defined by the condition that for large $|z|$

$$\frac{\sqrt{R_b(z)}}{\sqrt{R_a(z)}} P(z) + P^*(z) = O\left(\frac{1}{z}\right),$$

and hence finally

$$J(t_0) = \Omega^+(t_0) + \Omega^-(t_0) = P^*(t_0)$$

and by (89.3)

$$\varphi(t_0) = \frac{\sqrt{R_a(t_0)}}{\sqrt{R_b(t_0)}} P^*(t_0).$$

Here, use has been made of the following simple formula which is a direct result of Cauchy's residue theorem; if $f(z)$ is holomorphic in a region consisting of points of the plane lying outside the contours Λ_k and continuous up to these contours, and if for large $|z|$ the function $f(z) = P(z) + O(1/z)$, where $P(z)$ is a polynomial, then for z, lying outside the contours Λ_k,

$$\frac{1}{2\pi i}\int_{\Lambda}\frac{f(t)\,dt}{t-z}=f(z)-P(z).$$

But since, by hypothesis, $\varphi(t)$ must also be bounded at the ends b_k, then $P^*_k\ (b_k) = 0$, $k = 1, 2, \ldots, p$ and hence $P^*(t_0) \equiv 0$, and therefore $\varphi(t_0) \equiv 0$.

Now it is easily verified by direct substitution that the (unique) solution of the problem (89.1), belonging to the class H, is given by

$$\varphi(t_0)=\frac{\sqrt{R(t_0)}}{\pi i}\int_{L}\frac{f(t)dt}{\sqrt{R(t)}(t-t_0)}, \tag{89.4}$$

where, as before,

$$R(z)=\prod_{j=1}^{p}(z-a_j)(z-b_j),\quad \sqrt{R(t)}=[\sqrt{R(t)}]^+;$$

together with these a definite expression for the polynomial $P(t)$ will be found.

For this purpose put

$$\Phi(z)=\frac{1}{2\pi i}\int_{L}\frac{\varphi(t)dt}{t-z}, \tag{89.5}$$

where by $\varphi(t)$ will be understood the expression, given by (89.4); then, subject to verification, (89.1) assumes the form

$$\Phi^+(t_0)+\Phi^-(t_0)=f(t_0)+P(t_0), \tag{89.6}$$

where $P(t_0)$ is some polynomial of degree not greater than $p-1$.

In order to evaluate $\Phi(z)$, put

$$\Psi(z)=\frac{\sqrt{R(z)}}{2\pi i}\int_{L}\frac{f(t)\,dt}{\sqrt{R(t)}(t-z)}. \tag{89.7}$$

Obviously, from (89.7) and (89.4),

$$\varphi(t)=\Psi^+(t)-\Psi^-(t),$$

and therefore it is easily seen that

$$\Phi(z)=\frac{1}{2\pi i}\int_{L}\frac{\varphi(t)\,dt}{t-z}=\frac{1}{2\pi i}\int_{\Lambda}\frac{\Psi(t)\,dt}{t-z},$$

where Λ is the same as above. Replacing $\Psi(t)$ by the value obtained from (89.7), one deduces

$$\begin{aligned}\Phi(z) &= \frac{1}{2\pi i}\int_{\Lambda}\frac{\sqrt{R(t)}\,dt}{t-z}\,\frac{1}{2\pi i}\int_{L}\frac{f(t_1)dt_1}{\sqrt{R(t_1)}\,(t_1-t)} = \\ &= \frac{1}{2\pi i}\int_{L}\frac{f(t_1)\,dt_1}{\sqrt{R(t_1)}}\,\frac{1}{2\pi i}\int_{\Lambda}\frac{\sqrt{R(t)}dt}{(t-z)(t_1-t)}.\end{aligned} \tag{89.8}$$

Applying the Cauchy residue theorem to the inner integral and remembering that z lies outside the contours Λ_k, while t_1 is situated inside one of them (see above), it is found that

$$\frac{1}{2\pi i}\int_{\Lambda}\frac{\sqrt{R(t)}dt}{(t-z)(t_1-t)} = \frac{\sqrt{R(z)}}{t_1-z} + Q(z, t_1), \tag{89.9}$$

where $Q(z, t)$ is a polynomial of degree $p-1$, determined by the condition that for large $|z|$

$$\frac{\sqrt{R(z)}}{z-t} = Q(z, t) + O\left(\frac{1}{z}\right).$$

The polynomial $Q(z, t)$ may be defined as follows: let $Q(z)$ be a polynomial of degree p, determined by the condition that for large $|z|$

$$\sqrt{R(z)} = Q(z) + O\left(\frac{1}{z}\right); \tag{89.10}$$

then obviously

$$Q(z, t) = \frac{Q(z)-Q(t)}{z-t.} \tag{89.11}$$

Substituting (89.9) in (89.8), one obtains finally

$$\Phi(z) = \frac{\sqrt{R(z)}}{2\pi i}\int_{L}\frac{f(t_1)\,dt_1}{\sqrt{R(t_1)}(t_1-z)} + \tfrac{1}{2}P(z),$$

where $P(z)$ is the definite polynomial of degree not greater than $p-1$

$$P(z) = \frac{1}{\pi i}\int_{L}\frac{f(t)\,Q(z, t)\,dt}{\sqrt{R(t)}}. \tag{89.12}$$

It follows directly from the above expression for $\Phi(z)$ that the condition (89.6) is fulfilled.

Thus the proposition is proved; in addition, the definite expression (89.12) for the polynomial $P(z)$ has been obtained.

This latter expression makes it possible to establish (differently from the manner by which it was done in § 88 for the more general case) the necessary and sufficient conditions for the existence of the solution of the equation

$$\frac{1}{\pi i}\int_L \frac{\varphi(t)\,dt}{t-t_0} = f(t_0)$$

which remains bounded at all the ends; as before, $f(t)$ will be assumed to belong to the class H.

This condition obviously consists in satisfying

$$\int_L \frac{f(t)\,Q(z,t)\,dt}{\sqrt{R(t)}} \equiv 0, \tag{89.13}$$

where

$$Q(z,t) = Q_0(t)z^{p-1} + Q_1(t)z^{p-2} + \ldots + Q_{p-1}(t),$$

and the $Q_k(t)$ are certain polynomials which are easily determined by use of (89.10) and (89.11). In particular, it is obvious that $Q_0(t) = 1$ and that, in general, the highest term of the polynomial $Q_k(t)$ is t^k. Hence it follows that the system of functions $Q_0(t)$, $Q_1(t)$, ..., $Q_{p-1}(t)$ is equivalent to the system $1, t, t^2, \ldots, t^{p-1}$ in the sense that the functions of each of these systems are essentially linear combinations of the other functions.

The condition (89.13) is obviously equivalent to the p conditions

$$\int_L \frac{Q_k(t)\,f(t)\,dt}{\sqrt{R(t)}} = 0, \quad k = 0, 1, \ldots, p-1,$$

or, on the basis of what has just been stated, to p conditions of the simpler form

$$\int_L \frac{t^k f(t)\,dt}{\sqrt{R(t)}} = 0. \tag{89.14}$$

This result, as has already been pointed out, is a particular case of the results, obtained in § 88 by a different method.

The particular case $p = 1$ will be specially noted. In this case the polynomial $P(t)$ reduces to a constant and one arrives at the following:

Let it be required to find on the arc ab the function $\varphi(t)$ of the class H and the constant C such that

$$\frac{1}{\pi i}\int_L \frac{\varphi(t)\,dt}{t-t_0} = f(t_0) + C, \tag{89.15}$$

where $f(t)$ is a function of the class H given on ab. Then the (unique) solution is given by

$$\varphi(t_0) = \frac{\sqrt{(t_0-a)(t_0-b)}}{\pi i}\int_{ab}\frac{f(t)dt}{\sqrt{(t-a)(t-b)}(t-t_0)}, \quad (89.16)$$

$$C = \frac{1}{\pi i}\int_{ab}\frac{if(t)\,dt}{\sqrt{(t-a)(t-b)}}. \quad (89.17)$$

§ 90. Some variations of the inversion problem (continued).

The problem (89.15) is a particular case of the problem (89.1). The former problem can be generalized in another direction with a view to solving the following problem:

To find the function $\varphi(t)$ of the class H and the constants C_k, subject to the condition

$$\frac{1}{\pi i}\int_L\frac{\varphi(t)\,dt}{t-t_0} = f(t_0) + C_k \text{ (for } t_0 \in L_k,\ k = 1, 2, \ldots, p), \quad (90.1)$$

where $f(t)$ is a given function of the class H.

First of all it will be shown that the problem cannot have more than one solution, i.e., that, if

$$\frac{1}{\pi i}\int_L\frac{\varphi(t)\,dt}{t-t_0} = C_k \text{ (for } t_0 \in L_k,\ k = 1, 2, \ldots, p), \quad (90.2)$$

then necessarily $\varphi(t) \equiv 0$, $C_k = 0$.

In fact, let $\varphi(t)$ satisfy the above conditions. Put

$$\Phi(z) = \frac{1}{2\pi i}\int_L\frac{\varphi(t)\,dt}{t-z}. \quad (90.3)$$

By (90.2), one will have on L_k: $\Phi^+(t_0) + \Phi^-(t_0) = C_k$, i.e.,

$$\Phi^+(t_0) - \frac{C_k}{2} = -\left[\Phi^-(t_0) - \frac{C_k}{2}\right],$$

and hence it obviously follows that the function

$$\Psi(z) = \sqrt{(z-a_k)(z-b_k)}\left[\Phi(z) - \frac{C_k}{2}\right]$$

is holomorphic in some neighbourhood of the arc L_k (including this arc itself) and is zero at the points a_k, b_k, i.e.,

$$\Psi(z) = (z-a_k)(z-b_k)\Psi_0(z),$$

where $\Psi_0(z)$ is holomorphic in the neighbourhood of the arc L_k. Thus near L_k

$$\Phi(z) - \frac{C_k}{2} = \sqrt{(z - a_k)(z - b_k)}\,\Psi_0(z),$$

and consequently

$$\Phi'(z) = \frac{\Omega(z)}{\sqrt{(z - a_k)(z - b_k)}},$$

where $\Omega(z)$ is holomorphic in the neighbourhood of L_k. It follows from the above that

$$\Phi'(z)\sqrt{R(z)} = P(z)$$

is holomorphic in the whole plane. Since, furthermore, $P(z) = O(z^{p-2})$ for large $|z|$, then $P(z)$ is a polynomial of degree not greater than $p - 2$. Thus one obtains for $\Phi(z)$ the expression

$$\Phi(z) = \int_{\infty}^{z} \frac{P(t)\,dt}{\sqrt{R(t)}}, \tag{90.4}$$

where the integration is taken along any path, not intersecting L. In addition, $\Phi(z)$ is known to be holomorphic (and hence single-valued) in the entire plane, cut along L. Hence it is concluded that $\Phi(z) \equiv 0$.

Indeed, using the single-valuedness of $\Phi(z)$ in the cut plane, it is easily inferred from the representation (90.4) that $\Phi(z)$ assumes the completely defined finite values $\Phi(a_k)$, $\Phi(b_k)$ at the ends a_k, b_k; moreover, if Λ_k is a contour drawn in the clockwise direction and infinitely close to the arc L_k, then

$$0 = \int_{\Lambda_k} \frac{P(t)\,dt}{\sqrt{R(t)}} = 2\int_{L} \frac{P(t)\,dt}{\sqrt{R(t)}} = 2[\Phi(b_k) - \Phi(a_k)],$$

so that, putting

$$\Phi(z) = u + iv,$$

necessarily $u(a_k) = u(b_k)$, $v(a_k) = v(b_k)$.
Consider the integral

$$J = \int_{\Lambda} u\,dv,$$

where Λ is the union of the contours Λ_k. Obviously

$$J = \int_{L} (u^+dv^+ - u^-dv^-).$$

But it follows from $\Phi^+ + \Phi^- = C_k = \alpha_k + i\beta_k$ that $u^+ + u^- = \alpha_k$, $dv^- = - dv^+$, and therefore

$$J = \int_L (u^+ + u^-)dv^+ = \sum_{k=1}^{p} \alpha_k \int_{L_k} dv^+ = \sum_{k=1}^{p} \alpha_k[v(b_k) - v(a_k)] = 0.$$

On the other hand,

$$J = \iint\left[\left(\frac{\partial u}{\partial x}\right)^2 + \left(\frac{\partial u}{\partial y}\right)^2\right]dx\,dy,$$

where the double integral is taken over the entire cut plane. Hence the conclusion is drawn that $\Phi(z) = u + iv = 0$. Consequently $\varphi(t) = 0$ and the statement is proved.

The solution of the problem (90.1) will now be considered. The formula (89.14) gives the necessary and sufficient conditions which the constants C_k must satisfy, so that the equation (90.1) has a solution of the class H. These conditions may be written as

$$\sum_{k=1}^{p} a_{jk}C_k + A_j = 0, \quad (j = 0, 1, \ldots, p-1), \tag{90.5}$$

where

$$a_{jk} = \int_{L_k} \frac{t^j dt}{\sqrt{R(t)}}, \quad A_j = \int_L \frac{t^j f(t)\,dt}{\sqrt{R(t)}}. \tag{90.6}$$

The determinant of the matrix $\| a_{jk} \|$ is non-zero, for otherwise the homogeneous system, obtained from (90.5) for $f(t) \equiv 0$, would have the non-zero solutions $C_1, C_2, \ldots, C_p$, and so the problem (90.2) would have a non-zero solution which is impossible.

Therefore the system (90.5) has always a definite solution. This solution has obviously the following form:

$$C_k = \int_L \frac{\omega_k(t)f(t)\,dt}{\sqrt{R(t)}}, \tag{90.7}$$

where the ω_k are certain polynomials of degree not greater than $p-1$, with their coefficients depending only on the form of the line L; in actual fact the coefficients depend only on the position of the ends a_k, b_k and, from the topological point of view, on the relative disposition of the arcs L_k, i.e., in other words, the polynomials $\omega_k(t)$ remain invariant under arbitrary continuous deformations of the L_k, for which the ends of the arcs remain fixed and for which the arcs themselves do not intersect. The polynomials $\omega_k(t)$ are easily seen to be linearly independent.

Once the constants C_k have been determined, the function $\varphi(t)$ may be found using (89.4) which in the present case gives

$$\varphi(t_0) = \frac{\sqrt{R(t_0)}}{\pi i} \int_L \frac{f(t)dt}{\sqrt{R(t)}(t-t_0)} + \frac{\sqrt{R(t_0)}}{\pi i} \sum_{k=1}^{p} C_k \int_{L_k} \frac{dt}{\sqrt{R(t)}(t-t_0)}, \tag{90.8}$$

where by the C_k must be understood the expressions (90.7).

Substituting (90.7) in (90.8), one obtains the inversion formula

$$\varphi(t_0) = \frac{\sqrt{R(t_0)}}{\pi i} \int_L \frac{f(t)}{\sqrt{R(t)}} \left\{ \frac{1}{t-t_0} + \sum_{k=1}^{p} \omega_k(t) \int_{L_k} \frac{d\tau}{\sqrt{R(\tau)}\,(\tau - t_0)} \right\} dt. \tag{90.9}$$

A formula, being essentially the same as (90.9), was obtained independently by N. P. Vekua [1].

It will be seen that, as was to be expected, $\varphi(t)$ becomes zero at all ends.

In the case $p = 1$, one can easily verify directly that the formulae (89.16), (89.17) are obtained.

CHAPTER 12

EFFECTIVE SOLUTION OF SOME BOUNDARY PROBLEMS OF THE THEORY OF HARMONIC FUNCTIONS

The effective solution of a number of important boundary problems in the theory of harmonic functions which are encountered in hydrodynamics, the theory of elasticity and other fields of mathematical physics will be given in this chapter.

§ 91. Dirichlet and analogous problems for the plane with cuts distributed along a straight line.

Let S be the plane, cut along the segments $L_j = a_j b_j$ ($j = 1, 2, \ldots, p$) of the real axis Ox. The union of these segments will be denoted by L:

$$L = L_1 + L_2 + \ldots + L_p.$$

Three problems will be solved for the region S. The last of these (Problem C) will be the customary Dirichlet problem, while the other two will be modifications of that problem. The latter are of independent interest and also of assistance in the solution of the Dirichlet problem.

Problem A: *To find a function $\Phi(z) = u + iv$, sectionally holomorphic in S and vanishing at infinity, subject to the boundary condition*

$$u^+ = f^+(t), \quad u^- = f^-(t) \text{ on } L, \tag{91.1}$$

where $f^+(t)$, $f^-(t)$ are real functions, given on L and satisfying the H condition.

Using the notation of § 38, 1°, put

$$\Psi(z) = \frac{\Phi(z) + \bar{\Phi}(z)}{2}, \quad \Omega(z) = \frac{\Phi(z) - \bar{\Phi}(z)}{2}. \tag{91.2}$$

The sectionally holomorphic functions $\Psi(z)$ and $\Omega(z)$ must obviously satisfy the conditions

$$\bar{\Psi}(z) = \Psi(z), \; \Omega(z) = -\bar{\Omega}(z). \tag{91.3}$$

The boundary condition (91.1) is easily seen to lead to the following:

$$\Psi^+(t) + \Psi^-(t) = 2f(t), \tag{91.4a}$$

$$\Omega^+(t) - \Omega^-(t) = 2g(t) \tag{91.4b}$$

on L, where

$$2f(t) = f^+(t) + f^-(t),\ 2g(t) = f^+(t) - f^-(t). \tag{91.5}$$

In fact,

$$u^+(t) = \Re\Phi^+(t) = \tfrac{1}{2}[\Phi^+(t) + \overline{\Phi^+(t)}],$$

$$u^-(t) = \Re\Phi^-(t) = \tfrac{1}{2}[\Phi^-(t) + \overline{\Phi^-(t)}].$$

But [cf. (38.5a)]

$$\overline{\Phi^+(t)} = \bar{\Phi}^-(t),\ \overline{\Phi^-(t)} = \bar{\Phi}^+(t),$$

resulting in (91.4a), (91.4b).

The function $\Omega(z)$ is uniquely (§ 78) defined by (91.4b):

$$\Omega(z) = \frac{1}{\pi i}\int_L \frac{g(t)\,dt}{t-z}. \tag{91.6}$$

Using (38.10), the second condition of (91.3) is seen to be satisfied.

The function $\Omega(z)$ will be seen to be almost bounded (§ 77, 6°) near all ends [for the problem (91.4b) all ends a_j, b_j being special]; in addition, it will be bounded near those ends c at which

$$f^+(c) = f^-(c).$$

The determination of $\Psi(z)$, however, leads to one of the simplest particular cases of the Hilbert problem, already treated in § 84; here one has only to consider the additional condition $\bar{\Psi}(z) = \Psi(z)$.

Let $c_1, c_2, \ldots, c_{2p}$ be the points a_j, b_j, taken in any order. The fundamental function of the class $h(c_1, c_2, \ldots, c_q)$, i.e., the fundamental solution of this class of the homogeneous Hilbert problem

$$\Psi^+(t) + \Psi^-(t) = 0 \text{ on } L, \tag{91.7}$$

has, according to (84.3) with $C = 1$, the following form:

$$X(z) = \frac{\sqrt{R_1(z)}}{\sqrt{R_2(z)}} = \sqrt{\frac{R_1(z)}{R_2(z)}}, \tag{91.8}$$

where, as in § 84,

$$R_1(z) = \prod_{j=1}^{q} (z - c_j),\ R_2(z) = \prod_{j=q+1}^{2p} (z - c_j) \tag{91.9}$$

the meaning of the root having been discussed in the same section.

As has already been indicated there, for the problem (91.4a) all the ends a_k, b_k are non-special. The general solution of the class

$h(c_1, c_2, \ldots, c_q)$ of this problem, vanishing at infinity, has for $p \geqq q$ the form

$$\Psi(z) = \frac{1}{\pi i}\frac{\sqrt{R(z_1)}}{\sqrt{R_2(z)}}\int_L \frac{\sqrt{R_2(t)}f(t)dt}{\sqrt{R_1(t)}\,(t-z)} + \frac{\sqrt{R_1(z)}}{\sqrt{R_2(z)}}P_{p-q-1}(z), \quad (91.10)$$

where $P_{p-q-1}(z)$ is an arbitrary polynomial of degree not greater than $p-q-1(P_{p-q-1}(z) \equiv 0$ for $p = q)$ and the roots are again as in § 84. For $p < q$, a solution of the class considered of the problem (91.4a), vanishing at infinity, exists, provided only that the conditions (91.12) established below are satisfied; then again the solution is given by (91.10) with $P_{p-q-1}(z) \equiv 0$.

In order that the solution of the problem (91.4a), just established, should satisfy all the required conditions, one must still have by (91.3): $\overline{\Psi}(z) = \Psi(z)$. It will now be verified that this condition is satisfied.

First of all it is easily seen that

$$\overline{X}(z) = X(z). \quad (91.11)$$

Furthermore, using (38.10) and keeping in mind that in the present case [by (38.5a), (91.11) and (91.7)] $\overline{X^+(t)} = \overline{X}^-(t) = X^-(t) = -X^+(t)$ on L, it is easily verified that the first term on the right side of (91.10) satisfies the relation $\overline{\Psi}(z) = \Psi(z)$. Therefore the whole of the right side of (91.10) will also satisfy this condition, if and only if the coefficients in $P_{p-q-1}(z)$ are real numbers.

The above-mentioned conditions for the existence of solutions of the class $h(c_1, c_2, \ldots, c_q)$ of the problem (91.4a), vanishing at infinity, have for $p < q$ the form (§ 84)

$$\int_L \frac{\sqrt{R_2(t)}}{\sqrt{R_1(t)}}t^k f(t)dt = 0, \; k = 0, 1, \ldots, q-p-1. \quad (91.12)$$

The solutions $\Phi(z)$ of the original problem A will be called solutions of the class $h(c_1, c_2, \ldots, c_q)$, if they are almost bounded near the ends $c_1, c_2, \ldots, c_q$. Then it is easy to draw from the above the conclusion:

A solution of the class $h(c_1, c_2, \ldots, c_q)$ of the problem A is given for $p \geqq q$ by

$$\Phi(z) = \Psi(z) + \Omega(z) =$$

$$= \frac{1}{\pi i}\frac{\sqrt{R_1(t)}}{\sqrt{R_2(t)}}\int_L \frac{\sqrt{R_2(t)}}{\sqrt{R_1(t)}}\frac{f(t)dt}{t-z} + \frac{1}{\pi i}\int_L \frac{g(t)dt}{t-z} + \frac{\sqrt{R_1(z)}}{\sqrt{R_2(z)}}P_{p-q-1}(z), \quad (91.13)$$

where $P_{p-q-1}(z)$ is an arbitrary polynomial, of degree not greater than

$p-q-1$, *with real coefficients* ($P_{p-q-1}(z) \equiv 0$ for $p = q$); *for* $p < q$, *the (unique) solution exists, if and only if the conditions* (91.12) *are satisfied, and it is given by* (91.13) *for* $P_{p-q-1} \equiv 0$.

The result obtained is a generalization (in the sense of division of the solutions into classes and of the establishment of necessary and sufficient conditions of solubility) of valuable results obtained by M. V. Keldysh and L. I. Sedov [1].

NOTE. It is easily seen (cf. § 83) that all the above results and formulae remain true, if it is assumed that the given functions $f^+(t)$ and $f^-(t)$ belong to the class $h(c_1, c_2, \ldots, c_q)$, instead of satisfying the H condition on L. In particular, if $f^+(t)$ and $f^-(t)$ are arbitrary functions of the class h_0 (i.e., of the class H^*), the general solution of the class h_0 will be given by (91.13) for $q = 0$

$$\Phi(z) = \frac{1}{\pi i\sqrt{R(z)}}\int_L \frac{\sqrt{R(t)}f(t)dt}{t-z} + \frac{1}{\pi i}\int_L \frac{g(t)dt}{t-z} + \frac{P_{p-1}(z)}{\sqrt{R(z)}}, \tag{91.13a}$$

where, as before,

$$R(z) = \prod_{j=1}^{p} (z-a_j)(z-b_j). \tag{91.9a}$$

PROBLEM B: *To find a sectionally holomorphic function* $\Phi(z) = u + iv$, *vanishing at infinity and almost bounded near all ends, subject to the boundary condition*

$$u^+ = f^+(t) + C_k, \quad u^- = f^-(t) + C_k \text{ on } L_k, \quad k = 1, 2, \ldots, p, \tag{91.14}$$

where $f^+(t)$ *and* $f^-(t)$ *are given real functions satisfying the* H *condition and the* C_k *are real constants, not given beforehand.*

In essence, this problem coincides with the modified Dirichlet problem (§ 60), and this name will be retained here.

The principal difference between this and the modified Dirichlet problem stated in § 60 consists of the fact that there the function $u(x, y)$ had to be continuous from the left at all points of the boundary, while here the ends may be excluded; near these ends $u(x, y)$ need only be almost bounded. However, it will be seen that under the present conditions $u(x, y)$ is also bounded near the ends and even continuous at those ends c where $f^+(c) = f^-(c)$; clearly, if the last condition is not satisfied, the continuity at c is impossible.

Introducing the functions $\Psi(z)$ and $\Omega(z)$ by the formulae (91.2), one arrives at the following two Hilbert problems:

$$\Psi^+(t) + \Psi^-(t) = 2f(t) + 2C_k \text{ on } L_k, \quad k = 1, 2, \ldots, p, \tag{91.15a}$$

$$\Omega^+(t) - \Omega^-(t) = 2g(t) \text{ on } L, \tag{91.15b}$$

where $f(t)$ and $g(t)$ are defined by (91.5).

The second problem is solved as in the previous case; the solution will be almost bounded near all ends [and even bounded near those ends, where $f^+(t)$ and $f^-(t)$ take the same value].

There remains to find an almost bounded — and hence bounded [for the problem (91.15a) all ends are non-special so that every almost bounded solution will also be bounded (§ 83, 1°)] — solution of (91.15a), i.e., a solution of the class h_{2p}; the fundamental function of this class will be

$$X_{2p}(z) = \sqrt{R(z)},\quad R(z) = \prod_{k=1}^{p} (z - a_k)(z - b_k) \qquad (91.16)$$

[cf. (84.12); here $C = 1$].

The necessary and sufficient conditions for the existence of a solution of the class h_{2p} of the problem (91.15a), vanishing at infinity, has the form [from (84.18) for $q = 2p$]

$$\frac{1}{\pi i}\int_L \frac{t^k[f(t) + C(t)]dt}{\sqrt{R(t)}} = 0,\quad k = 0, 1, \ldots, p - 1,$$

where $C(t) = C_j$ on L_j. These conditions give the p linear equations for the determination of the C_j

$$\sum_{j=1}^{p} \gamma_{kj} C_j + \gamma_k = 0,\quad k = 0, 1, \ldots, p - 1, \qquad (91.17)$$

where

$$\gamma_{kj} = \frac{1}{\pi i}\int_{L_j} \frac{t^k dt}{\sqrt{R(t)}},\quad \gamma_k = \frac{1}{\pi i}\int_L \frac{t^k f(t)dt}{\sqrt{R(t)}}. \qquad (91.18)$$

The γ_{kj} and γ_k are easily seen to be real.

If the system (91.17) has a (real) solution, then the required solution of the problem (91.15a) will be given by

$$\Psi(z) = \frac{\sqrt{R(z)}}{\pi i}\int_L \frac{f(t) + C(t)}{\sqrt{R(t)}(t - z)}\,dt, \qquad (91.19)$$

where $C(t) = C_j$ on L_j and the C_j ($j = 1, \ldots, p$) satisfy the system (91.17). For this $\overline{\Psi}(z) = \overline{\Psi(z)}$.

Now it will be shown that the determinant of the system (91.17), i.e., the determinant of the matrix $\|\gamma_{ij}\|$, is non-zero. Actually, if it were zero, the homogeneous system obtained from (91.17) for $f(t) \equiv 0$ would have the solution $C_1, C_2, \ldots, C_p$, different from zero. The function, corresponding to this solution,

$$\Psi_0(z) = \frac{\sqrt{R(z)}}{\pi i}\int_L \frac{C(t)dt}{\sqrt{R(t)}(t - z)}$$

would be a function, vanishing at infinity, continuous up to and on all segments L_j, including the ends, while $[\Re\Psi_0(t)]^+ = [\Re\Psi_0(t)]^- = C_j$ on L_j. But then, by the theorem on the uniqueness of the solution of the modified Dirichlet problem proved in § 60 (the proof is easily seen to be applicable to the present case), it is necessary that $C_1 = C_2 = \ldots = C_p = 0$ which contradicts the condition. Hence the determinant of the system (91.17) is different from zero and this system always has a unique solution.

Thus there is always a unique solution of the problem and it is given by

$$\Phi(z) = \Psi(z)+\Omega(z) = \frac{\sqrt{R(z)}}{\pi i}\int_L \frac{f(t)+C(t)}{\sqrt{R(t)}}\,dt + \frac{1}{\pi i}\int_L \frac{g(t)dt}{t-z}, \quad (91.20)$$

where $C(t) = C_j$ on L_j, while the constants C_j are uniquely defined by the system (91.17).

NOTE. It is easily seen that *the harmonic functions* $u = \Re\Phi(z)$ *will not only be almost bounded, but also bounded near all ends.* This is obvious with regard to the first term on the right side of (91.20). The second term may be written near any end c in the form

$$\frac{1}{\pi i}\int_L \frac{g(t)dt}{t-z} = \mp \frac{g(c)}{\pi i}\log(z-c) + \frac{1}{\pi i}\int_L \frac{g(t)-g(c)}{t-z}\,dt,$$

where the upper sign is taken for the ends a_j and the lower for the ends b_j. Hence, near c,

$$u = \Re\Phi(z) = \mp g(c)\frac{\vartheta}{\pi} + u_0,$$

where u_0 is continuous at the point c and $\vartheta = \arg(z-c)$.

Furthermore, it is obvious that, if $g(c) = f^+(c) - f^-(c) = 0$, then u will be continuous up to and including c.

PROBLEM C (Dirichlet problem): *To find a function* $U(x, y)$, *harmonic and bounded everywhere in* S, *under the boundary condition*

$$U^+ = f^+(t),\ U^- = f^-(t) \text{ on } L, \quad (91.21)$$

where $f^+(t)$ *and* $f^-(t)$ *are given real functions, satisfying the H condition.*

Two methods of solution of this problem will be given. The first method reduces it to the problem B and the second to the problem A.

FIRST METHOD. The problem C may be reduced to the problem B in a number of ways; one of the simplest will be indicated.

Let $\omega_j(x, y)$, or simply $\omega_j(z)$, denote the following elementary harmonic functions:

$$\omega_j(z) = \Re\frac{(z-b_j)\log(z-b_j)-(z-a_j)\log(z-a_j)}{b_j-a_j}, \quad j = 1, \ldots, p; \quad (91.22)$$

$\omega_j(z)$ will be understood to be the branch of the function which is single-valued in the plane cut along L, such that for large $|z|$

$$\omega_j(z) = -\log|z| - 1 + O\left(\frac{1}{z}\right). \qquad (91.22a)$$

The function $\omega_j(z)$ is easily seen to take on the segments a_jb_j the same values from the right and from the left; these values will be denoted by $\omega_j(t)$.

Put

$$U(x, y) = u(x, y) + \alpha_0 + \sum_{j=1}^{p} \alpha_j\omega_j(z), \qquad (91.23)$$

where the $\alpha_0, \alpha_1, \ldots, \alpha_p$ are real constants the last p of which are connected by the relation

$$\alpha_1 + \alpha_2 + \ldots + \alpha_p = 0, \qquad (91.24)$$

ensuring that the sum

$$\sum_{j=1}^{p} \alpha_j \omega_j(z)$$

becomes zero at infinity; $u(x, y)$ is a new unknown harmonic function, vanishing at infinity.

From (91.21) one obtains the following boundary condition for u:

$$u^+ = f^+(t) - \omega(t), \quad u^- = f^-(t) - \omega(t) \text{ on } L, \qquad (91.25)$$

where

$$\omega(t) = \alpha_0 + \sum_{j=1}^{p} \alpha_j \omega_j(t). \qquad (91.26)$$

Instead of solving the problem (91.25), the problem B

$$u^+ = f^+(t) - \omega(t) + C_j, \quad u^- = f^-(t) - \omega(t) + C_j \text{ on } L_j \qquad (91.27)$$

will be solved, assuming the constants $\alpha_0, \alpha_1, \ldots, \alpha_p$ to be fixed and, as required in Problem B, $u(x, y)$ to be the real part of some sectionally holomorphic function $\Phi(z)$, vanishing at infinity and almost bounded near all ends.

The constants C_j are known to be completely determined for this case (cf. above solution of the problem B). It is clear that they will be linear combinations of the $\alpha_0, \alpha_1, \ldots, \alpha_p$:

$$C_j = \sum_{k=0}^{p} \gamma_{jk}\alpha_k + \gamma_j, \qquad (91.28)$$

where the γ_{jk} are certain constants, independent of the choice of $f^+(t)$, $f^-(t)$, and the γ_j are constants which are zero for $f^+(t) = f^-(t) \equiv 0$.

The constants $\alpha_0, \alpha_1, \ldots, \alpha_p$ will now be selected, so that the $C_j = 0$ $(j = 1, \ldots, p)$ and, in addition, the condition (91.24) is satisfied. One thus obtains the system of equations

$$\sum_{k=0}^{p} \gamma_{jk}\alpha_k + \gamma_j = 0, \quad j = 1, 2, \ldots, p,$$
$$\sum_{k=1}^{k} \alpha_k = 0 \tag{91.29}$$

for the determination of the constants α_k.

It will be shown below that this system always has a unique solution. If the constants α_k are selected in correspondence with (91.29), the solution of the original problem C will be given by (91.23), where by $u(x, y)$ is to be understood the solution of the problem (91.27). According to the Note to Problem B the function $u(x, y)$ will not only be almost bounded, but also bounded near all ends.

The determinant of the system (91.29) will now be shown to be different from zero. Let it be assumed that it is zero. Then the homogeneous system, obtained from (91.29) for $f^+(t) = f^-(t) = 0$ (i.e., all $\gamma_j = 0$), has the non-zero solutions $\alpha_0, \alpha_1, \ldots, \alpha_p$. For these values of the α_j, the function $U(x, y)$ defined by (91.23) will be a solution of the Dirichlet problem (91.21) for $f^+ = f^- = 0$. But then, as is known, $U \equiv 0$. However, if $U \equiv 0$, then necessarily $\alpha_0 = \alpha_1 = \ldots = \alpha_p = 0$. In fact, $\alpha_0 = 0$, since both terms on the right of (91.23), which are different from α_0, vanish at infinity. [The function $u(x, y)$ is a solution of the problem B, and hence, as has been shown, is the real part of a function, holomorphic in S and vanishing at infinity]. Further, let $\alpha_k \neq 0$ $(k \geqq 1)$. If z passes round the segment a_kb_k (and no other segment), the function conjugate to $u(x, y)$ reverts to the original value as well as the function conjugate to $\omega_j(z)$ $(j \neq k)$, while the function, conjugate to $\omega_k(z)$, is easily seen to suffer an increase. So, if $\alpha_k \neq 0$, the function, conjugate to the right side of (91.23), receives an increment, which is not possible since the left side $U \equiv 0$.

Thus the proposition is proved and the problem C can be considered solved.

THE SECOND METHOD of solution of the problem C, given by M. V. Keldysh and L. I. Sedov [1], consists of its reduction to the problem A; it is somewhat less general than the above, since it requires the given functions $f^+(t)$ and $f^-(t)$ to be differentiable, but in practice it often leads to simpler results.

This method will be stated here with some minor alterations and simplifications.

Let

$$\Phi(z) = U(x, y) + iV(x, y) \tag{91.30}$$

be an analytic function the real part of which is the unknown harmonic function $U(x, y)$. The function $V(x, y)$ conjugate to $U(x, y)$ will, in general, be multi-valued; hence the same will hold for $\Phi(z)$. But it is known that the function

$$\Phi'(z) = \frac{\partial U}{\partial x} + i\frac{\partial V}{\partial x}$$

will be single-valued (as a consequence of the univalence of U); for the function $V(x, y)$ receives, for a circuit over any contour enclosing one single segment $a_k b_k$ and no others, a constant increment which does not depend on the path; hence it is clear that $\Phi'(z)$ returns to the original value for the passage over such a path. Also, for large $|z|$,

$$\Phi'(z) = O\left(\frac{1}{z^2}\right) \tag{91.31}$$

as a consequence of U being bounded. For the harmonic function, single-valued and bounded near the point $z = \infty$, may be expanded for large $|z|$ in a series of the form

$$U(x, y) = \alpha_0 + \sum_{k=1}^{p} (\alpha_k \cos k\vartheta + \beta_k \sin k\vartheta) r^{-k} \quad (z = re^{i\vartheta}),$$

and hence for sufficiently large $|z|$

$$\Phi(z) = U + iV = \alpha_0 + i\beta_0 + \sum_{k=1}^{\infty} (\alpha_k + i\beta_k) z^{-k},$$

$$\Phi'(z) = O\left(\frac{1}{z^2}\right).$$

It will be assumed that the functions $f^+(t)$ and $f^-(t)$ take the same value at all ends a_k, b_k (i.e., the boundary values of the unknown function U will be assumed to be continuous) and that they have the derivatives $f^{+\prime}(t)$, $f^{-\prime}(t)$ belonging to the class H^*. The unknown function $\Phi'(z)$ will be assumed to be sectionally holomorphic. [This assumption will be justified 'a posteriori', since a solution of such a form will be seen to exist; then it will follow from the uniqueness of the solution of the Dirichlet problem that no other solutions have been lost].

The solution of the problem A for the function $\Phi'(z)$ will now be considered, where $f^{+\prime}(t)$, $f^{-\prime}(t)$ play the role of $f^+(t)$, $f^-(t)$. By the assumed conditions, the functions $f(t)$ and $g(t)$ of (91.5) are such that

$$f(a_k) = f^+(a_k) = f^-(a_k), \; f(b_k) = f^+(b_k) = f^-(b_k), \; g(a_k) = g(b_k) = 0, \quad k = 1, \ldots, p. \tag{91.32}$$

Using (91.13a), one obtains

$$\Phi'(z) = \frac{1}{\pi i \sqrt{R(z)}} \int_L \frac{\sqrt{R(t)}\, f'(t)dt}{t-z} + \frac{1}{\pi i} \int_L \frac{g'(t)dt}{t-z} + $$

$$+ \frac{C_1 z^{p-2} + \ldots + C_{p-1}}{\sqrt{R(z)}}, \qquad (91.33)$$

where, as before,

$$R(z) = \prod_{j=1}^{p} (z-a_j)(z-b_j)$$

and the roots are to be understood as in § 84; $C_1, \ldots, C_{p-1}$ are real constants. The polynomial in the numerator of the last term of (91.33) has been taken to be of degree $p-2$ and not $p-1$, in order to satisfy the condition (91.31), because the first two terms are easily seen to satisfy this condition. With regard to the first term this is obvious; for the second term the coefficient of z^{-1} in its expansion in decreasing powers of z near the point at infinity is

$$-\frac{1}{\pi i}\int_L g'(t)dt = -\frac{1}{\pi i}\sum_{k=1}^{p}\left[g(b_k) - g(a_k)\right],$$

and hence it is zero by (91.32).

Having found $\Phi'(z)$, $\Phi(z)$ is determined by

$$\Phi(z) = \int_0^z \Phi'(\zeta)d\zeta + C,$$

where by C may be understood a real constant.

For the determination of the constants $C_1, \ldots, C_{p-1}, C$, there are obviously the following p conditions [not forgetting that $f^+(a_k) = = f^-(a_k)$]:

$$\Re \int_0^{a_k} \Phi'(z)dz + C = f(a_k). \qquad (91.34)$$

The integration may be taken along any path, leading from O to a_k without intersecting L, e.g. along the left side of the Ox axis [when $\Phi^{+\prime}(t)dt$ has to be taken under the integral sign].

The conditions (91.34) are easily seen to give a system of p linear equations involving $C_1, C_2, \ldots, C_{p-1}, C$. It will be shown that this system always has a unique solution. The proof is easily obtained on the basis of the uniqueness theorem on the solutions of the Dirichlet problem, but a simple direct proof will be given here which is likewise due to M. V. Keldysh and L. I. Sedov [1].

The homogeneous system, corresponding to the system (91.34), will be considered; it will be obtained by putting $f(t) = g(t) = 0$, so that this homogeneous system has the form

$$\Re \int_0^{a_k} \frac{C_1 t^{p-2} + \ldots + C_{p-1}}{\sqrt{R(t)}} dt + C = 0, \quad k = 1, 2, \ldots, p, \quad (91.35)$$

and hence it is easily inferred that

$$\int_{b_k}^{a_{k+1}} \frac{C_1 t^{p-2} + \ldots + C_{p-1}}{\sqrt{R(t)}} dt = 0, \; k = 1, 2, \ldots, p-1. \quad (91.36)$$

Consequently, on each segment $b_k a_{k+1}$ (the number of such segments being $p-1$), the polynomial $C_1 z^{p-2} + C_2 z^{p-3} + \ldots + C_{p-1}$ necessarily changes sign at least once, so that it must have not less than $p-1$ roots; but this is only possible for $C_1 = C_2 = \ldots = C_{p-1} = 0$. Hence it is obvious that $C = 0$.

Thus the homogeneous system corresponding to (91.34) has no solutions different from zero. Consequently its determinant is not zero and the system (91.34) always has a unique solution.

§ 92. The Dirichlet and analogous problems for the plane with cuts distributed over a circle.

The problems A, B, C of § 91 can be solved quite similarly in the case of a region S, representing the plane cut along a finite number of arcs of some circle. This case can be reduced to the case of the last section by means of a simple inversion, but a direct solution, similar to the one of § 91, will lead more quickly to the result. The deduction of the corresponding formulae is similar to the above and will be left to the reader (cf. § 91).

§ 93. The Riemann-Hilbert problem for discontinuous coefficients.

This problem, important from the point of view of application, will be formulated as follows (cf. § 39). Let L be a simple contour and let S^+ (finite or infinite) be the part of the plane, bounded by L. It is required:

To find a function $\Phi(z) = u + iv$, holomorphic in S^+ and continuous from the left everywhere on L. except possibly at the points $c_1, c_2, \ldots, c_r$

of this contour, near which the inequalities

$$| \Phi(z) | < \frac{\text{const.}}{| z - c_j |^\alpha}, \qquad \alpha < 1 \tag{93.1}$$

hold, for the boundary condition

$$a(t)u^+ - b(t)v^+ = c(t); \tag{93.2}$$

here $a(t)$, $b(t)$, $c(t)$ *are real functions of the class* H_d (*cf.* § 85), *given on* L, *with the possible points of discontinuity* $c_1, c_2, \ldots, c_r$, *while* $a^2(t) + b^2(t) \neq 0$ *everywhere on* L; regarding the points of discontinuity c_j, this must be understood in the sense that

$$a^2(c_j \pm 0) + b^2(c_j \pm 0) \neq 0.$$

Further, it will be assumed *that the angle between the tangent to* L *and a constant direction satisfies the* H *condition.*

As in § 42, this problem can be reduced by means of conformal mapping to the case in which S^+ is the circle $|z| < 1$. Therefore *it will also be assumed in the sequel that* L *is the circle* $|z| = 1$ *and* S^+ *the circular region* $|z| < 1$; S^- will be the region $|z| > 1$. After what has been said in § 40, § 42 and § 85, the solution of the stated problem is clear. Therefore the problem will only be discussed briefly *).

As in § 40, the function $\Phi(z)$, unknown in S^+, will be continued into the region S^-, so that for all z in S^- and S^+

$$\Phi_*(z) = \bar{\Phi}(1/z) = \Phi(z). \tag{93.3}$$

As in § 40, this leads to the Hilbert problem corresponding to the Riemann-Hilbert problem (93.2)

$$(a + ib)\Phi^+(t) + (a - ib)\Phi^-(t) = 2c \tag{93.4}$$

or

$$\Phi^+(t) = G(t)\Phi^-(t) + g(t), \tag{93.5}$$

where

$$G(t) = -\frac{a - ib}{a + ib}, \quad g(t) = \frac{2c}{a + ib}. \tag{93.6}$$

As in § 85, all the solutions of the Riemann-Hilbert problem (93.2) will be divided into classes by putting into the class $h(c_1, c_2, \ldots, c_q)$ the solutions which remain bounded near $c_1, c_2, \ldots, c_q$.

---

*) The complete solution given in the text has not, to the knowledge of the Author, been stated elsewhere. One particular case has been considered by F. Noether [1], but even then no complete solution was obtained. The complete solution in another particular case was given by S. L. Sobolev [1] by a method differing from that used here.

The index $\varkappa$ of a given class will be defined by

$$\varkappa = \frac{1}{2\pi i}[\log G(t)]_L = \frac{1}{2\pi}[\arg G(t)]_L = \frac{1}{2\pi}\left[\arg \frac{a - ib}{a + ib}\right]_L, \quad (93.7)$$

where the discontinuities in the argument of $G(t)$ for the passage through the points c_j must be selected in correspondence with the class of solutions considered, as stated in § 85 subsequent to formula (85.2). The increments of the argument of the ratio $(a - ib)/(a + ib)$, i.e., $-G(t)$, will always be taken as the increments of the argument of $G(t)$.

In contrast to the case of continuous coefficients, the number $\varkappa$ may be even as well as odd. [It cannot be confirmed at this stage that the change in argument of the ratio $(a - ib)/(a + ib)$ is twice the change in argument of $a - ib$, because the increment of the latter argument for passage through the points c_j has not been fixed separately].

Let $X(z)$ be the fundamental function of the given class $h(c_1, c_2, \ldots, c_q)$ of the problem (93.5), satisfying the additional condition

$$X_*(z) = \overline{X}(1/z) = z^{\varkappa} X(z); \quad (93.8)$$

the function $X(z)$ is defined, apart from a constant real multiplier, and can be evaluated by use of (40.9)—(40.11), (40.13) and (40.14), if the expression

$$\log [t^{-\varkappa}G(t)] = i \arg \left[-t^{-\varkappa}\frac{a - ib}{a + ib}\right] = i\Theta(t) \quad (93.9)$$

is now understood, as stated in § 85 subsequent to (85.7).

The general solution of a given class of the homogeneous Riemann-Hilbert problem, obtained from (93.2) for $c(t) \equiv 0$, is given for $\varkappa \geqq 0$ by

$$\Phi(z) = X(z)(C_0 z^{\varkappa} + C_1 z^{\varkappa - 1} + \ldots + C_{\varkappa}), \quad (93.10)$$

where $C_0, C_1, \ldots, C_{\varkappa}$ are constants connected by the relations

$$C_k = \overline{C}_{\varkappa - k}, \quad (93.11)$$

but in other respects arbitrary as in § 40 (with the only difference that here $\varkappa$ can be an odd number). Accordingly, *the homogeneous problem has, for $\varkappa \geqq 0$, exactly $\varkappa + 1$ linearly independent solutions of a given class,* as in § 40 (linear independence being understood as in § 40); *the homogeneous problem has, for $\varkappa \leqq -1$, no non-zero solutions of a given class.*

The non-homogeneous problem (93.2) *is soluble (in a given class) for $\varkappa \geqq -1$ for every function $c(t)$; in particular, for $\varkappa = -1$, its solution is unique.*

This problem has, for $\varkappa \leqq -2$, a (unique) solution of a given class, only if the following $(-\varkappa-1)$ conditions are satisfied:
For $\varkappa$ even (in this case these formulae are the same as in § 40):

$$\begin{aligned}
&\int_0^{2\pi} \Omega(\vartheta)c(\vartheta)\cos k\vartheta\, d\vartheta = 0, \quad k = 0, 1, \ldots, -\frac{\varkappa}{2}-1,\\
&\int_0^{2\pi} \Omega(\vartheta)c(\vartheta)\sin k\vartheta\, d\vartheta = 0, \quad k = 1, 2, \ldots, -\frac{\varkappa}{2}-1.
\end{aligned} \tag{93.12}$$

For $\varkappa$ odd:

$$\begin{aligned}
&\int_0^{2\pi} \Omega(\vartheta)c(\vartheta)\cos (k+\tfrac{1}{2})\vartheta\, d\vartheta = 0,\\
&\int_0^{2\pi} \Omega(\vartheta)c(\vartheta)\sin (k+\tfrac{1}{2})\vartheta\, d\vartheta = 0,\\
&\qquad k = 0, 1, \ldots, -\frac{\varkappa+1}{2}-1.
\end{aligned} \tag{93.13}$$

Here $\Omega(\vartheta)$ is the real function defined as in § 40 [(40.22)], viz.

$$\Omega(\vartheta) = \frac{1}{\pm\sqrt{a^2(\vartheta)+b^2(\vartheta)}} \exp\left\{-\frac{1}{4\pi}\int_0^{2\pi} \Theta(\vartheta_1)\cot\frac{\vartheta_1-\vartheta}{2}\, d\vartheta_1\right\}, \tag{93.14}$$

with the following addition: in § 40, the root $\sqrt{a^2(\vartheta)+b^2(\vartheta)}$ was indifferent to the sign in front of it, because the expression varied continuously on L, without change of sign. In the present case this is not true. Therefore it must now be kept in mind that

$$\pm\sqrt{a^2(\vartheta)+b^2(\vartheta)} = (a+ib)\sqrt{\frac{a-ib}{a+ib}} = (a+ib)e^{i\frac{\omega}{2}},$$

where

$$\omega = \arg\frac{a-ib}{a+ib},$$

and for the passage through the points of discontinuity c_j the argument ω must be altered as indicated above, subsequent to (93.7).

Finally, the formulae giving the solution of the given class $h(c_1, c_2, \ldots, c_q)$ of the original non-homogeneous problem (93.2) will be written down. These formulae have the same form as in § 40.

For $\varkappa \geqq 0$, one of the particular solutions of the problem (93.2) is given by

$$\Phi(z) = \frac{X(z)}{2\pi i}\left\{\int_L \frac{c\,dt}{(a+ib)X^+(t)(t-z)} + z^\varkappa \int_L \frac{t^{-\varkappa}c\,dt}{(a+ib)X^+(t)(t-z)}\right\} - $$

$$- \frac{z^\varkappa X(z)}{2\pi i}\int_L \frac{t^{-\varkappa}c}{(a+ib)X^+(t)}\frac{dt}{t}; \qquad (93.15)$$

for $\varkappa \leqq -1$, the (unique) solution is given by

$$\Phi(z) = \frac{X(z)}{\pi i}\int_L \frac{c\,dt}{(a+ib)X^+(t)(t-z)}, \qquad (93.16)$$

provided that, for $\varkappa \leqq -2$, the conditions for the existence of the solution (93.12) or (93.13) are fulfilled.

The formula giving the (particular) solution for $\varkappa = 0$ will be noted separately, viz.

$$\Phi(z) = \frac{X(z)}{\pi i}\int_L \frac{c\,dt}{(a+ib)X^+(t)(t-z)} - \frac{X(z)}{2\pi i}\int_L \frac{c\,dt}{(a+ib)X^+(t)t}. \qquad (93.17)$$

§ 94. Particular cases: The mixed problem of the theory of holomorphic functions.

Let there be given on the contour L the disconnected arcs a_1b_1, $a_2b_2, \ldots, a_pb_p$ and let *it be required to find the function* $\Phi(z) = u + iv$, *holomorphic in* S^+, *satisfying the boundary condition*

$$u^+ = f(t) \text{ on } L', \quad v^+ = g(t) \text{ on } L'', \qquad (94.1)$$

where $f(t)$ *and* $g(t)$ *are functions of the class* H, *given on* L' *and* L'' *respectively*, while L' is the union of the arcs a_jb_j $(j = 1, 2, \ldots, p)$ and L'' is the remaining part of L, i.e., the union of the arcs b_ja_{j+1} $(j = 1, 2, \ldots, p)$; by a_{p+1} is to be understood a_1; as in the last section, $\Phi(z)$ will be assumed to be continuous up to and including L everywhere, with the possible exception of the points a_j, b_j near which

$$|\Phi(z)| < \frac{\text{const.}}{|z-a_j|^\alpha}, \quad |\Phi(z)| < \frac{\text{const.}}{|z-b_j|^\alpha}, \quad \alpha < 1. \qquad (94.2)$$

This problem is a particular case of the preceding one, i.e.,

$$a(t)u^+ - b(t)v^+ = c(t), \qquad (94.3)$$

where now

$$\begin{aligned} a(t) &= 1, \; b(t) = 0, \quad c(t) = f(t) \quad \text{on } L', \\ a(t) &= 0, \; b(t) = -1, \quad c(t) = g(t) \quad \text{on } L''. \end{aligned} \tag{94.4}$$

As in § 93, it can be reduced to the case of the circle by means of conformal transformation. Therefore L may be assumed to be the circle $|z| = 1$ and S^+ the circular region $|z| < 1$.

By the general method, stated in the last section, one may first of all find the fundamental solution $X(z)$ of a given class of the homogeneous Hilbert problem $(a + ib)\Phi^+ + (a - ib)\Phi^- = 0$ (on L), satisfying the additional condition

$$X_*(z) = z^{\varkappa} X(z). \tag{94.5}$$

In the present case, this last problem has the form

$$\Phi^+(t) + \Phi^-(t) = 0 \text{ on } L', \; \Phi^+(t) - \Phi^-(t) = 0 \text{ on } L''. \tag{94.6}$$

The second of the conditions (94.6) shows that L'' is not a line of discontinuity for the function $\Phi(z)$, so that the above problem reduces to the problem of determining a sectionally holomorphic function $\Phi(z)$ with the line of discontinuity $L' = a_1b_1 + \ldots + a_pb_p$ for the first of the conditions (94.6), i.e., to the problem

$$\Phi^+(t) + \Phi^-(t) = 0 \text{ on } L'. \tag{94.7}$$

This problem has already been solved in § 84.

For definiteness, the solution of the largest class, i.e., of the class h_0, will be sought. The corresponding fundamental solution of the problem (94.7) is (§ 84)

$$X(z) = \frac{C}{\sqrt{R(z)}},$$

where C is an arbitrary non-zero constant and

$$R(z) = \prod_{j=1}^{p} (z - a_j)(z - b_j); \tag{94.8}$$

in correspondence with the condition of § 84, it will be assumed that for large $|z|$

$$\sqrt{R(z)} = z^p + \alpha_1 z^{p-1} + \ldots. \tag{94.9}$$

Accordingly, by

$$\sqrt{R(0)} = \sqrt{\prod_{j=1}^{p} a_j b_j} \tag{94.10}$$

will be understood the value of the stated branch of $\sqrt{R(z)}$ for

$z = 0$. Taking into consideration that $\bar{a}_j = a_j^{-1}$, $\bar{b}_j = b_j^{-1}$, one obtains

$$R_*(z) = \bar{R}(1/z) = \prod_{j=1}^{p} (1/z - 1/a_j)(1/z - 1/b_j) = \frac{R(z)}{z^{2p}R(0)},$$

and hence, as is easily verified,

$$[\sqrt{R(z)}]_* = \frac{\sqrt{R(z)}}{z^p\sqrt{R(0)}},$$

where the roots must be understood as stated above. Hence

$$X_*(z) = \frac{\bar{C}}{[\sqrt{R(z)}]_*} = \frac{\bar{C}\sqrt{R(0)}}{\sqrt{R(z)}} z^p = \frac{\bar{C}}{C} \sqrt{R(0)}\, z^p X(z). \quad (94.11)$$

Since in the present case

$$\varkappa = \varkappa_0 = p \quad (94.12)$$

(because the degree of $X(z)$ at infinity is known to be $-\varkappa$), the relation (94.5) will be satisfied, if

$$\frac{\bar{C}}{C}\sqrt{R(0)} = 1.$$

Let

$$\sqrt{R(0)} = \sqrt{\prod_{j=1}^{p} a_j b_j} = e^{i\alpha}, \quad (94.13)$$

where α is a real constant. Then one can obviously take

$$C = e^{\frac{i\alpha}{2}} = \sqrt[4]{R(0)}. \quad (94.14)$$

Thus, finally, the fundamental solution of the class h_0 of the Hilbert problem (94.7), satisfying the condition (94.5), is given by

$$X(z) = \frac{\sqrt[4]{R(0)}}{\sqrt{R(z)}} = \frac{e^{\frac{i\alpha}{2}}}{\sqrt{R(z)}}; \quad (94.15)$$

this function is at the same time the fundamental solution of the required form of the homogeneous Hilbert problem (94.6).

Consequently, from the results of the last section, *the general solution of the class h_0 of the homogeneous mixed problem, obtained from (94.1) for $f(t) \equiv 0$, $g(t) \equiv 0$, is given by*

$$\Phi_0(z) = \frac{\sqrt[4]{R(0)}}{\sqrt{R(z)}}(C_0 z^p + C_1 z^{p-1} + \ldots + C_p) \quad (94.16)$$

where $C_0, C_1, \ldots, C_p$ *are (in general complex) constants, connected by the relations*

$$C_{p-j} = \overline{C}_j, \; j = 0, 1, \ldots, p \tag{94.17}$$

and otherwise arbitrary.

Having obtained $X(z)$ and the general solution (94.16) of the homogeneous problem, the general solution of the class h_0 of the original non-homogeneous problem can be written down immediately by use of (93.15); in fact, this formula gives some particular solution of the non-homogeneous problem; the required general solution of the class h_0 of the original problem is obtained by adding to the latter the general solution (94.16) of the homogeneous problem.

However, a simpler expression for the general solution is obtained, if a solution of one of those classes corresponding to the index zero is chosen instead of the particular solution.

For example, such a class will be the class $h(a_1, a_2, \ldots, a_p)$, i.e., the class of solutions bounded at the ends $a_1, a_2, \ldots, a_p$ and possibly unbounded at the ends $b_1, b_2, \ldots, b_p$. The corresponding fundamental solution of the homogeneous Hilbert problem (94.7) will obviously be

$$Z(z) = C\sqrt{\frac{R_a(z)}{R_b(z)}},$$

where

$$R_a(z) = \prod_{j=1}^{p} (z - a_j), \quad R_b(z) = \prod_{j=1}^{p} (z - b_j). \tag{94.18}$$

From the condition, adopted in § 84, the root above must be understood as the branch for which

$$\left[\sqrt{\frac{R_a(z)}{R_b(z)}}\right]_{z=\infty} = 1. \tag{94.19}$$

Again put

$$\sqrt{\frac{R_a(0)}{R_b(0)}} = \sqrt{\prod_{j=1}^{p} \frac{b_j}{a_j}} = e^{i\beta} \tag{94.20}$$

and

$$C = e^{-\frac{i\beta}{2}} = \sqrt[4]{\prod_{j=1}^{p} \frac{b_j}{a_j}}.$$

Then the fundamental function of the class $h(a_1, a_2, \ldots, a_p)$

$$Z(z) = e^{-\frac{i\beta}{2}}\sqrt{\frac{R_a(z)}{R_b(z)}} \tag{94.21}$$

will satisfy the condition

$$Z_*(z) = Z(z).$$

Correspondingly, one of the particular solutions $\Psi(z)$ of the original non-homogeneous problem will be [by (43.17)]

$$\Psi(z) = \frac{1}{\pi i}\frac{\sqrt{R_a(z)}}{\sqrt{R_b(z)}}\int_L \frac{\sqrt{R_b(t)}\,h(t)dt}{\sqrt{R_a(t)}(t-z)} - \\ - \frac{1}{2\pi i}\frac{\sqrt{R_a(z)}}{\sqrt{R_b(z)}}\int_L \frac{\sqrt{R_b(t)}}{\sqrt{R_a(t)}}\frac{h(t)\,dt}{t}, \tag{94.22}$$

where

$$h(t) = \begin{cases} f(t) \text{ on } L', \\ ig(t) \text{ on } L'', \end{cases} \tag{94.23}$$

and by $\sqrt{R_b(t)}/\sqrt{R_a(t)} = \sqrt{R_b(t)/\sqrt{R_a(t)}}$ must be understood the value taken by $\sqrt{R_b(z)}/\sqrt{R_a(z)} = 1/\sqrt{R_a(z)/R_b(z)}$ (from the left side) on L (i.e., from within the circle L).

The general solution of the class h_0 of the original problem will be

$$\Phi(z) = \Psi(z) + \Phi_0(z), \tag{94.24}$$

where $\Psi(z)$ is given by (94.22) and $\Phi_0(z)$ is determined by (94.16), (94.17).

Furthermore, note that the general solution of the class $h(a_1, \ldots, a_p)$ will obviously be given by

$$\Phi(z) = \Psi(z) + Ce^{-\frac{i\beta}{2}}\sqrt{\frac{R_a(z)}{R_b(z)}}, \tag{94.25}$$

where C is an arbitrary real constant.

§ 95. The mixed problem for the half-plane. Formula of M. V. Keldysh and L. I. Sedov.

The mixed problem for the half-plane, like the general Riemann-Hilbert problem for the half-plane, can be reduced to the same problem for the circle by means of a simple inversion. However, a direct solution will be given here.

Let there be given on the real axis Ox the disconnected finite segments a_jb_j ($j = 1, 2, \ldots, p$). Denote by L' the union of these segments and by L'' the remaining part of the real axis, so that L'' consists of the finite segments b_ja_{j+1} ($j = 1, 2, \ldots, p-1$) and the infinite "segment" b_pa_1, formed by the two semi-infinite straight lines $b_p < x < \infty$ and $-\infty < x < a_1$. In this case the mixed problem will be formulated in the following manner:

To find a function $\Phi(z) = u + iv$, holomorphic in the upper half-plane $y > 0$ and bounded at infinity, satisfying the boundary condition

$$u^+ = f(t) \text{ on } L', \quad v^+ = g(t) \text{ on } L''; \tag{95.1}$$

it will be assumed that near L ($L = L' + L''$ being the entire real axis) the unknown function $\Phi(z)$ satisfies the same conditions as the function $\Phi(z)$ of the last section and that the given functions $f(t)$, $g(t)$ satisfy the H condition on L' and L'' respectively. In addition, it will be assumed that $g(t)$ is for large $|t|$ subject to the condition

$$\lim_{t \to +\infty} g(t) = \lim_{t \to -\infty} g(t),$$

$$|g(t) - g(\infty)| < \frac{\text{const.}}{|t|^\alpha}, \quad \alpha > 0, \tag{95.2}$$

where $g(\infty) = \lim_{t \to \pm\infty} g(t)$ (cf. § 43, 1°).

The solution of the problem can be obtained by a method similar to that used in the last section. Therefore only brief remarks will be made here.

The homogeneous Hilbert problem corresponding to the present problem has the form

$$\Phi^+(t) + \Phi^-(t) = 0 \text{ on } L', \quad \Phi^+(t) - \Phi^-(t) = 0 \text{ on } L'' \tag{95.3}$$

(just as in § 94). The fundamental solution of the class h_0 of this problem is given, apart from a constant factor, by

$$X(z) = \frac{1}{\sqrt{R(z)}}, \tag{95.4}$$

where

$$R(z) = \prod_{j=1}^{p} (z - a_j)(z - b_j); \tag{95.5}$$

by $\sqrt{R(z)}$ is to be understood the branch, holomorphic in the plane cut along L', such that near $z = \infty$

$$\sqrt{R(z)} = z^p + \alpha_1 z^{p-1} + \ldots; \tag{95.6}$$

this is equivalent to the fact that $\sqrt{R(z)}$ takes on Ox for $x > b_p$ positive values.

It is easily seen that $X_*(z) = \overline{X}(z) = X(z)$. Accordingly, the general solution of the class h_0 of the problem, obtained from (95.1) for $f(t) = g(t) = 0$, is given by

$$\Phi_0(z) = \frac{C_0 z^p + C_1 z^{p-1} + \ldots + C_p}{\sqrt{R(z)}}, \tag{95.7}$$

where the $C_0, C_1, \ldots, C_p$ are arbitrary real constants.

As in the last section, the fundamental solution $Z(z)$ of the class $h(a_1, a_2, \ldots, a_p)$ of the problem (95.3) will now be found. Such a solution, apart from a constant multiplier, will obviously be

$$Z(z) = \sqrt{\frac{R_a(z)}{R_b(z)}}, \tag{95.8}$$

where

$$R_a(z) = \prod_{j=1}^{p} (z - a_j), \quad R_b(z) = \prod_{j=1}^{p} (z - b_j),$$

and by $\sqrt{R_a(z)/R_b(z)} = \sqrt{R_a(z)}/\sqrt{R_b(z)}$ is to be understood the branch, holomorphic in the plane cut along L', reducing to unity at infinity. Also in this case $Z_*(z) = Z(z)$.

One of the particular solutions $\Psi(z)$ of the original problem (95.1) is given by [cf. § 43, (43.13)]

$$\Psi(z) = \frac{1}{\pi i} \frac{\sqrt{R_a(z)}}{\sqrt{R_b(z)}} \int_L \frac{\sqrt{R_b(t)}}{\sqrt{R_a(t)}} \cdot \frac{h(t)dt}{t - z}, \tag{95.9}$$

where

$$h(t) = \begin{cases} f(t) & \text{on } L', \\ ig(t) & \text{on } L'', \end{cases} \tag{95.10}$$

and by $\sqrt{R_b(t)}/\sqrt{R_a(t)}$ is understood the value taken by $\sqrt{R_b(z)}/\sqrt{R_a(z)}$, when $z \to t$ from the upper half-plane; $\sqrt{R_b(z)}/\sqrt{R_a(z)}$, for its part, is the inverse of $\sqrt{R_a(z)}/\sqrt{R_b(z)}$.

The general solution of the class h_0 of the original problem (95.1) is thus given by

$$\Phi(z) = \Psi(z) + \Phi_0(z), \tag{95.11}$$

where $\Phi_0(z)$ and $\Psi(z)$ are defined by (95.7) and (95.9).

The general solution of the class $h(a_1, a_2, \ldots, a_p)$ of the same problem is given by

$$\Phi(z) = \Psi(z) + C\frac{\sqrt{R_a(z)}}{\sqrt{R_b(z)}},$$

where C is an arbitrary real constant.

The formulae, obtained in this section, are due to M. V. Keldysh and L. I. Sedov [1] (cf. also A. Signorini [1] and L. I. Sedov [1].

CHAPTER 13

EFFECTIVE SOLUTION OF THE PRINCIPAL PROBLEMS OF THE STATIC THEORY OF ELASTICITY FOR THE HALF-PLANE, CIRCLE AND ANALOGOUS REGIONS

In this chapter the solution of the principal problems of the plane theory of elasticity will be given for the half-plane, circle and also for the plane with straight cuts. The method stated was given by the Author in the papers [9], [10], [11] and applied to the half-plane and plane with straight cuts. This method was applied to the case of the circle by the post-graduate student I. N. Kartsivadze [1].

§ 96. General formulae of the plane theory of elasticity.

In order to facilitate reference, the fundamental formulae of the plane static theory of elasticity in the absence of body forces will be given here; their deduction may be found in the Author's book [1].

Let X_x, Y_y, X_y be the stress components and u, v the displacement components in some cartesian coordinate system. (In the plane problem, in the usual notation, $X_z = Y_z = w = 0$; the component Z_z is, in general, non-zero, but can be expressed directly in terms of X_x and Y_y).

These components can be expressed in terms of two analytic functions $\Phi(z)$, $\Psi(z)$ of the complex variable $z = x + iy$ in the following manner:

$$X_x + Y_y = 2[\Phi(z) + \overline{\Phi(z)}], \tag{96.1}$$

$$Y_y - X_x + 2iX_y = 2[\bar{z}\Phi'(z) + \Psi(z)], \tag{96.2}$$

$$2\mu(u + iv) = \varkappa\varphi(z) - z\overline{\varphi'(z)} - \overline{\psi(z)}, \tag{96.3}$$

where

$$\varphi(z) = \int \Phi(z)dz + \text{const}, \quad \psi(z) = \int \Psi(z)dz + \text{const}, \tag{96.4}$$

and

$$\varkappa = \frac{\lambda + 3\mu}{\lambda + \mu} > 1; \tag{96.5}$$

λ and μ are positive constants (Lamé's constants), characterising the elastic properties of the body. [The formulae (96.1)—(96.3) are G. V. Kolosov's known formulae in a different notation].

In particular, (96.1) and (96.2) give

$$Y_y - iX_y = \Phi(z) + \overline{\Phi(z)} + z\overline{\Phi'(z)} + \overline{\Psi(z)}. \qquad (96.6)$$

In the sequel, X_x, Y_y, X_y will be assumed to have continuous partial derivatives up to the second order in the entire region S, occupied by the elastic body under consideration. This condition will be expressed by saying that the state of stress is regular. Under these conditions $\Phi(z)$ and $\Psi(z)$ will be holomorphic in S.

For given X_x, Y_y, X_y, the function $\Psi(z)$ is defined completely and $\Phi(z)$ apart from a purely imaginary constant term.

The functions $\varphi(z)$ and $\psi(z)$, in their turn, are defined for given $\Phi(z)$ and $\Psi(z)$, apart from constant (complex) terms.

In the sequel, cases will have to be considered in which the region S extends to infinity. In all these cases it will be assumed that the stress components remain bounded at infinity.

First, the boundary of the infinite region will be assumed to lie entirely in the finite part of the plane. Then it can be shown that for large $|z|$ the following relations hold:

$$\begin{aligned} \Phi(z) &= \Gamma_0 - \frac{X+iY}{2\pi(1+\varkappa)}\cdot\frac{1}{z} + O\left(\frac{1}{z^2}\right), \\ \Psi(z) &= 2\overline{\Gamma} + \frac{\varkappa(X-iY)}{2\pi(1+\varkappa)}\cdot\frac{1}{z} + O\left(\frac{1}{z^2}\right) \end{aligned} \qquad (96.7)$$

(cf. N. I. Muskhelishvili [1], § 36 et. seq.), where (X, Y) is the resultant vector of the external forces, applied to the boundary S,

$$\Gamma_0 = \Gamma_0' + i\Gamma_0'', \quad \Gamma = \Gamma' + i\Gamma'' \qquad (96.8)$$

are constants which likewise can be expressed in terms of quantities, having simple physical meaning; in fact,

$$\Gamma_0' = \tfrac{1}{4}(N_1 + N_2), \ \Gamma_0'' = \frac{2\mu\varepsilon_0}{1+\varkappa}, \quad \Gamma = -\tfrac{1}{4}(N_1 - N_2)e^{2i\alpha}, \qquad (96.9)$$

where N_1, N_2 are the values of the principal stresses at infinity, α is the angle between the principal axis, corresponding to N_1, and the Ox axis, and ε_0 is the value of the "rotation" at infinity, i.e.,

$$\varepsilon_0 = \left[\frac{\partial v}{\partial x} - \frac{\partial u}{\partial y}\right]_{z=\infty} \qquad (96.10)$$

In the case where the stresses and the rotation vanish at infinity

$$\Phi(z) = -\frac{X+iY}{2\pi(1+\varkappa)}\cdot\frac{1}{z} + O\left(\frac{1}{z^2}\right),$$
$$\Psi(z) = -\frac{\varkappa(X-iY)}{2\pi(1+\varkappa)}\cdot\frac{1}{z} + O\left(\frac{1}{z^2}\right). \tag{96.7a}$$

It will be noted that (96.7), (96.7a) do not hold, generally speaking, if the boundary of the region S extends to infinity.

For example, let S be the "lower" or "upper" half-plane ($y < 0$ or $y > 0$) and let the resultant vector (X, Y) of the external forces, acting on the boundary Ox, be finite. If, in addition, the stresses and the rotation vanish at infinity, then (under known additional assumptions [cf. N. I. Muskhelishvili [1] § 79]) in both cases

$$\Phi(z) = -\frac{X+iY}{2\pi z} + O\left(\frac{1}{z^2}\right), \quad \Psi(z) = \frac{X-iY}{2\pi z} + O\left(\frac{1}{z^2}\right). \tag{96.11}$$

§ 97. The first, second and mixed boundary problems for an elastic half-plane. *)

Let the body under consideration occupy the lower half-plane S^- $(y < 0)$. Then the functions $\Phi(z)$, $\Psi(z)$ of the last section are holomorphic in this half-plane. As in § 38, 1°, the functions $\Phi_*(z) = \bar{\Phi}(z)$ and $\Psi_*(z) = \bar{\Psi}(z)$, holomorphic in the upper half-plane S^+, will be considered [where it will be recalled that, by definition, (§ 38, 1°) $\bar{\Phi}(z) = \overline{\Phi(\bar{z})}$; also the following formula, resulting from this, will be noted: $\bar{\Phi}(\bar{z}) = \overline{\Phi(z)}$,]; the definition of $\Phi(z)$ will be extended to this upper half-plane by putting

$$\Phi(z) = -\bar{\Phi}(z) - z\bar{\Phi}'(z) - \bar{\Psi}(z) \text{ for } z \in S^+; \tag{97.1}$$

of course, such an extension of $\Phi(z)$ to the upper half-plane is not chosen accidentally; the basis for such a choice is the fact that in their present meaning the functions $\Phi(z)$ are the analytic continuation of one another through the unloaded parts of the Ox axis (cf. below).

Now let z be a point of the lower half-plane; then $\bar{z}$ will be a point of the upper half-plane and in (79.1) $\bar{z}$ can be put in place of z. This gives

$$\Phi(\bar{z}) = -\bar{\Phi}(\bar{z}) - \bar{z}\bar{\Phi}'(\bar{z}) - \bar{\Psi}(\bar{z}) \text{ for } z \in S^-,$$

and hence, passing to the conjugate values,

$$\bar{\Phi}(z) = -\Phi(z) - z\Phi'(z) - \Psi(z)$$

*) This and the following sections reproduce with some unimportant additions the contents of the Author's paper [9].

or

$$\Psi(z) = -\Phi(z) - \bar{\Phi}(z) - z\Phi'(z) \text{ for } z \in S^-. \qquad (97.2)$$

Thus the function $\Psi(z)$ for z in S^- is expressed by $\Phi(z)$, extended by the stated method to the half-plane S^+. (If $\Phi(z)$ were not continued into S^+, then the expression $\bar{\Phi}(z) = \overline{\Phi(\bar{z})}$ would be meaningless for $z \in S^-$, and hence for $\bar{z} \in S^+$.)

From (97.2), the formula (96.6) can be written as

$$Y_y - iX_y = \Phi(z) - \Phi(\bar{z}) + (z - \bar{z})\overline{\Phi'(z)}. \qquad (97.3)$$

If the function $\Phi(z)$ is known, then (96.1) and (97.3) completely determine the stress components. Thus *the components of stress are expressed by a single unique function* $\Phi(z)$, defined both in the upper and lower half-planes.

Nothing remarkable would lie in this fact, if the values of $\Phi(z)$ in the upper and lower half-plane were not connected with one another, because then there would be essentially two different functions. But the fact is that, as will now be assumed, *the values of* $\Phi(z)$ *in the upper and lower half-planes are the analytic continuations of one another through the unloaded parts of the Ox axis*, so that in the case in which there are unstressed parts of the Ox axis, one is actually dealing with a single function $\Phi(z)$, holomorphic in the whole plane, cut along the stressed parts of the Ox axis.

In the sequel (unless stated otherwise) it will be assumed that only that part L of the boundary Ox is stressed, which consists of the segments $L_k = a_k b_k$ ($k = 1, 2, \ldots, n$), and that the remaining part of the boundary Ox is free from external stresses (i.e., that $Y_y = X_y = 0$ on Ox outside the segments L_k). In addition, it will be assumed that the limits $\Phi^+(t)$, $\Phi^-(t)$ exist for all points t of the Ox axis, except possibly at the points a_k, b_k, and that

$$\lim_{y \to 0} y\Phi'(z) = \lim_{y \to 0} y\Phi'(t + iy) = 0 \qquad (97.4)$$

for all t, not coinciding with a_k, b_k. These restrictions are imposed, in order to ensure that X_x, Y_y, X_y tend to definite limits as $z \to t$; the points a_k, b_k have been excluded, so that no solutions, important from the point of view of application, are lost.

For these conditions, as shown by (97.3), one will have on the unstressed parts of the Ox axis $\Phi^-(t) - \Phi^+(t) = 0$, and this also shows that the values of $\Phi(z)$ in S^+ and S^- are analytic continuations of one another through the unstressed parts.

In the whole of this section the stresses and the rotation will be assumed to vanish at infinity. Then it is easily established from (96.11) of the last section that for large $|z|$

$$\Phi(z) = -\frac{X+iY}{2\pi z} + O\left(\frac{1}{z^2}\right), \tag{97.5}$$

where (X, Y) is the resultant vector of the external forces, applied to the boundary; for the lower half-plane, this relation expresses the same as the first expression of (96.11); for the upper half-plane, it follows from (97.1) and from the second relation of (96.11).

Further, near the points a_k, b_k, where, by supposition, it may not be bounded, the function $\Phi(z)$ will be assumed to satisfy the following condition:

$$|\Phi(z)| < \frac{A}{|z-c|^\alpha}, \tag{97.6}$$

where A, α are constants, $0 \leqq \alpha < 1$, and c is any of the points a_k, b_k.

Note also the formula

$$2\mu(u'+iv') = \varkappa\Phi(z) + \Phi(\bar{z}) - (z-\bar{z})\overline{\Phi'(z)} \tag{97.7}$$

which follows from (96.3) and (97.2), where u' and v' are the partial derivatives of u and v with respect to x.

These formulae make it extraordinarily easy to solve the fundamental boundary problems for the half-plane. The solution of some of them will be given in this and the following sections; the problems considered in 1° and 2° of the present section have already been solved long ago by other methods (cf. for example N. I. Muskhelishvili [1]); but it was thought expedient to give their solution also by the new method, in order to demonstrate its effectiveness by means of simple examples.

1°. THE FIRST FUNDAMENTAL PROBLEM. In this problem the external stresses are given: the pressure $P(t) = -Y_y^-$ and the tangential stress $T(t) = X_y^-$, applied to the entire boundary Ox (where it must not be forgotten that the elastic body occupies the lower half-plane S^-, for which reason the boundary values Y_y and X_y are denoted by Y_y^- and X_y^-). For simplicity it will be assumed that for large $|t|$: $P(t) = O(t^{-2})$, $T(t) = O(t^{-2})$ [which restriction is easily seen to follow from (97.5) and the condition (97.4); however, the problem can be solved for more general conditions]; also let $P(t)$ and $T(t)$ satisfy the H condition.

By (97.3), the boundary condition of the problem has the form

$$\Phi^+(t) - \Phi^-(t) = P(t) + iT(t), \tag{97.8}$$

and hence it follows directly that

$$\Phi(z) = \frac{1}{2\pi i}\int_{-\infty}^{+\infty} \frac{P(t)+iT(t)}{t-z}\,dt. \tag{97.9}$$

The solution (97.9) is easily seen to satisfy all the imposed conditions; in particular, the condition (97.4) follows from the statements in § 21.

2°. THE SECOND FUNDAMENTAL PROBLEM. In this problem the boundary values of the displacement components u, v are given: $u^- = g_1(t)$, $v^- = g_2(t)$ on the entire boundary Ox. It will be assumed that the given functions $g_1(t)$ and $g_2(t)$ have the first derivatives $g_1'(t)$, $g_2'(t)$, satisfying the H condition and, in addition, for large $|t|$ the condition (cf. the similar restrictions above): $g'_1(t) = O(t^{-1})$, $g'_2(t) = O(t^{-1})$. The boundary condition, which can be written as $u^- + iv^- = g_1(t) + ig_2(t)$, gives $u^{-\prime} + iv^{-\prime} = g_1'(t) + ig_2'(t)$, and hence, using (97.7) and assuming that the boundary values of the partial derivatives

$$u' = \frac{\partial u}{\partial x}, \quad v' = \frac{\partial v}{\partial x}$$

are equal to the derivatives $u^{-\prime}$, $v^{-\prime}$ of the boundary values u^-, v^- [where the assumption that the boundary value of the derivatives equals the derivatives of the boundary values is easily justified after the solution has been obtained], one obtains

$$\Phi^+(t) + \varkappa\Phi^-(t) = 2\mu[g_1'(t) + ig_2'(t)]. \tag{97.10}$$

Let $\Omega(z)$ be the function, equal to $\Phi(z)$ for $y > 0$ and equal to $-\varkappa\Phi(z)$ for $y < 0$. Then (97.10) takes the form

$$\Omega^+(t) - \Omega^-(t) = 2\mu[g'_1(t) + ig'_2(t)], \tag{97.10a}$$

and hence

$$\Omega(z) = \frac{\mu}{2\pi i}\int_{-\infty}^{+\infty} \frac{g'_1(t) + ig_2'(t)}{t - z}\, dt, \tag{97.11}$$

so that, finally,

$$\Phi(z) = \begin{cases} \Omega(z) & \text{for } y > 0, \\ -\dfrac{1}{\varkappa}\Omega(z) & \text{for } y < 0, \end{cases} \tag{97.12}$$

where $\Omega(z)$ is given by (97.11).

3°. THE FUNDAMENTAL MIXED PROBLEM.

In the case $n = 1$, a fairly simple, but not effective solution of this problem (for $n = 1$ the problems **A** and **B**, formulated below, will coincide) was obtained by the Author several years ago [8]; two years later V. M. Abramov [1] gave a much more effective solution, likewise for the case $n = 1$, using Mellin integrals. The solution given here for arbitrary n, obtained by the Author in his paper [9], is incomparably simpler. A solution for the particular case of rectilinear foundations [$g'(t) \equiv 0$], somewhat more complicated, but essentially closely related to the one given below, was later obtained by N. I. Glagolev [1], [2] who was not acquainted with the Author's paper [9].

As before, let $L = L_1 + L_2 + \ldots + L_n$ be the union of a finite number of segments L_k of the boundary Ox and let the displacement components be given on L and the external stress components on the remaining part L' of the boundary. Having solved the first fundamental problem (1°), the problem considered here can obviously always be reduced to the case in which the external stress components Y_y, X_y, given on L', are zero.

Thus it will be assumed that the part L' of the boundary is free from external stresses, i.e., that

$$Y_y^- = X_y^- = 0 \text{ on } L'. \tag{97.13}$$

In view of the great practical interest of the present problem, two of its versions will be considered ("Problem A" and "Problem B"). In both versions the boundary condition on L will have the form

$$u^- + iv^- = g(t) + c(t) \text{ on } L, \tag{97.14}$$

where $g(t) = g_1(t) + ig_2(t)$ is a function, given on L.

In *Problem A* it will be assumed that $c(t) = c =$ const. on L and that the resultant vector (X, Y) of the external stresses, applied to L, is given. For example, without loss of generality, c can be assumed to be zero, because the value of c affects only the rigid body displacement of the whole system.

In *Problem B* it will be assumed that $c(t) = c_k$ on L_k, where the c_k are constants, given beforehand and different for different segments. In this case the principal vectors (X_k, Y_k) of the external stresses, applied to every segment L_k separately, will be assumed given. Without loss of generality, one of the constants c_k can be fixed arbitrarily, for example $c_1 = 0$.

These problems correspond to the somewhat generalized problem of the pressure of a system of rigid stamps of given form on the boundary of a half-plane in the absence of slipping, while it is assumed that the stamps can only move vertically (i.e., perpendicularly to the straight boundary). In Problem A the stamps are rigidly connected, in Problem B they can move vertically, independent of one another. [The stated problems are idealisations of the somewhat generalized problem of the pressure of a foundation on the soil in the presence of so much friction (or rather adhesion) that sliding is excluded and the base of the foundation remains in contact with the soil].

The case in which the stamps can tilt is easily reduced to the above (cf. below).

By ordinary methods each of thc cases A and B is easily shown not to have more than one solution, provided there is no rigid body displacement of the whole system (i.e., of the elastic half-plane and the stamps).

In the sequel, the given function $g(t)$ will be assumed to possess the first derivative $g'(t)$, satisfying the H condition on L.

In both the cases A and B the boundary condition (97.14) gives $u^{-\prime} + iv^{-\prime} = g'(t)$ on L, and hence from (97.7) [cf. remarks prior to (97.10)]

$$\Phi^+(t) + \varkappa\Phi^-(t) = 2\mu g'(t) \text{ on } L. \tag{97.15}$$

However, the boundary condition (97.13) is equivalent to the demand that $\Phi^+(t) - \Phi^-(t) = 0$ on L', i.e., that $\Phi(z)$ is holomorphic in the whole plane cut along L, as will now be assumed.

The condition (97.15) differs essentially from (97.10) in that L is now only part of the Ox axis.

The problem (97.15) is a very simple particular case of the Hilbert problem, considered in § 81. Applying the results of that section, one obtains easily the general solution (where the most general solution, i.e., the solution of the class h_0 is being considered)

$$\Phi(z) = \frac{[R_0(z)]^{i\gamma}}{\sqrt{R(z)}} \cdot \frac{\mu}{\pi i} \int_L \frac{\sqrt{R(t)}\,[R_0(t)]^{-i\gamma} g'(t)dt}{t - z} + \frac{[R_0(z)]^{i\gamma} P_{n-1}(z)}{\sqrt{R(z)}}. \tag{97.16}$$

Here

$$\gamma = \frac{\log \varkappa}{2\pi}, \tag{97.17}$$

$$R(z) = \prod_{k=1}^{n} (z - a_k)(z - b_k), \quad R_0(z) = \prod_{k=1}^{n} \frac{z - a_k}{z - b_k}, \tag{97.18}$$

and

$$P_{n-1}(z) = C_0 z^{n-1} + C_1 z^{n-2} + \ldots + C_{n-1}$$

is a polynomial of degree not greater than $n - 1$ with the arbitrary coefficients $C_0, C_1, \ldots, C_{n-1}$; further, by $\sqrt{R(z)}$ one must, as always, understand the branch, single-valued in the plane cut along L, such that $\sqrt{R(z)}/z^n \to 1$ for $z \to \infty$, and by $[R_0(z)]^{\pm i\gamma}$ a branch, single-valued in the same region and tending to 1 as $z \to \infty$. Finally, by $\sqrt{R(t)}$ and $[R_0(t)]^{\pm i\gamma}$ on L one has to understand the values which $\sqrt{R(z)}$ and $[R_0(z)]^{\pm i\gamma}$ take on the upper side of L.

There remains to determine the coefficients $C_0, C_1, \ldots, C_{n-1}$ of the polynomial $P_{n-1}(z)$, arising from the additional conditions of the problems A and B.

First the problem B will be considered; in this case the given resultant vectors (X_k, Y_k) will serve as additional conditions. In order to express these, the normal pressure $P = -Y_y^-$ and the tangential stress $T = X_y^-$, exerted on the boundary of the half-plane by the stamps, i.e., on L, will be calculated.

The formula (97.3) gives for points t_0 on L

$$P(t_0) + iT(t_0) = \Phi^+(t_0) - \Phi^-(t_0),$$

and hence, starting from (97.16) and using the Plemelj formulae, one easily obtains

$$P(t_0) + iT(t_0) = \frac{\mu(\varkappa - 1)}{\varkappa} g'(t_0) + \frac{\varkappa + 1}{\varkappa} \Phi_0(t_0) + \frac{\varkappa + 1}{\varkappa} \cdot \frac{[R_0(t_0)]^{i\gamma} P_{n-1}(t_0)}{\sqrt{R(t_0)}}, \tag{97.19}$$

where

$$\Phi_0(t_0) = \frac{[R_0(t_0)]^{i\gamma}}{\sqrt{R(t_0)}} \cdot \frac{\mu}{\pi i} \int_L \frac{\sqrt{R(t)}\,[R_0(t)]^{-i\gamma} g'(t)dt}{t - t_0}. \tag{97.20}$$

Expressing that

$$\int_{L_k} [P(t_0) + iT(t_0)]dt_0 = -Y_k + iX_k, \tag{97.21}$$

one obtains a system of n linear equations for the determination of the constants C_j; this system has a unique solution, as is easily seen from the uniqueness of the solution of the original problem.

Next the problem A will be considered. Since the above solution satisfies the condition $u'^- + iv'^- = u^{-\prime} + iv^{-\prime} = g'(t)$ on L (as is easily verified by direct substitution), one will have on the segments L_k

$$u^- + iv^- = g(t) + c_k,$$

where the c_k are constants. There remains to consider the condition

$$c_1 = c_2 = \ldots = c_n. \tag{97.22}$$

For this purpose the values of $u'^- + iv'^-$ on the unstressed parts of the boundary will be calculated. From (97.7) one has for the point t_0 of the Ox axis, not belonging to L,

$$2\mu(u'^- + iv'^-) = (\varkappa + 1)\Phi(t_0) = (\varkappa + 1)\Phi_0(t_0) + (\varkappa + 1)\frac{[R_0(t_0)]^{i\gamma} P_{n-1}(t_0)}{\sqrt{R(t_0)}}, \tag{97.23}$$

where $\Phi_0(t_0)$ is given by (97.20), while t_0 lies this time on Ox outside L. Obviously (97.22) leads to the following conditions:

$$g(a_{k+1}) - g(b_k) = \int_{b_k}^{a_{k+1}} (u'^- + iv'^-)dt_0, \quad k = 1, 2, \ldots, n-1. \quad (97.24)$$

Substituting here for $u'^- + iv'^-$ from (97.23), a system of $(n - 1)$ linear equations for the C_j is obtained. One more equation is given by the condition that the resultant vector (X, Y) of the external stresses, applied to L, is known. It is altogether simpler to express this condition by starting from (96.11) which gives

$$\lim_{z\to\infty} z\Phi(z) = -\frac{X + iY}{2\pi}.$$

Applying this result to (97.16), one obtains immediately that the coefficient C_0 of z^{n-1} in the polynomial $P_{n-1}(z)$ is given by

$$C_0 = -\frac{X + iY}{2\pi}. \quad (97.25)$$

Thus there only remains to determine the coefficients $C_1, C_2, \ldots, C_{n-1}$ for which one has the system of $(n - 1)$ linear equations, mentioned above. This system has always a unique solution, as in the preceding case.

In the particular case where $g'(t) = 0$ (stamps or foundations with straight bases, parallel to the Ox axis) the formulae (97.16) and (97.19) take a very simple form, since the integral becomes zero.

Hitherto it was assumed that the stamps can only move vertically. Now the case will be considered in which they can tilt (of course, in the plane).

In Problem A (where the stamps are interconnected) let ε be the angle by which the system of stamps tilts. Then $g(t)$ must be replaced in the boundary condition (97.14) by $g(t) + i\varepsilon t$ which entails the substitution of $g'(t) + i\varepsilon$ for $g'(t)$ in all the preceding formulae.

Correspondingly there appears in the expression for $\Phi(z)$ an additional term, containing ε as a factor.

The value of ε may not be given directly; instead of it there may, for example, be given the resultant moment M of the external forces about O, acting on the system on stamps. Then there will be the additional relation for the determination of ε

$$M = -\int_L t_0 P(t_0)\, dt_0. \quad (97.26)$$

In the case of the problem B the angles of rotation ε_k ($k = 1, 2, \ldots, n$) of the various stamps can be different; if they are not given directly and, say, the moments M_k of the external stresses are given, acting on the separate stamps, then there will be the n additional equations for the determination of the ε_k

$$M_k = -\int_{L_k} t_0 P(t_0)\, dt_0 \tag{97.27}$$

(cf. the analogous calculation in § 100).

§ 98. The problem of pressure of rigid stamps on the boundary of an elastic half-plane in the absence of friction.

This problem was first solved by M. A. Sadowsky [1] for one particular case. The general solution for the case $n = 1$ was given by the Author [1]. Later on A. I. Begiashvili [1] gave a simple solution for arbitrary n, starting from the Author's method and using Keldysh and Sedov's formula (§ 95). The much simpler solution, stated here, was given by the Author in his paper [9]; simultaneously (and independently) the post-graduate student Bitsadze found a solution which is in essence closely related to the last.

This problem is similar to the preceding one (i.e., 3° of § 97) with the difference that now

$$\begin{aligned} X_y^- &= 0 \text{ everywhere on } Ox, \\ Y_y^- &= 0 \text{ on } Ox \text{ outside } L, \end{aligned} \tag{98.1}$$

while on $L = a_1b_1 + \ldots + a_nb_n$ only the normal component of displacement is given

$$v^-(t) = f(t) + c(t); \tag{98.2}$$

in the last formula $f(t)$ is a function, given on L and characterising the shape of the bases of the stamps. In fact, $y = f(x)$ ($x \in L$) is the equation of the union of the base-profiles of the stamps prior to their displacement; $c(t)$ is defined thus: either $c(t) = c$ on L (rigidly connected stamps) or $c(t) = c_k$ on L_k (disconnected stamps), where the c and c_k are now real constants. Without loss of generality, one can assume $c = 0$ in the first case and, for example, $c_1 = 0$ in the second case; the other c_k cannot be given beforehand.

In addition, in the first case the resultant vector $(0, Y)$ of the external stresses applied to the whole system of stamps will be given; in the second case the resultant vectors $(0, Y_k)$ will be given for each stamp separately.

Here it has been assumed that the stamps can only move vertically;

the case where they can tilt is reduced to the above, as it was done in 3° of the last section.

It will be assumed that $f'(t)$ satisfies the H condition on each of the segments $L_k = a_k b_k$.

The boundary conditions (98.1) show, as in § 97, 3°, that $\Phi(z)$ is holomorphic in the plane cut along L. In addition, it is easily established that the first of the conditions (98.1) gives, using (97.3),

$$\Phi^+(t) + \bar{\Phi}^+(t) = \Phi^-(t) + \bar{\Phi}^-(t) \text{ everywhere on } Ox,$$

and hence it follows that the function $\Phi(z) + \bar{\Phi}(z)$ is holomorphic in the whole plane; further, since it vanishes at infinity, it is zero everywhere. Consequently

$$\bar{\Phi}(z) = -\Phi(z). \tag{98.3}$$

After this, the boundary condition (98.2), which will now be written as [cf. remarks prior to (97.11)]

$$v^{-\prime}(t) = f'(t) \text{ on } L, \tag{98.4}$$

gives, using (97.7),

$$\Phi^+(t) + \Phi^-(t) = \frac{4\mu i f'(t)}{\varkappa + 1} \text{ on } L. \tag{98.5}$$

The general solution (of the class h_0) of the above problem is given by (§ 84)

$$\Phi(z) = \frac{2\mu}{\pi(\varkappa + 1)\sqrt{R(z)}} \int_L \frac{\sqrt{R(t)}\, f'(t)\,dt}{t - z} + \frac{iP_{n-1}(z)}{\sqrt{R(z)}}, \tag{98.6}$$

where $iP_{n-1}(z)$ is an arbitrary polynomial of degree not greater than $n - 1$.

Now the condition (98.3) has to be satisfied. The first term on the right of (98.6) satisfies this condition, since

$$\bar{\Phi}(z) = \Phi_*(z) = \frac{2\mu}{\pi(\varkappa + 1)[\sqrt{R(z)}]_*} \int_L \frac{\overline{\sqrt{R(t)}}\, f'(t)\,dt}{t - z},$$

and

$$[\sqrt{R(z)}]_* = \sqrt{R(z)}, \quad \overline{\sqrt{R(t)}} = \overline{[\sqrt{R(t)}]^+} = [\sqrt{R(t)}]^- = -\sqrt{R(t)}.$$

The second term will likewise satisfy this condition, if and only if all the coefficients of the polynomial $P(z)$ are real numbers.

Thus the general solution of the original problem is given by (98.6), where

$$P_{n-1}(z) = D_0 z^{n-1} + D_1 z^{n-2} + \ldots + D_{n-1} \tag{98.7}$$

is a polynomial with real coefficients.

Using (97.3) and taking into consideration that $X_y^- = 0$ on the boundary, the pressure $P(t)$ of a stamp on the boundary of the half-plane is given by

$$P(t_0) = \Phi^+(t_0) - \Phi^-(t_0) =$$
$$= \frac{4\mu}{\pi(\varkappa + 1)\sqrt{R(t_0)}} \int_L \frac{\sqrt{R(t)}\, f'(t)\, dt}{t - t_0} + \frac{2iP_{n-1}(t_0)}{\sqrt{R(t_0)}}. \qquad (98.8)$$

The coefficients D_j of the polynomial $P_{n-1}(z)$ can be determined from the additional conditions mentioned in the statement of the problem above, just as it was done in the last section.

As an example, attention will be given to the case where the resultant vectors $(0, Y_k)$ of the forces applied to the different stamps are known. Then

$$-Y_k = \int_{L_k} P(t_0) dt_0, \quad k = 1, 2, \ldots, n, \qquad (98.9)$$

where by $P(t_0)$ must be understood the expression (98.8). Thus a system of n linear equations with the n unknowns $D_0, \ldots, D_{n-1}$ is obtained. It is easily shown, based on the uniqueness of the solutions of the problem, that this system has always a unique solution.

So far the stamps were assumed to move only vertically. The case where they can tilt is easily reduced to the above in the same way as this was done in § 97, 3°.

§ 99. The problem of pressure of rigid stamps on the boundary of an elastic half-plane in the absence of friction (continued).

The solution, obtained in the last section, will now be studied in somewhat greater detail. To simplify the investigation, the case of a single stamp will be considered which touches the Ox axis along a single continuous segment ab; the general case can be studied similarly.

In the present case

$$\Phi(z) = \frac{2\mu}{\pi(\varkappa + 1)\sqrt{(z - a)(b - z)}} \int_a^b \frac{\sqrt{(t - a)(b - t)}\, f'(t)\, dt}{t - z} +$$
$$+ \frac{D}{\sqrt{(z - a)(b - z)}} \qquad (99.1)$$

and

$$P(t_0) = \frac{4\mu}{\pi(\varkappa+1)\sqrt{(t_0-a)(b-t_0)}} \int_a^b \frac{\sqrt{(t-a)(b-t)}f'(t)\,dt}{t-t_0} + \\ + \frac{2D}{\sqrt{(t_0-a)(b-t_0)}} \qquad (99.2)$$

will take the place of (98.6) and (98.8) respectively, where D is a real constant. In these formulae $-R(z) = (z-a)(b-z)$ has been introduced instead of $R(z)$, as a result of which the expression for $P(t)$ is in the real form. By $\sqrt{(t-a)(b-t)}$ for $a < t < b$ one has to understand a positive quantity, and by $\sqrt{(z-a)(b-z)}$ the branch, holomorphic in the plane cut along ab and taking positive values on the upper side of ab. This branch is easily seen to be specified by the condition that for large $|z|$

$$\sqrt{(z-a)(b-z)} = -iz + O(1). \qquad (99.3)$$

In order that the obtained solution will be physically possible, one must obviously have $P(t) \geqq 0$ (for $a \leqq t \leqq b$). Thus, after having obtained the solution, it has to be verified that it satisfies this condition.

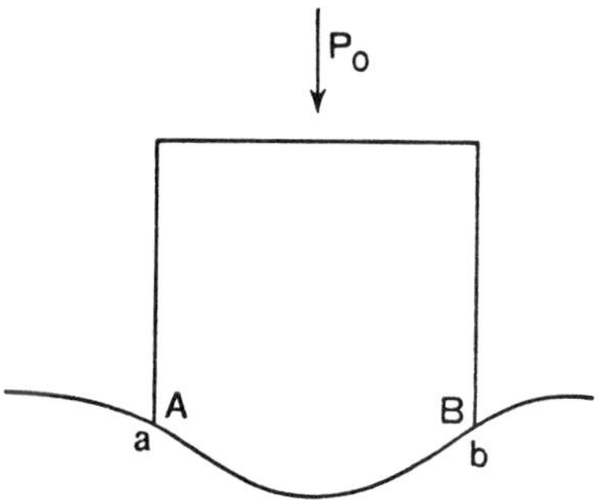

Fig. 11*a*.

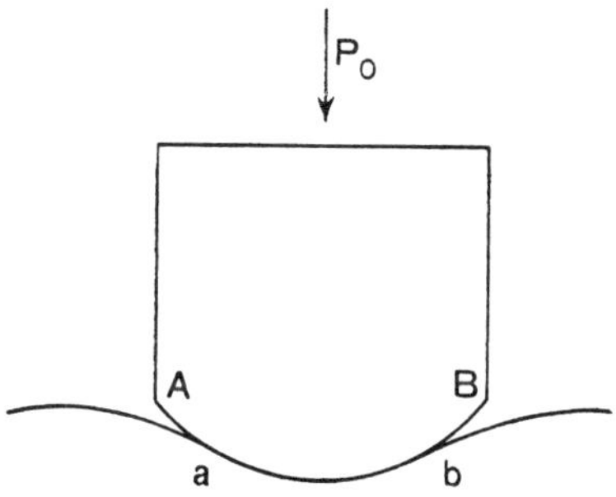

Fig. 11*b*.

For the solution of the problem it has been assumed that the segment of contact ab between the stamp and the elastic half-plane was given beforehand. This corresponds, for example, to the case in which the stamp has the form shown in fig. 11a and the force, applied to the stamp, is sufficiently large, so that the corners A and B of the stamp come into contact with the elastic body. The presence of the corners A, B likewise explains the occurrence of infinitely large stresses at the points a, b of the elastic body, coinciding with the corners A, B of the stamp.

Another case which is of considerable interest is that where the rigid profile, pressed into the elastic half-plane, has no corners (e.g. the case of the pressure of a circular disc) or where the applied force is not sufficiently large for the corners A and B to contact the elastic body, as in Fig. 11b.

In this case the ends a and b of the region of contact are unknown beforehand. However, also in this case the problem can be solved by the formulae obtained. In fact, the general formula (99.2) for the pressure $P(t_0)$ under the stamp will now contain the three initially unknown constants: D, a, b. For the determination of these three constants there are three relations; these are

$$P_0 = \int_a^b P(t_0)\,dt_0, \tag{99.4}$$

where $P_0 = -Y$ is the known force exerted on the stamp, and

$$P(a) = 0, \quad P(b) = 0, \tag{99.5}$$

expressing the condition that the pressure $P(t_0)$ tends continuously to zero, while the point t_0 leaves the region of contact. It will be noted that this last condition can be replaced by the more general (and, in addition, physically still more obvious) one, that the pressure $P(t_0)$ remains bounded at a, b, provided these points are not corner points of the profile of the stamp. Actually it is easily seen that, if the expression (99.2) is bounded at the points a, b, then it is necessarily zero there. In fact, putting for brevity $(t-a)(t-b) = -R(t) = Q(t)$, the formula (99.2) can also be written

$$P(t_0) = \frac{4\mu}{\pi(\varkappa+1)\sqrt{Q(t_0)}}\int_a^b \frac{Q(t)f'(t)\,dt}{\sqrt{Q(t)}(t-t_0)} + \frac{2D}{\sqrt{Q(t_0)}} =$$

$$= \frac{4\mu\sqrt{Q(t_0)}}{\pi(\varkappa+1)}\int_a^b \frac{f'(t)\,dt}{\sqrt{Q(t)}(t-t_0)} +$$

$$+ \frac{4\mu}{\pi(\varkappa+1)\sqrt{Q(t_0)}}\int_a^b \frac{Q(t)-Q(t_0)}{t-t_0}\,\frac{f'(t)\,dt}{\sqrt{Q(t)}} + \frac{2D}{\sqrt{Q(t_0)}} =$$

$$= \frac{4\mu\sqrt{Q(t_0)}}{\pi(\varkappa+1)}\int_a^b \frac{f'(t)\,dt}{\sqrt{Q(t)}(t-t_0)} + \frac{At_0+B}{\sqrt{Q(t_0)}}, \tag{99.2a}$$

where A and B are constants for which expressions are easily written down. The first term on the right side of (99.2a) becomes zero at the ends, by the results in § 29 (4°; here $\gamma = \frac{1}{2}$). Therefore, if $P(t_0)$ is bounded at the ends a, b, necessarily $A = B = 0$ and, consequently, $P(t_0)$ zero there. It will also be noted that in the present case $P(t_0)$ satisfies the H condition on ab, including the ends.

The constant D of the formula (99.2) is immediately determined from the condition (99.4):

$$D = \frac{P_0}{2\pi}. \tag{99.6}$$

This formula is directly determined by evaluation of the integral in (99.4) or, still simpler, by comparison of the coefficients of z^{-1} in (99.1) and (96.11) for large $|z|$.

Thus, for the determination of a and b, there are the two relations (99.5) which, generally speaking, determine them uniquely, provided the condition that $P(t_0) \geqq 0$ for $a \leqq t_0 \leqq b$ is taken into consideration.

For the determination of a and b the following method may still be used. Instead of writing down the most general solution (i.e., the solution of the class h_0) of the problem (98.5), the solution of the class $h(a, b)$ will be given, corresponding to the condition of boundedness of $P(t)$ at the ends a, b, i.e., the solution, remaining bounded at both ends. For, if the function $P(t)$ is bounded at the ends, then, as has been seen, it must satisfy the H condition on ab and become zero for $t = a$, $t = b$. But then likewise the function $\Phi(z)$ is bounded at the ends, as follows from

$$\Phi(z) = \frac{1}{2\pi i} \int_a^b \frac{P(t)\,dt}{t - z}$$

which is a direct consequence of $P(t) = \Phi^+(t) - \Phi^-(t)$.

In this case the index, corresponding to the problem (98.5), is equal to -1 (§ 84), and the corresponding fundamental function can be taken in the form

$$X(z) = \sqrt{(z - a)(b - z)}, \tag{99.7}$$

where the root is to be understood as above.

Therefore, according to the general theory [§ 84, (84.18)], a solution of the class $h(a, b)$ exists, if and only if

$$\int_a^b \frac{f'(t)\,dt}{\sqrt{(t - a)(b - t)}} = 0. \tag{99.8}$$

Provided this condition is satisfied, $\Phi(z)$ is given by

$$\Phi(z) = \frac{2\mu\sqrt{(z-a)(b-z)}}{\pi(\varkappa+1)} \int_a^b \frac{f'(t)\,dt}{\sqrt{(t-a)(b-t)}(t-z)}, \tag{99.9}$$

and the pressure by

$$P(t_0) = \Phi^+(t_0) - \Phi^-(t_0) =$$

$$= \frac{4\mu\sqrt{(t_0-a)(b-t_0)}}{\pi(\varkappa+1)} \int_a^b \frac{f'(t)\,dt}{\sqrt{(t-a)(b-t)}(t-t_0)}. \tag{99.10}$$

The condition (99.4) is easily expressed directly in terms of the given function $f'(t)$. This can be done either by evaluation of the integral in (99.4) or (as is simpler and leads to the same result) by comparison of the coefficients of z^{-1} in the expansions of the right sides of (99.9) and (96.11) for large $|z|$ which gives

$$\frac{4\mu}{\varkappa+1} \int_a^b \frac{tf'(t)\,dt}{\sqrt{(t-a)(b-t)}} = P_0. \tag{99.11}$$

The formulae (99.8) and (99.11) likewise serve for the determination of a and b.

The case, where the stamp moves vertically, has been considered; the case, where it can also tilt, is treated similarly.

EXAMPLE. The calculations, connected with the solution of the above problem, are very much simplified in the case (entirely sufficient in practice), where $f(t)$ is a polynomial or, more generally, a rational function, because then all the integrals can be evaluated in an elementary manner in finite form. It is sufficient to explain this by an example.

Let

$$f(t) = \frac{t^2}{2R}; \tag{99.12}$$

this may be interpreted as the case, where the loaded rigid profile is a circle of large radius R.

In view of the symmetry, one can take in advance $-a = b = l$, where l has to be determined. Then the condition (99.8) will be automatically satisfied. The condition (99.11), however, gives

$$\frac{4\mu}{R(\varkappa+1)} \int_{-l}^{+l} \frac{t^2\,dt}{\sqrt{l^2-t^2}} = P_0,$$

and hence

$$l^2 = \frac{R(\varkappa + 1)}{2\pi\mu} P_0. \tag{99.13}$$

The function $\Phi(z)$ is given by (99.9), where one has to take $a = -l$, $b = l$. The integral, appearing in this formula, can be evaluated in an elementary manner. In fact, it is obvious that

$$I = \frac{1}{\pi i} \int_{-l}^{+l} \frac{t\,dt}{\sqrt{l^2 - t^2}(t - z)} = \frac{1}{2\pi i} \int_{\Delta} \frac{t\,dt}{\sqrt{l^2 - t^2}(t - z)},$$

where Δ is any simple contour, surrounding the segment ab in clockwise direction and not containing z inside. Using (99.3), one obtains by Cauchy's theorem

$$I = \frac{z}{\sqrt{l^2 - z^2}} - i,$$

and hence

$$\Phi(z) = \frac{2\mu i z}{R(\varkappa + 1)} + \frac{2\mu}{R(\varkappa + 1)} \sqrt{l^2 - z^2} \tag{99.14}$$

and

$$P(t) = \Phi^+(t) - \Phi^-(t) = \frac{4\mu}{R(\varkappa + 1)} \sqrt{l^2 - t^2}. \tag{99.15}$$

§ 100. Equilibrium of a rigid stamp on the boundary of an elastic half-plane in the presence of friction.

This section reproduces with minor alterations the Author's paper [11]. Approximately at the same time N. I. Glagolev [1] published a paper in which he gave the solution of the problem, considered here, for the particular case of a stamp with a straight base. N. I. Glagolev's method of solution was somewhat more complicated (as he reduced the problem to a singular integral equation), but in essence it is closely related to the Author's method. In a recently published paper L. A. Galin [1] gave a somewhat different method of solution of the present problem; he also solved it for the case of stamps, sliding with finite constant velocity, and for the case of an anisotropic body.

The problem of the equilibrium of a rigid stamp on the boundary of an elastic half-plane was solved in the last sections for the two extreme cases, where the coefficient of friction is zero (§ 98) and where it is infinitely large (§ 97, 3°); in the last case more assumptions were made, namely, that the elastic material does not leave the stamp and that, thus, the presence of even arbitrarily large negative pressure is admissible.

Using the method, stated in the last sections, the solution for a finite frictional coefficient can also be obtained; this is the only possible case in practice. For this purpose consideration will be restricted only to the case where the stamp is at the verge of equilibrium; but it is obvious that the solution, obtained for this case, may likewise be applied to the case of slow sliding of the stamp on the boundary of the half-plane. More exactly, it will be assumed that under the stamp on the boundary of the elastic half-plane

$$T = kP,$$

where P and T are respectively the pressure and tangential stress, applied to points of the boundary of the half-plane, and k is the frictional coefficient.

1°. As before, the Ox axis will be directed along the boundary of the elastic half-plane and the Oy axis perpendicular to it, so that the elastic body occupies the lower half-plane $y \leq 0$. For such a choice of axes: $P = -Y_y^-$, $T = X_y^-$.

Moreover, the stamp will be assumed to be in contact with the elastic half-plane along a single continuous segment $L = ab$ (the result being easily generalized to the case, where the region of contact consists of a finite number of distinct segments: cf. the preceding sections).

Besides, it will be assumed for the time being that the stamp can only move vertically (i.e., parallel to the Oy axis).

The boundary conditions of the present problem have the form

$$T(t) = kP(t), \tag{100.1}$$

$$v^-(t) = f(t) + \text{const.} \tag{100.2}$$

on L, $T(t) = P(t) = 0$ on Ox outside L. As before, t is here the coordinate of a point on the Ox axis, v is the projection of the displacement on the Oy axis, $f(t)$ is a given function, determining the profile of the stamp; in fact, $y = f(x) + \text{const.}$ is the equation of this profile. The function $f(t)$ will be assumed to have a derivative, satisfying the H condition.

In addition, the quantity

$$P_0 = \int_L P(t)dt, \tag{100.3}$$

i.e., the overall pressure of the stamp on the half-plane, will be assumed known. The total tangential force will then obviously be $T_0 = kP_0$; thus the resultant vector $(X, Y) = (T_0, -P_0)$ of the external force, acting on the stamp, and of the balancing reaction of the elastic half-plane will be given.

In the notation and under the assumptions of § 97, the boundary conditions (100.1) and (100.2) of the present problem, using (97.3) and (97.7) respectively, can be written in the following form, where (as in § 98) the relation $v^{-\prime} = f'(t)$ will be used instead of (100.2):

$$(1-ik)\Phi^+(t) + (1+ik)\bar{\Phi}^+(t) = \\ = (1-ik)\Phi^-(t) + (1+ik)\bar{\Phi}^-(t), \tag{100.4}$$

$$\varkappa\Phi^-(t) + \Phi^+(t) - \varkappa\bar{\Phi}^+(t) - \bar{\Phi}^-(t) = 4i\mu f'(t) \tag{100.5}$$

on L, while the condition $P(t) = T(t) = 0$ on Ox outside L is equivalent to the condition that $\Phi(z)$ is holomorphic outside the segment L.

The formula (100.4) shows that the function $(1-ik)\Phi(z) + (1+ik)\bar{\Phi}(z)$ is holomorphic in the entire plane and, since it must vanish at infinity, that

$$(1-ik)\Phi(z) + (1+ik)\bar{\Phi}(z) = 0 \tag{100.6}$$

in the whole plane. Expressing $\bar{\Phi}(z)$ in terms of $\Phi(z)$ from (100.6) and substituting in (100.5), one obtains the boundary condition for $\Phi(z)$

$$\Phi^+(t) = A\Phi^-(t) + B(t) \quad \text{on } L, \tag{100.7}$$

where

$$A = -\frac{\varkappa+1+ik(\varkappa-1)}{\varkappa+1-ik(\varkappa-1)}, \quad B(t) = \frac{4i\mu(1+ik)}{\varkappa+1-ik(\varkappa-1)}f'(t).$$

The above expressions can be simplified by introduction of the constant γ, defined by

$$\tan\pi\gamma = k\frac{\varkappa-1}{\varkappa+1}, \quad 0 \leqq \gamma < \tfrac{1}{2} \tag{100.8}$$

(remembering that $\varkappa > 1$, $k \geqq 0$). Then

$$\varkappa+1 \pm ik(\varkappa-1) = \sqrt{(\varkappa+1)^2 + k^2(\varkappa-1)^2}\,e^{\pm\pi i\gamma} = \frac{(\varkappa+1)e^{\pm\pi i\gamma}}{\cos\pi\gamma},$$

and hence

$$A = -e^{2\pi i\gamma}, \quad B(t) = \frac{4i\mu(1+ik)e^{\pi i\gamma}\cos\pi\gamma}{\varkappa+1}f'(t). \tag{100.9}$$

Applying the method of § 81 for the solution of the problem (100.7) and noting that one may write

$$\log A = 2\pi i(\tfrac{1}{2}+\gamma),$$

one easily obtains

$$\Phi(z) = \frac{2\mu(1+ik)e^{\pi i\gamma}\cos\pi\gamma}{\pi(\varkappa+1)(z-a)^{\frac{1}{2}+\gamma}(b-z)^{\frac{1}{2}-\gamma}} \int_a^b \frac{(t-a)^{\frac{1}{2}+\gamma}(b-t)^{\frac{1}{2}-\gamma}f'(t)dt}{t-z} + \\ + \frac{C}{(z-a)^{\frac{1}{2}+\gamma}(b-z)^{\frac{1}{2}-\gamma}}, \tag{100.10}$$

where C is a constant and $(z-a)^{\frac{1}{2}+\gamma}(b-z)^{\frac{1}{2}-\gamma}$ means the branch, holomorphic outside the segment ab and taking on the upper side of this segment the real positive value $(t-a)^{\frac{1}{2}+\gamma}(b-t)^{\frac{1}{2}-\gamma}$; the indicated branch is easily seen to be characterised by the fact that

$$\lim_{z\to\infty} \frac{(z-a)^{\frac{1}{2}+\gamma}(b-z)^{\frac{1}{2}-\gamma}}{z} = -ie^{\pi i\gamma}. \tag{100.11}$$

The constant C is immediately determined from (96.11) which gives

$$\lim_{z\to\infty} z\Phi(z) = \frac{-T_0+iP_0}{2\pi} = \frac{iP_0(1+ik)}{2\pi},$$

and hence it follows by (100.11) that

$$C = \frac{P_0(1+ik)e^{\pi i\gamma}}{2\pi};$$

thus (100.10) takes the form

$$\Phi(z) = \frac{2\mu(1+ik)e^{\pi i\gamma}\cos\pi\gamma}{\pi(\varkappa+1)(z-a)^{\frac{1}{2}+\gamma}(b-z)^{\frac{1}{2}-\gamma}} \int_a^b \frac{(t-a)^{\frac{1}{2}+\gamma}(b-t)^{\frac{1}{2}-\gamma}f'(t)dt}{t-z} + \\ + \frac{P_0(1+ik)e^{\pi i\gamma}}{2\pi(z-a)^{\frac{1}{2}+\gamma}(b-z)^{\frac{1}{2}-\gamma}}. \tag{100.12}$$

All the conditions of the problem will be easily seen to be fulfilled, provided, as it was assumed, $f'(t)$ satisfies the H condition on L.

Thus the problem is solved, because $\Phi(z)$ completely specifies the state of stress.

The solution will, of course, only be physically possible, if the pressure $P(t)$ under the stamp, determined by use of (100.12), satisfies the condition $P(t) \geqq 0$. This pressure is easily found. In fact, using (97.3),

$$P(t_0) + iT(t_0) = P(t_0)(1+ik) = \Phi^+(t_0) - \Phi^-(t_0).$$

Evaluating the last difference by the help of the Plemelj formula, one easily obtains

$$P(t_0) = -\frac{4\mu \sin \pi\gamma \cos \pi\gamma}{\varkappa + 1} f'(t_0) +$$

$$+ \frac{4\mu \cos^2 \pi\gamma}{\pi(\varkappa + 1)(t_0 - a)^{\frac{1}{2}+\gamma} (b - t_0)^{\frac{1}{2}-\gamma}} \int_a^b \frac{(t - a)^{\frac{1}{2}+\gamma} (b - t)^{\frac{1}{2}-\gamma} f'(t)dt}{t - t_0} +$$

$$+ \frac{P_0 \cos \pi\gamma}{\pi(t_0 - a)^{\frac{1}{2}+\gamma}(b - t_0)^{\frac{1}{2}-\gamma}}. \tag{100.13}$$

For $k = 0$ (when also $\gamma = 0$), the solution for the idealized case in which there is no friction (§§ 98, 99) is again obtained.

2°. Now the case, most important from the point of view of application, will be considered, i.e., the case in which the base of the stamp is straight.

(a) STAMP WITH A STRAIGHT BASE, PARALLEL TO Ox.

In this case $f'(t) = 0$ and the formulae (100.12) and (100.13) give

$$\begin{aligned} \Phi(z) &= \frac{P_0(1 + ik)e^{\pi i\gamma}}{2\pi(z - a)^{\frac{1}{2}+\gamma}(b - z)^{\frac{1}{2}-\gamma}}, \\ P(t) &= \frac{P_0 \cos \pi\gamma}{\pi(t - a)^{\frac{1}{2}+\gamma}(b - t)^{\frac{1}{2}-\gamma}}. \end{aligned} \tag{100.14}$$

(b) STAMP WITH A STRAIGHT INCLINED BASE.

Let the base of the stamp form a small angle ε with the Ox axis. Then one can put $f(t) = \varepsilon t + \text{const.}$, $f'(t) = \varepsilon$.

Replacing $f'(t)$ in (100.12) by the constant ε and putting, for brevity, $a = -l$, $b = +l$, one obtains after some simple steps

$$\Phi(z) = \frac{(1 + ik)e^{\pi i\gamma}}{2\pi(\varkappa + 1)} \cdot \frac{P_0(\varkappa + 1) - 8\pi\mu\varepsilon\gamma l - 4\pi\mu\varepsilon z}{(l + z)^{\frac{1}{2}+\gamma}(l - z)^{\frac{1}{2}-\gamma}} +$$

$$+ \frac{2\mu\varepsilon i(1 + ik)}{(\varkappa + 1)}. \tag{100.15}$$

In order to obtain this result, one has to evaluate the integral

$$\Omega(z) = \int_{-l}^{+l} \frac{(l + t)^{\frac{1}{2}+\gamma}(l - t)^{\frac{1}{2}-\gamma}dt}{t - z} = \frac{1}{1 + e^{2\pi i\gamma}} \int_\Lambda \frac{(l + t)^{\frac{1}{2}+\gamma}(l - t)^{\frac{1}{2}-\gamma}dt}{t - z},$$

where Λ is a contour, surrounding the segment ab in clockwise direction and not containing z inside; the last equality is easily seen to be true by shrinking Λ into ab. On the other hand, the last integral is easily evaluated by Cauchy's theorem. In fact, using that for large $|z|$

$$(l + z)^{\frac{1}{2}+\gamma}(l - z)^{\frac{1}{2}-\gamma} = - ie^{\pi i\gamma}\,[z + 2l\gamma + O(1/z)],$$

one gets

$$\frac{1}{2\pi i}\int_{\Lambda}\frac{(l + t)^{\frac{1}{2}+\gamma}(l - t)^{\frac{1}{2}-\gamma}dt}{t - z} = (l + z)^{\frac{1}{2}+\gamma}(l - z)^{\frac{1}{2}-\gamma} + ie^{\pi i\gamma}(z + 2l\gamma).$$

The pressure $P(t)$ under the stamp is given by

$$P(t) = \frac{\Phi^+(t) - \Phi^-(t)}{1 + ik} = \frac{\cos\pi\gamma}{\pi(\varkappa + 1)}\cdot\frac{P_0(\varkappa + 1) - 8\pi\mu\varepsilon\gamma l - 4\pi\mu\varepsilon t}{(l + t)^{\frac{1}{2}+\gamma}(l - t)^{\frac{1}{2}-\gamma}}. \tag{100.16}$$

The solution will be physically possible, i.e., $P(t) \geqq 0$ on ab, provided

$$-\frac{P_0(\varkappa + 1)}{4\pi\mu l(1 - 2\gamma)} \leqq \varepsilon \leqq \frac{P_0(\varkappa + 1)}{4\pi\mu l(1 + 2\gamma)}. \tag{100.17}$$

3°. Hitherto the stamp was assumed to move only vertically. This assumption is not important. Actually, if ε is the angle of rotation of the stamp (of course in the plane Oxy), then the corresponding solution $\Phi(z)$ is obtained by replacement of $f'(t)$ by $f'(t) + \varepsilon$ in (100.12). As the result of this substitution, a term, equal to the term on the right of (100.15) containing ε, is added to $\Phi(z)$, determined by (100.12); as in (100.15), it will be assumed here that in (100.12) $a = -l$, $b = +l$. The quantity ε may be calculated from one or the other additional condition. In particular, ε will have a definite value, if it is assumed that also the resultant moment M of the external forces, acting on the stamp, about the origin of coordinates O is given. Then there is the following relation for the determination of ε:

$$M = -\int_{-l}^{l} tP(t)dt \tag{100.18}$$

(the positive direction of rotation, in the definition of the moment as also for the measuring of the angle ε, being assumed counterclockwise for the present choice of coordinate axes).

For the explanation of the above it is sufficient to consider the example of a stamp with a straight base. In this case $P(t)$ will be given by (100.16) and, as the result of a simple calculation (cf. remarks following (100.15)),

$$M = 2\gamma lP_0 + \frac{2\pi\mu(1 - 4\gamma^2)l^2}{(\varkappa + 1)}\,\varepsilon. \tag{100.19}$$

The last relation determines ε in terms of the given M. In particular, if $M = 0$, i.e., if the external forces, acting on the stamp, are equivalent to a single force acting through the centre of the base,

$$\varepsilon = -\frac{\gamma(\varkappa + 1)P_0}{\pi\mu l(1 - 4\gamma^2)}; \tag{100.20}$$

since $0 \leqq \gamma < \frac{1}{2}$, this value of ε corresponds by (100.17) to a physically possible case.

In the general case (i.e., where the base is not straight) ε will be determined quite similarly, since obviously the moment is given by an expression, differing from the right side of (100.19) by a term depending on the form of $f'(t)$ only and not on ε.

§ 101. Another method of solution of the boundary problems for the half-plane.

Instead of starting from the representation of the stress in terms of the functions $\Phi(z)$, $\Psi(z)$, one may use the functions $\varphi(z)$, $\psi(z)$ for that purpose. Also in this case the stresses (and displacements) may obviously be expressed in terms of the single function $\varphi(z)$, suitably continued into the upper half-plane.

One inconvenience will be that $\varphi(z)$ will, generally speaking, be multi-valued. However, this inconvenience can be removed by elimination of the multi-valued part of $\varphi(z)$ which is very simply done. On the other hand, the introduction of $\varphi(z)$ instead of $\Phi(z)$ has the advantage that for the setting up of the boundary conditions, containing the boundary values of the displacements, one has not to differentiate these values beforehand.

§ 102. The problem of contact of two elastic bodies (the generalized plane problem of Hertz).

The two elastic bodies S_1, S_2, in shape closely resembling half-planes and touching along a segment *ab* of their boundaries, will be considered (Fig. 12). The segment of contact *ab* is not given beforehand and is subject to definition. In advance there will be given: the shape of the boundaries (closely resembling straight lines) before the deformation and the resultant vector of the external force, applied, say, by the body S_2 to the body S_1. It will be assumed that there is no friction. Besides, the stresses and rotations (§ 96) of S_1 and S_2 at infinity will be assumed to be zero.

This problem, presenting great independent interest, is also important, because it solves the problem of the contact of two (plane) bodies of arbitrary shape, provided the line of contact is very short in comparison with the size of the bodies; then, if merely the stresses and deformations near the point of contact are of interest, it can be assumed without appreciable error that the body under consideration approximates in shape to a half-plane.

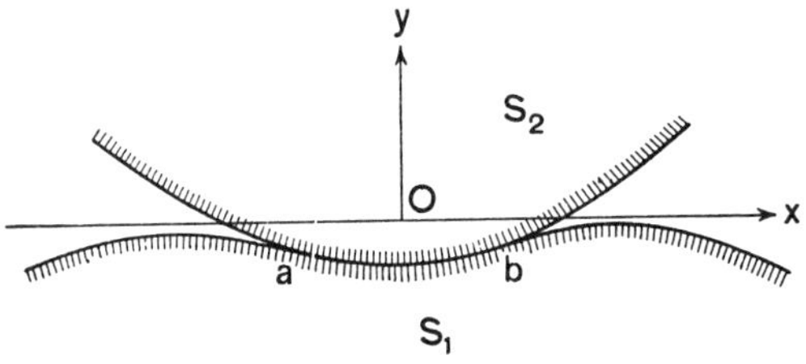

Fig. 12.

In the three-dimensional case, the problem of contact of two elastic bodies was first stated and solved by Hertz for some restricting assumptions, in particular, for the assumption that the region of contact is very small and that the equation of the undeformed surfaces near the point of contact can sufficiently accurately be represented in the form $z = Ax^2 + 2Bxy + Cy^2$ for a suitable choice of coordinate axes. (I. Ya. Shtaerman [2] reduced, under more general assumptions, the three-dimensional Hertz problem to an integral equation.)

Thus the above problem of contact of two bodies, approximating to half-planes, is the two-dimensional analogue of Hertz's problem, but somewhat generalized, since the line of contact was not assumed small and, accordingly, no assumptions were made with regard to the shape of the boundaries, except that they approximate to straight lines (and are sufficiently smooth).

This problem has recently been investigated by several authors. I. Ya. Shtaerman [1], [3] reduced it to the Fredholm equation of the first kind (written in the notation of this book)

$$\int_a^b P(t) \log |t - t_0| \, dt = f(t_0) + \text{const.}, \tag{*}$$

where $P(t)$ is the unknown pressure of the one body on the other at the point t of the line of contact of the bodies and $f(t)$ is a given function. The problem of the pressure of a rigid stamp on an elastic half-plane can be reduced to exactly the same equation (cf. N. I. Muskhelishvili [1], § 87); this equation is easily solved by quadrature for given a and b (cf. N. I. Muskhelishvili [1], § 88). A. V. Bitsadze [1] reduced the problem to a singular integral equation the obvious solution of which is found directly. Actually, Bitsadze's solution can be obtained by differentiating both sides of the equation (*).

The solution of the problem will be established below by a method, completely analogous to that by which the problem was solved for the case where one of the bodies in contact is absolutely rigid (§ 98).

The bodies S_1 and S_2 will be assumed to occupy the lower and the upper half-planes, i.e., S^- and S^+ respectively, and correspondingly the components of stress and displacement and the constants λ, μ, $\varkappa$, referring to S_1 and S_2, will be marked by the suffices 1 and 2.

Let $\Phi_1(z)$ be the sectionally holomorphic function, corresponding to the body S_1 and defined as in § 97; let $\Phi_2(z)$ be a similar function for the body S_2; these functions are holomorphic in the entire plane, with the exclusion of the segment ab of the Ox axis, because the boundaries of the bodies are free from external stresses outside this segment. Since, by assumption, there is no friction, $[X_y^-]_1 = 0$ on Ox; hence, as in § 98, the conclusion is drawn that $\bar{\Phi}_1(z) = -\Phi_1(z)$; similarly one concludes that $\bar{\Phi}_2(z) = -\Phi_2(z)$. Further, if $P(t)$ is the pressure at the point t of the one body on the other, then, as in § 98,

$$P(t) = \Phi_1^+(t) - \Phi_1^-(t); \tag{102.1}$$

similarly

$$P(t) = \Phi_2^-(t) - \Phi_2^+(t). \tag{102.2}$$

Comparing these equations, one finds that $[\Phi_1 + \Phi_2]^+ = [\Phi_1 + \Phi_2]^-$, and hence it follows that the sum $\Phi_1(z) + \Phi_2(z)$ is holomorphic in the whole plane; since it vanishes at infinity, $\Phi_1(z) + \Phi_2(z) = 0$. Thus there is the result

$$\bar{\Phi}_1(z) = -\Phi_1(z),\ \bar{\Phi}_2(z) = -\Phi_2(z),\ \Phi_2(z) = -\Phi_1(z). \tag{102.3}$$

If now

$$y = f_1(t) \text{ and } y = f_2(t)$$

are the equations of the boundaries of the bodies S_1 and S_2 before deformation [$f_1(t)$, $f_2(t)$ having to be small quantities], one will have after deformation on the line of contact ab

$$f_1(t) + v_1^-(t) = f_2(t) + v_2^+(t),$$

and hence

$$v_1^- - v_2^+ = f(t) \text{ on } L \tag{102.4}$$

or

$$v_1^{-\prime} - v_2^{+\prime} = f'(t), \tag{102.5}$$

where

$$f(t) = f_2(t) - f_1(t). \tag{102.6}$$

The function $f'(t)$ will be assumed to satisfy the H condition.

Expressing now the boundary condition (102.4) by applying (97.7) to S_1 and S_2 respectively, one easily obtains, on the basis of (102.3),

$$\Phi_1^+(t) + \Phi_1^-(t) = \frac{if'(t)}{M}, \tag{102.7}$$

where

$$M = \frac{\varkappa_1 + 1}{4\mu_1} + \frac{\varkappa_2 + 1}{4\mu_2}. \tag{102.8}$$

Thus the same mathematical problem has been reached which arose from the problem of the pressure of a completely rigid stamp on a half-plane [§ 98, (98.5)] in the case where the line of contact ab is not given beforehand, as in § 99.

In the present case, as in § 99, it is required to find a solution, vanishing at infinity and bounded near the ends. Using the formulae of § 99 or directly the results of § 84, one arrives at the following conclusion.

The function $\Phi_1(z)$ is given by

$$\Phi_1(z) = \frac{\sqrt{(z-a)(b-z)}}{2\pi M} \int_a^b \frac{f'(t)\,dt}{\sqrt{(t-a)(b-t)}(t-z)}. \tag{102.9}$$

There will be two relations for the determination of a and b. Firstly, the relation

$$\int_a^b \frac{f'(t)\,dt}{\sqrt{(t-a)(b-t)}} = 0, \tag{102.10}$$

expressing the condition of existence of a solution of the problem (102.7), bounded at the ends; secondly, the relation

$$\int_a^b \frac{tf'(t)\,dt}{\sqrt{(t-a)(b-t)}} = MP_0, \tag{102.11}$$

where P_0 is the known value of the resultant vector of the external forces, exerted by the body S_2 on S_1 (or S_1 on S_2). The relation (102.11) is obtained, as in § 99, by comparison of the coefficients of z^{-1} in the expansions of the right sides of (102.9) and (96.11). As in § 99, $\sqrt{(z-a)(b-z)}$ means a branch such that for large $|z|$

$$\sqrt{(z-a)(b-z)} = -iz + O(1); \tag{102.12}$$

$\sqrt{(t-a)(b-t)}$ is the positive value of the root.

The pressure $P(t_0) = \Phi_1^+(t_0) - \Phi_1^-(t_0)$ is given by

$$P(t_0) = \frac{\sqrt{(t_0 - a)(b - t_0)}}{\pi M} \int_a^b \frac{f'(t)\,dt}{\sqrt{(t-a)(b-t)}(t - t_0)}. \quad (102.13)$$

If the function $f(t)$ satisfies the condition

$$f(-t) = f(t), \quad (102.14)$$

then, by symmetry, one can take beforehand $-a = b = l$, where l has to be determined. In this case, the condition (102.10) is automatically satisfied, and there remains the relation

$$\int_0^l \frac{tf'(t)\,dt}{\sqrt{l^2 - t^2}} = \tfrac{1}{2} MP_0 \quad (102.11a)$$

for the determination of l.

For the assumption (102.14) and $a = -b$, the final formulae obtained coincide with those found by A. V. Bitsadze in the paper, mentioned above.

As already stated in § 99, the integrals in the above formulae can be evaluated in an elementary manner, if $f(t)$ is a rational function, in particular, a polynomial. Putting, for example,

$$f(t) = At^{2n},$$

where n is an integer, one obtains immediately I. Ya. Shtaerman's [1] solution. Putting

$$f(t) = \frac{t^2}{2}\left(\frac{1}{R_1} + \frac{1}{R_2}\right),$$

which corresponds to the case in which S_1 and S_2 are bounded by circles of radii S_1 and S_2 (which are large in comparison with the line of contact), one obtains the solution found by L. Föppl [1] by another method. (From the point of view of computation this is the same case as in the example at the end of § 99).

Several other examples can be found in I. Ya. Shtaerman's paper [3].

§ 103. The fundamental boundary problems for the plane with straight cuts. *)

1°. GENERAL FORMULAE. Let the region S, occupied by an elastic body, be the entire plane, cut along n segments $a_k b_k$ ($k = 1, 2, \ldots, n$) of the Ox axis; the union of these segments will be denoted by L.

*) This section reproduces almost without any alterations the contents of the Author's paper [10].

In this section the stresses will not be assumed to vanish at infinity, but only to be bounded there.

Then the functions $\Phi(z)$, $\Psi(z)$, appearing in (96.1), (96.2), will be holomorphic in S and have for large $|z|$ the form

$$\begin{aligned} \Phi(z) &= \Gamma_0 - \frac{X+iY}{2\pi(1+\varkappa)} \cdot \frac{1}{z} + O(1/z^2), \\ \Psi(z) &= 2\bar{\Gamma} + \frac{\varkappa(X-iY)}{2\pi(1+\varkappa)} \cdot \frac{1}{z} + O(1/z^2); \end{aligned} \tag{103.1}$$

in these formulae (X, Y) is the resultant vector of the external forces, exerted on the edges of the union of cuts L, $\Gamma_0 = \Gamma_0' + i\Gamma_0''$ and $\Gamma = \Gamma' + i\Gamma''$ are the constants

$$\Gamma_0' = \tfrac{1}{4}(N_1 + N_2),\ \Gamma_0'' = \frac{2\mu\varepsilon_0}{1+\varkappa},\ \Gamma = -\tfrac{1}{4}(N_1 - N_2)e^{2i\alpha}, \tag{103.2}$$

where N_1, N_2 are the values of the principal stresses at infinity, α is the angle between the principal axis, corresponding to N_1, and the Ox axis and ε_0 is the value of the "rotation" at infinity.

Using the notation of § 38, the function

$$\Omega(z) = \bar{\Phi}(z) + z\bar{\Phi}'(z) + \bar{\Psi}(z) \tag{103.3}$$

will be introduced which is likewise holomorphic in S and has for large $|z|$ the form

$$\Omega(z) = \bar{\Gamma}_0 + 2\Gamma + \frac{\varkappa(X+iY)}{2\pi(1+\varkappa)} \cdot \frac{1}{z} + O(1/z^2). \tag{103.4}$$

Substitution of $\Omega(z)$ instead of $\Psi(z)$ in (96.6) leads to

$$Y_y - iX_y = \Phi(z) + \Omega(\bar{z}) + (z - \bar{z})\bar{\Phi}'(\bar{z}). \tag{103.5}$$

In a similar manner (96.3) may be transformed, if $\psi(z)$ is replaced by the function

$$\omega(z) = \int \Omega(z)dz = z\bar{\Phi}(z) + \bar{\psi}(z) + \text{const.}, \tag{103.6}$$

determined, apart from an additive constant, by the functions $\Phi(z)$, $\Psi(z)$ and also $\varphi(z)$ and $\psi(z)$. Then (96.3) takes the form

$$2\mu(u + iv) = \varkappa\varphi(z) - \omega(\bar{z}) - (z - \bar{z})\overline{\Phi(z)} + \text{const.} \tag{103.7}$$

Thus the stresses and displacements are expressed in terms of the two functions $\Phi(z)$ and $\Omega(z)$.

In the scqucl it will be assumed that the functions $\Phi(z)$, $\Omega(z)$ are sectionally holomorphic in the sense of the definition in § 15, so that, in particular, near any of the ends a_k, b_k

$$|\Phi(z)| < \frac{A}{|z-c|^{\alpha}}, \quad |\Omega(z)| < \frac{A}{|z-c|^{\alpha}}, \tag{103.8}$$

where A, α are positive constants, $0 < \alpha < 1$, and c is the corresponding end. In addition, it will be assumed that for all t, lying on L but not coinciding with an end,

$$\lim_{z \to t} y\Phi'(z) = 0. \tag{103.9}$$

2°. Now the solution of the first fundamental problem will be considered, i.e., the quantities $Y_y{}^+$, $X_y{}^+$, $Y_y{}^-$ $X_y{}^-$ on L will be assumed given; by (+) and (—) are, as always, denoted the boundary values, taken respectively on the upper and lower edges of the cut. [A less simple solution of this problem was given by D. I. Sherman [3]. The Author uses this opportunity to make up for an omission in his paper [10]; unfortunately he did not know then about the paper by D. I. Sherman and therefore did not refer to the solution given in it].

In addition, the constants Γ_0' and Γ, i.e., the values of the stresses at infinity, will be assumed given. Since the distribution of stresses is being discussed, it can be assumed, without loss of generality, that $\Gamma_0'' = 0$, i.e., that

$$\Gamma_0 = \overline{\Gamma}_0 = \Gamma_0'.$$

By (103.5) and (103.9), the boundary conditions take the form

$$\Phi^+(t) + \Omega^-(t) = Y_y^+ - iX_y^+, \quad \Phi^-(t) + \Omega^+(t) = Y_y^- - iX_y^- \tag{103.10}$$

on L. Adding and subtracting, one obtains

$$[\Phi(t) + \Omega(t)]^+ + [\Phi(t) + \Omega(t)]^- = 2p(t), \tag{103.11}$$

$$[\Phi(t) - \Omega(t)]^+ - [\Phi(t) - \Omega(t)]^- = 2q(t) \tag{103.12}$$

on L, where $p(t)$, $q(t)$ are functions, given on L. The functions $p(t)$ and $q(t)$ will be assumed to satisfy the H condition on L.

Since $\Phi(\infty) - \Omega(\infty) = -2\Gamma$, the general solution of the boundary problem (103.12) is given by (§ 78)

$$\Phi(z) - \Omega(z) = \frac{1}{\pi i} \int_L \frac{q(t)\,dt}{t-z} - 2\Gamma. \tag{103.13}$$

Further, putting

$$R(z) = \prod_{k=1}^{n} (z - a_k)(z - b_k) \tag{103.14}$$

and using the results of § 84, one obtains the general solution (of the class h_0) of the boundary problem (103.11), bounded at infinity [where the last condition follows from (103.1) and (103.4)],

$$\Phi(z) + \Omega(z) = \frac{1}{\pi i \sqrt{R(z)}} \int_L \frac{\sqrt{R(t)}p(t)\,dt}{t-z} + \frac{2P_n(z)}{\sqrt{R(z)}}, \tag{103.15}$$

where $P_n(z) = C_0 z^n + C_1 z^{n-1} + \ldots + C_n$ is a polynomial of degree n; by $\sqrt{R(z)}$ the branch is to be understood which is holomorphic in the cut plane, such that $\sqrt{R(z)}/z^n \to 1$ for $z \to \infty$, and by $\sqrt{R(t)}$ the value of this branch, taken on the upper side of L.

The formulae (103.13) and (103.15) give

$$\Phi(z) = \Phi_0(z) + \frac{P_n(z)}{\sqrt{R(z)}} - \Gamma,$$
$$\Omega(z) = \Omega_0(z) + \frac{P_n(z)}{\sqrt{R(z)}} + \Gamma, \tag{103.16}$$

where

$$\Phi_0(z) = \frac{1}{2\pi i \sqrt{R(z)}} \int_L \frac{\sqrt{R(t)}p(t)\,dt}{t-z} + \frac{1}{2\pi i} \int_L \frac{q(t)\,dt}{t-z}, \tag{103.17}$$

$$\Omega_0(z) = \frac{1}{2\pi i \sqrt{R(z)}} \int_L \frac{\sqrt{R(t)}p(t)\,dt}{t-z} - \frac{1}{2\pi i} \int_L \frac{q(t)\,dt}{t-z}. \tag{103.18}$$

Under the conditions, imposed here on $p(t)$ and $q(t)$, (103.9) will be satisfied (cf. § 21). There remains to determine the polynomial $P_n(z)$. The coefficient C_0 is immediately given by the first of the formulae (103.16) and by the condition $\Phi(\infty) = \Gamma_0$, hence

$$C_0 = \Gamma_0 + \Gamma. \tag{103.19}$$

The remaining coefficients must be found from the condition of the univalence of the displacements. By (103.7), this condition demands that the expression $\varkappa\varphi(z) - \omega(\bar{z})$ must return to its original value, when the point (x, y) describes a contour Λ_k, surrounding the segment L_k. By shrinking the contours Λ_k into the L_k, it is easily seen that the condition of univalence is expressed by the following equations (where the terms $\Phi_0^+ - \Phi_0^-$ and $\Omega_0^+ - \Omega_0^-$ are easily evaluated by the Plemelj formulae):

$$2(\varkappa + 1)\int\limits_{L_k} \frac{P_n(t)dt}{\sqrt{R(t)}} + \varkappa \int\limits_{L_k} [\Phi_0^+(t) - \Phi_0^-(t)]dt +$$
$$+ \int\limits_{L_k} [\Omega_0^+(t) - \Omega_0^-(t)]dt = 0 \quad (k = 1, 2, \ldots, n) \qquad (103.20)$$

which are a system of n linear equations in $C_1, C_2, \ldots, C_n$.

This system is always soluble. In fact, the homogeneous system, obtained if $\Gamma_0 = \Gamma = 0$, $Y_y{}^+ = X_y{}^+ = Y_y{}^- = X_y{}^- = 0$, can have no other solutions except $C_1 = C_2 = \ldots = C_n = 0$, because the initial problem, as is easily verified, has in this case only the trivial solution $\Phi(z) \equiv 0, \Omega(z) \equiv 0$. Therefore the non-homogeneous system (103.20) has always a unique solution and the problem is solved.

In the particular case where the edges of the cuts are free from stresses (*problem of a thin plate weakened by cracks under tension*), $\Phi_0(z) \equiv \Omega_0(z) \equiv 0$, and the solution assumes the extraordinarily simple form

$$\Phi(z) = \frac{P_n(z)}{\sqrt{R(z)}} - \Gamma,\ \Omega(z) = \frac{P_n(z)}{\sqrt{R(z)}} + \Gamma, \qquad (103.21)$$

where the coefficients of the polynomial $P_n(z)$ are determined by

$$C_0 = \Gamma_0 + \Gamma,\ \int\limits_{L_k} \frac{P_n(t)dt}{\sqrt{R(t)}} = 0 \quad (k = 0, 1, \ldots, n).$$

For $n = 1$ (i.e., single cut), putting $a_1 = -a$, $b_1 = a$, one obtains the very simple formulae

$$\Phi(z) = \frac{(\Gamma_0 + \Gamma)z}{\sqrt{z^2 - a^2}} - \Gamma,\ \Omega(z) = \frac{(\Gamma_0 + \Gamma)z}{\sqrt{z^2 - a^2}} + \Gamma. \quad (103.22)$$

A (less simple) solution of the problem for this particular case ($n = 1$) is well known.

3°. THE SECOND FUNDAMENTAL PROBLEM will now be considered, i.e., it will be assumed that the values of the displacements are given on L: $u^+(t)$, $v^+(t)$ on the upper edge and $u^-(t)$, $v^-(t)$ on the lower, while, if $u(a_k), v(a_k)$ and $u(b_k), v(b_k)$ denote the (given) displacements of the points a_k, b_k,

$$\begin{aligned} u^+(a_k) &= u^-(a_k) = u(a_k),\ v^+(a_k) = v^-(a_k) = v(a_k), \\ u^+(b_k) &= u^-(b_k) = u(b_k),\ v^+(b_k) = v^-(b_k) = v(b_k). \end{aligned} \qquad (103.23)$$

In addition, the constants Γ_0 and Γ (where this time it is not assumed that $\Gamma_0'' = 0$) and also the resultant vector (X, Y) of the external forces, applied to L, will be assumed known.

In order not to have to consider directly the functions $\varphi(z)$, $\psi(z)$ which may be multi-valued, the boundary conditions will be set up, not starting from (103.7), but from the formulae obtained from them by differentiation with respect to x

$$2\mu(u' + iv') = \varkappa\Phi(z) - \Omega(\bar{z}) - (z - \bar{z})\overline{\Phi}'(\bar{z}), \qquad (103.24)$$

where u', v' are the partial derivatives of u, v with respect to x. According to the above formulae, the boundary conditions are written as

$$\begin{aligned} \varkappa\Phi^+(t) - \Omega^-(t) &= 2\mu(u'^+ + iv'^+), \\ \varkappa\Phi^-(t) - \Omega^+(t) &= 2\mu(u'^- + iv'^-). \end{aligned} \qquad (103.25)$$

Adding and subtracting, one obtains

$$[\varkappa\Phi(t) - \Omega(t)]^+ + [\varkappa\Phi(t) - \Omega(t)]^- = 2f(t), \qquad (103.26)$$

$$[\varkappa\Phi(t) + \Omega(t)]^+ - [\varkappa\Phi(t) + \Omega(t)]^- = 2g(t) \qquad (103.27)$$

on L, where $f(t)$, $g(t)$ are given functions. These functions will be assumed to satisfy the H condition on L.

Similarly to the above, the general solutions of the boundary problems (103.26) and (103.27) are given by

$$\varkappa\Phi(z) + \Omega(z) = \frac{1}{\pi i}\int_L \frac{g(t)dt}{t - z} + 2\Gamma + \varkappa\Gamma_0 + \overline{\Gamma}_0, \qquad (103.28)$$

$$\varkappa\Phi(z) - \Omega(z) = \frac{1}{\pi i\sqrt{R(z)}}\int_L \frac{\sqrt{R(t)}f(t)dt}{t - z} + \frac{2P_n(z)}{\sqrt{R(z)}}. \qquad (103.29)$$

These formulae determine the unknown functions $\Phi(z)$ and $\Omega(z)$, apart from the term involving the polynomial

$$P_n(z) = C_0 z^n + C_1 z^{n-1} + \ldots + C_n. \qquad (103.30)$$

The first two coefficients C_0 and C_1 of this polynomial are found directly from (103.29), taking into consideration that, by (103.1) and (103.4), for large $|z|$

$$\varkappa\Phi(z) - \Omega(z) = \varkappa\Gamma_0 - \overline{\Gamma}_0 - 2\Gamma - \frac{\varkappa(X - iY)}{\pi(\varkappa + 1)} \cdot \frac{1}{z} + O(1/z^2). \qquad (103.31)$$

It is easily seen from (103.23) and (103.25) that the displacements u, v, calculated from the functions $\Phi(z)$, $\Omega(z)$ by (103.7), will be single-valued. However, these displacements will take on the cuts L_k the given values, apart from some constant terms c_k which may be different on different cuts. The functions $\Phi(z)$ and $\Omega(z)$ will only satisfy the conditions of the problem, provided $c_1 = c_2 = \ldots = c_n$ (where, by means of a rigid body displacement, one can make $c_1 = c_2 = \ldots = c_n = 0$). It is easily seen from (103.24) that these conditions may be expressed as

$$\int_{b_k}^{a_{k+1}} [\varkappa\Phi(t) - \Omega(t)]dt = 2\mu \left\{ u(a_{k+1}) - u(b_k) + i[v(a_{k+1}) - v(b_k)] \right\},$$

$$k = 1,\ 2,\ \ldots,\ n-1, \qquad (103.32)$$

where on the right side appear the quantities given in (103.23).

Substituting on the left of (103.32) the expression (103.29), a system of $n - 1$ linear equations is obtained for the determination of the $n - 1$ coefficients $C_2, \ldots, C_n$ which are still undetermined; similarly as above, this system is easily seen to have always a unique solution. Thus the problem is solved. The solution for the particular case $n = 1$ was obtained earlier (cf. N. I. Muskhelishvili [1], § 72, where the case of an elliptic hole is considered which includes the single cut as a particular case).

As before, the problem may be solved in the case where the displacements are given apart from constant terms which may differ on different cuts, but then, in addition, the resultant vectors of the external forces, acting on each cut separately, must be given.

4°. Finally some remarks will be made on a problem, considered by D. I. Sherman [4]. In this problem the external stresses acting, say, on the upper edges and the displacements on the lower edges of the cuts are given. (Sherman solved the problem by reducing it to a system of singular integral equations). By (103.5) and (103.24) the boundary conditions may be written as

$$\Phi^+(t) + \Omega^-(t) = Y_y^+ - iX_y^+,$$
$$\varkappa\Phi^-(t) - \Omega^+(t) = 2\mu(u'^- + iv'^-) \qquad (103.33)$$

on L. Multiplying the second of these equations by $\pm i/\sqrt{\varkappa}$ and adding to the first (cf. D. I. Sherman [4], p. 333), one obtains on L

$$\left[\Phi(t) - \frac{i}{\sqrt{\varkappa}}\,\Omega(t)\right]^+ + i\sqrt{\varkappa}\left[\Phi(t) - \frac{i}{\sqrt{\varkappa}}\,\Omega(t)\right]^- = 2f(t), \qquad (103.34)$$

$$\left[\Phi(t) + \frac{i}{\sqrt{\varkappa}}\,\Omega(t)\right]^+ - i\sqrt{\varkappa}\left[\Phi(t) + \frac{i}{\sqrt{\varkappa}}\,\Omega(t)\right]^- = 2g(t), \qquad (103.35)$$

where $f(t)$ and $g(t)$ are functions, given on L. Thus the functions

$$\Phi(z) - \frac{i}{\sqrt{\varkappa}}\,\Omega(z), \quad \Phi(z) + \frac{i}{\sqrt{\varkappa}}\,\Omega(z)$$

are determined as the solutions of the boundary problems (103.34) and (103.35) which are very simple particular cases of the non-homogeneous Hilbert problem, solved in § 81. The solutions of these problems determine the above functions, and hence also the functions $\Phi(z), \Omega(z)$, apart from the terms containing the polynomials the coefficients of which are found by the help of some additional conditions of the problem, just as before.

§ 104. The boundary problems for circular regions. *)

The method of solution of the boundary problems for the half-plane, given in the preceding sections, is easily transferred to the case, where the region, occupied by the elastic body, is a circle or an infinite region with a circular inside boundary. For definiteness, the case of the circle will be considered.

Denote by S^+ the circular region under consideration, by Γ its boundary and by S^- the remaining part of the plane. Take the radius of the circle as unity and its centre at the origin of coordinates.

If polar coordinates are introduced, so that

$$z = x + iy = re^{i\vartheta},$$

then (cf. for example N. I. Muskhelishvili [1], § 39)

$$\widehat{rr} + \widehat{\vartheta\vartheta} = 2[\Phi(z) + \overline{\Phi(z)}], \tag{104.1}$$

$$\widehat{rr} + i\widehat{r\vartheta} = \Phi(z) + \overline{\Phi(z)} - e^{-2i\vartheta}[z\overline{\Phi'(z)} + \overline{\Psi(z)}], \tag{104.2}$$

where $\widehat{rr}$, $\widehat{r\vartheta}$, $\widehat{\vartheta\vartheta}$ are the components of stress in polar coordinates and $\Phi(z) = \varphi'(z)$, $\Psi(z) = \psi'(z)$ are the same holomorphic functions as in § 96. Rewrite the formula (96.3), viz.

$$2\mu(u + iv) = \varkappa\varphi(z) - z\overline{\varphi'(z)} - \overline{\psi(z)} \tag{104.3}$$

expressing the components u and v (in Cartesian coordinates).

Using the notation of § 38, 2°, the function $\Phi(z)$ will be continued into the region S^- by putting for $|z| > 1$

$$\Phi(z) = -\overline{\Phi}(1/z) + \frac{1}{z}\overline{\Phi}'(1/z) + \frac{1}{z^2}\overline{\Psi}(1/z). \tag{104.4}$$

Substituting in the above formula, which holds, by hypothesis, for $|z| > 1$, the quantity $1/\bar{z}$ for z, assuming now $|\bar{z}| = |z| < 1$, one obtains

$$\Phi(1/\bar{z}) = -\overline{\Phi}(\bar{z}) + \bar{z}\overline{\Phi}'(\bar{z}) + \bar{z}^2\overline{\Psi}(\bar{z}),$$

and hence, going to the conjugate values,

*) The results of this section are due to I. N. Kartsivadze (1].

$$\Psi(z) = \frac{1}{z^2}\Phi(z) + \frac{1}{z^2}\overline{\Phi}(1/z) - \frac{1}{z}\Phi'(z) \tag{104.5}$$

for $|z| < 1$.

Thus the components of stress and displacement can be expressed by a single function $\Phi(z)$, holomorphic both in S^+ and S^- (where the boundary Γ does not belong to S^+ nor S^-). That the function $\Phi(z)$ is holomorphic in S^- follows again from (104.4), because, by hypothesis, the functions $\Phi(z)$ and $\Psi(z)$, appearing on the right side, are holomorphic for $|z| < 1$. The same formula (104.4) also shows that *the function* $\Phi(z)$ *is bounded at infinity.* In addition, one more condition must be imposed regarding the behaviour of this function near the point at infinity, in order that it shall correspond to a certain regular state of stress (§ 96) of the elastic body under consideration. In fact, the function $\Phi(z)$ must be such that the function $\Psi(z)$, corresponding to it by (104.5), will be holomorphic in S^+; in other words, that the point $z = 0$ will not be a pole of the right side of (104.5).

Put

$$\begin{aligned} \Phi(z) &= A_0 + A_1 z + A_2 z^2 + \ldots \quad (|z| < 1), \\ \Phi(z) &= B_0 + \frac{B_1}{z} + \frac{B_2}{z^2} + \ldots \quad (|z| > 1). \end{aligned} \tag{104.6}$$

Then, as shown by (104.5), $\Psi(z)$ will not have a pole at $z = 0$, if and only if

$$A_0 + \overline{B}_0 = 0, \quad B_1 = 0. \tag{104.7}$$

In the sequel, this condition will be assumed to be fulfilled.

In order to express the components of stress by the function $\Phi(z)$, it is sufficient to replace on the right side of (104.2) the function $\Psi(z)$ by the expression (104.5). However, the formula (104.2) will first be transformed in a certain manner; in fact, noting that

$$e^{-2i\vartheta} = \bar{z}/z,$$

it may be written as

$$\widehat{rr} + i\widehat{r\vartheta} = \Phi(z) + \overline{\Phi(z)} - \bar{z}\,\overline{\Phi'(z)} - (\bar{z}/z)\overline{\Psi(z)};$$

but by (104.5)

$$\overline{\Phi(z)} - \bar{z}\,\overline{\Phi'(z)} = \bar{z}^2\,\overline{\Psi(z)} - \Phi(1/\bar{z}); \tag{*}$$

introducing this expression in the preceding equation, one obtains the formula, convenient for the present purpose,

$$\widehat{rr} + i\widehat{r\vartheta} = \Phi(z) - \Phi(1/\bar{z}) + \bar{z}(\bar{z} - 1/z)\overline{\Psi(z)}, \tag{104.8}$$

where one has to use for $\overline{\Psi(z)}$ the expression given by (104.5).

The formula for the displacements, convenient for the present purpose, will be obtained in the following manner; differentiating by parts both sides of (104.3) with respect to ϑ, where, for example,

$$\frac{\partial\varphi(z)}{\partial\vartheta} = \varphi'(z)\frac{\partial z}{\partial\vartheta} = ire^{i\vartheta}\varphi'(z) = iz\Phi(z),$$

one obtains

$$2\mu(u' + iv') = iz[\varkappa\Phi(z) - \overline{\Phi(z)} + \bar{z}\overline{\Phi'(z)} + (\bar{z}/z)\overline{\Psi(z)}],$$

where

$$u' = \frac{\partial u}{\partial\vartheta}, \quad v' = \frac{\partial v}{\vartheta\partial},$$

and hence, using again (*),

$$2\mu(u' + iv') = iz\{\varkappa\Phi(z) + \Phi(1/\bar{z}) + \bar{z}(1/z - \bar{z})\overline{\Psi(z)}\}, \quad (104.9)$$

where by $\overline{\Psi(z)}$ must again be understood the expression, given by (104.5).

In the sequel, it will be assumed that $\Phi(z)$ is continuous on Γ from S^+ and from S^-, with the possible exception of a finite number of points c_k on Γ near which

$$|\Phi(z)| < \frac{\text{const.}}{|z - c_k|^\alpha}, \quad 0 \leqq \alpha < 1; \quad (104.10)$$

in addition, it will be assumed that

$$\lim_{r\to 1}(1 - r)\Phi'(z) = 0, \quad z = re^{i\vartheta} \quad (104.11)$$

for all values of ϑ, except possibly at the points corresponding to the c_k. It is easily seen from (104.5), by the strength of these conditions, that also

$$\lim_{r\to 1}(\bar{z} - 1/z)\Psi(z) = \lim_{r\to 1} e^{-i\vartheta}(r - 1/r)\Psi(z) = 0. \quad (104.12)$$

If the part Γ' of Γ is unstressed, i.e., if on Γ' : $\widehat{rr} = \widehat{r\vartheta} = 0$, then, as shown by (104.8), $\Phi^+(t) - \Phi^-(t) = 0$ on Γ'. Consequently *the values of $\Phi(z)$ outside and inside Γ are analytic continuations of one another through the unstressed parts of the boundary*, as in the case of the half-plane. In fact, the definition (104.4) of the function $\Phi(z)$ in S^- was chosen with this property in view.

By use of the above formulae a series of fundamental boundary problems are easily solved for the circle, similarly as it was done in the preceding sections for the half-plane.

The solution of the fundamental mixed problem will be considered here for the case, when the displacements u, v on the arcs $a_k b_k$ ($k = 1, \ldots, n$) of the circle and the external stresses on the remaining part of Γ are given. Since the solution of the problem for the case, in which the external forces are given over the entire boundary, are well known (cf. for example N. I. Muskhelishvili [1]; also I. N. Kartsivadze [1] where the solution of this problem is given by the help of the formulae, deduced in the present section), the mixed problem under consideration may be reduced to the case, where the displacements are given on the arcs $a_k b_k$, while the remaining part of the boundary is free from external stresses. Thus let

$$u^+ + iv^+ = g(t) \text{ on } L, \tag{104.13}$$

$$\widehat{rr}^+ + i\widehat{r\vartheta}^+ = 0 \text{ on } L', \tag{104.14}$$

where L is the union of the arcs $a_k b_k$ ($k = 1, \ldots, n$) and L' is the remaining part of the boundary. With regard to the given function $g(t)$ it will be assumed that its derivative with respect to ϑ satisfies the H condition.

Using (104.9), one easily obtains from (104.13)

$$\varkappa\Phi^+(t) + \Phi^-(t) = 2\mu g'(t) \text{ on } L, \tag{104.15}$$

where

$$g'(t) = \frac{dg}{dt} = -\, ie^{-i\vartheta}\frac{dg}{d\vartheta}. \tag{104.16}$$

The condition (104.14) reduces, as was mentioned above, to $\Phi^+(t) - \Phi^-(t) = 0$ on L', expressing that *the function $\Phi(z)$ is holomorphic in the entire plane, cut along L.*

Thus the problem of determining $\Phi(z)$ reduces to finding a solution, bounded at infinity, of the Hilbert problem (104.15) which is a particular case of the one, solved in § 81. Again the most general solution, i.e., one of the class h_0, will be found. Applying the formulae of the above-mentioned section, one finds without difficulty (compare the completely analogous case of the half-plane, § 97, 3°)

$$\Phi(z) = \frac{[R_0(z)]^{i\gamma}\mu}{\pi i\varkappa\sqrt{R(z)}}\int_L \frac{\sqrt{R(t)}}{[R_0(t)]^{i\gamma}}\frac{g'(t)dt}{t-z} + \frac{[R_0(z)]^{i\gamma}P_n(z)}{\sqrt{R(z)}}, \tag{104.17}$$

where

$$R(z) = \prod_{k=1}^{n}(z-a_k)(z-b_k), \quad R_0(z) = \prod_{k=1}^{n}\frac{z-b_k}{z-a_k}, \tag{104.18}$$

$$\gamma = \frac{\log\varkappa}{2\pi}, \tag{104.19}$$

$$P_n(z) = C_0 z^n + C_1 z^{n-1} + \ldots + C_n, \tag{104.20}$$

and the $C_0, \ldots, C_n$ are constants, subject to determination from additional conditions (cf. below). By $\sqrt{R(z)}$ must be understood the branch, holomorphic in the plane cut along L, such that

$$\lim_{z \to \infty} z^{-n} \sqrt{R(z)} = 1,$$

and by $[R_0(z)]^{i\gamma}$ a branch, holomorphic in the same region, which tends to 1 for $z \to \infty$; finally, by $\sqrt{R(t)}$, $[R_0(t)]^{i\gamma}$ will be understood the values taken by the specified branches for $z \to t$ from S^+.

There remains to determine the constants $C_0, C_1, \ldots, C_n$ so that all the conditions, following from the statement of the original problem, will be satisfied. Firstly, the conditions (104.7) have to be satisfied and, secondly, the boundary condition (104.13), and not only the condition (104.15) obtained from the latter by differentiation with respect to ϑ. The last condition is easily seen to reduce to

$$\int_{b_k a_{k+1}} [\varkappa \Phi^+(t_0) + \Phi^-(t_0)] dt_0 = 2\mu [g(a_{k+1}) - g(b_k)],$$
$$k = 1, 2, \ldots, n \quad (a_{n+1} = a_1), \tag{104.21}$$

where $\Phi^+(t_0)$ and $\Phi^-(t_0)$ are the expressions, obtained from (104.17). Since on the arcs $b_k a_{k+1}$: $\Phi^+(t_0) = \Phi^-(t_0)$, the preceding formulae give

$$(\varkappa + 1) \int_{b_k a_{k+1}} \Phi_0(t_0) dt_0 + \sum_{j=1}^{n} A_{kj} C_j = 2\mu [g(a_{k+1}) - g(b_k)], \tag{104.22}$$

where

$$\Phi_0(t_0) = \frac{[R_0(t_0)]^{i\gamma} \mu}{\pi i \varkappa \sqrt{R(t_0)}} \int_L \frac{\sqrt{R(t)}}{[R_0(t)]^{i\gamma}} \frac{g'(t)\, dt}{t - t_0}, \tag{104.23}$$

$$A_{kj} = (\varkappa + 1) \int_{b_k a_{k+1}} \frac{[R_0(t)]^{i\gamma} t^{n-j}\, dt}{\sqrt{R(t)}}. \tag{104.24}$$

Hence n linear equations have been obtained involving $C_0, C_1, \ldots, C_n$. There remain still the conditions (104.7) to be satisfied. The second of these conditions is easily seen to be a consequence of the conditions (104.22), already obtained. Actually, it is easily seen from (104.22), equivalent to (104.21), that

$$\int_{\Gamma} [\varkappa\Phi^+(t_0) + \Phi^-(t_0)]\, dt_0 = 0,$$

where it must be kept in mind that, by (104.15),

$$\int_{a_k b_k} [\varkappa\Phi^+(t_0) + \Phi^-(t_0)]\, dt_0 = 2\mu[g(b_k) - g(a_k)].$$

But, since $\Phi(z)$ is holomorphic in S^+, the integral over the first term is zero; hence

$$\int_{\Gamma} \Phi^-(t_0) dt_0 = 0,$$

and this just means that the coefficient B_1 in the expansion of $\Phi(z)$ in decreasing powers of z near infinity is zero.

Thus there remains still to express the first of the conditions (104.7) which can be written as

$$\Phi(0) + \overline{\Phi(\infty)} = 0;$$

this gives by (104.17)

$$C_n + \frac{\sqrt{R(0)}}{[R_0(0)]^{i\gamma}}\overline{C}_0 + \frac{\mu}{\pi i \varkappa}\int_L \frac{\sqrt{R(t)}}{[R_0(t)]^{i\gamma}}\frac{g'(t)\, dt}{t} = 0. \qquad (104.25)$$

Eventually one has the system of $n+1$ linear equations (104.22) and (104.25) for the determination of the constants $C_0, C_1, \ldots, C_n$ or, more correctly, the system of $2n+2$ equations for the determination of the real and imaginary parts of these constants.

One has still to show that this system is always soluble. For this, in its turn, it is sufficient to show that the homogeneous system, obtained for $g(t) = \text{const.}$, has no other solution but $C_0 = C_1 = \ldots = C_n = 0$. However, this is a simple consequence of the known theorem on the uniqueness of the solution of the mixed problem.

§ 105. Certain analogous problems. Generalizations.

The methods of solution of the boundary problems of the theory of elasticity, studied in this chapter, may be extended to a series of analogous cases. First of all, as already stated, the results of the last section are easily transferred to the case where the region under consideration is the infinite plane with a circle as inside boundary. Besides, the problems for the case of the infinite plane, cut along arcs of one and the same circle, may be solved by similar methods (i.e., problems, analogous to those of § 103).

Further, the methods of this chapter can, without difficulty, be generalized to the case of a region, mapped on the circle by help of rational functions, i.e.,, of regions, considered by the Author in a series of earlier papers by the help of other methods (N. I. Muskhelishvili [1], [12], [13] and others). In this manner a series of new results can be obtained, as well as simpler solutions of some problems, already solved by other methods.

PART V

SINGULAR INTEGRAL EQUATIONS FOR THE CASE OF ARCS OR DISCONTINUOUS COEFFICIENTS AND SOME OF THEIR APPLICATIONS

In this Part the general theory of singular integral equations involving Cauchy integrals is given for the case in which the line of integration is the union of smooth arcs and the coefficients of the dominant part of the equation are continuous (Chapt. 14).

The theory for the case, where the line of integration is the union of smooth contours and the coefficients of the dominant part have first order discontinuities at a finite number of points, may be developed in a similar manner; therefore only short remarks will be made with reference to this case (Chapt. 15).

The theory may easily be generalized to the case, where the line of integration consists of arcs and contours and the coefficients of the dominant part have first order discontinuities at a finite number of points.

Among the numerous existing and possible applications of the theory of singular integral equations of the stated type only the solution of the Dirichlet problem (and its analogues) for the plane, cut along smooth arcs of arbitrary form (chapt. 16), and the solution of an integro-differential equation, occurring in the theory of wings with finite aspect ratio, will be given.

The important and interesting results, obtained by D. I. Sherman [1], [2] by means of the generalized method of Carleman, will not be mentioned here as the Author intends to reproduce a number of them in the third edition of his book [1] which is in preparation.

CHAPTER 14

SINGULAR INTEGRAL EQUATIONS FOR THE CASE OF ARCS AND CONTINUOUS COEFFICIENTS

The results stated in this chapter were obtained by the Author in his paper [2] and substantially extended in the paper by N. I. Muskhelishvili and D. A. Kveselava [1]. In this latter paper the concept of classes of solutions was introduced for the first time and the fundamental theorems of § 112 were proved. In addition, another method of investigation due to D. A. Kveselava [1] is given in § 115.

§ 106. Definitions.

Throughout this chapter (§§ 106—115) the terms and notation of § 77 will be used; in particular, by

$$L = L_1 + L_2 + \ldots + L_p$$

will be understood the union of the disconnected smooth arcs $L_k = a_k b_k$ ($k = 1, 2, \ldots, p$) with the positive direction running from a_k to b_k.

The symbols t, t_0, t_1 will always refer to points on L. In order to avoid repetition, the following convention will be introduced: the statement that $\varphi(t)$ satisfies the singular integral equation of the form

$$A(t_0)\varphi(t_0) + \frac{1}{\pi i}\int_L \frac{K(t_0, t)\varphi(t)\,dt}{t - t_0} = f(t_0)$$

will mean that the above equation must hold for all t_0 on L, with the possible exclusion of the ends.

As in § 44, singular operators will be defined by

$$\mathbf{K}\varphi \equiv A(t_0)\varphi(t_0) + \frac{1}{\pi i}\int_L \frac{K(t_0, t)\varphi(t)\,dt}{t - t_0} \tag{106.1}$$

or

$$\mathbf{K}\varphi \equiv A(t_0)\varphi(t_0) + \frac{B(t_0)}{\pi i}\int_L \frac{\varphi(t)\,dt}{t - t_0} + \frac{1}{\pi i}\int_L k(t_0, t)\varphi(t)dt, \tag{106.2}$$

where

$$B(t_0) = K(t_0, t_0),\ k(t_0, t) = \frac{K(t_0, t) - K(t_0, t_0)}{t - t_0}. \tag{106.3}$$

The functions $A(t_0)$ and $B(t_0)$ will always be assumed to belong to the class H on L (i.e., to satisfy the H condition on each of the closed arcs L_k); in addition, it will be assumed that the function $k(t_0, t)$ likewise satisfies the H condition with respect to both variables on each of the closed arcs L_k. Then obviously

$$K(t_0, t) = B(t_0) + (t - t_0)k(t_0, t)$$

also satisfies this condition.

The operator $\mathbf{K}$ will be said to be of the regular type, if the functions

$$S(t) = A(t) + B(t),\ D(t) = A(t) - B(t) \tag{106.4}$$

do not vanish anywhere on L. In the sequel it will be assumed that this condition is satisfied.

The operator $\mathbf{K}^0$, defined by

$$\mathbf{K}^0\varphi \equiv A(t_0)\varphi(t_0) + \frac{B(t_0)}{\pi i}\int_L \frac{\varphi(t)\,dt}{t - t_0}, \tag{106.5}$$

will be called the dominant part of the operator $\mathbf{K}$. If $\mathbf{k}$ denotes the Fredholm operator of the first kind

$$\mathbf{k}\varphi \equiv \frac{1}{\pi i}\int_L k(t_0, t)\varphi(t)\,dt, \tag{106.6}$$

the operator $\mathbf{K}$ may be represented as the sum of the operators $\mathbf{K}^0$ and $\mathbf{k}$, i.e.,

$$\mathbf{K}\varphi = \mathbf{K}^0\varphi + \mathbf{k}\varphi. \tag{106.7}$$

The operator $\mathbf{K}$ and the operator $\mathbf{K}'$, defined by

$$\mathbf{K}'\psi \equiv A(t_0)\psi(t_0) - \frac{1}{\pi i}\int_L \frac{K(t, t_0)\psi(t)\,dt}{t - t_0}, \tag{106.8}$$

will be said to be adjoint to one another, just as in the case where L consists of contours. In particular, the operator adjoint to $\mathbf{K}^0$ will be the operator $\mathbf{K}^{0\prime}$, defined by

$$\mathbf{K}^{0\prime}\psi \equiv A(t_0)\psi(t_0) - \frac{1}{\pi i}\int_L \frac{B(t)\psi(t)\,dt}{t - t_0}. \tag{106.9}$$

Accordingly

$$\mathbf{K}'\psi = \mathbf{K}^{0\prime}\psi + \mathbf{k}'\psi. \tag{106.10}$$

Thus the operator $\mathbf{K}'$ is decomposed into the sum of two operators $\mathbf{K}^{0\prime}$ and $\mathbf{k}'$, although this decomposition is different from that of $\mathbf{K}$, since an operator of the form (106.9) now plays the role of the dominant part, where the coefficient B appears in the form $B(t)$ under the integral sign and not in the form $B(t_0)$ outside it.

Of course, the operator $\mathbf{K}$ may likewise be decomposed in a similar manner, i.e., one may put

$$\mathbf{K}\varphi = \mathbf{K}^1\varphi + \mathbf{k}^1\varphi, \tag{106.11}$$

where

$$\mathbf{K}^1\varphi \equiv A(t_0)\varphi(t_0) + \frac{1}{\pi i}\int_L \frac{B(t)\varphi(t)\,dt}{t - t_0} \tag{106.12}$$

and

$$\mathbf{k}^1\varphi = \frac{1}{\pi i}\int_L k^1(t_0, t)\varphi(t)dt, \; k^1(t_0, t) = \frac{K(t_0, t) - K(t, t)}{t - t_0}. \tag{106.13}$$

This decomposition may serve the same purpose as that of (106.7); however, in order not to introduce new notation ($\mathbf{K}^1$ and $\mathbf{k}^1$), such a decomposition will, as a rule, only be used in connection with the operator $\mathbf{K}'$, adjoint to the given operator $\mathbf{K}$. This actually does not impose any restrictions, because the operator $\mathbf{K}$ can be considered as adjoint to the operator $\mathbf{K}'$.

The following important formula occurred in § 46:

$$\int_L \psi \mathbf{K}\varphi \, dt = \int_L \varphi \mathbf{K}'\psi \, dt, \tag{106.14}$$

where φ and ψ are arbitrary functions, satisfying the H condition. This formula is also easily seen to hold in the present case in which L is the union of arcs. However, it has now to be applied to a wider class of functions φ, ψ than those of the class H, in fact, to functions of the class H^*. If φ and ψ are arbitrary functions of the class H^*, then the integrals of (106.14) may turn out to be divergent. But in the sequel the formula (106.14) will only be used in cases where near every end one of the functions $\varphi(t)$, $\psi(t)$ belongs to the class H^*_ε. The formula (106.14) is easily seen to hold true in such cases.

The equation

$$\mathbf{K}\varphi = f(t_0), \tag{106.15}$$

where $f(t_0)$ is a given function of the class H^*, will be called a singular integral equation with a Cauchy type kernel or simply a singular equation. A solution $\varphi(t)$ of this equation will always be sought in the class H^*. The equation

$$\mathbf{K}^0\varphi = f(t_0) \tag{106.16}$$

will be called the dominant equation corresponding to the equation $\mathbf{K}\varphi = f$. Finally, the equations

$$\mathbf{K}\varphi = f, \; \mathbf{K}'\psi = g \tag{106.17}$$

will be called adjoint, whatever may be the right sides f and g (of the class H^*).

If, as it has been agreed, the operator $\mathbf{K}$ is of the regular type, i.e., if

$$A^2(t) - B^2(t) \neq 0 \text{ everywhere on } L \tag{106.18}$$

(including the ends), then the corresponding equation (106.15) will be called an equation of the regular type. Dividing both sides of the equation by

$$\sqrt{A^2(t_0) - B^2(t_0)},$$

one obtains an equation in which the coefficients A and B satisfy the condition

$$A^2(t_0) - B^2(t_0) = 1; \tag{106.19}$$

however, in spite of the fact that, with (106.19), the following formulae are somewhat simplified, the coefficients A and B will not be related by this condition.

§ 107. Solution of the dominant equation.

Similarly as in Part II a start will be made with the solution of the dominant equation

$$\mathbf{K}^0\varphi \equiv A(t_0)\varphi(t_0) + \frac{B(t_0)}{\pi i}\int_L \frac{\varphi(t)dt}{t - t_0} = f(t_0). \tag{107.1}$$

As already stated, it will be assumed that

$$A^2(t) - B^2(t) \neq 0 \text{ everywhere on } L. \tag{107.2}$$

In addition, $f(t)$ will be assumed for the present to belong to the class H; however, as already stipulated, the solution of the equation (107.1) will be sought in the class H^*.

The sectionally holomorphic function

$$\Phi(z) = \frac{1}{2\pi i}\int_L \frac{\varphi(t)dt}{t - z}, \tag{107.3}$$

vanishing at infinity, will be introduced into the consideration. Then

$$\varphi(t_0) = \Phi^+(t_0) - \Phi^-(t_0), \quad \frac{1}{\pi i}\int_L \frac{\varphi(t)dt}{t - t_0} = \Phi^+(t_0) + \Phi^-(t_0), \tag{107.4}$$

and hence it follows that the function $\Phi(z)$ must be a solution, vanishing at infinity, of the Hilbert problem

$$(A + B)\Phi^+ - (A - B)\Phi^- = f \tag{107.5}$$

or

$$\Phi^+(t_0) = G(t_0)\,\Phi^-(t_0) + \frac{f(t_0)}{A(t_0) + B(t_0)}, \tag{107.6}$$

where

$$G(t_0) = \frac{A(t_0) - B(t_0)}{A(t_0) + B(t_0)}. \tag{107.7}$$

Special and non-special ends corresponding to this problem will now be called special and non-special ends corresponding to the integral equation (107.1). As before, non-special ends will be denoted by $c_1, c_2, \ldots, c_m$.

The most general solution (of the class H^*) of the problem (107.6) having finite degree at infinity may, for example, be written in the form (§ 81, Note)

$$\Phi(z) = \frac{X(z)}{2\pi i} \int_L \frac{f(t)dt}{[A(t) + B(t)]X^+(t)(t - z)} + X_0(z)Q(z), \quad (107.8)$$

where $X(z)$ is the fundamental function of any class

$$h(c_1, c_2, \ldots, c_q),$$

corresponding to the problem (107.6), $X_0(z)$ is the fundamental function of the class h_0 of the same problem and $Q(z)$ is some polynomial.

The solution $\Phi(z)$ must also be subjected to the condition $\Phi(\infty) = 0$; this condition will be dealt with below. Some conclusions will now be deduced from the fact that, if solutions of the equation (107.1) exist, then all of them are necessarily given by the formula $\varphi(t_0) = \Phi^+(t_0) - \Phi^-(t_0)$, in which $\Phi(z)$ has the form (107.8).

The function $\varphi(t_0)$ will be determined. For this purpose introduce the notation

$$Z(t_0) = [A(t_0) + B(t_0)]X^+(t_0) = [A(t_0) - B(t_0)]X^-(t_0); \quad (107.9)$$

the functions $X^+(t_0)$ and $X^-(t_0)$ are defined by (79.7). The function $Z(t)$ will be called the fundamental function of a given class $h(c_1, c_2, \ldots, c_q)$, corresponding to the equation $\mathbf{K}^0\varphi = f$ or the operator $\mathbf{K}^0$.

In particular, the fundamental function $Z_0(t)$ of the class $h_0 = h(0)$, corresponding to the case $q = 0$, is determined by

$$Z_0(t_0) = [A(t_0) + B(t_0)]X_0^+(t_0) = [A(t_0) - B(t_0)]X_0^-(t_0). \quad (107.9a)$$

Introduce now the notation

$$A^*(t_0) = \frac{A(t_0)}{A^2(t_0) - B^2(t_0)}, \quad B^*(t_0) = \frac{B(t_0)}{A^2(t_0) - B^2(t_0)}. \quad (107.10)$$

Then, using the Plemelj formulae, one easily obtains (compare § 47)

$$\varphi(t_0) = \mathbf{K}^*f + B^*(t_0)Z_0(t_0)P(t_0), \quad (107.11)$$

where

$$\mathbf{K}^*f \equiv A^*(t_0)f(t_0) - \frac{B^*(t_0)Z(t_0)}{\pi i} \int_L \frac{f(t)dt}{Z(t)(t - t_0)} \quad (107.12)$$

and $P(t_0)$ is a polynomial.

From the definition (107.9) of $Z(t_0)$ and from the formulae (79.7) and (79.9) it is obvious that

$$Z(t_0) = \omega_0(t_0) \prod_{k=1}^{2p} (t_0 - c_k)^{\gamma_k}, \tag{107.13}$$

where $\omega_0(t_0)$ is a non-vanishing function of the class H and

$$0 < \Re\gamma_k < 1 \ (k=1, \ldots, q), \quad -1 < \Re\gamma_k < 0 \ (k=q+1, \ldots, m),$$
$$\Re\gamma_k = 0 \ \ (k = m+1, \ldots, 2p). \tag{107.14}$$

The same formulae will be true for $Z_0(t_0)$; one only has to assume in this case that $q = 0$.

Proceeding from the above the following conclusions are reached:

1°. *Near all special ends c_k ($k = m + 1, \ldots, 2p$) the solution $\varphi(t)$ belongs to the class H_ε^* and is bounded.* This is clear from § 29, 4° for those special ends for which $\gamma_k \neq 0$. However, if for a given special end $c_k : \gamma_k = 0$, then $G(c_k) = 1$, i.e., $B(c_k) = 0$; after this the statement becomes obvious.

2°. *If near any non-special end the solution $\varphi(t)$ is bounded, then it belongs necessarily near this end to the class H.*

In fact, let c_k be a non-special end near which $\varphi(t_0)$ is bounded. Introduce in (107.11) as fundamental function $Z(t)$ one which vanishes at c_k. Then the first term on the right side belongs near c_k to the class H (§ 29, 4°). Therefore, in order that $\varphi(t_0)$ will be bounded, it is necessary that the polynomial has the root c_k; after this the statement is obvious.

All possible solutions of the integral equation $K^0\varphi = f$ will now be divided into classes by taking into the class $h(c_1, c_2, \ldots, c_q)$, $0 \leqq q \leqq m$, all solutions $\varphi(t)$ which remain bounded near the non-special ends $c_1, c_2, \ldots, c_q$. Such solutions have been seen to belong near $c_1, c_2, \ldots, c_q$ to the class H; therefore the present definition of classes of solutions agrees with the definition of classes of functions $\varphi(t)$, used in § 82.

Clearly, a solution of a similar class of the Hilbert problem (107.6) will necessarily correspond to a solution of the class $h(c_1, c_2, \ldots, c_q)$ of the integral equation (107.1). Therefore, for the determination of all solutions of this class of the equation (107.1), it is sufficient to find all the solutions of the similar class of the Hilbert problem (107.6), vanishing at infinity.

Starting from the fact that the solution of this last problem is known (§ 81), the following conclusions are easily drawn.

The index of a similar class of the Hilbert problem (107.6) will be called the index $\varkappa$ of the given class $h(c_1, \ldots, c_q)$ of the equation $\mathbf{K}^0\varphi = f$ or of the operator $\mathbf{K}^0$. If by $Z(t)$ is understood the fundamental function of the class $h(c_1, \ldots, c_q)$, then, *for* $\varkappa \geqq 0$, *all the solutions of the class* $h = h(c_1, c_2, \ldots, c_q)$ *of the equation* $\mathbf{K}^0\varphi = f$ *are given by*

$$\varphi(t_0) = \mathbf{K}^* f + B^*(t_0)Z(t_0)P_{\varkappa-1}(t_0), \tag{107.15}$$

where $P_{\varkappa-1}(t_0)$ *is an arbitrary polynomial of degree not greater than* $\varkappa - 1$ $[P_{\varkappa-1}(t_0) \equiv 0$ for $\varkappa = 0)]$;

for $\varkappa < 0$, *the (unique) solution exists, provided the following (nesessary and sufficient) conditions are satisfied:*

$$\int_L \frac{t^k f(t)}{Z(t)}\, dt = 0, \; k = 0, 1, \ldots, -\varkappa - 1. \tag{107.16}$$

It is then given by the same formula (107.15) *in which* $P_{\varkappa-1}(t_0) \equiv 0$.

From the above it also follows that for $\varkappa \leqq 0$ the homogeneous equation $\mathbf{K}^0\varphi = 0$ has no non-zero solutions of the class h; for $\varkappa > 0$, it has exactly $\varkappa$ linearly independent solutions of the class h, the sum of which is given by

$$\varphi(t) = B^*(t)Z(t)P_{\varkappa-1}(t), \tag{107.17}$$

where $P_{\varkappa-1}(t)$ is an arbitrary polynomial of degree not greater than $\varkappa - 1$.

The preceding results are easily seen to remain true, if the function $f(t)$, instead of belonging to the class H, belongs to the class $h(c_1, c_2, \ldots, c_q)$. But in this case the solutions may be unbounded near the special ends, if $f(t)$ is unbounded near these ends.

NOTE. The equation

$$\mathbf{K}_1\varphi \equiv A(t_0)\varphi(t_0) - \frac{B(t_0)}{\pi i}\int_L \frac{\varphi(t)\, dt}{t - t_0} = f(t_0), \tag{107.18}$$

obtained from (107.1) by replacing $B(t_0)$ by $-B(t_0)$, will now be considered. This equation will not be adjoint to the equation (107.1), except when $B(t_0) = $ const. (the adjoint equation will be considered in the next section). The Hilbert problem corresponding to (107.18) is obtained from (107.6) by replacing $B(t_0)$ by $-B(t_0)$, i.e., it will have the form

$$\Psi^+(t_0) = [G(t_0)]^{-1}\Psi^-(t_0) + \frac{f(t_0)}{A(t_0) - B(t_0)}, \tag{107.19}$$

because, when B is replaced by $-B$, the function G^{-1} takes the place of G. Thus the homogeneous Hilbert problems, corresponding to (107.6) and (107.19), are associate (§ 80). Therefore, by the results of § 80, if $X(z)$ and $\varkappa$ are the fundamental function and the index of the class h corresponding to the problem (107.6), then $[X(z)]^{-1}$ and $-\varkappa$ are the fundamental function and the index of the class h', adjoint to h, corresponding to the problem (107.19).

Now, in view of the formula (107.9) defining the fundamental function $Z(t)$ corresponding to the equation (107.1) and of the same formula constructed for the equation (107.19), the following conclusion is reached:

All the formulae and results of the present section will remain true, if

$$B(t), \quad Z(t), \quad \varkappa, h$$

are replaced by

$$-B(t), \quad \frac{A^2(t)-B^2(t)}{Z(t)}, \quad -\varkappa, h'$$

respectively, where h' is the class, associate to the class h.

§ 108. Solution of the equation adjoint to the dominant equation.

The equation

$$\mathbf{K}^{0\prime}\psi \equiv A(t_0)\psi(t_0) - \frac{1}{\pi i}\int_L \frac{B(t)\psi(t)\,dt}{t-t_0} = g(t_0), \qquad (108.1)$$

adjoint to the equation $\mathbf{K}^0\varphi = f$, will now be considered. The function $g(t)$ will be assumed to belong to the class H and, as always, a solution $\psi(t)$ belonging to the class H^* will be sought.

The sectionally holomorphic function

$$\Psi(z) = \frac{1}{2\pi i}\int_L \frac{B(t)\psi(t)\,dt}{t-z},$$

vanishing at infinity, will be introduced. In view of the relations

$$B(t_0)\psi(t_0) = \Psi^+(t_0) - \Psi^-(t_0),$$
$$\frac{1}{\pi i}\int_L \frac{B(t)\psi(t)dt}{t-t_0} = \Psi^+(t_0) + \Psi^-(t_0), \qquad (108.2)$$

it may be shown, similarly as in the case of § 48, that the solution of the equation (108.1) is equivalent to the following boundary problem: to find a function $\psi(t)$ of the class H^* and a sectionally holomorphic function $\Psi(z)$, vanishing at infinity, subject to the conditions

$$A(t_0)\psi(t_0) = \Psi^+(t_0) + \Psi^-(t_0) + g(t_0),$$
$$B(t_0)\psi(t_0) = \Psi^+(t_0) - \Psi^-(t_0). \qquad (108.3)$$

However, these are equivalent to the conditions

$$(A+B)\psi = 2\Psi^+ + g, \quad (A-B)\psi = 2\Psi^- + g$$

or $$\psi = \frac{2\Psi^+}{A+B} + \frac{g}{A+B}, \quad \psi = \frac{2\Psi^-}{A-B} + \frac{g}{A-B}. \qquad (108.4)$$

Comparing the right sides, one is led to the Hilbert problem

$$\Psi^+(t_0) = [G(t_0)]^{-1}\Psi^-(t_0) + \frac{B(t_0)g(t_0)}{A(t_0) - B(t_0)}, \tag{108.5}$$

where, as in the last section,

$$G(t_0) = \frac{A(t_0) - B(t_0)}{A(t_0) + B(t_0)} \tag{108.6}$$

and it is required to find a solution $\Psi(z)$, vanishing at infinity.

Having solved this problem, the function $\psi(t)$ may be found from either of the formulae (108.4).

The homogeneous Hilbert problem

$$\Psi^+(t_0) = [G(t_0)]^{-1}\Psi^-(t_0) \tag{108.7}$$

is associate to the homogeneous problem

$$\Phi^+(t_0) = G(t_0)\Phi^-(t_0), \tag{108.8}$$

corresponding to the problem (107.6) [cf. § 80; compare also the Note in the last section]. Therefore, if $X(z)$ is the fundamental solution of the class $h = h(c_1, c_2, \ldots, c_q)$ of this latter problem, then by the results of § 80 $[X(z)]^{-1}$ will be the fundamental solution of the adjoint class $h' = h(c_{q+1}, \ldots, c_m)$ of the problem (108.7). Therefore, as is easily seen, the general solution of the problem (108.5), having finite degree at infinity, can be written in the following form:

$$\Psi(z) = \frac{[X(z)]^{-1}}{2\pi i}\int_L \frac{X^+(t)B(t)g(t)dt}{[A(t) - B(t)](t - z)} + [X_m(z)]^{-1}Q(z), \tag{108.9}$$

where $Q(z)$ is a polynomial and $X_m(z)$ is the fundamental solution of the class $h_m = h(c_1, c_2, \ldots, c_m)$ of the homogeneous problem (108.8), and hence $[X_m(z)]^{-1}$ is the fundamental solution of the class h_0 of the problem (108.7).

The condition $\Psi(\infty) = 0$ has still to be expressed. Without concentrating on this condition, some consequences will be deduced, in a similar manner as it was done in the last section, from the fact that each solution of the integral equation (108.1) is necessarily given by (108.4), where $\Psi(z)$ is an expression of the form (108.9). Determining $\psi(t_0)$ with the help of one of the formulae (108.4), one obtains, after some simple manipulations similar to those of § 107 (cf. also § 48),

$$\psi(t_0) = \mathbf{K}^{*\prime}g + \frac{P(t_0)}{Z_m(t_0)}, \tag{108.10}$$

where $P(t_0)$ is a polynomial, $\mathbf{Z}_m(t_0)$ is the fundamental function of the class $h_m = h(c_1, c_2, \ldots, c_m)$, corresponding to the equation (107.1), and

$$\mathbf{K}^{*\prime}g \equiv A^*(t_0)g(t_0) + \frac{1}{\pi i Z(t_0)} \int_L \frac{Z(t)B^*(t)g(t)dt}{t - t_0}, \qquad (108.11)$$

so that $\mathbf{K}^{*\prime}$ is the operator, adjoint to $\mathbf{K}^*$ of the last section.

Near all special ends each solution $\psi(t_0)$ is readily seen to belong to the class H_ε^* and to be bounded (cf. § 107).

Furthermore, just as in the last section, if a solution is bounded near any non-special end, then it belongs necessarily to the class H near this end.

Now all possible solutions of the equation (108.1) will be divided into classes, just as it has been done in the last section.

The index of the corresponding homogeneous Hilbert problem (108.7) of a similar class will be called the index of a given class of the equation (108.1) or of the operator $\mathbf{K}^{0\prime}$. (In § 109, the concept of the index will be given for a singular integral equation of the general form; the definition for the equation of the particular form (108.1) is easily verified to lead to the same value of the index).

If $X(z)$ is the fundamental solution of the class $h = h(c_1, \ldots, c_q)$ of the problem (108.8), then, as stated, $[X(z)]^{-1}$ will be the fundamental solution of the adjoint class $h' = h(c_{q+1}, \ldots, c_m)$ of the problem (108.7). Hence the indices $\varkappa$ and $\varkappa'$ of adjoint classes h and h' of the adjoint equations (107.1) and (108.1) are equal in magnitude and opposite in sign:

$$\varkappa' = -\varkappa.$$

On the basis of the above the following proposition will be easily established:

if $\varkappa' = -\varkappa \geqq 0$, *then all solutions of the class* $h' = h(c_{q+1}, \ldots, c_m)$ *of the equation* (108.1) *are given by*

$$\psi(t_0) = \mathbf{K}^{*\prime}g + \frac{P_{\varkappa'-1}(t_0)}{Z(t_0)}, \qquad (108.12)$$

where $P_{\varkappa'-1}(t_0)$ *is an arbitrary polynomial of degree not greater than* $\varkappa' - 1$ $[P_{\varkappa'-1}(t_0) \equiv 0$ for $\varkappa' = 0]$, *and* $Z(t_0)$ *is the fundamental function of the class h, adjoint to h', for the adjoint equation* (107.1);

if $\varkappa' = -\varkappa < 0$, *then a solution of the class h' exists only, provided the conditions*

$$\int_L Z(t)B^*(t)t^j g(t)dt = 0, \quad j = 0, 1, \ldots, -\varkappa' - 1 \qquad (108.13)$$

are satisfied, in which case the (unique) solution is again given by (108.12) *with* $P_{\varkappa'-1}(t_0) \equiv 0$.

The deduction of (108.12) is just like that of (108.10); in addition, one must take into consideration the condition $\Phi(\infty) = 0$, which also shows that the degree of the arbitrary polynomial must not exceed $\varkappa' - 1$ and that for $\varkappa' < 0$ the conditions (108.13) must hold true.

From the above it likewise follows that, for $\varkappa' \leqq 0$, the homogeneous equation $\mathbf{K}^{0\prime}\psi = 0$ has no solution of the class h', different from zero; for $\varkappa' > 0$, it has exactly $\varkappa'$ linearly independent solutions of the class h' the sum of which is given by

$$\psi(t) = \frac{P_{\varkappa'-1}(t)}{Z(t)}. \tag{108.14}$$

The preceding results will remain true, if $g(t)$ belongs to the class h' (and not necessarily to the class H), except that in this case the solutions may be unbounded near special ends.

NOTE 1. Comparing the conditions of solubility of the adjoint equations

$$\mathbf{K}^0\varphi = f, \quad \mathbf{K}^{0\prime}\psi = g,$$

the following result is easily reached: for the solubility of the equation $\mathbf{K}^0\varphi = f$ in a given class h it is necessary and sufficient that

$$\int_L f\psi_j dt = 0, \quad j = 1, 2, \ldots, \varkappa',$$

where ψ_j $(j = 1, \ldots, \varkappa')$ is a complete system of linearly independent solutions of the adjoint class h' of the adjoint homogeneous equation $\mathbf{K}^{0\prime}\psi = 0$; similarly, it is necessary and sufficient for the solubility of the equation $\mathbf{K}^{0\prime}\psi = g$ in a class h' that

$$\int_L g\varphi_j dt = 0, \quad j = 1, 2, \ldots, \varkappa,$$

where φ_j $(j = 1, 2, \ldots, \varkappa)$ is a complete system of linearly independent solutions of the class h, adjoint to h', of the adjoint homogeneous equation $\mathbf{K}^0\varphi = 0$.

It will still be noted that, if k and k' respectively are the numbers of linearly independent solutions of the classes h and h' of the homogeneous equations $\mathbf{K}^0\varphi = 0$ and $\mathbf{K}^{0\prime}\psi = 0$, then (for $\varkappa \geqq 0$: $k = \varkappa$, $k' = 0$; for $\varkappa \leqq 0$: $k = 0$, $k' = -\varkappa$)

$$k - k' = \varkappa,$$

where $\varkappa$ is the index of the class h for the operator $\mathbf{K}^0$.

The above results are particular cases of the important theorems which will be proved in § 112 below.

NOTE 2. The Note at the end of the last section obviously also remains true here.

§ 109. Reduction of the singular equation $\mathbf{K}\varphi = \mathbf{f}$.

The results of § 107 permit a very simple reduction of the singular integral equation $\mathbf{K}\varphi = f$ to a Fredholm equation, just as it was done in § 57.

The method of reduction to a Fredholm equation, studied in the present section, represents the development of an idea, initiated by T. Carleman [1], in the same direction as the treatment applied by I. N. Vekua to the case of contours (cf. § 57). Carleman considered the case where L is a segment of the real axis and [in the present notation] $\varkappa_0 = 1$.

The equation $\mathbf{K}\varphi = f$ will be written as

$$\mathbf{K}\varphi \equiv \mathbf{K}^0\varphi + \mathbf{k}\varphi = f, \tag{109.1}$$

where, as before,

$$\mathbf{K}^0\varphi \equiv A(t_0)\varphi(t_0) + \frac{B(t_0)}{\pi i}\int_L \frac{\varphi(t)dt}{t-t_0} \tag{109.2}$$

and

$$\mathbf{k}\varphi \equiv \frac{1}{\pi i}\int_L k(t_0, t)\varphi(t)dt. \tag{109.3}$$

It will be noted that, if, as has been assumed, the function $k(t_0, t)$ satisfies the H condition with respect to both variables and $\varphi(t)$ is a function of the class H^*, then $\mathbf{k}\varphi$ is easily seen to be a function of t_0 of the class H.

For simplicity, the function $f(t)$ will be assumed to belong to the class H. The equation (109.1) may be rewritten

$$\mathbf{K}^0\varphi = f - \mathbf{k}\varphi \tag{109.4}$$

and the right side (belonging to the class H) will temporarily be considered as a known function.

Just as in § 107, the division of all possible solutions of the present equation into classes $h = h(c_1, c_2, \ldots, c_q)$ can be established. The fundamental function and the index (of the same class) of the corresponding dominant equation $\mathbf{K}^0\varphi = f$ or the operator $\mathbf{K}^0$ will be called the fundamental function $Z(t)$ and the index $\varkappa$ of the class h of the equation (109.1) or of the operator $\mathbf{K}$.

As in § 107, the operator defined by

$$\mathbf{K}^*f = A^*(t_0)f(t_0) - \frac{B^*(t_0)Z(t_0)}{\pi i}\int_L \frac{f(t)\,dt}{Z(t)(t-t_0)} \tag{109.5}$$

will be denoted by $\mathbf{K}^*$, the notation being the same as in § 107.

Now, applying to (109.4) the results of § 107, the following conclusions are easily reached.

Let it be required to find all the solutions of the class $h(c_1, c_2, \ldots, c_q) = h$ of the equation (109.1) and let $Z(t)$ and $\varkappa$ be the corresponding fundamental function and index. Then:

for $\varkappa \geqq 0$, the equation (109.1) is equivalent (in the sense that solutions of the class h are being sought) to the Fredholm equation

$$\varphi(t_0) + \mathbf{K}^*\mathbf{k}\varphi = f_0(t_0), \tag{109.6}$$

where

$$f_0(t_0) = \mathbf{K}^*f + B^*(t_0)Z(t_0)P_{\varkappa-1}(t_0) \tag{109.7}$$

and $P_{\varkappa-1}(t_0)$ is an arbitrary polynomial of degree not greater than $\varkappa - 1$ $[P_{-1}(t_0) \equiv 0]$;

for $\varkappa < 0$, the equation (109.1) is equivalent (in the same sense) to the Fredholm equation (109.6), where one must put $P_{\varkappa-1}(t_0) \equiv 0$, and to the following set of additional conditions:

$$\int_L \frac{t^j\mathbf{k}\varphi(t)\,dt}{Z(t)} = \int_L \frac{t^j f(t)\,dt}{Z(t)}, \quad j = 0, 1, \ldots, -\varkappa - 1. \tag{109.8}$$

The last conditions originating from (107.16) can obviously be written as

$$\int_L \varrho_j(t)\varphi(t)\,dt = \int_L \frac{t^j f(t)\,dt}{Z(t)}, \quad j = 0, 1, \ldots, -\varkappa - 1, \tag{109.8a}$$

where

$$\varrho_j(t) = \mathbf{k}'\left[\frac{t^j}{Z(t)}\right] = \frac{1}{\pi i}\int_L \frac{k(t_1, t)t_1^{\,j}\,dt_1}{Z(t_1)} \tag{109.9}$$

is a completely defined function, easily seen to belong to the class H.

It will be noted that the function $f_0(t_0)$, defined by (109.7), belongs to the class $h(c_1, c_2, \ldots, c_q)$ and, in addition, near special ends to the class H_ε^*, and that it remains bounded there; this follows from the fact that $f(t)$, by assumption, belongs to the class H.

§ 110. Reduction of the singular equation $\mathbf{K}'\psi = g$.

The method of reduction, stated in the last section, can be somewhat modified, i.e., instead of starting from the decomposition of the operator **K** following (106.7), one can start from the decomposition given by (106.11). But in order not to introduce new notation, this alternative method will be applied to the equation

$$\mathbf{K}'\psi \equiv \mathbf{K}'^0\psi + \mathbf{k}'\psi = g, \tag{110.1}$$

adjoint to the given one which, as has been pointed out already in § 106, virtually does not imply a restriction, because the equation $\mathbf{K}'\psi = g$ can, of course, be considered independently.

Applying the same method as in the last section and using the results of § 108, one arrives at the following conclusion:

for $\varkappa' \geqq 0$, *the equation* (110.1) *is equivalent, in the sense that solutions of the class* $h'(c_{q+1}, \ldots, c_m) = h'$ are *being sought, to the Fredholm equation*

$$\psi(t_0) + \mathbf{K}^{*\prime}\mathbf{k}'\psi = g_0(t_0), \tag{110.2}$$

where

$$g_0(t_0) = \mathbf{K}^{*\prime}g + \frac{P_{\varkappa'-1}(t_0)}{Z(t_0)}, \tag{110.3}$$

$P_{\varkappa'-1}(t_0)$ *is an arbitrary polynomial of degree not greater than* $\varkappa - 1$ $[P_{-1}(t_0) \equiv 0]$ *and* $Z(t_0)$ *is the fundamental function of the class* $h(c_1, \ldots, c_q) = h$, *adjoint to* h', *for the adjoint equation* $\mathbf{K}\varphi = 0$;

for $\varkappa' < 0$, *the equation* (110.1) *is equivalent* (*in the same sense*) *to the Fredholm equation* (110.2), *where one has to assume* $P_{\varkappa'-1}(t_0) \equiv 0$, *and to the following set of additional conditions*:

$$\int_L \sigma_j(t)\psi(t)dt = \int_L Z(t)B^*(t)t^j g(t)dt, \quad j = 0, 1, \ldots, -\varkappa' - 1, \tag{110.4}$$

where

$$\sigma_j(t) = \mathbf{k}[Z(t)B^*(t)t^j] = \frac{1}{\pi i}\int_L k(t, t_1)Z(t_1)B^*(t_1)t_1^{\,j}\,dt_1 \tag{110.5}$$

are fully determined functions, belonging to the class H.

Here $\mathbf{K}^{*\prime}$ is the operator, adjoint to $\mathbf{K}^*$, just as in § 108 [(108.11)]; $B^*(t)$ is defined by (107.10). It will be noted that the equation (110.2) is not, generally speaking, adjoint to the equation (109.6), because the operator, adjoint to $\mathbf{K}^*\mathbf{k}$, is $\mathbf{k}'\mathbf{K}^{*\prime}$ and not $\mathbf{K}^{*\prime}\mathbf{k}'$.

Finally, it will be seen that the function $g_0(t_0)$ belongs to the class $h'(c_{q+1}, \ldots, c_m)$ and also to the class $\overset{*}{H}_\varepsilon$ near special ends and remains bounded near these; this follows from the fact that, by assumption, $g(t)$ belongs to the class H.

§ 111. Investigation of the equation, resulting from the reduction.

The equation (109.6) will now be studied more closely; it will be written in the form

$$\varphi(t_0) + \mathbf{K}^*\mathbf{k}\varphi \equiv \varphi(t_0) + \frac{1}{\pi i}\int_L N(t_0, t)\varphi(t)dt = f_0(t_0), \quad (111.1)$$

where $f_0(t_0)$ is a function of the class $h(c_1, c_2, \ldots, c_q)$, belonging to the class $H_\varepsilon{}^*$ at special ends and bounded near them, and

$$N(t_0, t) = A^*(t_0)k(t_0, t) - \frac{B^*(t_0)Z(t_0)}{\pi i}\int_L \frac{k(t_1, t)\,dt_1}{Z(t_1)(t_1 - t_0)}. \quad (111.2)$$

Together with this equation the corresponding homogeneous equation

$$\varphi(t_0) + \mathbf{K}^*\mathbf{k}\varphi = 0 \quad (111.3)$$

and the adjoint homogeneous equation

$$\psi(t_0) + \mathbf{k}'\mathbf{K}^{*\prime}\psi \equiv \psi(t_0) + \frac{1}{\pi i}\int_L N(t, t_0)\psi(t)dt = 0 \quad (111.4)$$

will be considered, where, by (111.2),

$$N(t, t_0) = A^*(t)k(t, t_0) - \frac{B^*(t)Z(t)}{\pi i}\int_L \frac{k(t_1, t_0)dt_1}{Z(t_1)(t_1 - t)}. \quad (111.5)$$

In § 109, the equation (111.1) was called a Fredholm equation, although its kernel $N(t_0, t)$ does not belong to that type of kernel which as a rule is called regular.

In spite of this, all the fundamental Fredholm theorems are applicable to the equation (111.1), if they are stated in a suitable manner. This will be shown by reduction of (111.1) to a Fredholm equation with a bounded kernel.

By (107.13)

$$Z(t) = \omega_0(t) \prod_{j=1}^{2p} (t - c_j)^{\gamma_j}, \quad (111.6)$$

where $\omega_0(t)$ is a certain non-zero function of the class H,

$$0 < \Re\gamma_j < 1 \ (j = 1, \ldots, q), \ -1 < \Re\gamma_j < 0 \ (j = q + 1, \ldots, m),$$
$$\Re\gamma_j = 0 \ (j = m + 1, \ldots, 2p).$$

Let

$$\prod_{j=q+1}^{m} (t - c_j)^{\gamma_j} = T(t), \quad (111.7)$$

so that

$$Z(t) = T(t)\Omega(t), \tag{111.8}$$

where $\Omega(t)$ is a bounded function of the class H_{ε}^{*}, belonging, in addition, to the class H near the ends $c_1, c_2, \ldots, c_q$ (and vanishing there), and put

$$\varphi(t) = T(t)\varphi_0(t). \tag{111.9}$$

Then the equation (111.1) will have the form

$$\varphi_0(t_0) + \frac{1}{\pi i}\int_L \frac{N(t_0, t)T(t)\varphi_0(t)\,dt}{T(t_0)} = \frac{f_0(t_0)}{T(t_0)}. \tag{111.10}$$

It is easily seen that the expression

$$n(t_0, t) = \frac{1}{\pi i}\frac{N(t_0, t)}{T(t_0)} \tag{111.11}$$

is, with respect to the variable t_0, a bounded function of the class $H_{\varepsilon}{}^{*}$, belonging, in addition, to the class H near all non-special ends; however, with respect to the variable t, the function $n(t_0, t)$ belongs to the class H. The right side of (111.10) is easily seen to be a bounded function of the class $H_{\varepsilon}{}^{*}$ and to belong to the class H near all non-special ends.

The variable t will now be replaced by a new variable τ by letting

$$T(t)dt = d\tau \tag{111.12}$$

or, for definiteness,

$$\tau = \int_{a_j}^{t} T(t)dt \text{ on } L_j \ (j = 1, \ldots, p). \tag{111.12a}$$

Denoting $\varphi_0(t)$ simply by $\varphi_0(\tau)$ and dealing similarly with regard to the other functions, the equation (111.10) will in this way be reduced to the form

$$\varphi(\tau_0) + \int_{\Lambda} n(\tau_0, \tau)\varphi_0(\tau)d\tau = \frac{f_0(\tau_0)}{T(\tau_0)}, \tag{111.13}$$

i.e., to an equation with a bounded kernel having the above-mentioned properties [cf. the remarks following (111.11)]; Λ denotes here the union of the arcs Λ_k in the auxiliary ζ plane, corresponding to the arcs L_k in the z plane. (The arcs Λ_k may now intersect one another, but this is of no importance here).

For the present purpose the solutions of the equation (111.1) must be sought in the class $h(c_1, c_2, \ldots, c_q)$. But, in the sequel. by solutions of the equation (111.1) will be understood any absolutely integrable solutions, because all such solutions are easily seen to belong necessarily to the class $h(c_1, c_2, \ldots, c_q)$.

In fact, let the unknown function $\varphi(t)$ be absolutely integrable [under which assumption the integral on the left of (111.1) is easily seen to retain the definite (ordinary) meaning]; then also the function $\varphi_0(t)$ has this property. On the other hand, as is easily seen from (111.13), everyone of its absolutely integrable solutions will be a bounded function of the class $H_\varepsilon{}^*$, belonging, in addition, to the class H near all non-special ends; each such solution gives, by (111.9), a solution of the class $h(c_1, c_2, \ldots, c_q)$ of the equation (111.1), which proves the proposition.

Thus a solution of (111.1) in the class $h(c_1, c_2, \ldots, c_q)$ is equivalent to a solution of the same equation in the class of absolutely integrable functions, and likewise to a solution of the Fredholm equation (111.13) in the ordinary sense, i.e., when its solution is being sought in the class of bounded (and, of course, integrable) functions.

Next, the homogeneous equation (111.4), adjoint to (111.1), will be considered. If the change of variable (111.12) is made in this equation, then it takes the form (in the earlier notation)

$$\psi(\tau_0) + \int_{\Lambda} n(\tau, \tau_0)\psi(\tau)d\tau = 0, \tag{111.14}$$

i.e., it becomes a Fredholm equation with a bounded kernel, adjoint to the equation (111.13). On the basis of what has been said above with respect to the function $n(\tau, \tau_0)$, it is easily seen that each absolutely integrable solution of (111.14) belongs necessarily to the class H and is bounded. Hence, when dealing with solutions of (111.4), bounded functions will always be implied.

Consider now the fundamental Fredholm theorems on the equality of the numbers of linearly independent solutions of adjoint homogeneous equations and on the conditions of solubility of the non-homogeneous equation.

The first of these theorems is directly applicable to the homogeneous equation, corresponding to (111.13), and to its adjoint equation (111.14), since the kernels of these equations are bounded.

For application to the equations (111.3) and (111.4), however, this theorem may obviously be formulated as follows:

The number of linearly independent (absolutely integrable) solutions of the homogeneous equation

$$\varphi(t_0) + \mathbf{K}^*\mathbf{k}\varphi = 0,$$

which solutions necessarily belong to the class $h(c_1, c_2, \ldots, c_q)$ and, besides, to the class H_ε^* at special ends, *is finite and equal to the number of linearly independent (bounded) solutions of its adjoint homogeneous equation*

$$\psi(t_0) + \mathbf{k}'\mathbf{K}^{*\prime}\psi = 0$$

(which solutions necessarily belong to the class H).

Next consider the second theorem. When applied to the equations (111.13) and (111.14), this theorem states that the necessary and sufficient conditions for the solubility of the equation (111.13) are

$$\int_\Lambda \frac{f_0(\tau)}{T(\tau)}\,\omega_j(\tau)d\tau = 0,\ j = 1,\ 2,\ \ldots,\ n, \qquad (111.15)$$

where $\omega_j(\tau)$ is a complete system of linearly independent solutions of the homogeneous equation (111.14). Reverting to the former variable t, the above conditions will be of the form

$$\int_L f_0(t)\omega_j(t)dt = 0,\ j = 1,\ 2,\ \ldots,\ n. \qquad (111.16)$$

Thus, for direct application to the equations (111.1) and (111.4), this theorem will be formulated as follows:

In order that the equation

$$\varphi(t_0) + \mathbf{K}^*\mathbf{k}\varphi = f_0(t_0) \qquad (111.1)$$

will be soluble (in the class of absolutely integrable functions), it is necessary and sufficient that its right side satisfies the conditions (111.16), *where $\omega_j(t)$ is a complete system of linearly independent (bounded) solutions of the homogeneous equation* (111.4), *adjoint to the preceding one. All (absolutely integrable) solutions of* (111.1) *belong necessarily to the class $h(c_1, c_2, \ldots, c_q)$ and, in addition, to the class H_ε^* near special ends, remaining bounded near these.*

It will be remembered that $f_0(t_0)$ is a function, belonging to the class $h(c_1, c_2, \ldots, c_q)$ and to the class H_ε^* at special ends and remaining bounded near these.

If the conditions of solubility of the Fredholm equation (111.13) are fulfilled, then its general solution, by the results of § 52, is of the form

$$\varphi_0(\tau_0) = \frac{f_0(\tau_0)}{T(\tau_0)} + \int_\Lambda \gamma(\tau_0, \tau)\frac{f_0(\tau)}{T(\tau)}\,d\tau + \sum_{j=1}^{n} C_j\chi_{0j}(\tau_0), \qquad (111.17)$$

where $\gamma(\tau_0, \tau)$ is the generalized resolvent (cf. § 52), χ_{0j} ($j = 1, 2, \ldots, n$) is a complete system of linearly independent solutions of the corresponding homogeneous equation and the C_j are arbitrary constants.

It is easily deduced from the functional equations, given at the end of § 52 [(52.13) and (52.14)], that the resolvent $\gamma(\tau_0, \tau)$ has near ends the same character as the kernel $n(\tau_0, \tau)$, i.e., it is with respect to τ_0 a bounded function of the class $H_\varepsilon{}^*$, belonging, besides, to the class H near all non-special ends, while with respect to τ it is a function of the class H.

Reverting now, by use of the formulae (111.9) and (111.12), to the original unknown function $\varphi(t)$ and the variable t, one deduces that the most general (absolutely integrable) solution of the equation (111.1), which is known to belong necessarily to the class $h(c_1, c_2, \ldots, c_q)$, is given by

$$\varphi(t_0) = \mathbf{\Gamma} f_0 + \sum_{j=1}^{n} C_j \chi_j(t), \tag{111.18}$$

where $\chi_j(t)$ is a complete system of linearly independent (absolutely integrable) solutions of the homogeneous equation (111.3) which are known to belong necessarily to the class $h(c_1, c_2, \ldots, c_q)$; $\mathbf{\Gamma}$ is the operator, defined by

$$\mathbf{\Gamma} f \equiv f(t_0) + \int_L T(t_0)\gamma(t_0, t) f(t) dt. \tag{111.19}$$

From the remarks, made above with regard to the properties of the function $\gamma(t_0, t)$, it is easily deduced that *the operator* $\mathbf{\Gamma}$ *transforms each function* $f(t)$ *of the class* $h(c_1, c_2, \ldots, c_q)$ *into a function of the same class, and that its adjoint operator* $\mathbf{\Gamma}'$, *defined by*

$$\mathbf{\Gamma}' g \equiv g(t_0) + \int_L T(t)\gamma(t, t_0) g(t) dt, \tag{111.20}$$

transforms each function $g(t)$ *of the class* $h' = h(c_{q+1}, \ldots, c_m)$ *into a function of the same class.*

The equation (110.2) of the last section, viz.,

$$\psi(t_0) + \mathbf{K}^{*\prime}\mathbf{k}'\psi = g_0(t_0), \tag{111.21}$$

obtained by reduction of the equation $\mathbf{K}'\psi = g$, and the homogeneous equation

$$\omega(t_0) + \mathbf{k}\mathbf{K}^*\omega = 0, \tag{111.22}$$

adjoint to (111.21), can be treated similarly; however, this will not be done here.

§ 112. Solution of a singular equation. Fundamental theorems.

The problem of the solution of the singular integral equation

$$\mathbf{K}\varphi \equiv \mathbf{K}^0\varphi + \mathbf{k}\varphi = f \tag{112.1}$$

in a given class $h(c_1, c_2, \ldots, c_q)$ will again be considered. Reducing the above equation by the method stated in § 109, one obtains the Fredholm equation

$$\varphi + \mathbf{K}^*\mathbf{k}\varphi = f_0(t_0) \tag{112.2}$$

and the additional conditions

$$\int_L \varrho_j(t)\varphi(t)dt = \int_L \frac{t^j f(t)dt}{Z(t)}, \; j = 0, 1, \ldots, -\varkappa - 1. \tag{112.3}$$

The original singular equation is equivalent, in the sense that solutions of the class $h(c_1, c_2, \ldots, c_q)$ are being sought, to the combination of the equation (112.2) and the conditions (112.3).

The notation of § 109 will be used here; in particular,

$$f_0(t_0) = \mathbf{K}^*f + B^*(t_0)Z(t_0)P_{\varkappa-1}(t_0), \tag{112.4}$$

where $P_{\varkappa-1}(t_0)$ is an arbitrary polynomial of degree not greater than $\varkappa - 1$ which will be written as

$$P_{\varkappa-1}(t_0) = A_1 t^{k_1} + A_2 t^{k_2} + \ldots + A_\varkappa t^{k_\varkappa}, \tag{112.5}$$

where the $k_1, k_2, \ldots, k_\varkappa$ denote the numbers $0, 1, \ldots, \varkappa - 1$, taken in any order, and the $A_1, A_2, \ldots, A_\varkappa$ are arbitrary constants.

In the case $\varkappa \geqq 0$, the additional conditions (112.3) vanish, while in the case $\varkappa \leqq 0$ one has to assume $P_{\varkappa-1}(t_0) \equiv 0$.

One direct consequence of the above will be indicated; in fact, *the number of linearly independent solutions (of any class) of the homogeneous equation* $\mathbf{K}\varphi = 0$ *is finite*, because all the solutions of a given class $h(c_1, c_2, \ldots, c_q)$ are at the same time solutions of the homogeneous equation $\varphi + \mathbf{K}^*\mathbf{k}\varphi = 0$.

First, the question of the solution of (112.1) for the case $\varkappa \geqq 0$ will be considered. In this case the equation (112.1) is equivalent, in the sense of finding solutions of the class $h(c_1, c_2, \ldots, c_q)$, to the equation (112.2). The conditions of solubility of the latter have the form (cf. § 111)

$$\int_L \omega_j(t) f_0(t)dt = 0, \; j = 1, 2, \ldots, n, \tag{112.6}$$

where $\omega_j(t)$ is the complete system of linearly independent solutions of the homogeneous equation

$$\omega + \mathbf{k}'\mathbf{K}^{*\prime}\omega = 0, \tag{112.7}$$

adjoint to (112.2); the functions $\omega_j(t)$ are known to belong to the class H.

Substituting in (112.6) the expression (112.4) for the function $f_0(t_0)$ and introducing the notation

$$\delta_i = \int_L \omega_i \mathbf{K}^* f dt, \tag{112.8}$$

it is seen that the conditions (112.6) have the form

$$\sum_{j=1}^{\varkappa} \gamma_{ij} A_j = \delta_i, \quad i = 1, 2, \ldots, n, \tag{112.9}$$

where the γ_{ij} are fully determined constants which do not depend on $f(t)$. The constants δ_i can still be presented in the form

$$\delta_i = \int_L \overset{*}{\omega_i} f dt, \quad i = 1, 2, \ldots, n, \tag{112.10}$$

where

$$\overset{*}{\omega_i} = \mathbf{K}^{*\prime} \omega_i. \tag{112.11}$$

The functions $\overset{*}{\omega_i}$ *belong to the class* $h(c_{q+1}, \ldots, c_m) = h'$, *adjoint to* $h(c_1, c_2, \ldots, c_q)$; this follows from the fact that the ω_j belong to the class H, and from the form of the operator $\mathbf{K}^{*\prime}$. Further, it is easily seen that *the functions* $\omega_j{}^*$ *are linearly independent.* In fact, it follows from (112.7) that $\omega_j + \mathbf{k}'\mathbf{K}^{*\prime}\omega_j = 0$, and hence that $\omega_j = - \mathbf{k}'\omega_j{}^*$; therefore, if the functions $\omega_j{}^*$ were linearly dependent, then also the functions ω_j would be linearly dependent, which contradicts their definition.

Let the rank of the matrix $\| \gamma_{ij} \|$ be r ($r \leqq n$, $r \leqq \varkappa$). Without loss of generality, the determinant of the matrix

$$\| \gamma_{ij} \| \quad (i, j = 1, 2, \ldots, r)$$

may be assumed different from zero. Then it is known that the conditions of solubility of the system (112.9) for $A_1, \ldots, A_\varkappa$ are

$$\begin{Vmatrix} \gamma_{11} & \gamma_{12} & \cdots & \cdots & \gamma_{1r} & \delta_1 \\ \gamma_{21} & \gamma_{22} & \cdots & \cdots & \gamma_{2r} & \delta_2 \\ \cdots & \cdots & \cdots & \cdots & \cdots & \cdots \\ \cdots & \cdots & \cdots & \cdots & \cdots & \cdots \\ \gamma_{r1} & \gamma_{r2} & \cdots & \cdots & \gamma_{rr} & \delta_r \\ \gamma_{r+j,1} & \gamma_{r+j,2} & \cdots & \cdots & \gamma_{r+j\ r} & \delta_{r+j} \end{Vmatrix} = 0, \; j = 1, 2, \ldots, n - r, \tag{112.12}$$

or

$$\delta_{r+j} + \sum_{i=1}^{r} a_{ji}\delta_i = 0, \quad j = 1, 2, \ldots, n - r, \qquad (112.13)$$

where the a_{ji} are fully determined constants which are independent of the function $f(t)$.

Substituting in the above equations the expressions (112.10) for the δ_j, it is seen that *the conditions of solubility of the equation* (112.2) *have the form*

$$\int_L \lambda_j f(t)\, dt = 0, \; j = 1, 2, \ldots, n - r, \qquad (112.14)$$

where the $\lambda_j = \lambda_j(t)$ are fully defined linearly independent functions of the class $h' = h(c_{q+1}, \ldots, c_m)$, and, in fact,

$$\lambda_j(t) = \omega^*_{r+j}(t) + \sum_{i=1}^{r} a_{ji}\omega^*_i(t), \; j = 1, \; \ldots, n - r. \quad (112.15)$$

The conditions (112.14) will be assumed satisfied; then the equation (112.2) is soluble. Its general solution will be constructed. Since, by assumption, the conditions (112.14) and hence (112.13) are fulfilled, the system (112.9) is soluble with respect to $A_1, A_2, \ldots, A_r$; its most general solution will be found by retaining the arbitrary constants $A_{r+1}, \ldots, A_\varkappa$ and solving the first r equations of the system (112.9) for $A_1, \ldots, A_r$. The general solution of (112.9) will therefore have the form

$$A_j = B_{jr+1}A_{r+1} + \ldots + B_{j\varkappa}A_\varkappa + \Gamma_{j1}\delta_1 + \ldots + \Gamma_{jr}\delta_r, \quad (112.16)$$

$j = 1, 2, \ldots, r$, where the B_{ji}, Γ_{ji} are certain constants, not depending on $f(t)$.

If now the above values of the constants $A_1, \ldots, A_r$ are introduced on the right side of the equation (112.4), then the resulting integral equation will be soluble for any values of the arbitrary constants $A_{r+1}, \ldots, A_\varkappa$. Solving this equation with the help of (111.18) and taking into consideration (112.8), it is easily deduced that the general solution of (112.2) has the form

$$\varphi(t_0) = \mathbf{\Gamma^* K^*} f + C_1\chi_1 + C_2\chi_2 + \ldots + C_n\chi_n + C_{n+1}\chi_{n+1} + \cdots + \\ + C_{\varkappa+n-r}\chi_{\varkappa+n-r}, \qquad (112.17)$$

where the $\chi_1, \chi_2, \ldots, \chi_n$ and $C_1, C_2, \ldots, C_n$ are the same as in (111.18); by $C_{n+1}, \ldots, C_{\varkappa+n-r}$ are denoted, for uniformity, the arbitrary constants $A_{r+1}, \ldots, A_\varkappa$, and by $\chi_{n+1}, \ldots, \chi_{\varkappa+n-r}$ the fully determined functions, not depending on $f(t)$ and belonging, as also the functions $\chi_1, \ldots, \chi_n$, to the class $h(c_1, c_2, \ldots, c_q)$; these functions will be discussed below. Finally, $\mathbf{\Gamma}^*$ is the operator, defined by

$$\mathbf{\Gamma}^* g \equiv g(t_0) + \int_L \Gamma^*(t_0, t) g(t)\, dt, \tag{112.18}$$

where $\Gamma^*(t_0, t)$ differs from $T(t_0)\gamma(t_0, t)$ of (111.19) by certain terms which are easily written down; if these simple manipulations are carried out, it will become obvious that the operator $\mathbf{\Gamma}^*$ has the property of $\mathbf{\Gamma}$ at the end of the last section; in fact, the operator $\mathbf{\Gamma}^*$ transforms a function of the class $h = h(c_1, c_2, \ldots, c_q)$ into a function of the same class, and the adjoint operator $\mathbf{\Gamma}^{*\prime}$ transforms a function of the class $h' = h(c_{q+1}, \ldots, c_m)$ into a function of the same class.

In particular, the case will be considered where the original singular equation (112.1) is homogeneous, i.e., where $f(t) \equiv 0$. In this case, (112.1) is equivalent to the following (in general, non-homogeneous) equation

$$\varphi(t_0) + \mathbf{K}^*\mathbf{k}\varphi = B^*(t_0) Z(t_0) P_{\varkappa-1}(t_0). \tag{112.19}$$

In the present case all $\delta_i = 0$ and the conditions of solubility (112.13) of the system (112.9) are satisfied. However, the relations (112.16) will take the form

$$A_j = B_{jr+1} A_{r+1} + \ldots + B_{j\varkappa} A_\varkappa \quad (j = 1, 2, \ldots, r). \tag{112.16a}$$

If these expressions for $A_1, A_2, \ldots, A_r$ are introduced on the right side of (112.19), it takes the form

$$\varphi + \mathbf{K}^*\mathbf{k}\varphi = C_{n+1}\sigma_{n+1} + \ldots + C_{\varkappa+n-r}\sigma_{\varkappa+n-r}; \tag{112.20}$$

again $C_{n+1}, \ldots, C_{\varkappa+n-r}$ have been written instead of $A_{r+1}, \ldots, A_\varkappa$; by $\sigma_{n+1}, \ldots, \sigma_{\varkappa+n-r}$ have been denoted certain linearly independent functions of the class $h(c_1, c_2, \ldots, c_q)$ for which expressions are easily written down.

The general solution of the homogeneous equation $\mathbf{K}\varphi = 0$, equivalent to the equation (112.20), for arbitrary $C_{n+1}, \ldots, C_{\varkappa+n-r}$ has, by the general formula (112.17), the form

$$\varphi(t_0) = C_1\chi_1(t_0) + \ldots + C_n\chi_n(t_0) + C_{n+1}\chi_{n+1}(t_0) + \ldots + \\ + C_{\varkappa+n-r}\chi_{\varkappa+n-r}(t_0), \tag{112.21}$$

where $C_1, C_2, \ldots, C_{\varkappa+n-r}$ are arbitrary constants. Thus the functions $\chi_j(t)$ are solutions of the homogeneous equation $\mathbf{K}\varphi = 0$; the first n of them are at the same time linearly independent solutions of the equation $\varphi + \mathbf{K}^*\mathbf{k}\varphi = 0$. The functions $\chi_j(t)$ for $j \geqq n$, however, are obviously solutions of the equations

$$\varphi + \mathbf{K}^*\mathbf{k}\varphi = \sigma_j \quad (j = n + 1, \ldots, \varkappa + n - r), \tag{112.22}$$

obtained from (112.20) by taking the C_i equal to zero, except for $C_j = 1$. Now it is easily seen that all the functions $\chi_j(t)$ ($j = 1, 2, \ldots, \varkappa + n - r$) are linearly independent.

In fact, for certain values of the C_j, let the function φ, defined by (112.21), be identically zero. Then obviously

$$0 \equiv \varphi + \mathbf{K}^*\mathbf{k}\varphi \equiv C_{n+1}\sigma_{n+1} + \ldots + C_{\varkappa+n-r}\sigma_{\varkappa+n-r}$$

which is only possible for $C_{n+1} = \ldots = C_{\varkappa+n-r} = 0$, because the functions σ_j are linearly independent. But then it follows from (112.21) for $\varphi \equiv 0$ that also $C_1 = C_2 = \ldots = C_n = 0$.

Thus, the homogeneous equation $\mathbf{K}\varphi = 0$ has exactly $\varkappa + n - r$ linearly independent solutions of the class $h(c_1, \ldots, c_q)$.

Since, further, $n \geqq r$, it may be stated that, *for* $\varkappa \geqq 0$, *the homogeneous equation* $\mathbf{K}\varphi = 0$ *has at least* $\varkappa$ *linearly independent solutions of the class* $h(c_1, \ldots, c_q)$.

Now the case will be considered where $\varkappa$ is negative. In this case one has to assume $P_{\varkappa-1}(t_0) \equiv 0$ in equation (112.2) and the conditions of solubility (112.6) of this equation reduce to the conditions $\delta_j = 0$, $j = 1, 2, \ldots, n$, i.e., again to conditions of the form

$$\int_L \lambda_j(t)f(t)\,dt = 0, \quad j = 1, 2, \ldots, n, \tag{112.23}$$

where the $\lambda_j(t)$ are certain definite linearly independent functions of the class $h' = h(c_{q+1}, \ldots, c_m)$.

Let the conditions (112.23) be satisfied; then the equation (112.2) is soluble; however, this still does not mean that the original equation (112.1) is soluble in a given class $h(c_1, \ldots, c_q)$, because in the present case ($\varkappa < 0$) the ($-\varkappa$) additional conditions (112.3) must be satisfied. Introduce the general solution (112.17) on the left side of (112.3); of course, one must then assume in (112.17) $C_{n+1} = \ldots = C_{n+\varkappa-r} = 0$. Thus one obtains for the determination of $C_1, C_2, \ldots, C_n$ a system of ($-\varkappa$) linear equations analogous to the system (112.9), the conditions of solubility of which lead again to some set of conditions of the form

$$\int_L \lambda_j(t)f(t)\,dt = 0, \quad j = n+1, n+2, \ldots, n+s, \tag{112.24}$$

where $s \leqq -\varkappa$. If the corresponding elementary calculations are carried out and the formula

$$\int_L \varrho_j \mathbf{\Gamma}^*\mathbf{K}^* f\,dt = \int_L f\mathbf{K}^{*\prime}\mathbf{\Gamma}^{*\prime}\varrho_j\,dt \tag{112.25}$$

and also the property of the operator $\mathbf{\Gamma}^{*\prime}$ stated above are taken into consideration, then it is easily seen that the functions $\lambda_j(t)$ $(j = n+1, \ldots, n+s)$ belong, just as the functions $\lambda_j(t)$ of the formula (112.23), to the class $h' = h(c_{q+1}, \ldots, c_m)$.

The conditions (112.23), together with the conditions (112.24), give a system of necessary and sufficient conditions for the solubility of the equation (112.1) in the class $h = h(c_1, \ldots, c_q)$.

Now the proof will be given of the general theorems, analogous to those of § 53; the actual proofs presented here are the results of generalization of those, given by I. N. Vekua [4] for the case of contours and continuous coefficients. These theorems are:

THEOREM I. *The necessary and sufficient conditions of solubility in a given class* $h = h(c_1, \ldots, c_q)$ *of the equation* $\mathbf{K}\varphi = f$ are

$$\int_L f\psi_j dt = 0, \quad j = 1, 2, \ldots, k', \tag{112.26}$$

where ψ_j $(j = 1, \ldots, k')$ *is a complete system of linearly independent solutions of the adjoint class* $h' = h(c_{q+1}, \ldots, c_m)$ *of the adjoint homogeneous equation* $\mathbf{K}'\psi = 0$.

THEOREM II. *If* k *is the number of linearly independent solutions of the class* h *of the homogeneous equation* $\mathbf{K}\varphi = 0$, $\varkappa$ *is the index of this class, and* k' *is the number of linearly independent solutions of the adjoint class* h' *of the adjoint homogeneous equation* $\mathbf{K}'\psi = 0$, *then*

$$k - k' = \varkappa. \tag{112.27}$$

PROOF OF THEOREM I. The necessity of (112.26) is obvious from (106.14) which is easily seen to be applicable in the present case.

Now the sufficiency of the conditions (112.26) will be proved. It has already been seen that the necessary and sufficient conditions for the solubility of the equation $\mathbf{K}\varphi = f$ in the class h have the form

$$\int_L \lambda_j(t)f(t)dt = 0, \quad j = 1, 2, \ldots, N, \tag{112.28}$$

where the $\lambda_j(t)$ are definite functions of the class h', adjoint to h, and N is some positive integer or zero. Theorem I will obviously be proved, if it can be shown that the conditions (112.28) follow from the conditions (112.26).

Let $g(t)$ be an arbitrary function of the class H, vanishing at all ends, so that $\mathbf{K}g$ belongs to the class H. The equation $\mathbf{K}\varphi = \mathbf{K}g$ is soluble in the class h (and even in the class H), because it has the solution $\varphi = g$. Hence, necessarily,

$$0 = \int_L \lambda_j \mathbf{K}g dt = \int_L g\mathbf{K}'\lambda_j dt = 0.$$

Due to the arbitrariness of the function $g(t)$, it follows from the preceding equation that $\mathbf{K}'\lambda_j = 0$. Thus the λ_j are solutions of the class h' of the homogeneous equation $\mathbf{K}'\psi = 0$ and hence are linear combinations of the functions $\psi_1, \psi_2, \ldots, \psi_{k'}$; hence the conditions (112.28) follow from (112.26) and Theorem I is proved.

Proof of theorem II. First the case $\varkappa \geqq 0$ will be considered. In this case the necessary and sufficient conditions for solubility in the class h have the form (112.14), where $\lambda_1, \lambda_2, \ldots, \lambda_{n-r}$ are linearly independent functions of the class h'. On the other hand, as has just been proved, the conditions (112.26) are necessary and sufficient for solubility (in the same class). Thus, if any function f of the class H satisfies (112.14), then it likewise satisfies (112.26) and vice versa. Hence it follows (cf. Appendix 3) that the functions $\lambda_1, \lambda_2, \ldots, \lambda_{n-r}$ are essentially linear combinations of the functions $\psi_1, \psi_2, \ldots, \psi_{k'}$, and conversely. Hence

$$k' = n - r.$$

Further, the number of linearly independent solutions of the class h of the equation $\mathbf{K}\varphi = 0$ is, by (112.21), equal to $\varkappa + n - r$, so that

$$k = \varkappa + n - r;$$

the equation (112.27) follows from the last two relations.

Next the case $\varkappa < 0$ will be examined. This case can be reduced to the preceding one by exchange of the roles of the operators $\mathbf{K}$ and $\mathbf{K}'$ and the classes h and h'. For this purpose two ways lie open: one may either apply the method of reduction of § 109 or that of § 110 to the singular equation $\mathbf{K}'\psi = g$. In the latter case, as is easily seen, the reasoning and results are quite analogous to those which correspond to the first method of reduction. In both cases the relation $k' - k = \varkappa'$ is obtained, where $\varkappa'$ is the index of the class h' for the operator $\mathbf{K}'$, which is found to be equal to $(-\varkappa)$ (compare the Note at the end of § 107). Therefore the equation (112.27) is again obtained.

Thus Theorem II has been proved for all cases.

Note 1. The above theorems and results are seen to remain true, if the function $f(t)$, instead of belonging to the class H, belongs to the class $h(c_1, \ldots, c_q)$ and, at the same time, to the class $H_\varepsilon{}^*$ near special ends, remaining bounded near these. In neither case will there arise any essential changes, if $f(t)$, belonging to the class $h(c_1, \ldots, c_q)$, belongs to the class H^* near special ends; the above theorems remain true in this case.

Note 2. If, after suitable introduction of a selected real variable, for example, of the arc coordinate s, the singular integral equation under consideration becomes real, then the fundamental theorems proved above also hold true in the case, where consideration is restricted to solutions in the real field. This is analogous to what was stated in § 54. For this purpose, one must understand by the homogeneous equation, adjoint to a given equation, the likewise real adjoint equation, constructed from the given equation which was reduced to the real form (cf. § 54).

§ 113. Application to the dominant equation of the first kind.

In § 88 the equation

$$\frac{1}{\pi i}\int_L \frac{\varphi(t)dt}{t-t_0} = f(t_0) \tag{113.1}$$

has been solved. It will now be called the dominant equation of the first kind; it is obtained from the equation $\mathbf{K}^0\varphi = f$, considered in § 107, by putting $A(t_0) = 0$, $B(t_0) = 1$.

The fundamental function $X(z)$ of the class $h(c_1, c_2, \ldots, c_q)$ for the Hilbert problem (107.6), corresponding to the equation (113.1), is given by (§ 88 and § 84)

$$X(z) = C\,\frac{\sqrt{R_1(z)}}{\sqrt{R_2(z)}}, \tag{113.2}$$

where C is an arbitrary constant, different from zero, and

$$R_1(z) = \prod_{j=1}^{q} (z-c_j), \quad R_2(z) = \prod_{j=q+1}^{2p} (z-c_j); \tag{113.3}$$

the roots are to be understood as in § 84; in fact,

$$\frac{\sqrt{R_1(z)}}{\sqrt{R_2(z)}} = \sqrt{\frac{R_1(z)}{R_2(z)}}, \quad \frac{\sqrt{R_2(z)}}{\sqrt{R_1(z)}} = 1 : \sqrt{\frac{R_1(z)}{R_2(z)}}, \tag{113.4}$$

where by the root on the right side must be understood the branch, holomorphic in the plane cut along L; as in § 84, it will be assumed, for definiteness, that for large $|z|$

$$\sqrt{\frac{R_1(z)}{R_2(z)}} = z^{q-p} + \alpha_1 z^{q-p-1} + \ldots. \tag{113.5}$$

In the present case all ends are non-special. Since the degree of $X(z)$ at infinity is $q-p$, the index of the class $h(c_1, \ldots, c_q)$ is

$$\varkappa = p - q. \tag{113.6}$$

The fundamental function $Z(t)$ of the class $h(c_1, \ldots, c_q)$ for the equation (113.1) is given by (107.9) for $A = 0$, $B = 1$, so that

$$Z(t) = X^+(t) = -X^-(t) = C\frac{\sqrt{R_1(t)}}{\sqrt{R_2(t)}} = C\sqrt{\frac{R_1(t)}{R_2(t)}}, \quad (113.7)$$

where one must understand by the last root the value taken by the function $\sqrt{R_1(z)/R_2(z)}$ on L from the left.

Accordingly, by (107.12),

$$\mathbf{K}^*f \equiv \frac{\sqrt{R_1(t_0)}}{\pi i\sqrt{R_2(t_0)}}\int_L \frac{\sqrt{R_2(t)}f(t)dt}{\sqrt{R_1(t)}(t-t_0)}. \quad (113.8)$$

A solution of the class $h(c_1, c_2, \ldots, c_q)$ of the equation (113.1) will, by the general formula of § 107, have the form

$$\varphi(t_0) = \frac{\sqrt{R_1(t_0)}}{\pi i\sqrt{R_2(t_0)}}\int_L \frac{\sqrt{R_2(t)}f(t)dt}{\sqrt{R_1(t)}(t-t_0)} + \frac{\sqrt{R_1(t_0)}}{\sqrt{R_2(t_0)}}P_{p-q-1}(t_0), \quad (113.9)$$

where $P_{p-q-1}(t_0)$ is an arbitrary polynomial of degree not greater than $p - q - 1$ [$P_{p-q-1}(t_0) \equiv 0$ for $p \leqq q$]; for $p < q$, it is necessary and sufficient for the existence of a solution that

$$\int_L \frac{\sqrt{R_2(t)}}{\sqrt{R_1(t)}} t^j f(t)dt = 0, \quad j = 0, 1, \ldots, q-p-1. \quad (113.10)$$

Thus the result of § 88 has again been obtained. It will still be noted that (113.1) is self-adjoint and that the conditions of solubility (113.10) express nothing else than Theorem I of the last section, if it is applied to the present particular case.

§ 114. Reduction and solution of an equation of the first kind.

The equation of the first kind of the general form

$$\frac{1}{\pi i}\int_L \frac{K(t_0, t)\varphi(t)dt}{t-t_0} = f(t_0) \quad (114.1)$$

or, in the former notation,

$$\frac{B(t_0)}{\pi i}\int_L \frac{\varphi(t)dt}{t-t_0} + \frac{1}{\pi i}\int_L k(t_0, t)\varphi(t)dt = f(t_0), \quad (114.2)$$

where

$$B(t_0) = K(t_0, t_0), \quad k(t_0, t) = \frac{K(t_0, t) - K(t_0, t_0)}{t-t_0}, \quad (114.3)$$

will now be considered. As always, the equation under consideration will be assumed to be of the normal type; in the present case this leads to the condition $B(t) \neq 0$ everywhere on L. Therefore, without loss of generality, one may assume

$$B(t) = 1. \tag{114.4}$$

In correspondence with these the equation considered takes the form

$$\mathbf{K}\varphi \equiv \frac{1}{\pi i}\int_L \frac{\varphi(t)\,dt}{t-t_0} + \frac{1}{\pi i}\int_L k(t_0, t)\varphi(t)dt = f(t_0). \tag{114.5}$$

As before, $k(t_0, t)$ will be assumed to belong to the class H with respect to t_0 and to t; in addition, it will be assumed for the present that $f(t)$ also belongs to the class H.

The equation (114.1) is a particular case of the equation (106.1), and therefore all the results of the preceding sections may be applied to it. The same results and formulae may be obtained by applying directly to (114.5) the method of reduction, stated in § 109, and by using the solution of the dominant equation (113.1), stated in the last section. Some of these results will be briefly reproduced.

1°. The fundamental function of the class $h(c_1, c_2, \ldots, c_q)$, corresponding to (114.5), is given by

$$\mathbf{Z}(t) = C\frac{\sqrt{R_1(t)}}{\sqrt{R_2(t)}}, \tag{114.6}$$

where C is an arbitrary constant, different from zero; the index of the class $h(c_1, c_2, \ldots, c_q)$ is

$$\varkappa = p - q. \tag{114.7}$$

2°. For $\varkappa = p - q \geqq 0$, the equation (114.5) is equivalent, in the sense of seeking a solution of the class $h(c_1, c_2, \ldots, c_q)$, to the Fredholm equation

$$\varphi(t_0) + \mathbf{K}^*\mathbf{k}\varphi = \mathbf{K}^*f + \frac{\sqrt{R_1(t_0)}}{\sqrt{R_2(t_0)}}P_{p-q-1}(t_0), \tag{114.8}$$

where $P_{p-q-1}(t_0)$ is an arbitrary polynomial of degree not greater than $p-q-1$ $[P_{p-q-1}(t_0) \equiv 0$ for $p = q]$.

3°. For $\varkappa = p - q < 0$, the equation (114.5) is equivalent (in the same sense) to the equation

$$\varphi(t_0) + \mathbf{K}^*\mathbf{k}\varphi = \mathbf{K}^*f \tag{114.9}$$

and additional conditions of the form

$$\int_L \varrho_j\varphi\,dt = \int_L \frac{\sqrt{R_2(t)}t^j f(t)\,dt}{\sqrt{R_1(t)}}, \quad j = 0, 1, \ldots, q-p-1. \tag{114.10}$$

In the above formulae

$$\mathbf{K}^*f \equiv \frac{\sqrt{R_1(t_0)}}{\pi i\sqrt{R_2(t_0)}}\int_L \frac{\sqrt{R_2(t)}f(t)\,dt}{\sqrt{R_1(t)}(t-t_0)}, \quad \mathbf{k}\varphi \equiv \frac{1}{\pi i}\int_L k(t_0, t)\varphi(t)dt, \tag{114.11}$$

corresponding to which

$$\mathbf{K}^*\mathbf{k}\varphi = \frac{1}{\pi i}\int_L N(t_0, t)\varphi(t)\,dt, \tag{114.12}$$

where

$$N(t_0, t) = \frac{\sqrt{R_1(t_0)}}{\pi i\sqrt{R_2(t_0)}}\int_L \frac{\sqrt{R_2(t_1)}k(t_1, t)\,dt_1}{\sqrt{R_1(t_1)}(t_1 - t_0)} \tag{114.13}$$

and

$$\varrho_j(t) = \frac{1}{\pi i}\int_L \frac{\sqrt{R_2(t_1)}t_1^j k(t_1, t)dt_1}{\sqrt{R_1(t_1)}}. \tag{114.14}$$

4°. Attention will now be given to the fact that all the solutions of the equation (114.5), remaining bounded near a given end c, necessarily vanish there.

This property of the solutions was already noted in § 88 for one particular case, namely, for an equation of the form (113.1). However, the general case reduces to this particular case, because the equation (114.5) may be written

$$\frac{1}{\pi i}\int_L \frac{\varphi(t)\,dt}{t - t_0} = f(t_0) - \mathbf{k}\varphi;$$

if the solution φ is bounded near the end c, then $\mathbf{k}\varphi$ belongs to the class H near this end, and so does f; hence follows the proposition.

5°. Finally, it will be noted that all the preceding formulae and results will remain true, if the function $f(t)$ is assumed to belong to the class $h(c_1, c_2, \ldots, c_q)$, and not necessarily to the class H.

§ 115. An alternative method for the investigation of singular equations.

In § 55 a very effective method of investigation of singular equations for the case of contours (which was stated by I. N. Vekua) was discussed. (Here, the method stated in I. N. Vekua's paper [5] must be kept in mind; it differs from that, given by the same author in his paper [4]) An analogous method which is of interest here will also be applied in the case of arcs. In this case the problem is somewhat complicated, since, when choosing the reducing operator, one has to take care of the fact that the behaviour of the kernel of the Fredholm equation, obtained as the result of the reduction and equivalent to the original singular equation, should be sufficiently simple near the ends.

Such reducing operators can easily be found on the basis of the following consideration.

It is known that, in the notation of § 107, a solution of a given class h of the dominant equation

$$\mathbf{K}^0\varphi = f$$

is given, for $\varkappa \geqq 0$, by

$$\varphi(t_0) = \mathbf{K}^*f + B^*(t_0)Z(t_0)P_{\varkappa-1}(t_0),$$

where $P_{\varkappa-1}(t_0)$ is an arbitrary polynomial of degree not greater than $\varkappa - 1$. Thus, for $\varkappa \geqq 0$, one has the identity

$$\mathbf{K}^0\mathbf{K}^*f \equiv f,$$

so that $\mathbf{K}^0\mathbf{K}^* = \mathbf{E}$, where $\mathbf{E}$ is the unit operator. It will be seen that the operator $\mathbf{K}^0$ is, in the present case, a reducing operator with respect to $\mathbf{K}^*$, and the result of the reduction is found to be very simple.

It is easily predicted that similar circumstances will hold for $\varkappa < 0$, and likewise if the roles of $\mathbf{K}^0$ and $\mathbf{K}^*$ are interchanged. Actually, it can be established by a simple reasoning that

$$\mathbf{K}^*\mathbf{K}^0\varphi = \varphi(t_0) + \frac{B^*(t_0)Z(t_0)}{\pi i}\int\limits_L P_{\varkappa-1}(t_0, t)\varphi(t)dt,$$

$$\mathbf{K}^0\mathbf{K}^*\varphi = \varphi(t_0) - \frac{B(t_0)}{\pi i}\int\limits_L \frac{Q_{\varkappa'-1}(t_0, t)\varphi(t)dt}{Z(t)},$$

where $\varphi(t)$ is an arbitrary function of the class h and $P_{\varkappa-1}(t_0, t)$ and $Q_{\varkappa'-1}(t_0, t)$ are fully determined polynomials of degree not greater than $\varkappa - 1$ and $\varkappa' - 1$ respectively ($\varkappa' = -\varkappa$), while $P_{\varkappa-1}(t_0, t) \equiv 0$ for $\varkappa \leqq 0$, $Q_{\varkappa'-1}(t_0, t) \equiv 0$ for $\varkappa' \leqq 0$. If, in particular, $\varkappa = \varkappa' = 0$, then $\mathbf{K}^*\mathbf{K}^0 = \mathbf{K}^0\mathbf{K}^* = \mathbf{E}$, so that the operators $\mathbf{K}^0$ and $\mathbf{K}^*$ are the inverse of one another. Although in the general case, when $\varkappa \neq 0$, these operators are not inverse to one another, they have the essential properties of such operators. In addition, it is not difficult to prove the following property of the operator $\mathbf{K}^*$.

For $\varkappa \geqq 0$, the homogeneous equation $\mathbf{K}^*\omega = 0$ has no non-zero solutions of the class h.

For $\varkappa \leqq 0$, the equation $\mathbf{K}^*\omega = f$ is soluble in the class h for each function f belonging to the class h; the corresponding homogeneous equation has exactly $\varkappa' = -\varkappa$ linearly independent solutions of the class h.

The operator $\mathbf{K}^*$ may therefore play the same role as the operator $\mathbf{M}$ in the equivalence theorem by I. N. Vekua, proved in § 55. Thus also in the present case an equivalence theorem, analogous to that in § 55, has been obtained.

Using this theorem one can, just as in § 55, obtain the fundamental theorems proved by another method in § 112; in the present case, the reasoning will be somewhat involved as compared with that in § 55, since here one has, in addition, to consider the behaviour of the solutions near the ends.

The method of investigation set out in the present section is due to D. A. Kveselava [1] in whose paper the reader will find the proofs of the above propositions. In particular, D. A. Kveselava considers a case somewhat different from the one above which is, in fact, that stated in the next section. But the arguments in both cases are quite similar.

CHAPTER 15

SINGULAR INTEGRAL EQUATIONS IN THE CASE OF DISCONTINUOUS COEFFICIENTS

§ 116. Definitions.

In this chapter (§§ 116, 117) the terminology and notation of § 85 will be used. In particular, L will be a simple smooth contour, $d_1, d_2, \ldots, d_r$ some fixed points on L, taken in the same order in which they are encountered on traversing L in the positive direction, and $c_1, c_2, \ldots, c_r$ will be the same points, taken in any order. The classes of functions H^*, $H_\varepsilon{}^*$, H_d, given on L, will refer to these points.

By a singular integral equation will be understood an equation of one of the following two types:

$$\mathbf{K}\varphi \equiv A(t_0)\varphi(t_0) + \frac{B(t_0)}{\pi i}\int_L \frac{\varphi(t)dt}{t-t_0} + \frac{1}{\pi i}\int_L k(t_0, t)\varphi(t)dt = f(t_0) \quad \text{(I)}$$

and

$$\mathbf{K}'\psi \equiv A(t_0)\psi(t_0) - \frac{1}{\pi i}\int_L \frac{B(t)\psi(t)dt}{t-t_0} + \frac{1}{\pi i}\int_L k(t, t_0)\psi(t)dt = g(t_0), \quad \text{(II)}$$

where $A(t)$ and $B(t)$ are given functions of the class H_d and $k(t_0, t)$ is likewise a given function, belonging to the class H_d with respect to each of the variables t and t_0. The functions $f(t)$, $g(t)$ are given functions which for the present will be assumed to belong to the class H_d. The equations (I) and (II) will be called adjoint.

It will always be assumed that the equations under consideration are of the regular type, i.e., that $A(t) + B(t) \neq 0$, $A(t) - B(t) \neq 0$ everywhere on L; for the points c_j this will be understood to mean: $A(c_j \pm 0) + B(c_j \pm 0) \neq 0$, $A(c_j \pm 0) - B(c_j \pm 0) \neq 0$.

In contrast to what was true in the case considered in the last chapter the equations (I) and (II) belong to different types which, generally speaking, are not related to one another. In fact, if one tried to rearrange the equation (II) by separation of a "dominant part" of the operator $\mathbf{K}'$, as has been done above, one would obtain the equation

$$A(t_0)\psi(t_0) - \frac{B(t_0)}{\pi i}\int_L \frac{\psi(t)dt}{t-t_0} -$$

$$-\frac{1}{\pi i}\int_L \left\{\frac{B(t)-B(t_0)}{t-t_0} - k(t, t_0)\right\}\psi(t)dt = g(t_0),$$

which, in general, does not belong to the type (I), because the term $[B(t)-B(t_0)]/(t-t_0)$ is, generally speaking, unbounded like $(t-t_0)^{-1}$, when the points t and t_0 lie near, but on opposite sides of one of the points c_j.

In spite of this the equation (II) will not be considered independently, but as the adjoint equation of (I); this will be done solely for the purpose of avoiding the introduction of further notation.

§ 117. Reduction and solution of singular equations in the case of discontinuous coefficients.

The theory of the equations (I) and (II) of the last section can be developed similarly as the theory for the case already considered, where L consisted of disconnected arcs and the functions $A(t)$, $B(t)$, $k(t_0, t)$ belonged to the class H on L.

The dominant equation

$$\mathbf{K}^0\varphi \equiv A(t_0)\varphi(t_0) + \frac{B(t_0)}{\pi i}\int_L \frac{\varphi(t)dt}{t-t_0} = f(t_0), \tag{I°}$$

corresponding to (I), and the analogous equation

$$\mathbf{K}^{0\prime}\psi \equiv A(t_0)\psi(t_0) - \frac{1}{\pi i}\int_L \frac{B(t)\psi(t)dt}{t-t_0} = g_0(t_0) \tag{II°}$$

are solved in the same manner as the equations (107.1) and (108.1) treated in § 107 and § 108. In fact, the solution of (I°) leads to the solution of the Hilbert problem (compare § 107)

$$\Phi^+(t_0) = G(t_0)\Phi^-(t_0) + \frac{f(t_0)}{A(t_0)+B(t_0)}, \tag{117.1}$$

and the solution of (II°) to the solution of the Hilbert problem (compare § 108)

$$\Psi^+(t_0) = [G(t_0)]^{-1}\Psi^-(t_0) + \frac{B(t_0)g(t_0)}{A(t_0)-B(t_0)}, \tag{117.2}$$

where

$$G(t_0) = \frac{A(t_0)-B(t_0)}{A(t_0)+B(t_0)}.$$

Quite as in §§ 107, 108, the division of the solutions of (I°) and (II°) into classes $h(c_1, c_2, \ldots, c_q)$ can be set up and the indices $\varkappa$ and $\varkappa'$ of given classes of (I°) and (II°) defined; the indices $\varkappa$ and $\varkappa'$ of adjoint classes of the adjoint equations (I°) and (II°) will be equal in magnitude and opposite in sign.

The solutions of (I°) and (II°) are given by exactly the same formulae as the solutions of the equations (107.1) and (108.1) in § 107 and § 108; only that now one has, of course, to understand by $X(z)$ the function defined in § 85 [(85.9) and (85.8)]. It must still be remembered that in the present case functions of the class H_d play the role of functions of the class H.

The further deductions and formulae are completely analogous to those which were obtained in the last chapter and therefore no time will be spent on them.

All these results are easily extended to the case where L consists of any (finite) number of contours.

Furthermore, the results will remain unaltered, if, for the finding of the solutions of a given class, it is assumed that the functions $f(t)$ or $g(t)$ on the right side of (I) or (II) belong to the same class and, in addition, are bounded functions of the class $H_\varepsilon{}^*$ near the special ends. No essential changes will result in the case, where $f(t)$ or $g(t)$ are arbitrary functions of the same class in which the solutions are sought.

A direct investigation of the equations (I) and (II) was recently carried out by D. A. Kveselava [1]. In this investigation the transference of the results, referring to the case of arcs and continuous coefficients, to the case of contours and discontinuous coefficients is not essentially new, but the application of the method is new, as was briefly mentioned in § 115. Such a transference was also carried out by F. D. Gakhov [3] who made use of the results of the paper by N. I. Muskhelishvili and D. A. Kveselava [1]. If only the results which are studied in this Part are considered, then the last method can hardly be considered simpler than that used here which, as already mentioned, was given in the Author's paper [2] and in the paper [1] by the Author and D. A. Kveselava. By the help of this method a number of new results may be obtained, similar to those stated in Part II (§§ 58,59) and this has been done to some degree by D. A. Kveselava.

CHAPTER 16

APPLICATION TO THE DIRICHLET AND SIMILAR PROBLEMS *)

The results, studied in the preceding chapters, can be successfully applied to the solution of many important problems of mathematical physics.

In this chapter, one of the simpler and at the same time typical applications will be given, i.e., the application to the Dirichlet problem for the plane, cut along a finite number of arcs of arbitrary shape, and to certain of its variations. Such applications play an important part in aerodynamics (e.g., in the theory of wings).

§ 118. The Dirichlet and similar problems for the plane, cut along arcs of arbitrary shape.

Let $L = ab$ be a simple smooth arc. In addition, it will be assumed that the curvature of L satisfies the H condition, which means that the coordinates x and y of the point t on L have second derivatives with respect to the arc coordinate, satisfying the H condition.

Let S be the plane cut along L. The following problem will be considered:

Problem A. To find a function $\Phi(z)$, holomorphic in S, vanishing at infinity, continuous on L from the left and from the right with the possible exception of the ends, near which it may be unbounded, subject to the condition

$$|\Phi(z)| < \frac{\text{const.}}{|z-c|^{\alpha}}, \quad 0 \leqq \alpha < 1 \quad (c = a \text{ or } b)$$

(i.e., in other words, $\Phi(z)$ must be a sectionally holomorphic function with the line of discontinuity L, vanishing at infinity), and satisfying the following boundary condition:

$$\Re\Phi^+(t_0) = \Re\Phi^-(t_0) = f(t_0) \text{ on } L, \tag{118.1}$$

---

*) The results of this chapter which refer to the case of a single contour (§ 118, 119) were obtained by the Author in his paper [2], where, however, only the problem A_1 was solved (§ 118).

The generalization to the case of an arbitrary number of contours (§ 120) was given by N. I. Vekua [1]; cf. also the Author's paper [2].

except possibly at the ends, where $f(t_0)$ is a real function of the class H, given on L.

The problem A may be subdivided into the following cases:

PROBLEM A_1: The function $\Phi(z)$ has, in addition, to be bounded near both ends a, b.

PROBLEM A_2: The function $\Phi(z)$ has, in addition, to be bounded near one of the ends a, b.

PROBLEM A_3: The function $\Phi(z)$ is unbounded near both ends a, b.

First the question of the existence of non-zero solutions of the homogeneous problem corresponding to the problem A will be considered, i.e., the problem, corresponding to the case where $f(t) \equiv 0$. This homogeneous problem will be called the problem A^0, and the homogeneous problems, corresponding to A_1, A_2, A_3, the problems $A_1{}^0$, $A_2{}^0$, $A_3{}^0$ respectively. For the problem A^0 the boundary condition (118.1) takes the form

$$\Re\Phi^+(t_0) = \Re\Phi^-(t_0) = 0 \text{ on } L. \tag{*}$$

Firstly, the particular case will be considered, where $L = ab$ is part of the real axis and where the condition $\Phi(\infty) = 0$ is replaced by the more general condition $\Phi(z_0) = 0$, z_0 being a given point not on L (in particular $z_0 = \infty$ may be taken); the function $\Phi(z)$ will be understood to be bounded at infinity.

Introduce into the consideration the sectionally holomorphic function (with the line of discontinuity L), bounded at infinity,

$$\Omega(z) = \Phi(z) - \bar{\Phi}(z);$$

here, use has been made of the notation of § 38, 1°, i.e., $\bar{\Phi}(z)$ is the function defined by

$$\bar{\Phi}(z) = \overline{\Phi(\bar{z})}.$$

It will be remembered that

$$\bar{\Phi}^+(t) = \overline{\Phi^-(t)},\ \bar{\Phi}^-(t) = \overline{\Phi^+(t)} \text{ on } L. \tag{**}$$

From (*) and (**) it is easily seen that $\Omega^+(t) - \Omega^-(t) = 0$ on L, and hence that $\Omega(z) = \text{const.} = 2C$. Thus $\bar{\Phi}(z) = \Phi(z) - 2C$, where C is a constant. Therefore, as is easily seen from (*) and (**),

$$\Phi^+(t) + \Phi^-(t) = 2C \text{ on } L.$$

Thus one has one of the simplest cases of the Hilbert problem for the determination of $\Phi(z)$. The general solution of this problem, bounded at infinity, is given by

$$\Phi(z) = C + \frac{Az + B}{\sqrt{(z-a)(b-z)}}, \tag{***}$$

where A and B are arbitrary constants.

Now, expressing that the function $\Phi(z)$ must satisfy the boundary condition (*) of the problem A^0, it is easily concluded that the constants A, B, C must be imaginary. Furthermore, expressing the additional conditions required in the problems $A_1{}^0$ and $A_2{}^0$, and likewise the condition $\Phi(z_0) = 0$, the following conclusions are arrived at by elementary means:

The problems $A_1{}^0$ and $A_2{}^0$ do not have non-zero solutions. The most general solution of the problem $A_3{}^0$ has the form

$$\Phi(z) = k\omega(z),$$

where k is a real arbitrary constant and $\omega(z)$ is a fully determined sectionally holomorphic function, bounded at infinity.

The following fact will still be noted. If in the problem A the condition $\Phi(z_0) = 0$ is omitted, then, as is easily deduced from (***), the unique solution of this problem will be $\Phi(z) = ki$, where k is a real constant, so that $\Re\Phi(z) = 0$.

The case where $L = ab$ is a straight segment has been considered. The general case, however, will be reduced to the preceding one by conformal transformation of the region S on the region Σ which is the ζ plane, cut along some part $\alpha\beta$ of the real axis; it is assumed here that the ends a and b are thus transformed into the ends α and β respectively. On the basis of known properties of conformal transformations (cf. for example S. Warschawski [1]) it is easily seen that for the present conditions, imposed on L, the function $\Phi(z)$, satisfying near the end c the condition

$$|\,\Phi(z)\,| < \frac{\text{const.}}{|\,z - c\,|^{\mu}}, \quad 0 \leqq \mu < 1,$$

is transformed by conformal mapping into a function, satisfying in the ζ plane the same condition near the corresponding end. Thus the results, obtained above for the solution of the problem A^0, also remain true for the general case.

The solution of the problem A will now be considered. This solution will be written in the form (cf. § 65)

$$\Phi(z) = \frac{1}{\pi}\int_L \frac{\mu(t)\,dt}{t - z}, \tag{118.2}$$

where $\mu(t)$ is an unknown real function of the class H^*; in the sequel, a solution of the problem A (the problems A_1, A_2, A_3, respectively) will, for the present, be a solution of the form (118.2).

The boundary condition (118.1) leads to a real singular equation of the first kind

$$\frac{1}{\pi}\int_L \mu(t)\frac{dr}{r(t_0, t)} \equiv \frac{1}{\pi}\int_L \mu(t)\frac{\cos\alpha\,(t_0, t)}{r(t_0, t)}ds = f(t_0), \tag{118.3}$$

where, as in § 64, $r(t_0, t) = |t - t_0|$ and $\alpha(t_0, t)$ is the angle between the positive tangent at t and the vector $\overrightarrow{t_0 t}$. The above equation can still be written as (cf. § 64)

$$\frac{1}{\pi} \int_L \frac{\mu(t)dt}{t - t_0} + \frac{1}{\pi i} \int_L \mu(t) \frac{d\vartheta}{ds} ds = f(t_0), \tag{118.4}$$

where $\vartheta = \arg (t - t_0)$. Under the conditions, introduced for L, this equation is of the type considered in § 114. Actually, by the result at the end of § 8, $d\vartheta/ds$ satisfies the H condition for t and t_0.

Since the equation (118.3) is real, *the whole investigation will be carried out in the real field* (cf. Note 2 at the end of § 112).

The homogeneous equation, adjoint to (118.3), viz.,

$$\int_L \nu(t) \frac{\cos \alpha(t, t_0)}{r(t_0, t)} ds = 0 \tag{118.5}$$

has a simple physical meaning. In fact, let it be required to find the density $\nu(t)$ of the potential of a simple layer, distributed on L and taking a constant value there; then

$$\int_L \nu(t) \log \frac{1}{r(t_0, t)} ds = \text{const.} \tag{118.6}$$

("problem of distribution of charge" on a conductor L). Differentiating both sides of the above equation with respect to s_0, equation (118.5) is obtained.

Now consider separately each of the problems A_1, A_2, A_3.

1°. For the solution of Problem A_1 one has obviously to find a solution of the equation (118.3) of the class $h(a, b)$, i.e., a solution, bounded at both ends a, b. (By § 114, 4°, such a solution necessarily vanishes at the ends). The index $\varkappa$ of this class is equal to -1, on the basis of the results of § 114, where now $p = 1$ and $q = 2$.

The homogeneous equation corresponding to (118.3) has no solutions of the class $h(a, b)$, different from zero; this follows from the fact that the problem $A_1{}^0$ has no non-zero solution. Consequently, the adjoint homogeneous equation (118.5) has exactly one linearly independent solution of the class h_0, adjoint to $h(a, b)$, which will be called $\nu_0(t)$.

It is easily seen that

$$\int_L \nu_0(t) ds \neq 0. \tag{118.7}$$

Otherwise the function

$$U(x, y) = \int_L \nu_0(t) \log \frac{1}{r(t_0, t)} ds$$

would be harmonic in S, continuous on L from both sides including the ends, constant on L and vanishing at infinity. Consequently, $U(x, y) \equiv 0$, and hence it would follow that $\nu_0(t) = 0$ [cf. § 65, (65.11)], which contradicts the condition.

It is necessary and sufficient for the solubility of (118.3) that

$$\int_L f(t)\nu_0(t)ds = 0. \tag{118.8}$$

If this condition is not satisfied, then the constant C may always be chosen such that

$$\int_L [f(t) + C]\nu_0(t)ds = 0; \tag{118.9}$$

this follows from (118.7). Therefore, by solving the problem

$$\Re\Phi^+(t_0) = \Re\Phi^-(t_0) = f(t_0) + C \tag{118.10}$$

and putting

$$U = \Re\Phi(z) - C, \tag{118.11}$$

one obtains the solution of the ordinary Dirichlet problem

$$U^+ = U^- = f(t_0), \tag{118.12}$$

bounded at infinity; in fact, it takes the value $-C$ at infinity.

2°. For the solution of the problem A_2 one has to find a solution of (118.3) of the class $h(a)$; for definiteness it will be assumed that the solution must be bounded at the end a. The index $\varkappa$ of the class $h(a)$ is zero (§ 114; here $p = 1$, $q = 1$). The homogeneous equation, corresponding to (118.3), does not have any non-zero solution of the class $h(a)$; this follows from the fact that the problem $A_2{}^0$ has no solution different from zero. Consequently, the adjoint equation (118.5) also has no solution of the class $h(b)$ different from zero. This means that the equation (118.3) always has one and only one solution of the class $h(a)$ and the problem A_2 always has a unique solution.

3°. For the solution of problem A_3 a solution of the equation (118.3) of the class h_0 has to be found. The index $\varkappa$ of this class is equal to unity (§ 114; here $p = 1$, $q = 0$). The adjoint homogeneous equation (118.5) does not have any non-zero solutions of the adjoint class $h(a, b)$, because, from 2° above, it has no solutions different from zero, even in the wider class $h(b)$. Consequently, the equation (118.3) is always soluble and the solution contains linearly one arbitrary constant (because the corresponding homogeneous equation has a single linearly independent solution).

Thus the problem A_3 is always soluble and its general solution contains linearly one (real) arbitrary constant.

So far a solution of the problem A always implied a solution, representable in the form (118.2). There still remains to show that the above condition, imposed in the process of solution of the problem, in actual fact does not restrict the generality of the results obtained.

In the case of the problem A_2 this is obvious, since it has been proved that a solution, representable in the form (118.2), always exists; other solutions of the problem A_2 do not exist, because the problem A_2^0 has solutions different from zero.

In the case of the problem A_1 only the following doubt may arise; in fact, whether the condition (118.8) remains necessary for the solubility of the problem, if the representation of the solution in the form (118.2) is abandoned. The condition (118.8) is easily shown to be likewise necessary in this case. Let

$$\int_L f(t)\nu_0(t)ds \neq 0$$

and let the problem A_1, in spite of this, have the solution $\Phi_1(z)$, vanishing at infinity. Now the real constant C will be chosen such that the condition (118.9) holds true, and a solution $\Phi_2(z)$ will be constructed

$$\Re\Phi_2^+(t_0) = \Re\Phi^-(t_0) = f(t_0),$$

as in 1° above, such that $\Re\Phi_2(\infty) = -C \neq 0$. But then the function $\Phi(z) = \Phi_2(z) - \Phi_1(z)$, bounded everywhere, will satisfy the condition $\Re\Phi^+(t_0) = \Re\Phi^-(t_0) = 0$, and hence, by a remark above (page 360), $\Re\Phi(z) = 0$ everywhere, which contradicts the fact that $\Re\Phi(\infty) = \Re\Phi_2(\infty) - \Re\Phi_1(\infty) = -C \neq 0$.

In the case of the problem A_3 doubt may only arise due to the fact that, in seeking solutions in the form (118.2), the most general solution will not be found. But this doubt is easily dispersed, if the result in 3° is compared with that referring to the solution of the homogeneous problem A_3^0, stated above (page 360).

NOTE. The investigation of the non-homogeneous equation

$$\frac{1}{\pi}\int_L \nu(t)\frac{\cos\alpha(t, t_0)}{r(t, t_0)}\,ds = g(t_0), \tag{118.13}$$

adjoint to the equation (118.3), is quite similar to that of the last equation; it is sufficient merely to interchange the roles of the equations (118.3) and (118.13).

One very important problem of hydromechanics leads to the equation (118.13), namely the problem of finding the flow around an arc of a given form. From the point of view of application, the greatest interest lies in the determination of a solution of the class $h(a)$ or, what comes to the same thing, of the class $h(b)$. (This is connected with the so called postulate of S. A. Chaplygin, cf. M. A. Lavrentjev [1]). The work of M. A. Lavrentjev [1], 1932, was devoted to this problem. Not having the general theory of singular equations at his disposal, this author gave a number of results of great interest and stated a method of approximate solution. The further development of this and similar methods of approximate solution of singular integral equations is, in the Author's opinions, one of the important future problems of the theory of these equations.

§ 119. Reduction to a Fredholm equation. Examples.

The singular integral equation (118.4) obtained in the last section can also be written as

$$\frac{1}{\pi}\int_L \frac{\mu(t)dt}{t-t_0} + \frac{1}{\pi i}\int_L \frac{\sin\alpha(t_0, t)e^{-i\alpha(t_0, t)}\mu(t)dt}{t-t_0} = f(t_0). \quad (119.1)$$

Depending on the class h in which the solutions are being sought, i.e., depending on which of the problems A_1, A_2 or A_3 is to be solved, this equation can be reduced to different Fredholm equations by the method of § 114. For this purpose it is, of course, possible and desirable to vary this method, depending on the concrete case which is of interest.

For definiteness, the problem A_1 will be considered, i.e., the case where a solution of the class $h(a, b)$ of the equation (119.1) has to be found. If one follows the method of § 114, then in this case, where the index $\varkappa = -1$, one will obtain, as the result of the reduction, a Fredholm equation and one supplementary condition of the form (114.10).

But it will be assumed that the solution of the Dirichlet problem in its ordinary form is of interest, i.e., that a function $U(x, y)$, bounded everywhere and harmonic in S, has to be determined for the boundary condition

$$U^+ = U^- = f(t_0) \text{ on } L. \quad (119.2)$$

Putting

$$U = \Re\Phi(z) - C, \quad (119.3)$$

where $\Phi(z)$ is the integral (118.2), vanishing at infinity, and C is a constant, not given beforehand, one arrives at an integral equation which is obtained from (119.1) by replacing $f(t_0)$ by $f(t_0) + C$, i.e., at the equation

$$\frac{1}{\pi}\int_L \frac{\mu(t)dr}{r(t_0, t)} \equiv \frac{1}{\pi}\int_L \frac{\mu(t)dt}{t-t_0} + \frac{1}{\pi i}\int_L \frac{\sin\alpha(t_0, t)e^{-i\alpha(t_0,t)}\mu(t)dt}{t-t_0} = \\ = f(t_0) + C. \tag{119.1a}$$

Now it will be remembered that, by what has been said at the end of § 89, the general (unique) solution of the class $h(a, b)$, i.e., of the class H, of the equation

$$\frac{1}{\pi}\int_L \frac{\mu(t)dt}{t-t_0} = g(t_0) + C, \tag{119.4}$$

where C is a constant not given beforehand, has the form

$$\mu(t_0) = -\frac{\sqrt{(t_0-a)(b-t_0)}}{\pi}\int_L \frac{g(t)dt}{\sqrt{(t-a)(b-t)}(t-t_0)}, \tag{119.5}$$

while C is fully determined by

$$C = -\frac{1}{\pi}\int_L \frac{g(t)dt}{\sqrt{(t-a)(b-t)}}. \tag{119.6}$$

Here, use has been made of the formulae (89.16) and (89.17) in a slightly modified form; in fact, $\sqrt{(t-a)(b-t)}$ has been written instead of $\sqrt{(t-a)(t-b)}$, since

$$\sqrt{(t-a)(t-b)} = i\sqrt{(t-a)(b-t)}; \tag{119.7}$$

in the case where $L = ab$ is a part of the real axis, $\sqrt{(t-a)(b-t)}$ will take real values on L.

Now, shifting the second term on the left side of (119.1a) to the right side and solving the resulting equation by the help of the above formulae, as if the right side were a given function (known apart from a constant), i.e., proceeding similarly as in the reduction of a singular equation in the general case, one obtains the Fredholm equation

$$\mu(t_0) + \frac{1}{\pi i}\int_L N(t_0, t_1)\mu(t_1)dt_1 = f_0(t_0), \tag{119.8}$$

equivalent to the original equation (119.1a) in the sense that a solution of the class H is being sought and containing no undetermined constant; here the following notation has been introduced:

$$N(t_0, t_1) = -\frac{\sqrt{(t_0 - a)(b - t_0)}}{\pi} \int_L \frac{\sin \alpha(t, t_1) e^{-i\alpha(t, t_1)} dt}{\sqrt{(t - a)(b - t)(t_1 - t)(t - t_0)}}, \tag{119.9}$$

$$f_0(t_0) = -\frac{\sqrt{(t_0 - a)(b - t_0)}}{\pi} \int_L \frac{f(t)dt}{\sqrt{(t - a)(b - t)(t - t_0)}}. \tag{119.10}$$

It is easily seen that under the present conditions the equation (119.8) is an ordinary regular Fredholm integral equation and that each of its (continuous) solutions belongs to the class H and vanishes at the ends.

The homogeneous equation, corresponding to (119.8), is easily shown to have no non-zero solutions of the class $h(a, b)$. In fact, let $\mu(t)$ be a solution of this homogeneous equation. Then, by (119.1a),

$$\frac{1}{\pi} \int_L \frac{\mu(t)dr}{r(t_0, t)} = \text{const.} \tag{*}$$

It may be assumed that $\mu(t)$ is a real function, because in the opposite case the real and imaginary parts of $\mu(t)$ could be considered separately. From the above equation it follows that the function $\Phi(z)$, defined for $\mu(t)$ by (118.2) and continuous on L from the left and from the right, including the ends [because $\mu(t)$ vanishes at the ends], satisfies the condition $\Re\Phi^+ = \Re\Phi^- = 0$ on L, including the ends, and vanishes at infinity; but such a function is necessarily zero, and therefore $\mu(t) \equiv 0$.

In the same way the equation (119.8) is readily shown to have only real solutions, assuming, of course, that $f(t_0)$ is a real function. In fact, the imaginary part of the solution is easily seen to satisfy (*) and therefore it is necessarily zero.

Since the homogeneous equation, corresponding to the Fredholm equation (119.8), does not have any non-zero solutions, the equation (119.8) is always soluble; the solution $\mu(t)$ is a real function and the function $U = \Re\Phi(z)$, corresponding to this solution, satisfies the boundary condition

$$U^+ = U^- = f(t_0) + \text{const.},$$

and the problem is solved by the help of the Fredholm equation.

It is somewhat inconvenient that the equation (119.8) is of complex form, while the Dirichlet problem refers to the field of real functions. However, this inconvenience is easily removed. In fact, separating in (119.8) real and imaginary parts [assuming $\mu(t)$ real], the following two real Fredholm equations of the second and first kind respectively are obtained:

$$\mu(s_0) + \int_L M(s_0, s_1)\mu(s_1)ds_1 = F(s_0), \tag{119.11}$$

$$\int_L M_1(s_0, s_1)\mu(s_1)ds_1 = F_1(s_0), \tag{119.12}$$

where s_0, s_1 are the arc coordinates of the points t_0 and t_1,

$$(M + iM_1)ds_1 = \frac{1}{\pi i}N(t_0, t_1)dt_1, \quad F + iF_1 = f_0(t_0).$$

It is easily shown that each (real) solution of the equation (119.11) will also be a solution of the equation (119.8), i.e., that (119.12) is a consequence of (119.11) and that the original equation reduces to the real Fredholm equation (119.11).

In fact, let μ' be any real solution of (119.11) and let μ, as before, be a solution of (119.8) which is necessarily real. Then obviously $\mu'' = \mu - \mu'$ will satisfy the homogeneous equation

$$\mu''(s_0) + \int_L M(s_0, s_1)\mu''(s_1)ds_1 = 0; \tag{*}$$

it is easily seen that μ'' belongs to the class H and vanishes at the ends.

Now put

$$\frac{1}{\pi}\int_L \mu''(t)\frac{dr}{r(t_0, t)} = \chi(t_0). \tag{**}$$

Applying to (**) the same process of reduction by which (119.1a) became (119.8) and taking (*) into consideration, one obtains

$$\chi_0(t_0) \equiv -\frac{\sqrt{(t_0 - a)(b - t_0)}}{\pi}\int_L \frac{\chi(t)dt}{\sqrt{(t - a)(b - t)(t - t_0)}} = i\nu(t_0),$$

where $\nu(t_0)$ is a certain real function of the class H; hence, on the strength of the equivalence of the equations (119.4), (119.5), the conclusion is drawn that

$$\chi(t_0) = \frac{1}{\pi i}\int_L \frac{i\nu(t)\,dt}{t - t_0} + \text{const.};$$

separating in the integral the imaginary part and noting that $\chi(t)$ is a real function, one obtains

$$\int_L \nu(t)\frac{dr}{r(t_0, t)} = \text{const.},$$

and hence it follows that $\nu(t) \equiv 0$. Consequently, $\chi(t) = \text{const.}$ and finally, by (**), $\mu''(t) \equiv 0$, which proves the proposition.

EXAMPLES:

1°. Let $L = ab$ be part of a straight line. Then $\sin\alpha(t, t_1) \equiv 0$, $N(t_0, t_1) \equiv 0$ and consequently, by (119.8) and (119.10),

$$\mu(t_0) = -\frac{\sqrt{(t_0 - a)(b - t_0)}}{\pi}\int_L \frac{f(t)dt}{\sqrt{(t-a)(b-t)}(t-t_0)}. \tag{119.13}$$

2°. Let $L = ab$ be an arc of a circle with its centre at the origin of coordinates. It is easily verified that also in this case $N(t_0, t_1) \equiv 0$, and therefore $\mu(t_0)$ will be given by exactly the same formula (119.13), as in the case of a straight segment.

§ 120. The Dirichlet problem for the plane, cut along a finite number of arcs of arbitrary shape.

The solution of the problem A of § 118 can, without difficulty, be generalized to the case where L consists of any (finite) number of arcs $L_j = a_j b_j$ which do not intersect one another:

$$L = a_1b_1 + a_2b_2 + \ldots + a_pb_p = L_1 + L_2 + \ldots + L_p;$$

as in § 118, it will be assumed that the curvature of the arcs L_j satisfies the H condition. The plane cut along L will be denoted by S.

For definiteness, the Dirichlet problem in its ordinary form will be considered, i.e., the problem of finding a function $U(x, y)$, harmonic in S, bounded at infinity and continuous on L from the left and from the right, including the ends, subject to the boundary condition

$$U^+ = U^- = f(t_0) \text{ on } L, \tag{120.1}$$

where $f(t_0)$ is a given real function of the class H.

As in the case of contours, the solution of the modified Dirichlet problem will be considered first, i.e., the problem of finding a function $\Phi(z)$, holomorphic in S, vanishing at infinity, continuous on L from the left and from the right, including the ends, for the boundary condition

$$\Re\Phi^+(t_0) = \Re\Phi^-(t_0) = f(t_0) + C_k \text{ on } L_k,\ k = 1, 2, \ldots, p, \tag{120.2}$$

where $f(t_0)$ is a given real function of the class H and the C_k are real constants, not given beforehand and likewise subject to definition.

This problem cannot have more than one solution, since the proof in § 60 is easily seen to be applicable here. As in § 118, the function $\Phi(z)$ will be sought in the form

$$\Phi(z) = \frac{1}{\pi} \int_L \frac{\mu(t)\, dt}{t - z}, \tag{120.3}$$

where $\mu(t)$ is a real unknown function of the class H. The boundary condition (120.2) leads to the integral equation (compare § 119)

$$\frac{1}{\pi} \int_L \frac{\mu(t)\, dt}{t - t_0} + \frac{1}{\pi i} \int_L \frac{\sin \alpha(t_0, t) e^{-i\alpha(t_0, t)} \mu(t)\, dt}{t - t_0} = f(t_0) + C_j$$

$$\text{for } t_0 \in L_j,\ j = 1, 2, \ldots, p, \tag{120.4}$$

which contains the undetermined constants C_j.

Now it will be remembered that, by what has been said in § 90, the solution of the integral equation

$$\frac{1}{\pi} \int_L \frac{\mu(t)\, dt}{t - t_0} = f(t_0) + C_j \text{ for } t_0 \in L_j,\ j = 1, \ldots, p, \tag{120.5}$$

where the constants C_j are not given in advance, is given by [cf. (90.9)]

$$\mu(t_0) = -\frac{\sqrt{R(t_0)}}{\pi} \int_L \frac{f(t)}{\sqrt{R(t)}} \left\{ \frac{1}{t - t_0} + \sum_{j=1}^{p} \omega_j(t) \int_{L_j} \frac{dt_1}{\sqrt{R(t_1)}(t_1 - t_0)} \right\} dt, \tag{120.6}$$

where the $\omega_j(t)$ are certain polynomials of degree not greater than $p - 1$,

$$R(t) = \prod_{j=1}^{p} (t - a_j)(t - b_j); \tag{120.7}$$

the meaning of the root $\sqrt{R(t)}$ has been repeatedly explained.

By the help of the inversion formula (120.6), the equation (120.4) may, quite as it was done in § 119, be reduced to a Fredholm equation, equivalent in the sense that solutions of the class H are being sought and not containing any undetermined constants C_j; this equation will not be written down here. Just as in the last section, it is easily shown that the Fredholm equation thus obtained always has one and only one solution which belongs necessarily to the class H and vanishes at the ends. Thus the modified Dirichlet problem can be considered solved.

The solution $U(x, y)$ of the Dirichlet problem (120.1) is now also easily obtained. For this purpose it is, of course, sufficient to find a solution, satisfying the boundary condition

$$U^+ = U^- = f(t_0) + C, \tag{120.1a}$$

where C is a certain constant, not given beforehand. With this object in view the following procedure may be adopted. Put

$$U = u + \sum_{j=1}^{p} \alpha_j u_j, \tag{120.8}$$

where u is a harmonic function expressible in the form

$$u = \Re \frac{1}{\pi} \int_L \frac{\mu(t)\, dt}{t - z}, \tag{120.9}$$

with $\mu(t)$ a real unknown function of the class H, the α_j are so far undetermined constants and the u_j are potentials of simple layers, distributed over the L_j,

$$u_j = \int_{L_j} \sigma_j(t) \log \frac{1}{r}\, ds, \tag{120.10}$$

where the $\sigma_j(t)$ are to be understood as arbitrarily chosen real functions such that the values of u_j on L satisfy the H condition and that

$$e_j = \int_{L_j} \sigma_j\, ds \neq 0. \tag{120.11}$$

For this purpose it is seen to be sufficient that the $\sigma_j(t)$ should be continuous; for example, $\sigma_j(t) = 1$ can be taken.

In the Author's paper [2] an incorrect method for the choice of the functions u_j, denoted there by ω_j, has been stated; this method of selection will only be useful in the case, where L consists of straight segments.

In order that the function U will be bounded at infinity, the constants α_j must fulfil the condition

$$\sum_{j=1}^{p} \alpha_j e_j = 0 \tag{120.12}$$

(when U will obviously be a function vanishing at infinity).

Now u will be defined as the solution of the modified Dirichlet problem, corresponding to the boundary condition

$$u^+ = u^- = f(t) - \sum_{j=1}^{p} \alpha_j u_j(t) + C_j \text{ on } L_j,\ j = 1, \ldots, p \tag{120.13}$$

for arbitrarily fixed α_j.

By solving this last problem, certain values will be obtained for the constants C_j which will be easily seen to be linear functions of the constants α_j.

Now these constants α_j will be chosen in such a way that

$$C_1 = C_2 = \ldots = C_p \tag{120.14}$$

and that the condition (120.12) is also satisfied; this can always be done uniquely, because the conditions (120.12), (120.14) represent a system of p linear equations in the α_j which, as it is not difficult to show, always has a unique solution.

For the above values of the α_j the solution of the Dirichlet problem will be given by (120.8).

CHAPTER 17

SOLUTION OF THE INTEGRO-DIFFERENTIAL EQUATION OF THE THEORY OF AIRCRAFT WINGS OF FINITE SPAN*)

In this book the general theory of singular integro-differential equations will not be examined. However, it has been considered expedient to deal here with one singular integro-differential equation, firstly, because it is of great practical interest, and secondly, because some of the above results may be directly used for the investigation of its solution.

§ 121. The integro-differential equation of the theory of aircraft wings of finite span.

In the theory of wings of finite span an important role is played by the following integro-differential equation (Prandtl; cf. for example V. Kármán and Burgers [1] p. 167; V. V. Golubev [1] p. 164):

$$\frac{\Gamma(t_0)}{B(t_0)} - \frac{1}{\pi}\int_{-a}^{+a} \frac{\Gamma'(t)dt}{t - t_0} = f(t_0), \qquad (121.1)$$

where $\Gamma(t)$ is the unknown function, $\Gamma'(t) = \dfrac{d\Gamma(t)}{dt}$, $B(t)$ and $f(t)$ are the given functions

$$B(t) = \frac{mb(t)}{8}, \quad f(t) = 4V\alpha(t). \qquad (121.2)$$

The physical meaning of the quantities in the above formulae is as follows: $2a$ is the span of the wing which is assumed symmetrical with respect to some plane OYZ, where the direction of the OZ axis coincides with the direction of the air flow at infinity. If OX is the axis of the abscissae, perpendicular to the OYZ plane, then $b(x)$ is the "chord of the profile" which corresponds to the abscissa x, $\Gamma(x)$ is the circulation of the airflow around this profile, $\alpha(x)$ is the "geometrical angle of incidence" and, finally, V is the velocity of the airflow at infinity. By m is denoted a constant which,

*) In this chapter the Author made use of papers by N. I. Vekua [11] and by L. G. Magnaradze [1] (reproducing different parts almost literally).

as a rule, is taken equal to 2π; more precisely, the value of m is 5.5 (cf. Karman and Burgers [1] p. 166).

By symmetry,

$$\Gamma(t) = \Gamma(-t),\ B(t) = B(-t),\ f(t) = f(-t); \qquad (121.3)$$

in addition, it will be assumed, as it is usually done, that

$$\Gamma(-a) = \Gamma(a) = 0. \qquad (121.4)$$

A large amount of work which has been published recently is devoted to the study and solution of the equation (121.1). In addition to the literature, quoted in the books by V. V. Golubev [1] and by Kármán and Burgers [1], the following papers by K. Schröder [2] and J. Weissinger [1] will be mentioned (in which likewise references to the most important works will be found). However, the Author believes that the most effective method of solution was given in the works of I. N. Vekua and L. G. Magnaradze, mentioned above. This method will be stated in the following sections.

§ 122. Reduction to a regular Fredholm equation.

In the sequel it will be assumed that the function $B(t)$ is nowhere zero, with the possible exception of the ends, and that $1/B(t)$ belongs to the class H^*. Furthermore, the function $\alpha(t)$ will be assumed to satisfy the H condition. As regards the solution $\Gamma(t)$, it will be required that its derivative belongs to the class H^*.

Removing the first term on the left of (121.1) to the right side and applying the inversion formula of § 88, one obtains

$$\Gamma'(t_0) = -\frac{1}{\pi\sqrt{a^2-t_0^2}} \int_{-a}^{+a} \frac{\sqrt{a^2-t^2}\,\Gamma(t)}{B(t)(t-t_0)}\,dt + F(t_0), \qquad (122.1)$$

where

$$F(t_0) = \frac{1}{\pi\sqrt{a^2-t_0^2}} \int_{-a}^{+a} \frac{\sqrt{a^2-t^2}\,f(t)\,dt}{t-t_0} \qquad (122.2)$$

and by the root is meant its positive value.

For the present case one has in (88.9) $p = 1$, $q = 0$, because $\Gamma'(t)$ has been assumed to belong to the class H^*, i.e., h_0; the root $\sqrt{t^2-a^2}$ has been replaced by $\sqrt{a^2-t^2} = -i\sqrt{t^2-a^2}$ which is positive for $-a < t < a$.

The term of the form

$$\frac{C}{\sqrt{a^2-t_0^2}}$$

which should be added to the right side in the present case is easily seen to vanish, on the basis of (121.3), if one takes into consideration that from $\Gamma(t) = \Gamma(-t)$ follows $\Gamma'(t) = -\Gamma'(-t)$, and hence $\Gamma'(0) = 0$.

Integrating both sides of (122.1) with respect to t_0 from $t_0 = 0$ to a variable value which will again be denoted by t_0, one obtains

$$\Gamma(t_0) - \frac{1}{\pi}\int_{-a}^{+a} \log\left|\frac{i(t_0-t)+\sqrt{a^2-t_0^2}-\sqrt{a^2-t^2}}{i(t_0-t)+\sqrt{a^2-t_0^2}+\sqrt{a^2-t^2}}\right|\frac{\Gamma(t)}{B(t)}\,dt =$$

$$= \int_0^{t_0} F(t)dt + C_0, \qquad (122.3)$$

where C_0 is a constant, for the present undetermined; putting

$$t = -a\cos\omega, \quad t_0 = -a\cos\omega_0,$$

this equation becomes

$$\Gamma(\omega_0) - \frac{a}{\pi}\int_0^{\pi} \frac{\sin\omega}{B(\omega)} \log\left|\frac{\sin\dfrac{\omega-\omega_0}{2}}{\sin\dfrac{\omega+\omega_0}{2}}\right| \Gamma(\omega)d\omega = G(\omega_0) + C_0, \quad (122.4)$$

where $G(\omega_0)$ is a completely determined function, the expression for which will not be written down here; $\Gamma(t)$, $B(t)$ have been denoted by $\Gamma(\omega)$, $B(\omega)$.

The equation (122.4) is a quasi-regular Fredholm equation; since it is well-known in the literature, no time will be spent on it, and it will be transformed into another Fredholm equation more convenient from the point of view of application.

For this purpose the equation (122.1) will again be considered and it will be written in the following manner:

$$B(t_0)\Gamma'(t_0) = -\frac{1}{\pi}\int_{-a}^{+a}\frac{\Gamma(t)dt}{t-t_0} + B(t_0)F(t_0) -$$

$$-\frac{B(t_0)}{\pi\sqrt{a^2-t_0^2}}\int_{-a}^{+a} R(t_0, t)\Gamma(t)dt, \qquad (122.5)$$

where

$$R(t_0, t) = \frac{1}{t-t_0}\left\{\frac{\sqrt{a^2-t^2}}{B(t)} - \frac{\sqrt{a^2-t_0^2}}{B(t_0)}\right\}. \qquad (122.6)$$

For simplicity, it will be assumed that

$$P(t) = \frac{\sqrt{a^2 - t^2}}{B(t)}, \quad -a \leqq t \leqq a \tag{122.7}$$

has a continuous first derivative. $R(t_0, t)$ will be a continuous function of both arguments.

Now, noting that under the present conditions [1])

$$\frac{d}{dt_0} \int_{-a}^{+a} \frac{\Gamma(t)dt}{t - t_0} = \int_{-a}^{+a} \frac{\Gamma'(t)dt}{t - t_0}, \tag{*}$$

one obtains from (122.5)

$$\frac{d}{dt_0} [B(t_0)\Gamma'(t_0)] + \frac{1}{\pi} \int_{-a}^{+a} \frac{\Gamma'(t)dt}{t - t_0} = g(t_0), \tag{122.8}$$

where

$$g(t_0) = \frac{d}{dt_0} \left\{ B(t_0)F(t_0) - \frac{B(t_0)}{\pi\sqrt{a^2 - t_0^2}} \int_{-a}^{+a} R(t_0, t)\Gamma(t)dt \right\}. \tag{122.9}$$

From (121.1) and (122.8) follows

$$B(t_0) \frac{d}{dt_0} [B(t_0)\Gamma'(t_0)] + \Gamma(t_0) = B(t_0)[f(t_0) + g(t_0)]. \tag{122.10}$$

If, for the present, the right side of (122.10) is considered as a known function, then this equation will be a second order linear differential equation for $\Gamma(t)$. Integrating this equation, one obtains

$$\Gamma(t_0) = C_1 \cos \tau(t_0) + C_2 \sin \tau(t_0) + \\ + \int_0^{t_0} \{f(t) + g(t)\} \sin \{\tau(t_0) - \tau(t)\}dt, \tag{122.11}$$

---

[1]) Integration by parts is easily seen to give

$$\int_{-a}^{+a} \frac{\Gamma(t)dt}{t - t_0} = \int_{-a}^{+a} \Gamma(t)d \log |t - t_0| = - \int_{-a}^{+a} \Gamma'(t) \log |t - t_0| \, dt, \tag{**}$$

and hence, differentiating both sides with respect to t_0 and remembering a result of § 13, one obtains (*). For the deduction of (**) it has been taken into consideration that, by assumption, $\Gamma(a) = \Gamma(-a) = 0$.

where

$$\tau(t_0) = \int_0^{t_0} \frac{dt}{B(t)} \tag{122.12}$$

and C_1, C_2 are for the present undetermined constants. Putting in (122.11) $t_0 = 0$, one obtains $C_1 = \Gamma(0)$; furthermore, it is easily seen that, by (121.3), $C_2 = 0$.

Now, inserting on the right of (122.11) instead of $g(t)$ the expression given by (122.9), one obtains, after some simple transformations, the integral equation

$$\Gamma(t_0) + \frac{1}{\pi} \int_{-a}^{+a} K(t_0, t)\Gamma(t)\,dt = h(t_0), \tag{122.13}$$

where

$$K(t_0, t) = \int_0^{t_0} \frac{R(t_1, t)}{\sqrt{a^2 - t_1^2}} \cos\,[\tau(t_0) - \tau(t_1)]\,dt_1, \tag{122.14}$$

$$h(t_0) = \Gamma(0) \cos \tau(t_0) + \int_0^{t_0} \Bigg\{ \sin\,[\tau(t_0) - \tau(t_1)]f(t_1) + \\ + \frac{\cos\,[\tau(t_0) - \tau(t_1)]}{\pi\sqrt{a^2 - t_1^2}} \int_{-a}^{+a} \frac{\sqrt{a^2 - t^2}\, f(t)\,dt}{t - t_1} \Bigg\} dt_1. \tag{122.15}$$

If one does not consider the unknown constant $\Gamma(0)$ on the right side of (122.13), this equation is a regular Fredholm equation.

It was first obtained by I. N. Vekua [11] in a different manner under the assumption that the relation (122.7) represented an analytic function on the segment $(-a, +a)$ [this assumption was essential for I. N. Vekua's method]. The result obtained by I. N. Vekua was soon generalized by L. G. Magnaradze [1] who started from assumptions, corresponding approximately to those made here. The method of obtaining the equation (122.13), studied in this section, is a reproduction of the essential part of L. G. Magnaradze's paper. [Magnaradze did not impose the conditions $\Gamma(-t) = \Gamma(t)$, $\Gamma(-a) = \Gamma(a) = 0$, and therefore the equation obtained by him contains several arbitrary constants.]

At first sight, the equation (122.13) seems to be more complicated than the equation (122.4). As a matter of fact, however, it is much simpler from the point of view of application. Firstly, its kernel is regular. But the chief advantage is that this equation may be solved in quite an elementary manner in many cases, important in practice. In particular, if the relation $P(t)$, defined by (122.7),

$$P(t) = \frac{\sqrt{a^2 - t^2}}{B(t)}$$

is a rational function, i.e., if the chord of the profile $b(t)$ is representable in the form

$$b(t) = \sqrt{1 - \frac{t^2}{a^2}}\, p(t), \tag{122.16}$$

where $p(t)$ is a rational function, then the equation (122.13) is easily seen to be very simply solved in closed form (for particulars consult the quoted papers by I. N. Vekua and L. G. Magnaradze). On the other hand, in the majority of cases of practical interest, the function $b(t)$ may, with sufficient accuracy, be represted in the form (122.16), where $p(t)$ is a rational fraction containing only a small number of parameters.

For example, putting $p(t) = b_0 =$ const., i.e.,

$$b(t) = b_0\sqrt{1 - \frac{t^2}{a^2}}, \tag{122.17}$$

one obtains the elliptic wing. In this case, obviously, $R(t_0, t) \equiv 0$, $K(t_0, t) \equiv 0$ and (122.13) immediately gives $\Gamma(t)$, and the determination of the constant $\Gamma(0)$, occuring on the right side, presents no difficulty.

In the case

$$b(t) = b_0 \sqrt{1 - \frac{t^2}{a^2}\,\frac{1 + \nu\dfrac{t^2}{a^2}}{1 + \mu\dfrac{t^2}{a^2}}}, \tag{122.18}$$

where b_0, μ, ν are constants, $b_0 > 0$, $\mu > -1$, $\nu > -1$ (I. N. Vekua [11]), the solution is also very simply found. Varying the constants μ and ν, a large number of practically important cases may be solved. For example, putting $\mu = 0$, $\nu = 0.9$, an almost rectangular wing will be obtained, as will be seen from the following table:

$\dfrac{t}{a} =$	0.1	0.2	0.3	0.4	0.5	0.6	0.7	0.8	0.9
$\dfrac{b(t)}{b_0} =$	1.00	1.02	1.03	1.05	1.06	1.06	1.03	0.95	0.75

The case

$$b(t) = b_0 \sqrt{1 - \frac{t^2}{a^2}} \frac{1 + \nu_1 \frac{t^2}{a^2} + \ldots + \nu_n \frac{t^{2n}}{a^{2n}}}{1 + \mu_1 \frac{t^2}{a^2} + \ldots + \mu_n \frac{t^{2n}}{a^{2n}}}, \qquad (122.19)$$

where b_0, μ_j, ν_j are constants, likewise leads to a fairly simple result. Wings of the particular shapes, corresponding to all $\mu_j = 0$, were considered by H. Schmidt [1] whose solution, however, was very complicated.

The reader will find a more detailed study of the equation (**122.13**) and a detailed solution in the special cases just stated in the papers by L. G. Magnaradze and I. N. Vekua, quoted above, especially in the paper by the latter author.

§ 123. Certain generalizations.

The equation (**121.1**) is a particular case of an equation of the form

$$\sum_{r=0}^{m} \alpha_r(t_0)\varphi^{(r)}(t_0) - \frac{1}{\pi} \int_L \sum_{j=0}^{n} \frac{K_j(t_0, t)\varphi^{(j)}(t)\, dt}{t - t_0} = f(t_0), \qquad (123.1)$$

where $\varphi^{(j)}(t)$ are derivatives of the order j. Equations of that form occur for a large number of practically important problems. A general study of equations of the form (**123.1**) was carried out by L. G. Magnaradze; a brief statement of the results was published in his paper [3]. It will be noted that in (**123.1**) the line of integration L may consist of a finite number of contours and arcs.

PART VI

THE HILBERT PROBLEM FOR SEVERAL UNKNOWN FUNCTIONS AND SYSTEMS OF SINGULAR INTEGRAL EQUATIONS

This Part is devoted to the generalization of the results, concerning the Hilbert boundary problem and singular integral equations with Cauchy type kernels, to the case of several unknown functions. Here the case of contours and continuous coefficients will be considered (regarding the other cases, only reference to the literature will be made).

If one is satisfied with the definition of the index of the system of singular integral equations as the difference between the number of solutions of the given homogeneous system and the number of solutions of the adjoint system, then the transference of the results referring to a single singular equation to the case of a system of such equations may be carried out almost automatically by the introduction of a convenient notation, and this is actually done in §§ 130, 131 below.

This is not the case, however, as far as the deduction of a clear expression for the index is concerned. This expression, as for a single equation, is closely connected with the solution of the corresponding Hilbert problem. In contrast to the case of a single unknown function, the Hilbert problem for several unknown functions may not be solved in closed form, provided, of course, the general case is being considered.

The first chapter of this Part (Chapt. 18) will be devoted to the solution of the Hilbert problem for several unknown functions.

CHAPTER 18

THE HILBERT PROBLEM FOR SEVERAL UNKNOWN FUNCTIONS *)

§ 124. Definitions.

1°. In the sequel, by L will be understood the union of simple smooth contours in the z plane which have no common points and bound some finite connected region S^+; the boundary L will not belong to S^+. By S^- will be denoted the (generally not connected) region which is the complement of $S^+ + L$. The positive direction on L will be assumed to leave the region S^+ on the left.

2°. A set of n functions $\Phi_1, \Phi_2, \ldots, \Phi_n$ will have to be considered, given in one or the other of the regions. Such a set of functions will be called a vector and denoted by the single letter Φ; this vector will be written

$$\Phi = (\Phi_1, \Phi_2, \ldots, \Phi_n), \tag{124.1}$$

and $\Phi_1, \Phi_2, \ldots, \Phi_n$ will be called the components of the vector Φ.

Consider the linear transformation

$$\begin{array}{lllll} \Psi_1 = A_{11}\Phi_1 & + A_{12}\Phi_2 & + \ldots & + A_{1n}\Phi_n, \\ \Psi_2 = A_{21}\Phi_1 & + A_{22}\Phi_2 & + \ldots & + A_{2n}\Phi_n, \\ \ldots \quad \ldots & \ldots & \ldots & \ldots \\ \ldots \quad \ldots & \ldots & \ldots & \ldots \\ \Psi_n = A_{n1}\Phi_1 & + A_{n2}\Phi_2 & + \ldots & + A_{nn}\Phi_n, \end{array}$$

or

$$\Psi_\alpha = \sum_{\beta=1}^{n} A_{\alpha\beta}\Phi_\beta \quad (\alpha = 1, 2, \ldots, n), \tag{124.2}$$

where the $A_{\alpha\beta}$ are functions, given in the same regions as Φ. The substitution (124.2) will be written

$$\Psi = A\Phi, \tag{124.3}$$

where $\Phi = (\Phi_1, \ldots, \Phi_n)$ and $\Psi = (\Psi_1, \ldots, \Psi_n)$ are vectors and A is the matrix

$$A = \| A_{\alpha\beta} \| = \left\| \begin{array}{cccc} A_{11} & A_{12} & \ldots & A_{1n} \\ A_{21} & A_{22} & \ldots & A_{2n} \\ \ldots & \ldots & \ldots & \ldots \\ \ldots & \ldots & \ldots & \ldots \\ A_{n1} & A_{n2} & \ldots & A_{nn} \end{array} \right\| \tag{124.4}$$

with the components $A_{\alpha\beta}$ ($\alpha, \beta = 1, 2, \ldots, n$). The determinant of

*) This chapter is a reproduction of part of the paper [1] by N. I. Muskhelishvili and N. P. Vekua [1].

the matrix A will be denoted by $\det A = \det \| A_{\alpha\beta} \|$. If $\det A \neq 0$, the matrix A will be called regular.

The inner product of the two vectors $\Phi = (\Phi_1, \ldots, \Phi_n)$ and $\Psi = (\Psi_1, \ldots, \Psi_n)$ will be defined as the sum $\Phi_1\Psi_1 + \ldots + \Phi_n\Psi_n$; it will be denoted by $\Phi\Psi$ or $\Psi\Phi$, so that

$$\Phi\Psi = \Psi\Phi = \Phi_1\Psi_1 + \Phi_2\Psi_2 + \ldots + \Phi_n\Psi_n. \tag{124.5}$$

As usual, the matrix $C = \| C_{\alpha\beta} \|$, where

$$C_{\alpha\beta} = \sum_{\gamma=1}^{n} A_{\alpha\gamma}B_{\gamma\beta} \quad (\alpha,\ \beta = 1,\ 2,\ \ldots,\ n),$$

will be called the product $C = AB$ of the two matrices $A = \| A_{\alpha\beta} \|$ and $B = \| B_{\alpha\beta} \|$.

The matrix, obtained from A by interchanging rows and columns, will be called the transposed matrix and it will be denoted by $A' = \| A'_{\alpha\beta} \|$, so that $A'_{\alpha\beta} = A_{\beta\alpha}$. It is easily verified that for any two vectors Φ, Ψ.

$$\Psi A\Phi = \Phi A'\Psi, \tag{124.6}$$

where the inner product of the vectors Ψ and $A\Phi$ (and analogously for the right side) must be understood to be denoted by $\Psi A\Phi$.

Note also the well known relations

$$(AB)' = B'A', \tag{124.7}$$

$$(AB)^{-1} = B^{-1}A^{-1},\ (A^{-1})' = (A')^{-1}; \tag{124.8}$$

in (124.8), the matrices A and B are assumed regular, i.e., their determinants are different from zero. In future, A'^{-1} will often be written instead of $(A')^{-1}$.

3°. In the sequel, when stating that certain matrices are continuous and satisfy the H condition etc., this will mean that all their components satisfy the conditions mentioned. The same will be true for vectors.

In particular, the vector $\Phi(z) = (\Phi_1, \ldots, \Phi_n)$ will be called sectionally holomorphic, if the $\Phi_\alpha = \Phi_\alpha(z)$ $(\alpha = 1, \ldots, n)$ are sectionally holomorphic functions. If all these functions have finite degree at infinity, then the vector $\Phi(z)$ will be said to have finite degree at infinity. The highest degree of any of the components $\Phi_\alpha(z)$ will be called the degree k at infinity of such a vector.

If in the neighbourhood of the point at infinity

$$\Phi_\alpha(z) = \gamma_\alpha(z) + O(1/z),$$

where $\gamma_\alpha(z)$ is a polynomial, then the vector

$$\gamma(z) = (\gamma_1, \gamma_2, \ldots, \gamma_n)$$

will be called the principal part of $\Phi(z)$ at infinity.

§ 125. Auxiliary theorems.

Although the transference of the Plemelj formulae and of the results studied in § 24 to the case where one has to deal with vectors and not with ordinary functions is quite obvious, some of these formulae and results will be reproduced here for the sake of later reference.

Let $\varphi(t)$ be a vector, given on L and satisfying the H condition, and let $\Phi(z)$ be the sectionally holomorphic vector, defined by

$$\Phi(z) = \frac{1}{2\pi i}\int_L \frac{\varphi(t)dt}{t-z}. \tag{125.1}$$

Then, by the Plemelj formulae,

$$\Phi^+(t_0) = \tfrac{1}{2}\varphi(t_0) + \frac{1}{2\pi i}\int_L \frac{\varphi(t)dt}{t-t_0}, \tag{125.2}$$

$$\Phi^-(t_0) = -\tfrac{1}{2}\varphi(t_0) + \frac{1}{2\pi i}\int_L \frac{\varphi(t)dt}{t-t_0}. \tag{125.3}$$

Further, let $\Phi(z)$ be a vector holomorphic in S^+ and continuous on L from the left (S^+). Then, by Cauchy's theorem,

$$\Phi(z) = \frac{1}{2\pi i}\int_L \frac{\Phi^+(t)dt}{t-z} \quad \text{for } z \text{ in } S^+, \tag{125.4}$$

$$0 = \frac{1}{2\pi i}\int_L \frac{\Phi^+(t)dt}{t-z} \quad \text{for } z \text{ in } S^-. \tag{125.5}$$

The equation (125.5) is the necessary and sufficient condition for the continuous vector $\Phi^+(t)$, given on L, to be the boundary value of a certain vector $\Phi(z)$, holomorphic in S^+ and continuous on L from the left.

This condition is equivalent to the following one:

$$0 = -\tfrac{1}{2}\Phi^+(t_0) + \frac{1}{2\pi i}\int_L \frac{\Phi^+(t)dt}{t-t_0}. \tag{125.6}$$

Similarly, let $\Phi(z)$ be a vector, holomorphic in S^-, continuous on L from the right and of finite degree at infinity,

$$\Phi(z) = \gamma(z) + O(1/z), \tag{125.7}$$

where $\gamma(z) = (\gamma_1, \gamma_2, \ldots, \gamma_n)$ is the principal part of the vector $\Phi(z)$

at infinity. Then

$$-\Phi(z) = \frac{1}{2\pi i}\int_L \frac{\Phi^-(t)dt}{t-z} - \gamma(z) \text{ for } z \text{ in } S^-, \tag{125.8}$$

$$0 = \frac{1}{2\pi i}\int_L \frac{\Phi^-(t)dt}{t-z} - \gamma(z) \text{ for } z \text{ in } S^+. \tag{125.9}$$

The equation (125.9) is the necessary and sufficient condition for the vector $\Phi^-(t)$, continuous on L, to be the boundary value of a vector $\Phi(z)$, holomorphic in S^-, continuous on L from the right and having the principal part $\gamma(z)$ at infinity. The condition (125.9) is equivalent to

$$0 = \tfrac{1}{2}\Phi^-(t_0) + \frac{1}{2\pi i}\int_L \frac{\Phi^-(t)dt}{t-t_0} - \gamma(t) \text{ on } L. \tag{125.10}$$

§ 126. The homogeneous Hilbert problem.

The homogeneous Hilbert problem for the n unknown functions $\Phi_1, \Phi_2, \ldots, \Phi_n$ will be formulated as follows:

To find a sectionally holomorphic vector $\Phi(z) = (\Phi_1, \Phi_2, \ldots, \Phi_n)$, having finite degree at infinity, for the boundary condition

$$\Phi_\alpha^+(t_0) = G_{\alpha 1}(t_0)\Phi_1^-(t_0) + G_{\alpha 2}(t_0)\Phi_2^-(t_0) + \ldots + G_{\alpha n}(t_0)\Phi_n^-(t_0)$$
$$(\alpha = 1, \ldots, n)$$

on L, or, more briefly, for the condition

$$\Phi^+(t_0) = G(t_0)\Phi^-(t_0), \tag{I}$$

where

$$G(t_0) = \| G_{\alpha\beta}(t_0) \|$$

is a matrix which is given on L, satisfies the H condition and is regular everywhere on L, i.e., so that its determinant $\det G(t_0) \neq 0$ everywhere on L.

In the sequel, when speaking of solutions of the problem (I), solutions other than the trivial solution $\Phi(z) \equiv 0$ will be considered.

An ingenious and complete solution of this homogeneous problem was given by Plemelj in his paper [2]. Plemelj's investigation will be studied in this and at the beginning of the next section with certain (sometimes essential) additions and simplifications.

In proceeding to the solution of the problem, it will be noted from the very beginning that, if the $\overset{1}{\Phi}(z), \overset{2}{\Phi}(z), \ldots, \overset{k}{\Phi}(z)$ are any particular solutions of the problem (I), then obviously the expression

$$\Phi(z) = P_1(z)\overset{1}{\Phi}(z) + P_2(z)\overset{2}{\Phi}(z) + \ldots + P_k(z)\overset{k}{\Phi}(z), \quad (126.1)$$

where the $P_1(z)$, $P_2(z)$, ..., $P_k(z)$ are arbitrary polynomials, will also be a solution of the same problem.

In the next section the complete solution of the stated problem will be given and it will be proved that the boundary values $\Phi^+(t)$, $\Phi^-(t)$ of each solution of this problem necessarily satisfy the H condition. This will be a consequence of the assumed condition that the $G_{\alpha\beta}(t)$ satisfy the H condition.

For the time being, only those solutions will be sought which have this property, and thus (if not particularly stated to the contrary), when saying that $\Phi(z)$ is a solution of (I), it will be implied that $\Phi^+(t)$ and $\Phi^-(t)$ satisfy the H condition. It follows from (I) that, if one of the vectors Φ^+, Φ^- satisfies the H condition, then the other will also have this property.

First a solution $\Phi(z)$ will be sought, having a given principal part at infinity which will be denoted by

$$\gamma(z) = (\gamma_1, \gamma_2, \ldots, \gamma_n), \quad (126.2)$$

where the $\gamma_\alpha = \gamma_\alpha(z)$ ($\alpha = 1, \ldots, n$) are given polynomials.

The problem (I) is then obviously equivalent to the following one: to find a vector $\Phi^-(t_0)$, defined on L and satisfying the H condition, such that

1° the vector $\Phi^-(t_0)$ is the boundary value of a vector $\Phi(z)$, holomorphic in S^-, continuous on L from the right and having at infinity the given principal part $\gamma(z)$;

2° the vector $\Phi^+(t_0)$, related to $\Phi^-(t_0)$ by the linear transformation

$$\Phi^+(t_0) = G(t_0)\Phi^-(t_0),$$

is the boundary value of a vector $\Phi(z)$, holomorphic in S^+ and continuous on L from the left.

By the results of the last section, the condition 1° is equivalent to

$$\tfrac{1}{2}\Phi^-(t_0) + \frac{1}{2\pi i}\int_L \frac{\Phi^-(t)dt}{t - t_0} = \gamma(t_0), \quad (126.3)$$

and the condition 2° to

$$-\tfrac{1}{2}G(t_0)\Phi^-(t_0) + \frac{1}{2\pi i}\int_L \frac{G(t)\Phi^-(t)dt}{t - t_0} = 0. \quad (126.4)$$

Thus the vector $\Phi^-(t)$ must simultaneously satisfy the two equations (126.3), (126.4) which are singular integral equations of a particular form, or, using the usual terminology, systems of singular equations with the n unknowns $\Phi_1^-(t), \ldots, \Phi_n^-(t)$. (In future, in similar cases, the terms "equation" and "system of equations" will be used indiscriminately).

By writing (126.4) in the form

$$-\tfrac{1}{2}\Phi^-(t_0) + \frac{1}{2\pi i}\int_L \frac{G^{-1}(t_0)G(t)\Phi^-(t)dt}{t-t_0} = 0$$

and subtracting from (126.3), one obtains the new integral equation

$$\Phi^-(t_0) - \frac{1}{2\pi i}\int_L \frac{G^{-1}(t_0)G(t)-E}{t-t_0}\Phi^-(t)dt = \gamma(t_0), \qquad (126.5)$$

where E is the unit matrix.

The equation (126.5) is an ordinary (quasi-regular) system of Fredholm integral equations of the second kind. Actually, it is easily seen that on the basis of the assumed conditions

$$\frac{G^{-1}(t_0)G(t)-E}{t-t_0} = \frac{K(t_0, t)}{|t-t_0|^\alpha}, \quad 0 \leqq \alpha < 1, \qquad (126.6)$$

where $K(t_0, t)$ is a certain matrix satisfying the H condition.

Two questions have now to be answered: under what conditions is the equation (126.5) soluble and does each solution of (126.5) lead to a solution of the original problem?

The answer to the first question will be given later. As regards the second, it will first of all be noted that, as is easily verified from (126.6), *every continuous solution of* (126.5) *satisfies the H condition.*

Let $\Phi^-(t)$ be any solution of the equation (126.5). This solution obviously leads to a solution of the Hilbert problem (I), if and only if $\Phi^-(t)$ satisfies simultaneously the conditions (126.3) and (126.4), the first of which expresses that $\Phi^-(t)$ is the boundary value of the vector $\Phi(z)$, holomorphic in S^- and having the principal part $\gamma(z)$ at infinity, and the second that $G(t)\Phi^-(t)$ is the boundary value of the vector $\Phi(z)$, holomorphic in S^+. These conditions will be given a somewhat different form. For this purpose introduce the sectionally holomorphic vector $\Psi(z)$, defined in the following manner:

$$\begin{aligned} \Psi(z) &= \frac{1}{2\pi i}\int_L \frac{\Phi^-(t)dt}{t-z} - \gamma(z) \text{ for } z \text{ in } S^+, \\ \Psi(z) &= \frac{1}{2\pi i}\int_L \frac{G(t)\Phi^-(t)dt}{t-z} \quad \text{for } z \text{ in } S^-; \end{aligned} \qquad (126.7)$$

the vector $\Psi(z)$ will vanish at infinity.

In this notation the conditions (126.3) and (126.4) are equivalent to

$$\Psi^+(t_0) = 0, \ \Psi^-(t_0) = 0 \qquad \text{on } L \tag{126.8}$$

which, in turn, are equivalent to

$$\Psi(z) = 0 \tag{126.9}$$

in the entire plane.

Now the equation (126.5) may be rewritten

$$\Psi^+(t_0) = G^{-1}(t_0)\Psi^-(t_0). \tag{II}$$

Thus, if the vector $\Psi(z)$ connected with the solution $\Phi^-(t)$ of the integral equation (126.5) by the relation (126.7) is identically zero, then $\Phi(z)$ will give a solution of the original problem. If, however, the vector $\Psi(z)$ is not identically zero, then it will be a (non-trivial) solution of the problem (II), vanishing at infinity.

Following Plemelj, the problem (II) will be called the accompanying (begleitende) problem of (I). Thus there is the following important result:

Lemma 1: *If the problem* (II), *accompanying the problem* (I), *has no (non-trivial) solutions, vanishing at infinity, then each solution of the integral equation* (126.5) *gives a solution of the original problem* (I).

In the way in which the Fredholm integral equation (126.5) was constructed for the problem (I), a similar equation may be obtained for the problem (II). However, it will be preferred to find a Fredholm equation containing $\Psi^+(t)$ and not $\Psi^-(t)$. This equation can immediately be written down, if it is noted that the condition (II) is equivalent to $\Psi^-(t_0) = G(t_0)\Psi^+(t_0)$, differing from (I) only in that Φ^+ and Φ^- are replaced by Ψ^- and Ψ^+ respectively. Further, if only solutions of the problem (II) are being sought (vanishing at infinity) and if it is noted that, by replacing Φ^- by Φ^+ and vice versa, the condition (125.6) is replaced by (125.10) and conversely, which leads simply to a change in sign in front of the integral (assuming now that $\gamma = 0$, since the solutions are to vanish at infinity), then it will become clear that the required Fredholm integral equation, corresponding to the problem (II), will have the form

$$\Psi^+(t_0) + \frac{1}{2\pi i}\int_L \frac{G^{-1}(t_0)G(t) - E}{t - t_0}\Psi^+(t)dt = 0. \tag{126.10}$$

Thus, the Fredholm equation for the accompanying problem (if solutions of the latter, vanishing at infinity, are being sought) is obtained from the Fredholm equation for the given problem by replacing Φ^- by Ψ^+, putting the right side equal to zero and changing the sign in front of the integral. Now the question of solubility of the equation (126.5) will be considered. For this purpose introduce into the consideration the homogeneous equation, adjoint to (126.5), viz.

$$\Psi'^{+}(t_0) + \frac{1}{2\pi i}\int_L \frac{G'(t_0)G'^{-1}(t_0) - E}{t - t_0}\Psi'^{+}(t)dt = 0; \quad (126.11)$$

the kernel of this equation is obtained from that of (126.5) by replacing the matrix

$$\frac{G^{-1}(t_0)G(t) - E}{t - t_0}$$

by the transposed matrix and by interchanging simultaneously t and t_0; the somewhat complicated notation $\Psi'^{+}(t)$ has been introduced for the unknown vector of (126.11) for a reason which will become clear from what follows.

The integral equation (126.11) is likewise easily connected with a certain Hilbert problem. In fact, consider the problem

$$\Phi'^{+}(t_0) = G'^{-1}(t_0)\Phi'^{-}(t_0), \quad (\mathrm{I}')$$

differing from the problem (I) by the substitution of G'^{-1} for G; the problem (I′) will be called the Hilbert problem, associate to the original problem (I). The problem

$$\Psi'^{+}(t_0) = G'(t_0)\Psi'^{-}(t_0), \quad (\mathrm{II}')$$

accompanying the associate problem (I′), will also be considered. The Fredholm integral equation for the problem (I′), constructed in the same way as the equation (126.5) for the problem (I), has the form

$$\Phi'^{-}(t_0) - \frac{1}{2\pi i}\int_L \frac{G'(t_0)G'^{-1}(t) - E}{t - t_0}\Phi'^{-}(t)dt = 0, \quad (126.12)$$

if consideration is restricted to solutions of the problem (I′), vanishing at infinity.

However, the Fredholm integral equation, constructed for the problem (II′) in the same way as the equation (126.10) was obtained for the problem (II), will have exactly the form (126.11).

Thus, *the homogeneous integral equation* (126.11), *adjoint to* (126.5), *is the Fredholm integral equation corresponding to the problem* (II′), *i.e., to the problem accompanying the associate one.*

In this way the associate Hilbert problem does not correspond to the adjoint Fredholm equation and a question may arise as to the suitability of the choice of the term "associate" for the problems (I) and (I′). However, it will be seen below (§ 129) that in the theory of singular integral equations definite adjoint systems of singular integral equations correspond (in the known sense) to the associate problems (I) and (I′).

On the basis of the above the following proposition is easily proved:

LEMMA 2. *If the boundary problem* (I) *is such that neither its accompanying nor its associate problems have* (*non-trivial*) *solutions vanishing at infinity, then the Fredholm integral equation* (126.5) *is soluble for any right side* (*being itself a vector with polynomials as components*) *and every solution of this equation gives a solution of the original problem.*

The second part of this Lemma has already been proved above (Lemma 1). There remains to prove the first part. For this purpose it will be noted that the problem (I′) is the accompanying problem of (II′). Further, since by assumption the problem (I′) has no solutions vanishing at infinity, by Lemma 1 each solution $\Psi'^{+}(t)$ of the integral equation (126.11), corresponding to the problem (II′), gives a solution of the latter [of course (126.11) may not have any non-trivial solutions at all; in this case (126.5) is soluble for any right side]; hence, in particular, $\Psi'^{+}(t)$ is necessarily the boundary value of some vector $\Psi'(z) = (\Psi'_1, \ldots, \Psi'_n)$, holomorphic in S^+.

However, for the solubility of the integral equation (126.5), it is known to be necessary and sufficient that the function $\gamma(t)$ on the right side should satisfy the condition

$$\int_L \gamma(t)\Psi'^{+}(t)dt = \int_L (\gamma_1\Psi_1'^{+} + \gamma_2\Psi_2'^{+} + \ldots + \gamma_n\Psi_n'^{+})dt = 0, \tag{126.13}$$

where $\Psi'^{+}(t) = (\Psi_1'^{+}, \Psi_2'^{+}, \ldots, \Psi_n'^{+})$ is any solution of (126.11). But, as has just been seen, the vector $\Psi_1'^{+}, \ldots, \Psi_n'^{+}$ is the boundary value of a function, holomorphic in S^+; in addition, by assumption, the $\gamma_1, \gamma_2, \ldots, \gamma_n$ are polynomials. Hence, by Cauchy's theorem, the above condition is always fulfilled and the proposition is proved.

At first sight, the case in which the conditions of Lemma 2 are satisfied is a very particular one. However, as will be seen later, the most general case can also be reduced to this particular case; owing to this, as will be shown, it is sufficient for the setting up of the general solution to find all the solutions, remaining bounded at infinity.

Thus it will be assumed temporarily that the conditions of Lemma 2 are fulfilled, and solutions of the problem (I), bounded at infinity, will be sought. By Lemma 2, all these solutions will be given by solutions of the integral equation (126.5), on the right side of which one must now introduce the arbitrary constant vector $\gamma = (\gamma_1, \gamma_2, \ldots, \gamma_n)$.

The vector γ will be the value of the unknown vector $\Phi(z)$ at infinity: $\Phi(\infty) = \gamma$.

By the same Lemma 2, the equation (126.5) is soluble for any values of the arbitrary constants $\gamma_1, \gamma_2, \ldots, \gamma_n$. The following n vectors will be taken successively in the capacity of the vector γ: $(1, 0, 0, \ldots, 0)$, $(0, 1, 0, \ldots, 0)$, $\ldots$, $(0, 0, \ldots, 1)$. To each of them there will correspond some solution of (126.5); these solutions will be denoted by

$$\begin{aligned} \overset{1}{\Phi}{}^-(t) &= (\overset{1}{\Phi}{}_1^-, \overset{1}{\Phi}{}_2^-, \ldots, \overset{1}{\Phi}{}_n^-), \\ \overset{2}{\Phi}{}^-(t) &= (\overset{2}{\Phi}{}_1^-, \overset{2}{\Phi}{}_2^-, \ldots, \overset{2}{\Phi}{}_n^-), \\ &\ldots \quad \ldots \quad \ldots \quad \ldots \quad \ldots \\ \overset{n}{\Phi}{}^-(t) &= (\overset{n}{\Phi}{}_1^-, \overset{n}{\Phi}{}_2^-, \ldots, \overset{n}{\Phi}{}_n^-). \end{aligned} \tag{126.14}$$

To these n solutions there will correspond the n solutions of the Hilbert problem

$$\begin{aligned} \overset{1}{\Phi}(z) &= (\overset{1}{\Phi}_1, \overset{1}{\Phi}_2, \ldots, \overset{1}{\Phi}_n), \\ \overset{2}{\Phi}(z) &= (\overset{2}{\Phi}_1, \overset{2}{\Phi}_2, \ldots, \overset{2}{\Phi}_n), \\ &\ldots \quad \ldots \quad \ldots \quad \ldots \\ \overset{n}{\Phi}(z) &= (\overset{n}{\Phi}_1, \overset{n}{\Phi}_2, \ldots, \overset{n}{\Phi}_n) \end{aligned} \tag{126.15}$$

which have the property that

$$\overset{\beta}{\Phi}_\alpha(\infty) = \delta_{\alpha\beta} = \begin{cases} 1 \text{ for } \alpha = \beta, \\ 0 \text{ for } \alpha \neq \beta. \end{cases} \tag{126.16}$$

The general solution of the integral equation (126.5) for arbitrary γ will obviously have the form

$$\Phi^-(t) = \gamma_1 \overset{1}{\Phi}{}^-(t) + \ldots + \gamma_n \overset{n}{\Phi}{}^-(t) + \gamma_{n+1} \overset{n+1}{\Phi}{}^-(t) + \ldots + \gamma_m \overset{m}{\Phi}{}^-(t), \tag{126.17}$$

where $\overset{n+1}{\Phi}{}^-(t), \ldots, \overset{m}{\Phi}{}^-(t)$ is a complete system of linearly independent solutions of the homogeneous equation, obtained from (126.5) for $\gamma = 0$, and the $\gamma_1, \gamma_2, \ldots, \gamma_n, \gamma_{n+1}, \ldots, \gamma_m$ are arbitrary constants, the first n of which are the components of the vector γ.

Corresponding to this, a solution of the Hilbert problem, remaining bounded at infinity, will have the form

$$\Phi(z) = \gamma_1 \overset{1}{\Phi}(z) + \ldots + \gamma_n \overset{n}{\Phi}(z) + \gamma_{n+1} \overset{n+1}{\Phi}(z) + \ldots + \gamma_m \overset{m}{\Phi}(z),$$

where $\overset{1}{\Phi}(z), \ldots, \overset{n}{\Phi}(z)$ are the same as above and $\overset{n+1}{\Phi}(z), \ldots, \overset{m}{\Phi}(z)$

are solutions of the problem (I), vanishing at infinity (because they correspond to the vector γ of zero value). All the $\overset{\beta}{\Phi}(z)$ ($\beta = 1, 2, \ldots, m$) are easily seen to be linearly independent.

Now the general case will be considered, where the conditions of Lemma 2 may not be satisfied, and a start will be made with the following simple proposition: *a number* $s \geqq 0$ *may be found such that the order of the zero at infinity of any solution of the problem* (I) *does not exceed* s.

Let the homogeneous equation, corresponding to the Fredholm equation (126.5), have s linearly independent solutions and let $\Phi(z)$ be any solution of the problem (I). It will be shown that the order k of the zero at infinity of $\Phi(z)$ cannot exceed s.

In fact, if $\Phi(z)$ has a zero of order k at infinity, then the vectors $\Phi(z), z\Phi(z), \ldots, z^{k-1}\Phi(z)$ likewise will be solutions of the problem (I), vanishing at infinity; consequently, their boundary values $\Phi^-(t), t\Phi^-(t), \ldots, t^{k-1}\Phi^-(t)$ will be k solutions of the homogeneous equation, obtained from (126.5) for $\gamma = 0$. Since these solutions are obviously linearly independent, then necessarily $k \leqq s$, as had to be proved.

This result will now be applied to the problems (II) and (I′), i.e., to the accompanying and the associate problems. By the above, an integer $r \geqq 0$ may be found such that neither of the problems (II) and (I′) admits a solution which would have a zero of order greater than r at infinity.

Now consider the problem (I) and endeavour to determine all its solutions having a degree not greater than r (i.e., a pole of order not greater than r) at infinity, where r is the number, just stated.

Let $\Phi(z)$ be such a solution. Then the vector $\overset{*}{\Phi}(z)$, defined as follows:

$$\overset{*}{\Phi}(z) = \Phi(z) \text{ for } z \text{ in } S^+, \quad \overset{*}{\Phi}(z) = \frac{\Phi(z)}{(z-a)^r} \text{ for } z \text{ in } S^-, \tag{126.18}$$

where a is an arbitrarily fixed point in the region S^+, will obviously be a solution of the Hilbert problem

$$\overset{*}{\Phi}{}^+(t_0) = (t_0 - a)^r G(t_0)\overset{*}{\Phi}{}^-(t_0), \tag{$\overset{*}{\mathrm{I}}$}$$

remaining bounded at infinity. The problem ($\overset{*}{\mathrm{I}}$) is similar to the original problem and differs from it only in that the matrix $G(t_0)$ is replaced by the matrix $(t_0 - a)^r G(t_0)$. Thus the determination of solutions of the problem (I), having a degree not greater than r at infinity, is reduced to the task of finding solutions of the problem

($\overset{*}{\mathrm{I}}$), remaining bounded at infinity. Now the problem ($\overset{*}{\mathrm{I}}$) will be shown to satisfy the conditions of Lemma 2. In fact, the problem, accompanying the problem ($\overset{*}{\mathrm{I}}$), has the form

$$\overset{*}{\Psi}{}^{+}(t_0) = (t_0 - a)^{-r} G^{-1}(t_0) \overset{*}{\Psi}{}^{-}(t_0). \qquad (\overset{*}{\mathrm{II}})$$

If this latter problem would have a solution, vanishing at infinity, then the problem (II) would admit the solution

$$\Psi(z) = \overset{*}{\Psi}(z) \qquad \text{for } z \text{ in } S^+,$$
$$\Psi(z) = (z - a)^{-r} \overset{*}{\Psi}(z) \quad \text{for } z \text{ in } S^-,$$

having at infinity a zero of order greater than r, which contradicts the definition of r. Similarly, it is proved that the problem ($\overset{*}{\mathrm{I}}{}'$), associate to the problem ($\overset{*}{\mathrm{I}}$), has no solutions, vanishing at infinity.

Applying to the problem ($\overset{*}{\mathrm{I}}$) the results obtained above for the case where the conditions of Lemma 2 are fulfilled and returning again to the solution $\Phi(z)$ of the original problem by means of (126.18), the following conclusion is easily reached:

THEOREM. *Each solution of the problem* (I) *having a degree not greater then r at infinity, where r is a given sufficiently large number, is of the form*

$$\Phi(z) = \gamma_1 \overset{1}{\Phi}(z) + \ldots + \gamma_n \overset{n}{\Phi}(z) + \gamma_{n+1} \overset{n+1}{\Phi}(z) + \ldots + \gamma_m \overset{m}{\Phi}(z), \qquad (126.19)$$

where the $\gamma_1, \gamma_2, \ldots, \gamma_m$ *are arbitrary constants, the* $\overset{1}{\Phi}(z), \ldots, \overset{m}{\Phi}(z)$ *are definite, linearly independent particular solutions, the first n of which have the property*

$$\lim_{z \to \infty} z^{-r} \overset{\beta}{\Phi}_{\alpha}(z) = \delta_{\alpha\beta} \quad (\alpha, \beta = 1, 2, \ldots, n), \qquad (126.20)$$

and the remaining ones (if $m = n$, there will be none) *have a degree less than r at infinity.*

To a certain extent this theorem solves the homogeneous Hilbert problem stated above, since r may be given a sufficiently large value and, thus, all solutions, having at infinity a degree not greater than an arbitrarily given value, may be obtained.

But such a solution is unsatisfactory in the following sense: for a change of r, problem ($\overset{*}{\mathrm{II}}$) and hence also the corresponding

Fredholm integral equation will change; the number m likewise will vary together with r and it is easily seen that it will increase without a bound for unlimited increase of r.

In the next section a solution will be stated which is free from these defects. In fact, it will be shown that a system of n particular solutions may always be found which will be called the fundamental system and which has the property that each solution of the problem (I), having finite degree at infinity, is a linear combination of these n particular solutions with polynomials as coefficients.

§ 127. The fundamental system of solutions of the homogeneous Hilbert problem and its general solution.

Let the integer r satisfy the condition of the theorem of the last section; it will be assumed fixed in all further reasonings.

By this theorem all solutions $\Phi(z)$ of the problem (I), having a degree not greater than r at infinity, are representable in the form (126.19). First of all, one almost obvious property of the first n solutions, appearing in this formula, will be noted.

The solutions

$$\overset{1}{\Phi}(z),\ \overset{2}{\Phi}(z),\ \ldots,\ \overset{n}{\Phi}(z)$$

are not connected by any relations of the form

$$Q_1(z)\overset{1}{\Phi}(z) + Q_2(z)\overset{2}{\Phi}(z) + \ldots + Q_n(z)\overset{n}{\Phi}(z) = 0, \qquad (127.1)$$

where the $Q_1(z), \ldots, Q_n(z)$ are polynomials, not all identically zero at the same time. In fact, (127.1) is equivalent to the n relations

$$Q_1(z)\overset{1}{\Phi}_\alpha(z) + Q_2(z)\overset{2}{\Phi}_\alpha(z) + \ldots + Q_n(z)\overset{n}{\Phi}_\alpha(z) = 0,\ \alpha = 1, 2, \ldots, n,$$

and the proposition follows from the fact that the determinant

$$\det \| \overset{\beta}{\Phi}_\alpha \|,\ (\alpha, \beta = 1, 2, \ldots, n)$$

is not identically zero, which is, in its turn, obvious from (126.20).

Now the construction of a particular form of the system of solutions will be undertaken which will play a principal part in the sequel. Among the solutions (126.19), the degree of which at infinity does not exceed r, there exist some which have the lowest possible degree at infinity (because the order of the zero at infinity of any solution is known not to be greater than a definite number). This lowest possible degree will be denoted by $(-\varkappa_1)$ and one of the solutions, having this degree, by $\overset{1}{\chi}(z)$.

[If it is assumed that the particular solutions, appearing on the right side of (126.19), have already been constructed, then an algorithm is easily found by the help of which the above solution and those following from it will be found, but this will not be considered here. It should be noted that Plemelj gave a somewhat different (more complicated) construction of these solutions, in consequence of which his construction does not have an algorithmic character; it carries the character of a pure existence proof.]

Further, let $(-\varkappa_2)$ be the lowest possible degree (where sometimes instead of "degree at infinity" simply "degree" will be written) of those solutions (126.19) which are not related to $\overset{1}{\chi}(z)$ by any relations of the form

$$\Phi(z) = P_1(z)\overset{1}{\chi}(z),$$

where $P_1(z)$ is a polynomial; clearly $\varkappa_1 \geqq \varkappa_2$. Let $\overset{2}{\chi}(z)$ be one of those functions, having the degree $(-\varkappa_2)$. By $(-\varkappa_3)$ will be denoted the lowest possible degree of those solutions (126.19) which are not connected with $\overset{1}{\chi}(z)$ and $\overset{2}{\chi}(z)$ by any relations of the form

$$\Phi(z) = P_1(z)\overset{1}{\chi}(z) + P_2(z)\overset{2}{\chi}(z),$$

where $P_1(z)$ and $P_2(z)$ are polynomials; denote one of the solutions, having the degree $(-\varkappa_3)$, by $\overset{3}{\chi}(z)$, etc. It will be shown that this process may be continued until some solution $\overset{n}{\chi}(z)$ with the number n is reached.

In fact, assume that the k solutions $\overset{1}{\chi}(z), \overset{2}{\chi}(z), \ldots, \overset{k}{\chi}(z)$ have been constructed by this method; if $k < n$, then there will exist among the solutions (126.19) some, which are not connected with the above by any relations of the form

$$\Phi(z) = P_1(z)\overset{1}{\chi}(z) + P_2(z)\overset{2}{\chi}(z) + \ldots + P_k(z)\overset{k}{\chi}(z), \quad (127.2)$$

where the $P_1(z), \ldots, P_k(z)$ are polynomials. Actually, if all the solutions (126.19) would be connected with the $\overset{1}{\chi}(z), \ldots, \overset{k}{\chi}(z)$ by similar relations, then this would, in particular, be true for the n solutions $\overset{1}{\Phi}(z), \ldots, \overset{n}{\Phi}(z)$. But then, as it is easily seen, the latter would be found to be connected with one another by relations of the form (127.1), which is impossible.

Thus, for $k < n$, there exist solutions, not connected with the $\overset{1}{\chi}(z), \ldots, \overset{k}{\chi}(z)$ by relations of the form (127.2), and, consequently, the process may be continued.

In such a manner the n solutions

$$\overset{1}{\chi}(z),\ \overset{2}{\chi}(z),\ \ldots,\ \overset{n}{\chi}(z) \quad (127.3)$$

will be obtained, having the degrees $-\varkappa_1, -\varkappa_2, \ldots, -\varkappa_n$, where

$$\varkappa_1 \geqq \varkappa_2 \geqq \ldots \geqq \varkappa_n. \qquad (127.4)$$

It will be shown later that the process cannot be continued further, i.e., that for $k = n$ each solution (126.19) is representable in the form (127.2); moreover, it will be shown that, in general, each solution of the problem (I) is representable in this way. For the present only the following remark will be made.

Let $\chi(z)$ be any solution, having a degree less than $(-\varkappa_k)$; then $\chi(z)$ can necessarily be represented in the form

$$\chi(z) = P_1(z)\overset{1}{\chi}(z) + P_2(z)\overset{2}{\chi}(z) + \ldots + P_{k-1}\overset{k-1}{\chi}(z), \qquad (127.5)$$

where the $P_1(z), \ldots, P_{k-1}(z)$ are polynomials. Actually, in the opposite case, when constructing by the above method the solution succeeding $\overset{k-1}{\chi}(z)$, the solution $\overset{k}{\chi}(z)$ would not be obtained, because a solution $\chi(z)$ of lower order would exist, not connected with the preceding ones by a relation of the form (127.5).

Now the following important property of the solutions (127.3) will be proved: *the expression*

$$\chi(z) = a_1\overset{1}{\chi}(z) + a_2\overset{2}{\chi}(z) + \ldots + a_n\overset{n}{\chi}(z), \qquad (127.6)$$

where not all the constants $a_1, a_2, \ldots, a_n$ are zero, cannot vanish at a finite point of the plane.

Let $\chi(z)$ vanish at some point c, not on L. Then obviously $\chi(z) = (z - c)\Phi(z)$, i.e.,

$$a_1\overset{1}{\chi}(z) + a_2\overset{2}{\chi}(z) + \ldots + a_n\overset{n}{\chi}(z) = (z - c)\Phi(z), \qquad (127.7)$$

where $\Phi(z)$ is a sectionally holomorphic vector which obviously is some solution of the problem (I). Let a_k be the last of the coefficients $a_1, a_2, \ldots$ which is not zero. Then the degree of the solution $\Phi(z)$ at infinity is less than the degree of the solution $\overset{k}{\chi}(z)$, and hence, by (127.5), a relation of the form

$$\Phi(z) = P_1(z)\overset{1}{\chi}(z) + \ldots + P_{k-1}(z)\overset{k-1}{\chi}(z)$$

will hold. But then it follows from (127.7) that $\overset{k}{\chi}(z)$ is connected with the $\overset{1}{\chi}(z), \ldots, \overset{k-1}{\chi}(z)$ by a similar relation, which contradicts the condition, and the proposition is proved for points not on L.

Now the case where c lies on L will be treated. When saying that $\chi(z)$ vanishes at the point c on L, this will mean that $\chi^+(c) = \chi^-(c) = 0$; one of these relations will follow from the other by the condition by $\chi^+(c) = G(c)\chi^-(c)$. The above reasoning will remain valid also for the point c, lying on L, if it can be proved that the relation

$$\chi(z) = (z - c)\Phi(z)$$

follows from $\chi^+(c) = \chi^-(c) = 0$, where $\Phi(z)$ is a sectionally holomorphic vector with its boundary values Φ^+ and Φ^- satisfying the H condition.

That this is the case will be shown by the following argument. Consider

$$\Phi(z) = \frac{\chi(z)}{z - c}. \tag{127.8}$$

This vector will be holomorphic in S^+ and S^-, it will have a finite degree at infinity and be continuous in each of the regions S^+ and S^- up to L, with the possible exclusion of the point c of the boundary; when $z \to c$ along any path, remaining in S^+ or in S^-, then $\chi(z) \to 0$.

Further, since $\chi^+(t)$ and $\chi^-(t)$ satisfy the H condition and $\chi^+(c)$, $\chi^-(c)$ are zero, it is clear that on L

$$\Phi^+(t) = \frac{\overset{0}{\Phi}{}^+(t)}{|t - c|^\alpha}, \quad \Phi^-(t) = \frac{\overset{0}{\Phi}{}^-(t)}{|t - c|^\alpha}, \tag{127.9}$$

where $0 \leqq \alpha \leqq 1$ and $\overset{0}{\Phi}{}^+(t)$, $\overset{0}{\Phi}{}^-(t)$ are continuous vectors. Therefore, as is easily seen, the formulae (125.4), (125.5), (125.8), (125.9) and likewise (125.6) and (125.10) are applicable to the vector $\Phi(z)$, if in the latter $t_0 \neq c$ is assumed. Further, since $\Phi(z)$ satisfies the boundary condition (I) of the Hilbert problem everywhere except possibly at the point $t_0 = c$, it is easily concluded from what has been said that the boundary value $\Phi^-(t_0)$ satisfies everywhere, except possibly at the point $t_0 = c$, the Fredholm integral equation (126.5) for a suitable choice of $\gamma(t_0)$. Starting from this and taking into consideration the second of the formulae (127.9), it is not difficult to prove by a simple estimate that $\Phi^-(t_0)$ necessarily satisfies the H condition everywhere on L, if this function is given a suitable value at the point c; the same thing will obviously be true for $\Phi^+(t_0) = G(t_0)\Phi^-(t_0)$. Hence it easily follows that $\Phi(z)$ possesses the required properties. Thus the statement is proved. [Plemelj did not consider the particular case where c lies on L (and, in general, this case is not assumed), in consequence of which his proof (reproduced in this book for c not on L) can hardly be recognized as complete].

From the above proposition follows:

Property 1°: *The determinant*

$$\Delta(z) = \det \| \overset{\beta}{\chi}_\alpha(z) \| \quad (\alpha, \beta = 1, \ldots, n) \tag{127.10}$$

does not vanish anywhere in the finite part of the plane.

In fact, if $\Delta(c) = 0$, then the constants $a_1, a_2, \ldots, a_n$, not simultaneously zero, can always be chosen in such a way that

$$a_1\overset{1}{\chi}(c) + a_2\overset{2}{\chi}(c) + \ldots + a_n\overset{n}{\chi}(c) = 0,$$

which was shown to be impossible.

The point $z = \infty$ may represent an exception; the determinant $\Delta(z)$ may have at this point either a pole or a zero (or may remain bounded). The behaviour of $\Delta(z)$ at infinity will be determined by the following proposition:

PROPERTY 2°. *Let* $(-\varkappa_\beta)$ *be the degree of the solution* $\overset{\beta}{\chi}(z)$ *at infinity; if one puts*

$$\overset{\beta}{\chi}{}^0(z) = z^{\varkappa_\beta}\overset{\beta}{\chi}(z) \quad (\beta = 1, 2, \ldots, n), \tag{127.11}$$

then the determinant

$$\Delta^0(z) = \det \| \overset{\beta}{\chi}{}^0_\alpha(z) \| \tag{127.12}$$

has a finite non-zero value at infinity.

That $\Delta^0(\infty)$ has a finite value, follows from the fact that $\overset{\beta}{\chi}(z)$ has the degree $(-\varkappa_\beta)$ at infinity. It will be proved now that $\Delta^0(\infty) \neq 0$. In fact, in the opposite case, the constants $a_1, a_2, \ldots, a_n$, not all simultaneously zero, could be chosen in such a way that for $z \to \infty$

$$a_1 z^{\varkappa_1}\overset{1}{\chi}(z) + \ldots + a_n z^{\varkappa_n}\overset{n}{\chi}(z) = O(1/z).$$

Let a_k be the last of the coefficients $a_1, a_2, \ldots$ which is not zero. Then

$$\chi(z) = a_1 z^{\varkappa_1-\varkappa_k}\overset{1}{\chi}(z) + a_2 z^{\varkappa_2-\varkappa_k}\overset{2}{\chi}(z) + \ldots + a_k\overset{k}{\chi}(z) = O(z^{-\varkappa_k-1});$$

consequently the solution $\chi(z)$ would have a degree less than $(-\varkappa_k)$ at infinity and therefore it could be represented in the form (127.5); but then also $\overset{k}{\chi}(z)$ could be represented in a similar manner, which is impossible.

One consequence of Property 2° will be noted which will be used often. Consider a sum of the form

$$P_1(z)\overset{1}{\chi}(z) + P_2(z)\overset{2}{\chi}(z) + \ldots + P_n(z)\overset{n}{\chi}(z), \tag{127.13}$$

where the $P_1(z), P_2(z), \ldots, P_n(z)$ are polynomials of order $m_1, m_2, \ldots, m_n$ respectively. The degrees of the different terms are $m_1 - \varkappa_1$, $m_2 - \varkappa_2, \ldots, m_n - \varkappa_n$. It follows directly from the fact that $\Delta^0(\infty) \neq 0$ that *the degree of* (127.13) *is equal to the degree of those terms which have the highest degree*, i.e., it is equal to the largest of the numbers $m_1 - \varkappa_1, \ldots, m_n - \varkappa_n$; in other words, it follows that in (127.13) the terms of higher degree cannot cancel.

In future, any n solutions of the homogeneous Hilbert problem

$$\overset{1}{\chi}(z),\ \overset{2}{\chi}(z),\ \ldots,\ \overset{n}{\chi}(z),$$

possessing the properties 1° and 2°, will be called a fundamental system of solutions of this problem. One such fundamental system has actually been constructed.

The matrix

$$X(z) = \| \overset{\beta}{\chi}_{\alpha}(z) \| = \left\| \begin{matrix} \overset{1}{\chi}_1, & \overset{2}{\chi}_1, & \ldots, & \overset{n}{\chi}_1 \\ \overset{1}{\chi}_2, & \overset{2}{\chi}_2, & \ldots, & \overset{n}{\chi}_2 \\ \ldots & \ldots & \ldots & \ldots \\ \overset{1}{\chi}_n, & \overset{2}{\chi}_n, & \ldots, & \overset{n}{\chi}_n \end{matrix} \right\| \tag{127.14}$$

the columns of which consist of the different solutions of the fundamental system will be called the fundamental matrix, corresponding to the homogeneous Hilbert problem.

By the boundary condition (I) it is easily seen that

$$X^+(t) = G(t)X^-(t), \tag{127.15}$$

and hence follows the formula which will often be used

$$G(t) = X^+(t)[X^-(t)]^{-1}. \tag{127.16}$$

Now the following fundamental theorem will be proved:

THEOREM. *All the solutions $\Phi(z)$ of the homogeneous Hilbert problem* (having finite degree at infinity) *are given by*

$$\Phi(z) = P_1(z)\overset{1}{\chi}(z) + P_2(z)\overset{2}{\chi}(z) + \ldots + P_n(z)\overset{n}{\chi}(z), \tag{127.17}$$

where $\overset{1}{\chi}(z), \overset{2}{\chi}(z), \ldots, \overset{n}{\chi}(z)$ is a fundamental system of solutions and the $P_1(z), P_2(z), \ldots, P_n(z)$ are polynomials.

[The proof given below differs essentially from that of Plemelj. The results established in the remaining part of this section were not obtained by Plemelj. In particular, he did not introduce the concept of indices.]

Now the condition may be abandoned that the boundary values of the unknown functions $\Phi(z)$ must satisfy the H condition.

By (I),

$$\Phi^+(t) = G(t)\Phi^-(t).$$

Replacing $G(t)$ from (127.16), one obtains

$$[X^+(t)]^{-1}\Phi^+(t) = [X^-(t)]^{-1}\Phi^-(t).$$

Hence it follows that the vector $[X(z)]^{-1}\Phi(z)$ is holomorphic in the entire plane; further, since by hypothesis $\Phi(z)$ has a finite degree at infinity, also $[X(z)]^{-1}\Phi(z)$ will have a finite degree, and thus

$$[X(z)]^{-1}\Phi(z) = P(z),$$

where

$$P(z) = (P_1,\ P_2,\ \ldots,\ P_n) \tag{127.18}$$

is a vector the components of which, i.e., $P_1 = P_1(z), \ldots, P_n = P_n(z)$, are polynomials.

Hence

$$\Phi(z) = X(z)P(z); \tag{127.19}$$

however, the last relation is easily seen to express the same as (127.17).

Conversely, it is obvious that (127.17) or, what is the same thing, (127.19) gives a solution of the boundary problem (I) for any choice of the polynomials $P_1(z), \ldots, P_n(z)$, which proves the statement above.

It will be noted that *it is sufficient to assume for the deduction of the conclusion regarding the representation of any solution in the form* (127.17) *that the system of solutions forming the matrix* $X(z)$ *has only Property* 1°. However, Property 2° of the fundamental system facilitates the choice of the polynomials, when it is required to find a solution having a given degree at infinity. In fact, on the basis of the consequence of Property 2° stated above, it is clear that all the solutions of the problem (I), having a degree not greater than a given integer k at infinity, are given by

$$\Phi(z) = P_{k+\varkappa_1}(z)\overset{1}{\chi}(z) + \ldots + P_{k+\varkappa_n}(z)\overset{n}{\chi}(z), \tag{127.20}$$

where the $P_{k+\varkappa_j}$ are arbitrary polynomials of degree not greater than $k + \varkappa_j$; for this one has to assume $P_{k+\varkappa_j} \equiv 0$, if $k + \varkappa_j < 0$. The solution $\Phi(z)$ will have exactly the degree k, if the degree of at least one of the polynomials $P_{k+\varkappa_j}$ attains $k + \varkappa_j$.

If all the $k + \varkappa_j < 0$, then the problem has no (non-trivial) solution, having a degree not greater than k at infinity.

Now, proceeding to the problem regarding the connection between two different fundamental systems of solutions of one and the same Hilbert problem, only a proof will be given of the fact that *the numbers* $\varkappa_1, \varkappa_2, \ldots, \varkappa_n$ *are always the same for all fundamental systems.*

Let

$$\overset{1}{\chi}(z),\ \overset{2}{\chi}(z),\ \ldots,\ \overset{n}{\chi}(z) \tag{*}$$

and

$$\overset{1}{\zeta}(z),\ \overset{2}{\zeta}(z),\ \ldots,\ \overset{n}{\zeta}(z) \tag{**}$$

be any two fundamental systems of solutions and let $-\varkappa_1$, $-\varkappa_2$, $\ldots$, $-\varkappa_n$ and $-\lambda_1$, $-\lambda_2$, $\ldots$, $-\lambda_n$ respectively be the degrees of the solutions (*) and (**) at infinity. These solutions will be assumed numbered in such a way that $\varkappa_1 \geqq \varkappa_2 \geqq \ldots \geqq \varkappa_n$ and $\lambda_1 \geqq \lambda_2 \geqq \geqq \ldots \geqq \lambda_n$. It has to be proved that $\lambda_1 = \varkappa_1, \ldots, \lambda_n = \varkappa_n$.

From the principal property of a fundamental system it follows that the solutions (**) may be expressed in terms of the solutions (*) by formulae of the form

$$\zeta = \overset{n}{P}_{\alpha 1}\overset{1}{\chi} + P_{\alpha 2}\overset{2}{\chi} + \ldots + P_{\alpha n}\overset{n}{\chi}, \tag{127.21}$$

where the $P_{\alpha\beta}$ are polynomials; by similar formulae the solutions (*) will be expressed in terms of the solutions (**).

For the sake of generality, it will now be assumed that

$$\varkappa_1 = \varkappa_2 = \ldots = \varkappa_k > \varkappa_{k+1}, \; \lambda_1 = \lambda_2 = \ldots = \lambda_l > \lambda_{l+1},$$

and it will be shown that $\varkappa_1 = \lambda_1$, $k = l$. In fact, formula (127.21), applied to the first l of the solutions (**), shows, if the degrees of the left and right sides are compared, that $-\lambda_1 \geqq -\varkappa_1$, because the degree of the right side of (127.21) at infinity cannot be less than $(-\varkappa_1)$. Interchanging the roles of the systems (*) and (**), one similarly obtains $-\varkappa_1 \geqq -\lambda_1$; hence $\varkappa_1 = \lambda_1$.

Similarly, a comparison of the degrees of the left and right sides of (127.21), applied to the first l of the solutions (**), shows that the right side can only contain the first k terms, i.e., that, for $\alpha = 1, 2, \ldots, l$,

$$\overset{\alpha}{\zeta} = P_{\alpha 1}\overset{1}{\chi} + \ldots + P_{\alpha k}\overset{k}{\chi}, \tag{127.22}$$

where obviously in the present case the $P_{\alpha 1}, \ldots, P_{\alpha k}$ are constants, but for uniformity (with a view to the sequel) they will be considered as particular cases of polynomials.

If now l would be larger than k, i.e., $l > k$, then the existence of at least one relation of the form

$$Q_1\overset{1}{\zeta} + Q_2\overset{2}{\zeta} + \ldots + Q_l\overset{l}{\zeta} = 0$$

would follow from the formulae (127.22), where the Q_j are polynomials [compare the remark subsequent to (127.22)], not all zero, which is impossible for solutions of fundamental systems. In exactly the same way, interchanging the roles of (*) and (**), it is shown that the converse inequality, i.e., $k < l$ cannot be true, and hence $l = k$.

Thus $\varkappa_1 = \lambda_1, \ldots, \varkappa_k = \lambda_k$, $\varkappa_k > \varkappa_{k+1}$, $\lambda_k > \lambda_{k+1}$. Now let $\varkappa_{k+1} = \ldots = \varkappa_{k+r} > \varkappa_{k+r+1}$, $\lambda_{k+1} = \ldots = \lambda_{k+s} > \lambda_{k+s+1}$. It will be shown that $\varkappa_{k+1} = \lambda_{k+1}$, $r = s$. Applying (127.21) to the solutions

$$\overset{k+1}{\zeta}, \; \overset{k+2}{\zeta}, \; \ldots, \; \overset{k+s}{\zeta},$$

the conclusion is drawn that the right sides of the formulae (127.21) cannot only consist of the first k terms, since in the opposite case there would be relations of the form (127.22) for $\alpha = 1, 2, \ldots, k + s$, and at least one relation of the form

$$Q_1\overset{1}{\zeta} + Q_2\overset{2}{\zeta} + \ldots + Q_k\overset{k}{\zeta} + Q_{k+1}\overset{k+1}{\zeta} + \ldots + Q_{k+s}\overset{k+s}{\zeta} = 0 \qquad (***)$$

would hold true, where the Q_j are polynomials, not all zero; and this is impossible.

From the above it follows that necessarily $-\lambda_{k+1} \geqq -\varkappa_{k+1}$. Similarly, it is concluded that $-\varkappa_{k+1} \geqq -\lambda_{k+1}$ and, consequently, $\varkappa_{k+1} = \lambda_{k+1}$. Applying (127.21) to the first $k + s$ solutions (**), one obviously will have formulae of the form

$$\overset{\alpha}{\zeta} = P_{\alpha 1}\overset{1}{\chi} + P_{\alpha 2}\overset{2}{\chi} + \ldots P_{\alpha, k+r}\overset{k+r}{\chi} \quad (\alpha = 1, 2, \ldots, k + s) \qquad (127.23)$$

(where the first k of the formulae coincide with those of (127.22), but this is not important here).

If now s would be greater than r, i.e., $s > r$, then at least one relation of the form (***) would follow from (127.23), which is impossible. Similarly, it is concluded that $r > s$ is not possible. Hence $r = s$.

The further process of reasoning is obvious and the proposition will be assumed proved.

On the basis of the above reasoning, a method of construction of all fundamental systems is easily indicated, once one of them has been found. But this will not be considered here.

The integers $\varkappa_1, \varkappa_2, \ldots, \varkappa_n$, not depending on the choice of the fundamental system, will be called the component indices of the corresponding homogeneous Hilbert problem and their sum

$$\varkappa_1 + \varkappa_2 + \ldots + \varkappa_n = \varkappa \qquad (127.24)$$

the total index or simply the index of this problem.

It is important that *the total index may be directly* (and besides in an elementary manner) *obtained from the matrix* $G(t)$, *characterising the Hilbert problem.*

In fact, it follows from (127.15) that

$$\Delta^+(t) = \det G(t) \,.\, \Delta^-(t),$$

where, as before, $\Delta(z) = \det X(z)$. Hence

$$[\log \Delta^+(t)]_L = [\log \det G(t)]_L + [\log \Delta^-(t)]_L,$$

where $[\ ]_L$ denotes the increase of the function in the brackets for a complete cycle of L in the positive direction.

The function $\Delta(z)$ is holomorphic in S^+, continuous on the boundary from the left and non-vanishing. Consequently

$$[\log \Delta^+(t)]_L = 0.$$

Further, the function $\Delta(z)$ is holomorphic in S^-, except possibly at the point $z = \infty$, non-vanishing in the finite part of the plane and continuous on L from the right. In the neighbourhood of the point at infinity the function $\Delta(z)$ has, by (127.11) and (127.12), the form

$$\Delta(z) = \frac{\Delta^0(z)}{z^\varkappa}, \tag{127.25}$$

where $\Delta^0(z)$ is holomorphic near $z = \infty$, and $\Delta^0(\infty) \neq 0$. Hence it follows that

$$[\log \Delta^-(t)]_L = -2\pi i\varkappa.$$

Thus the following important formula has been obtained:

$$\varkappa = \frac{1}{2\pi i} [\log \det G(t)]_L = \frac{1}{2\pi} [\arg \det G(t)]_L. \tag{127.26}$$

Finally, the close connection between the solutions of the associate problems

$$\Phi^+(t) = G(t)\Phi^-(t) \tag{I}$$

and

$$\Psi^+(t) = G'^{-1}(t)\Psi^-(t) \tag{I'}$$

will be noted; in fact, it follows from $X^+(t) = G(t)X^-(t)$ that

$$[X'^+(t)]^{-1} = G'^{-1}[X'^-(t)]^{-1},$$

and hence it is easily inferred that the matrix $[X'(z)]^{-1}$ plays the same role with regard to the problem (I′) as does the matrix $X(z)$ with respect to the problem (I). For this it is sufficient to show that the matrix $[X'(z)]^{-1}$ possesses the properties of a fundamental matrix. But this is easily seen by a simple check. The presence of Property 1° is obvious, since $\det [X'(z)]^{-1} = 1/\Delta(z)$.

Property 2° will now be proved. Put

$$[X'(z)]^{-1} = \| \overset{\beta}{\zeta}_\alpha(z) \|,$$

so that

$$\overset{\beta}{\zeta}_\alpha(z) = \frac{\overset{\beta}{\Delta}_\alpha(z)}{\Delta(z)}, \tag{127.27}$$

where $\overset{\beta}{\Delta}_\alpha(z)$ is the cofactor of the element $\overset{\beta}{\chi}_\alpha$ in the determinant $\Delta(z)$.

Hence it follows from (127.11) and (127.12) that

$$\overset{\beta}{\zeta}_\alpha(z) = \frac{\overset{\beta}{\Delta}{}^0_\alpha(z)}{\Delta^0(z)}\, z^{\varkappa_\beta}, \tag{127.28}$$

where $\overset{\beta}{\Delta}{}^0_\alpha(z)$ is the cofactor of $\overset{\beta}{\chi}{}^0_\alpha$ in the determinant $\Delta^0(z)$, and therefore that the degree of the solutions $\overset{\beta}{\zeta}(z)$ at infinity is exactly $\varkappa_\beta$ and that

$$\det \| z^{-\varkappa_\beta} \overset{\beta}{\zeta}_\alpha \| = \frac{1}{\Delta^0(z)}$$

does not vanish at infinity. This is Property 2° for the matrix $[X'(z)]^{-1}$.

Thus there is the following proposition: *if $X(z)$ is a fundamental matrix of solutions of the problem* (I), then $[X'(z)]^{-1}$ *is a fundamental matrix of solutions of the associate problem* (I'); *if $\varkappa_1, \varkappa_2, \ldots, \varkappa_n$ and $\varkappa$ are the component indices and the total index of the problem* (I), *then the component indices and the total index of the associate problem will be $-\varkappa_1, -\varkappa_2, \ldots, -\varkappa_n$ and $-\varkappa$.*

Let

$$\varkappa_1 \geqq \varkappa_2 \geqq \ldots \geqq \varkappa_m \geqq 0 > \varkappa_{m+1} \geqq \varkappa_{m+2} \geqq \ldots \geqq \varkappa_n \tag{127.29}$$

(if all the component indices are non-negative, then $m = n$; if all of them are negative, then $m = 0$).

Further, put

$$\lambda = \varkappa_1 + \varkappa_2 + \ldots + \varkappa_m, \quad -\mu = \varkappa_{m+1} + \ldots + \varkappa_n, \tag{127.30}$$

so that

$$\lambda - \mu = \varkappa. \tag{127.31}$$

From (127.20) the general solution of the problem (I), vanishing at infinity (i.e., having a degree not greater than $k = -1$), is given by

$$\Phi(z) = P_{\varkappa_1 - 1}(z)\overset{1}{\chi}(z) + \ldots + P_{\varkappa_m - 1}(z)\overset{m}{\chi}(z), \tag{127.32}$$

where $P_{\varkappa_j - 1}(z)$ is a polynomial of degree not greater than $\varkappa_j - 1$ with arbitrary coefficients ($P_{\varkappa_j - 1}(z) \equiv 0$, if $\varkappa_j = 0$).

Thus, this solution contains $\varkappa_1 + \varkappa_2 + \ldots + \varkappa_m = \lambda$ arbitrary constant coefficients of the polynomials $P_{\varkappa_j - 1}(z)$. These constants will (in any order) be denoted by $\alpha_1, \alpha_2, \ldots, \alpha_\lambda$. Then the general solution $\Phi(z)$ may be written in the form

$$\Phi(z) = \alpha_1 \overset{1}{\omega}(z) + \alpha_2 \overset{2}{\omega}(z) + \ldots + \alpha_\lambda \overset{\lambda}{\omega}(z), \tag{127.33}$$

where the $\overset{1}{\omega}(z), \ldots, \overset{\lambda}{\omega}(z)$ are certain particular solutions of the problem (I) which are obviously linearly independent. Thus, *the problem* (I) *has exactly* λ *linearly independent solutions, vanishing at infinity.*

Taking into consideration what has been said above with regard to the connection between the solutions of the problems (I) and (I′), it may be concluded that the problem (I′) has exactly μ linearly independent solutions, vanishing at infinity. Further, since $\lambda - \mu = \varkappa$, the following proposition results:

The difference between the numbers of linearly independent solutions, vanishing at infinity, of a given homogeneous Hilbert problem and of its associate problem is equal to the total index of the given problem.

NOTE. From the above it is easy to determine the numbers λ and μ, without having to calculate the component indices $\varkappa_j$. In fact, obviously $\lambda \leqq s$, where s is the number of linearly independent solutions of the homogeneous Fredholm equation, corresponding to the equation (126.5); it will follow then from (127.31) that $\mu \leqq s - \varkappa$. The number s, as is known (cf. for example E. Hellinger and O. Toeplitz [1], p. 1382), can, in its turn, be evaluated from the above without having to solve that integral equation. It will still be noted that it obviously follows from the preceding inequality that

$$\varkappa_j \leqq s \quad (j = 1, 2, \ldots, m), \quad -\varkappa_j \leqq s - \varkappa \quad (j = m + 1, \ldots, n).$$

§ 128. The non-homogeneous Hilbert problem.

As has been mentioned earlier, a complete, and besides elementary solution of this problem for the case of a single unknown function was given by F. D. Gakhov. In his paper [2] Gakhov also considered the non-homogeneous problem for several unknown functions and constructed directly a system of Fredholm integral equations for this case, similar to the system given by Plemelj for the homogeneous problem (cf. the beginning of § 126). However, Gakhov did not manage in this way to obtain complete solution. It will be shown in this section that the solution of the non-homogeneous problem is easily reduced to the solution of the corresponding homogeneous problem; this will be done by a method, similar to that which was used by F. D. Gakhov in the case of a single unknown function.

The non-homogeneous Hilbert problem for several unknown functions will be stated thus:

To find a sectionally holomorphic vector $\Phi(z) = (\Phi_1, \Phi_2, \ldots, \Phi_n)$, *having finite degree at infinity, for the boundary condition*

$$\Phi^+(t_0) = G(t_0)\Phi^-(t_0) + g(t_0) \text{ on } L, \tag{128.1}$$

where $G(t_0)$ *is a matrix, given on* L, *satisfying the* H *condition and* *not being singular anywhere on L, and $g(t_0)$ is a vector, given on L, likewise satisfying the H condition.*

The solution of this problem may be obtained in the following manner. As before, let $X(z)$ be a fundamental matrix, corresponding to the homogeneous problem obtained from (128.1) for $g(t_0) \equiv 0$. Then, by (127.16), $G(t_0) = X^+(t_0)[X^-(t_0)]^{-1}$. Substituting this expression in (128.1), one obtains

$$[X^+(t_0)]^{-1}\Phi^+(t_0) - [X^-(t_0)]^{-1}\Phi^-(t_0) = [X^+(t_0)]^{-1}g(t_0).$$

It follows from this formula and the Plemelj formulae (125.2) (125.3) that the vector

$$P(z) = [X(z)]^{-1}\Phi(z) - \frac{1}{2\pi i}\int_L \frac{[X^+(t)]^{-1}g(t)dt}{t-z},$$

holomorphic in each of the regions S^+ and S^-, takes the same boundary values as any point t_0 on L is approached from both sides. Consequently, the vector $P(z)$ is holomorphic in the entire plane; further, since it obviously must have a finite degree at infinity, all its components must be polynomials.

Therefore the unknown vector will necessarily have the form

$$\Phi(z) = \frac{X(z)}{2\pi i}\int_L \frac{[X^+(t)]^{-1}g(t)dt}{t-z} + X(z)P(z), \tag{128.2}$$

where

$$P(z) = (P_1, P_2, \ldots, P_n) \tag{128.3}$$

with the $P_\alpha = P_\alpha(z)$ being some polynomials.

Conversely, it is clear that the vector $\Phi(z)$, defined by (128.2), will satisfy the conditions of the problem for a quite arbitrary choice of the polynomials P_α.

Thus (128.2) gives the general solution of the non-homogeneous Hilbert problem.

From the point of view of future applications special interest will lie in the determination of solutions vanishing at infinity.

As before, let

$$\varkappa_1 \geqq \varkappa_2 \geqq \ldots \geqq \varkappa_m \geqq 0 > \varkappa_{m+1} \geqq \ldots \geqq \varkappa_n \tag{128.4}$$

and

$$\lambda = \varkappa_1 + \varkappa_2 + \ldots + \varkappa_m, \; \mu = -\varkappa_{m+1} - \ldots - \varkappa_n. \tag{128.5}$$

Introduce the notation

$$[X^+(t)]^{-1}g(t) = h(t) = (h_1, h_2, \ldots, h_n); \tag{128.6}$$

then (128.2) may obviously be written

$$\Phi(z) = \overset{1}{\chi}(z)\left\{\frac{1}{2\pi i}\int_L \frac{h_1(t)dt}{t-z} + P_1(z)\right\} + \ldots +$$

$$+ \overset{n}{\chi}(z)\left\{\frac{1}{2\pi i}\int_L \frac{h_n(t)dt}{t-z} + P_n(z)\right\}. \tag{128.7}$$

Noting that in the neighbourhood of the point at infinity

$$\int_L \frac{h_\alpha(t)dt}{t-z} = -z^{-1}\int_L h_\alpha(t)dt - z^{-2}\int_L th_\alpha(t)dt -$$

$$-z^{-3}\int_L t^2 h_\alpha(t)dt - \ldots \tag{128.8}$$

and calculating the degrees of the various terms on the right side of (128.7), the conclusion is easily reached that it is necessary and sufficient for the existence of a solution, vanishing at infinity, that $g(t)$ should satisfy the $\mu = -\varkappa_{m+1} - \varkappa_{m+2} - \ldots - \varkappa_n$ conditions

$$\int_L t^j h_\alpha(t)dt = 0 \tag{128.9}$$

$$(j = 0, 1, \ldots, -\varkappa_\alpha - 1, \quad \alpha = m+1, m+2, \ldots, n),$$

and that, provided these conditions are satisfied, the general solution of the required form is given by (128.2) in which one has now to consider

$$P(z) = (P_{\varkappa_1-1}, P_{\varkappa_2-1}, \ldots, P_{\varkappa_m-1}, 0, \ldots, 0),$$

where the $P_{\varkappa_j-1}$ are arbitrary polynomials of degree not greater than $\varkappa_j - 1$ (while $P_{\varkappa_j-1} \equiv 0$, if $\varkappa_j = 0$).

The set of conditions (128.9) can be written in the form of a single expression; in fact, multiplying the equations (128.9) by arbitrary constants and adding, one obtains

$$\int_L Q(t)h(t)dt = 0, \tag{128.10}$$

where $Q(t)$ is a vector, defined by

$$Q = (0, 0, \ldots, 0, Q_{-\varkappa_{m+1}-1}, \ldots, Q_{-\varkappa_n-1})$$

with the $Q_{-\varkappa_j-1}$ being polynomials with arbitrary coefficients of degree not greater than $-\varkappa_j - 1$.

In future, for uniformity, write

$$P(z) = (P_{\varkappa_1-1}, P_{\varkappa_2-1}, \ldots, P_{\varkappa_n-1}), \tag{128.11}$$

$$Q(z) = (Q_{-\varkappa_1-1}, Q_{-\varkappa_2-1}, \ldots, Q_{-\varkappa_n-1}), \tag{128.12}$$

stipulating that by P_α, Q_β will be understood polynomials of the degrees α and β respectively, while $P_\alpha \equiv 0$, $Q_\beta \equiv 0$, if $\alpha < 0, \beta < 0$ respectively.

In this notation the above result may be written:

For the existence of solutions of the problem (128.1), *vanishing at infinity, it is necessary and sufficient that the function* $g(t)$ *should satisfy the condition*

$$\int_L Q(t)[X^+(t)]^{-1}g(t)dt = 0, \tag{128.13}$$

where $Q(t)$ *is an arbitrary vector of the form* (128.12); *provided this condition is satisfied, the general solution of the required form is given by* (128.2), *where* $P(z)$ *is an arbitrary vector of the form* (128.11).

The numbers $\varkappa_1, \varkappa_2, \ldots, \varkappa_n$ and $\varkappa$, being the component indices and the total index of the homogeneous problem, obtained from (128.1) for $g(t_0) \equiv 0$, will also be called the component indices and the total index of the problem (128.1).

§ 129. Supplement to the solution of a dominant system of singular integral equations and of its associate system.

The particular system of singular integral equations

$$\sum_{\beta=1}^{n} A_{\alpha\beta}(t_0)\varphi_\beta(t_0) + \sum_{\beta=1}^{n} \frac{B_{\alpha\beta}(t_0)}{\pi i}\int_L \frac{\varphi_\beta(t)dt}{t-t_0} = f_\alpha(t_0) \; (\alpha = 1, \ldots, n) \tag{129.1}$$

will now be considered, where the $A_{\alpha\beta}(t_0)$, $B_{\alpha\beta}(t_0)$, $f_\alpha(t_0)$ are functions, given on L and satisfying the H condition there, and the $\varphi_\alpha(t)$ ($\alpha = 1, \ldots, n$) are unknown functions, which will likewise be required to satisfy the H condition.

A system of singular equations of the form (129.1) will be called a dominant system. Introducing the vectors

$$\varphi(t) = (\varphi_1, \varphi_2, \ldots, \varphi_n), \; f(t) = (f_1, f_2, \ldots, f_n) \tag{129.2}$$

and the matrices

$$A(t) = \| A_{\alpha\beta}(t) \|, \; B(t) = \| B_{\alpha\beta}(t) \|, \tag{129.3}$$

the system (129.1) may be written in the form of the single

equation

$$\mathbf{K}^0\varphi \equiv A(t_0)\varphi(t_0) + \frac{B(t_0)}{\pi i}\int_L \frac{\varphi(t)dt}{t-t_0} = f(t_0). \tag{129.4}$$

As will be seen later, the matrices

$$S = A + B, \quad D = A - B, \tag{129.5}$$

play the principal role in the theory of the equation (129.4), i.e., of the system of equations (129.1); they will be called the principal matrices of the equation (129.4) or of the operator $\mathbf{K}^0$. In the sequel it will be assumed that

$$\det S \neq 0, \ \det D \neq 0 \ \text{(everywhere on } L). \tag{129.6}$$

The equation (129.4) can be solved, using the results of the last sections, as will now be shown.

The sectionally holomorphic vector

$$\Phi(z) = \frac{1}{2\pi i}\int_L \frac{\varphi(t)dt}{t-z} \tag{129.7}$$

will be introduced. Then, by (125.2) and (125.3),

$$\varphi(t_0) = \Phi^+(t_0) - \Phi^-(t_0), \ \frac{1}{\pi i}\int_L \frac{\varphi(t)dt}{t-t_0} = \Phi^+(t_0) + \Phi^-(t_0). \tag{129.8}$$

Substituting these values in (129.4) and using (129.5), one obtains

$$S(t_0)\Phi^+(t_0) = D(t_0)\Phi^-(t_0) + f(t_0)$$

or

$$\Phi^+(t_0) = G(t_0)\Phi^-(t_0) + g(t_0), \tag{129.9}$$

where

$$G = S^{-1}D, \ g = S^{-1}f, \tag{129.10}$$

Thus the equation (129.4) is reduced to the non-homogeneous Hilbert problem (129.9) in the sense that a definite solution, vanishing at infinity, of the problem (129.9) corresponds, by (129.7), to each solution of the equation (129.4), and that a definite solution of (129.4) corresponds, by the first of the formulae (129.8), to each such solution of the problem (129.9).

By the results, obtained at the end of the last section, the problem (129.9) will admit a solution, vanishing at infinity, if and only if the condition (128.13) is fulfilled in which case the solution will be given by (128.2), where $P(z)$ has the form (128.11).

It follows from (128.2), on the basis of the general formulae (125.2) (where now $-\frac{1}{2}P$ will be written instead of P which, of course, does not affect the issue), that

$$\Phi^+(t_0) = X^+(t_0)\left\{\tfrac{1}{2}[X^+(t_0)]^{-1}g(t_0) + \frac{1}{2\pi i}\int\limits_L \frac{[X^+(t)]^{-1}g(t)dt}{t-t_0}\right\} - \\ -\tfrac{1}{2}X^+(t_0)P(t_0), \qquad (129.11)$$

$$\Phi^-(t_0) = X^-(t_0)\left\{-\tfrac{1}{2}[X^+(t_0)]^{-1}g(t_0) + \frac{1}{2\pi i}\int\limits_L \frac{[X^+(t)]^{-1}g(t)dt}{t-t_0}\right\} - \\ -\tfrac{1}{2}X^-(t_0)P(t_0),$$

and hence, by the formula $\varphi(t_0) = \Phi^+(t_0) - \Phi^-(t_0)$, an expression can be obtained for the unknown solution $\varphi(t)$. In order to simplify the last expression, introduce the further notation [where in (129.12) and for the deduction of (129.14) use will be made of the relations $X^+ = GX^-$, $G = S^{-1}D$]

$$Z(t) = S(t)X^+(t) = D(t)X^-(t), \qquad (129.12)$$

$$A^*(t) = \tfrac{1}{2}[S^{-1}(t) + D^{-1}(t)],\quad B^*(t) = -\tfrac{1}{2}[S^{-1}(t) - D^{-1}(t)]; \qquad (129.13)$$

then

$$\varphi(t_0) = A^*(t_0)f(t_0) - \frac{B^*(t_0)Z(t_0)}{\pi i}\int\limits_L \frac{[Z(t)]^{-1}f(t)dt}{t-t_0} + \\ + B^*(t_0)Z(t_0)P(t_0). \qquad (129.14)$$

The condition (128.13), necessary and sufficient for the existence of a solution, becomes in this notation

$$\int\limits_L Q(t)[Z(t)]^{-1}f(t)dt = 0. \qquad (129.15)$$

It will be remembered that in (129.14), (129.15)

$$\begin{aligned} P(t) &= (P_{\varkappa_1-1},\ P_{\varkappa_2-1},\ \ldots,\ P_{\varkappa_n-1}),\\ Q(t) &= (Q_{-\varkappa_1-1},\ Q_{-\varkappa_2-1},\ \ldots,\ Q_{-\varkappa_n-1}), \end{aligned} \qquad (129.16)$$

where the P_α, Q_α are polynomials with arbitrary coefficients of degree not greater than α ($P_\alpha \equiv 0$, $Q_\alpha \equiv 0$ for $\alpha < 1$). The formula (129.15) is equivalent to

$$\int_L f(t)[Z'(t)]^{-1}Q(t)dt = 0, \tag{129.17}$$

where, as always, Z' is the transposed matrix Z.

Let, as before,

$$\varkappa_1 \geqq \varkappa_2 \geqq \ldots \geqq \varkappa_m \geqq 0 > \varkappa_{m+1} \geqq \ldots \geqq \varkappa_n, \tag{129.18}$$
$$\lambda = \varkappa_1 + \varkappa_2 + \ldots + \varkappa_m, \; \mu = -\varkappa_{m+1} - \ldots - \varkappa_n.$$

The right side of (129.14) will actually contain the λ arbitrary constants $C_1, C_2, \ldots, C_\lambda$ as coefficients of the polynomials $P_\alpha(t)$. Obviously

$$P(t) = C_1\overset{1}{P}(t) + C_2\overset{2}{P}(t) + \ldots + C_\lambda\overset{\lambda}{P}(t), \tag{129.19}$$

where the $\overset{1}{P}(t), \ldots, \overset{\lambda}{P}(t)$ are certain linearly independent vectors the components of each of which are zero, except for one which is a non-negative integral power of t. Accordingly in (129.14)

$$B^*(t_0)Z(t_0)P(t_0) = C_1\overset{1}{\gamma}(t_0) + \ldots + C_\lambda\overset{\lambda}{\gamma}(t_0), \tag{129.20}$$

where the

$$\overset{\alpha}{\gamma}(t_0) = B^*(t_0)Z(t_0)\overset{\alpha}{P}(t_0) \tag{129.21}$$

are definite vectors, satisfying the H condition; they are easily seen to be linearly independent. In fact, if the right side (129.20) is identically zero for some values of the constants $C_1, C_2, \ldots, C_\lambda$, then

$$B^*(t_0)Z(t_0)P(t_0) = -\tfrac{1}{2}[X^+(t_0) - X^-(t_0)]P(t_0) = 0,$$

and hence it follows that the vector $X(z)P(z)$ is holomorphic in the entire plane. Since it vanishes at infinity, necessarily $X(z)P(z) \equiv 0$ in the whole plane. But this is only possible for $P(z) \equiv 0$, since $\det X(z) \neq 0$. Hence, necessarily, $C_1 = C_2 = \ldots = C_\lambda = 0$, and the statement is proved.

The condition (129.17) represents the condensed expression of conditions of the form

$$\int_L f(t)\overset{\alpha}{\psi}(t)dt = 0, \; \alpha = 1, 2, \ldots, \mu, \tag{129.22}$$

where the $\overset{\alpha}{\psi}(t)$ are certain linearly independent vectors, satisfying the H condition.

In order to obtain (129.22) from (129.17), it is sufficient to represent the vector $Q(t)$ in a form similar to (129.19)

$$Q(t) = D_1\overset{1}{Q}(t) + \ldots + D_\mu \overset{\mu}{Q}(t);$$

substituting in (129.17) and equating to zero the coefficients of the arbitrary constants $D_1, \ldots, D_\mu$, one obtains (129.22), where

$$\overset{\alpha}{\psi}(t) = [Z'(t)]^{-1}\overset{\alpha}{Q}(t), \quad \alpha = 1, 2, \ldots, \mu. \tag{129.23}$$

Thus there is the following result:

The necessary and sufficient conditions for the solubility of the system (129.1) *or, what is the same thing, of the equation* (129.4) *are given by* (129.17), *which is equivalent to the* μ *relations* (129.22). *Provided these conditions are satisfied, the general solution is given by* (129.14), *containing linearly* λ *arbitrary constants.*

The numbers $\varkappa_1, \varkappa_2, \ldots, \varkappa_n$ and $\varkappa$ will now be called the component indices and the total index of the system (129.1) or, what is the same thing, of the equation (129.4), or of the operator $\mathbf{K}^0$. The total index will also be called simply the index.

It will be noted that, if all the component indices are non-negative, then the condition of solubility is always satisfied and the solution contains $\varkappa$ arbitrary constants.

Now the homogeneous equation, obtained from (129.4) for $f \equiv 0$, will be considered. In this case the condition of solubility (129.17) is fulfilled, and there is the following result:

The homogeneous equation

$$\mathbf{K}^0\varphi = 0$$

has exactly λ *linearly independent solutions.* In particular, if $\lambda = 0$, i.e., if all the component indices are non-positive, the homogeneous equation has no solutions, different from zero.

Next consider a system of singular integral equations, similar to the system (129.1) and having in matrix notation the form

$$\mathbf{K}^{0\prime}\psi \equiv A'(t_0)\psi(t_0) - \frac{1}{\pi i}\int_L \frac{B'(t)\psi(t)dt}{t - t_0} = g(t_0); \tag{129.24}$$

here $A'(t)$, $B'(t)$ are matrices, given on L and satisfying the H condition such that the determinants of the matrices

$$S' = A' + B', \; D' = A' - B' \tag{129.25}$$

do not vanish anywhere on L [where in the sequel the main interest will be in the case where A' and B' are matrices, obtained by transposition from A and B of (129.4); however, for the present, A' and B' will be understood to be arbitrary matrices, satisfying the conditions stated above]; $g(t)$ is a vector, given on L and satisfying the H condition, and $\psi(t)$ is an unknown vector, which will also be subjected to the H condition.

The integral equation (129.24) is just as easily reduced to a certain Hilbert problem as is the equation (129.4), but a somewhat different method must be used.

For this purpose introduce the sectionally holomorphic vector

$$\Psi(z) = \frac{1}{\pi i}\int_L \frac{B'(t)\psi(t)dt}{t-z}, \tag{129.26}$$

vanishing at infinity. Taking into consideration the formulae

$$B'(t_0)\psi(t_0) = \tfrac{1}{2}[\Psi^+(t_0) - \Psi^-(t_0)],$$

$$\frac{1}{\pi i}\int_L \frac{B'(t)\psi(t)dt}{t-t_0} = \tfrac{1}{2}[\Psi^+(t_0) + \Psi(t_0)],$$

the conclusion is drawn that (129.24) is equivalent to the following problem: *To find a vector $\psi(t)$, defined on L and satisfying the H condition, and a sectionally holomorphic function $\Psi(z)$, vanishing at infinity, subject to the conditions*

$$2A'(t_0)\psi(t_0) = \Psi^+(t_0) + \Psi^-(t_0) + 2g(t_0),$$
$$2B'(t_0)\psi(t_0) = \Psi^+(t_0) - \Psi^-(t_0).$$

However, the last conditions, in their turn, are equivalent to the conditions (obtained from the above by addition and subtraction)

$$\begin{aligned} S'(t_0)\psi(t_0) &= \Psi^+(t_0) + g(t_0),\\ D'(t_0)\psi(t_0) &= \Psi^-(t_0) + g(t_0), \end{aligned} \tag{129.27}$$

or

$$\begin{aligned} \psi(t_0) &= S'^{-1}(t_0)\Psi^+(t_0) + S'^{-1}(t_0)g(t_0),\\ \psi(t_0) &= D'^{-1}(t_0)\Psi^-(t_0) + D'^{-1}(t_0)g(t_0). \end{aligned} \tag{129.28}$$

Comparing the right sides of the preceding equations, one arrives at the Hilbert problem

$$\Psi^+(t_0) = S'(t_0)D'^{-1}(t_0)\Psi^-(t_0) + [S'(t_0)D'^{-1}(t_0) - E]g(t_0), \tag{129.29}$$

where it is required to find a solution $\Psi(z)$, vanishing at infinity; having solved the problem (129.29), the solution of the original integral equation (129.24) will be found by the help of one of the formulae (129.28). Thus, the method of solution of (129.24) is essentially not different from that of (129.4).

Of particular interest here is the case where the matrices A' and B' of the equation (129.24) are the transposed matrices A and B of the equation (129.4).

In this case the operators $\mathbf{K}^0$ and $\mathbf{K}^{0\prime}$ will be called adjoint (a general definition of adjoint operators will be given in the next section); in correspondence with this the equations (systems of equations) (129.4), (129.24) will also be called adjoint, whatever may be the functions f and g on the right.

In the case considered the formula (129.29) assumes the form

$$\Psi^+(t_0) = G'^{-1}(t_0)\Psi^-(t_0) + [G'^{-1}(t_0) - E]g(t_0), \quad (129.30)$$

where G' is the matrix obtained by transposition of the matrix G in (129.9).

Thus *the homogeneous Hilbert problems*

$$\Phi^+ = G\Phi^-, \ \Psi^+ = G'^{-1}\Psi^-, \quad (129.31)$$

corresponding to the non-homogeneous problems (129.9) *and* (129.30), *are seen to be associate problems in the sense of the definition of* § 126.

It is known that, if $X(z)$ is a fundamental matrix corresponding to the first of the problems (129.31), then $[X'(z)]^{-1}$ will be a fundamental matrix, corresponding to the second problem. Further, since the principal difficulty of the solution of the Hilbert problem, homogeneous or non-homogeneous, lies in the determination of the corresponding fundamental matrix, it can be said that the problems of the solution of the adjoint integral equations (129.4) and (129.24) are equivalent problems.

In particular, the homogeneous equation

$$\mathbf{K}^{0\prime}\psi = 0,$$

adjoint to $\mathbf{K}^0\varphi = 0$, will be considered.

The Hilbert problem (129.30), corresponding to this, becomes the homogeneous problem $\Psi^+(t_0) = G'^{-1}(t_0)\Psi^-(t_0)$; further, since $[X'(z)]^{-1}$ is a fundamental matrix of solutions for this problem, its general solution is given by

$$\Psi(z) = [X'(z)]^{-1}Q(z), \quad (129.32)$$

where, in the former notation,

$$Q(z) = (Q_{-\varkappa_1-1}, \ldots, Q_{-\varkappa_n-1}).$$

However, the unknown general solution of the equation $\mathbf{K}^{0\prime}\psi = 0$ will be found by the help of either of the formulae (129.28) in which, of course, one has to take $g \equiv 0$; if the formula (129.12) is used, the general solution is obtained in the form

$$\psi(t) = [\mathbf{Z}'(t)]^{-1}\mathbf{Q}(t). \tag{129.33}$$

Now it will be seen that the μ vectors $\overset{\alpha}{\psi}(t)$, entering into the conditions of solubility (129.22) of the non-homogeneous equation $\mathbf{K}^0\varphi = f$, are *a complete system of linearly independent solutions of the adjoint homogeneous equation* $\mathbf{K}^{0\prime}\psi = 0$.

This fact is a particular case of Theorem I of § 131. Further, remembering that the number of linearly independent solutions of the homogeneous equation $\mathbf{K}^0\varphi = 0$ is λ and that $\lambda - \mu = \varkappa$, the following important result is obtained:

The difference of the numbers of linearly independent solutions of the adjoint homogeneous equations $\mathbf{K}^0\varphi = 0$ and $\mathbf{K}^{0\prime}\psi = 0$ *is equal to the total index of the equation* $\mathbf{K}^0\varphi = 0$.

This result is a particular case of Theorem III of § 131.

CHAPTER 19

SYSTEMS OF SINGULAR INTEGRAL EQUATIONS WITH CAUCHY TYPE KERNELS AND SOME SUPPLEMENTS

As has already been mentioned in the introduction to this Part, the generalization of the fundamental theorems proved in Part II to the case of a system of singular equations with Cauchy type kernels is easily done. This generalization is given in the two following sections (§§ 130, 131), being a reproduction without any essential alterations of the Author's paper [3] and its supplement [3a].

Of the work, done prior to the above-mentioned papers, that due to N. P. Vekua should be mentioned first (cf. § 132); next to it there is G. Giraud's paper [2], in which he gives the construction of a reducing operator coinciding essentially with that which will be given in § 130 [(130.20), (130.21)] (although somewhat less general) and proves a theorem coinciding with Theorem I of § 131. Giraud's paper [2] contains also a number of other results referring to the case of a single equation as well as to that of a system. These results concern mainly the dependence of the solution on the parameter λ; they were mentioned in § 59, 3°.

§ 130. Definitions. Auxiliary theorems. *)

In the sequel, if not stated to the contrary, L will mean the union of a finite number of one another not intersecting simple smooth contours.

Systems of singular integral equations of the following form will be considered:

$$\sum_{\beta=1}^{n} A_{\alpha\beta}(t_0)\varphi(t_0) + \frac{1}{\pi i}\int_L \sum_{\beta=1}^{n} \frac{K_{\alpha\beta}(t_0, t)\varphi_\beta(t)dt}{t-t_0} = f_\alpha(t_0),\ \alpha = 1, \ldots, n, \tag{130.1}$$

where t_0, t are the coordinates of points on L, $A_{\alpha\beta}(t)$, $K_{\alpha\beta}(t_0, t)$, $f_\alpha(t)$ are functions given on L and satisfying the H condition and the $\varphi_\alpha(t)$

*) cf. §§ 45, 46.

are unknown functions which will be required to satisfy the H condition.

By $\varphi(t)$ and $f(t)$ will be denoted the vectors with the components $\varphi_1(t), \varphi_2(t), \ldots, \varphi_n(t)$ and $f_1(t), f_2(t)., \ldots, f_n(t)$

$$\varphi = (\varphi_1, \varphi_2, \ldots, \varphi_n), \ f = (f_1, f_2, \ldots, f_n), \qquad (130.2)$$

by $A(t)$ and $K(t_0, t)$ the matrices with the elements $A_{\alpha\beta}(t)$ and $K_{\alpha\beta}(t_0, t)$; further, the matrix $B(t) = K(t, t)$ will be considered, so that, by definition,

$$A(t) = \| A_{\alpha\beta}(t) \|, \ B(t) = \| B_{\alpha\beta}(t) \|, \ K(t_0, t) = \| K_{\alpha\beta}(t_0, t) \|, \ (130.3)$$

where

$$B_{\alpha\beta}(t) = K_{\alpha\beta}(t, t). \qquad (130.4)$$

In this notation, the system (130.1) can be written

$$\mathbf{K}\varphi \equiv A(t_0)\varphi(t_0) + \frac{1}{\pi i}\int_L \frac{K(t_0, t)\varphi(t)dt}{t - t_0} = f(t_0) \qquad (130.5)$$

or

$$\mathbf{K}\varphi \equiv \mathbf{K}^0\varphi + \mathbf{k}\varphi = f(t_0), \qquad (130.6)$$

where

$$\mathbf{K}^0\varphi \equiv A(t_0)\varphi(t_0) + \frac{B(t_0)}{\pi i}\int_L \frac{\varphi(t)dt}{t - t_0}$$

and

$$\mathbf{k}\varphi \equiv \frac{1}{\pi i}\int_L k(t_0, t)\varphi(t)dt, \qquad (130.7)$$

with

$$k(t_0, t) = \frac{K(t_0, t) - K(t_0, t_0)}{t - t_0} = \frac{K(t_0, t) - B(t_0)}{t - t_0}. \qquad (130.8)$$

It is easily seen that

$$k(t_0, t) = \frac{k^*(t_0, t)}{|t - t_0|^\alpha}, \ 0 \leqq \alpha < 1, \qquad (130.9)$$

where $k^*(t_0, t)$ is a matrix satisfying the H condition (i.e., a matrix, the elements of which satisfy the H condition).

The operator $\mathbf{K}$ will be called a singular operator with Cauchy type kernel or, briefly, a singular operator. This operator transforms any vector $\varphi(t)$, satisfying the H condition, into a vector $\mathbf{K}\varphi$, likewise satisfying the H condition.

The operator $\mathbf{K}^0$ will be called the dominant part of the operator $\mathbf{K}$.

The equation $\mathbf{K}\varphi = f$ will be called a singular equation; this equation is equivalent to the system of equations (130.1) and in future, if convenient, the terms "equation" and "system of equations" will be used indiscriminately.

The equation $\mathbf{K}^0\varphi = f$ will be called the dominant equation, corresponding to the equation $\mathbf{K}\varphi = f$.

The matrices

$$S = A + B, \ D = A - B \tag{130.10}$$

will be called the principal matrices of the operator $\mathbf{K}$ or $\mathbf{K}^0$, and likewise of the equations $\mathbf{K}\varphi = f$ or $\mathbf{K}^0\varphi = f$; this concept has already been introduced in § 129 with regard to the operator $\mathbf{K}^0$.

If

$$\det S \neq 0, \ \det D \neq 0 \tag{130.11}$$

everywhere on L, the operator $\mathbf{K}$ or the equation $\mathbf{K}\varphi = f$ will be said to be of the regular type. In the sequel, only operators of the regular type will be considered.

In particular, if $B(t) = 0$, and hence $S = D$, then the equation $\mathbf{K}\varphi = f$ will be an ordinary (quasi-regular) system of Fredholm equations of the second kind. Therefore, in the case $S = D$, the operator $\mathbf{K}$ will be called a Fredholm operator.

The operator $\mathbf{K}$ and the operator $\mathbf{K}'$, defined by

$$\mathbf{K}'\psi \equiv A'(t_0)\psi(t_0) - \frac{1}{\pi i}\int_L \frac{K'(t, t_0)\psi(t)dt}{t - t_0}, \tag{130.12}$$

will be called adjoint operators; in (130.12) $A'(t_0)$ is the matrix obtained from $A(t_0)$ by transposition, and $K'(t, t_0)$ is the matrix, obtained by transposition of $K(t_0, t)$ and simultaneous interchange of the variables t and t_0; so that, if

$$A'(t) = \| A'_{\alpha\beta}(t) \|, \ K'(t, t_0) = \| K'_{\alpha\beta}(t, t_0) \|,$$

then

$$A'_{\alpha\beta}(t) = A_{\beta\alpha}(t), \ K'_{\alpha\beta}(t, t_0) = K_{\beta\alpha}(t, t_0).$$

Just as in § 46, the general formula

$$\int_L \psi\mathbf{K}\varphi dt = \int_L \varphi\mathbf{K}'\psi dt \tag{130.13}$$

is easily established, where $\varphi = \varphi(t)$, $\psi = \psi(t)$ are arbitrary vectors, satisfying the H condition.

Next, the question of multiplication of two singular operators will be considered. Let

$$\mathbf{K}_1\varphi \equiv A_1(t_0)\varphi(t_0) + \frac{1}{\pi i}\int_L \frac{K_1(t_0, t)\varphi(t)dt}{t - t_0}$$

and (130.14)

$$\mathbf{K}_2\psi \equiv A_2(t_0)\psi(t_0) + \frac{1}{\pi i}\int_L \frac{K_2(t_0, t)\psi(t)dt}{t - t_0}.$$

Denoting by $B_1, B_2, S_1, S_2, D_1, D_2$ the matrices, connected with the operators $\mathbf{K}_1$ and $\mathbf{K}_2$ in the same way as the matrices B, S, D with the operator $\mathbf{K}$, one easily obtains, as in § 45,

$$\mathbf{K}_1[\mathbf{K}_2\psi] \equiv \mathbf{K}^*\psi,$$

where $\mathbf{K}^*$ is a singular operator of the same type as $\mathbf{K}_1$ and $\mathbf{K}_2$, namely

$$\begin{aligned}\mathbf{K}^*\psi &\equiv [A_1(t_0)A_2(t_0) + B_1(t_0)B_2(t_0)]\psi(t_0) + \\ &+ \frac{1}{\pi i}\int_L \frac{[A_1(t_0)K_2(t_0, t) + K_1(t_0, t)A_2(t)]\psi(t)dt}{t - t_0} + \\ &+ \frac{1}{(\pi i)^2}\int_L \left[\int_L \frac{K_1(t_0, t_1)K_2(t_1, t)dt_1}{(t_1 - t_0)(t - t_1)}\right]\psi(t)dt.\end{aligned} \tag{130.15}$$

The only difference from the corresponding formula of § 45 is that here $A_1, B_1, K_1, A_2, B_2, K_2$ are matrices and that therefore the order of factors appearing in the above expression is not indifferent.

The operator $\mathbf{K}^*$, obtained by multiplication of $\mathbf{K}_1$ and $\mathbf{K}_2$ (in the order stated), will be denoted by $\mathbf{K}_1\mathbf{K}_2$:

$$\mathbf{K}^* = \mathbf{K}_1\mathbf{K}_2. \tag{130.16}$$

Denoting by S^* and D^* the principal matrices of the operator $\mathbf{K}^*$, one obtains, as in § 45,

$$S^* = S_1S_2, \quad D^* = D_1D_2. \tag{130.17}$$

It follows from these formulae that the operator $\mathbf{K}^*$ will be of the regular type, provided the operators $\mathbf{K}_1$ and $\mathbf{K}_2$ are so, and that the dominant part of $\mathbf{K}^*$ will be completely determined by the dominant parts of $\mathbf{K}_1$ and $\mathbf{K}_2$.

It also follows from above that, if the dominant parts of two of the three operators $\mathbf{K}_1$, $\mathbf{K}_2$, $\mathbf{K}^*$ are given, then the dominant part of the third will be uniquely determined. For example, if the dominant parts of $\mathbf{K}_2$ and $\mathbf{K}^*$ are given, one obtains for the principal matrices of the operator $\mathbf{K}_1$

$$S_1 = S^* S_2^{-1}, \; D_1 = D^* D_2^{-1}, \tag{130.18}$$

and the corresponding matrices A_1 and B_1 will be given by

$$A_1 = \tfrac{1}{2}[S^* S_2^{-1} + D^* D_2^{-1}], \; B_1 = \tfrac{1}{2}[S^* S_2^{-1} - D^* D_2^{-1}]. \tag{130.19}$$

In particular, *for a given operator* $\mathbf{K}_2$, *an operator* $\mathbf{K}_1$ *may be chosen in an infinite number of ways such that* $\mathbf{K}_1\mathbf{K}_2 = \mathbf{K}^*$ *will be a Fredholm operator.*

In fact, it is necessary and sufficient for this that $S^* = D^*$; hence

$$S_1 = S^* S_2^{-1}, \; D_1 = S^* D_2^{-1}, \tag{130.20}$$

$$A_1 = \tfrac{1}{2} S^* (S_2^{-1} + D_2^{-1}), \; B_1 = \tfrac{1}{2} S^* (S_2^{-1} - D_2^{-1}), \tag{130.21}$$

where S^* is an arbitrary matrix, satisfying the H condition on L such that $\det S^* \neq 0$ everywhere on L. In particular, $S^* = D^* = E$ may be taken, where E is the unit matrix; then *both operators* $\mathbf{K}_1\mathbf{K}_2$ *and* $\mathbf{K}_2\mathbf{K}_1$ *will be easily seen to be Fredholm operators.*

The integer

$$\varkappa = \frac{1}{2\pi i}\,[\log \det S^{-1} D]_L = \frac{1}{2\pi}\left[\arg \frac{\det (A - B)}{\det (A + B)}\right]_L \tag{130.22}$$

will be called the index of the singular operator $\mathbf{K}$ or the equation $\mathbf{K}\varphi = f$.

It follows from the above definition that the index depends only on the dominant part of $\mathbf{K}$; the concept of the index of singular operators was already introduced in § 129 and applied to the operator of the particular form $\mathbf{K}^0$.

It follows from (130.17) that, if $\varkappa'$ and $\varkappa''$ are the indices of $\mathbf{K}_1$ and $\mathbf{K}_2$, then the index $\varkappa^*$ of the operator $\mathbf{K}^* = \mathbf{K}_1\mathbf{K}_2$ equals the sum of $\varkappa'$ and $\varkappa''$:

$$\varkappa^* = \varkappa' + \varkappa''. \tag{130.23}$$

The index of a Fredholm operator is obviously equal to zero.

§ 131. Reduction of a system of singular equations. Fundamental theorems.

Let there be given a system of singular equations (130.1) which, as in the last section, will be written in the form of the single equation

$$\mathbf{K}\varphi = f. \tag{131.1}$$

It is known, from what has been said in the last section, that a singular operator **M** may always be chosen in an infinite number of ways such that the operator **MK** will be a Fredholm operator. An operator **M**, possessing this property, will be said to reduce **K**. From this it does not necessarily follow that also **KM** will be a Fredholm operator (however, as has been pointed out in the last section, **M** can be chosen such that both **MK** and **KM** will be Fredholm operators, but this is not essential for the results stated below. In the case of a single singular equation (§ 45), if **MK** is a Fredholm operator, then also **KM** has this property; however, in the present case when actually dealing with systems of equations, it has been seen that this does not always hold; therefore a difference may be made between operators, reducing a given operator from the left and from the right).

Let **M** be any reducing operator. Then all the solutions of (131.1) will be at the same time solutions of the Fredholm equation

$$\mathbf{MK}\varphi = \mathbf{M}f. \tag{131.2}$$

The converse conclusion is not always true and therefore the Fredholm equation (131.2) will not always be equivalent to the original singular equation. However, having solved (131.2), the general solution of the original equation (131.1) can always be found, just as in § 53.

In particular, it follows directly from the above that *the number of linearly independent solutions of the homogeneous equation* $\mathbf{K}\varphi = 0$ *is finite.*

Repeating word for word the reasoning given in § 53, the following fundamental theorems will be arrived at:

THEOREM I. *The necessary and sufficient conditions for the solubility of the equation*

$$\mathbf{K}\varphi = f$$

are

$$\int_L f(t)\overset{\alpha}{\psi}(t)dt = 0, \tag{131.3}$$

where $\overset{\alpha}{\psi}(t)$, $\alpha = 1, 2, \ldots, k'$, *is a complete system of linearly independent solutions of the homogeneous equation* $\mathbf{K}'\psi = 0$, *adjoint to the given equation.*

It will be remembered that in the present case

$$f = (f_1, f_2, \ldots, f_n), \quad \overset{\alpha}{\psi} = (\overset{\alpha}{\psi}_1, \overset{\alpha}{\psi}_2, \ldots, \overset{\alpha}{\psi}_n)$$

are vectors and that

$$f\overset{\alpha}{\psi} = f_1\overset{\alpha}{\psi}_1 + \ldots + f_n\overset{\alpha}{\psi}_n. \tag{131.4}$$

THEOREM II. *The difference between the number* k *of linearly independent solutions of the homogeneous equation* $\mathbf{K}\varphi = 0$ *and the number* k' *of linearly independent solutions of the adjoint homogeneous equation* $\mathbf{K}'\psi = 0$ *depends only on the dominant part of the operator* $\mathbf{K}$.

It is seen that these theorems are proved quite independently of the results, connected with the solution of the Hilbert problem.

However, if use is made of the results at the end of § 129 which are essentially based on the solution of the Hilbert problem, then the following important theorem is obtained:

THEOREM III. *The difference, entering into Theorem II above, is equal to the index* $\varkappa$ *of the operator* $\mathbf{K}$, i.e.,

$$k - k' = \varkappa. \tag{131.5}$$

It will be remembered that, by (130.22),

$$\varkappa = \frac{1}{2\pi i}\,[\log \det (A - B) - \log \det (A + B)]_L =$$

$$= \frac{1}{2\pi}\,[\arg \det (A - B) - \arg \det (A + B)]_L. \tag{131.6}$$

§ 132. Other methods of reduction and the investigation of systems of singular equations.

1°. The solutions of the dominant systems of singular equations and of their adjoint systems, obtained in § 129, lead to a method of reduction of systems of singular equations, quite similar to that stated in § 57 for the case of a single equation.

In particular, starting from these methods, the fundamental theorems proved in the last section by other means may be obtained. In fact, these theorems were first proved in this way (not considering Theorem I, first proved by G. Giraud) in a paper by N. P. Vekua which has only been used here for the general results. This paper, somewhat simplified and supplemented by the Author of this book (the most essential addition being the expression for the index $\varkappa$), was published as the combined paper: N. I. Muskhelishvili and N. P. Vekua [1]; part of it has been reproduced in Chapter 18.

The methods of reduction just referred to are in the general case less effective than the method stated in the last section, since they are linked with the Hilbert problem for several unknown functions, which cannot, generally speaking, be solved in closed form [in contrast to the case of one unknown function (cf. also 2° below)].

However, in a number of particular instances which are of practical interest, the Hilbert problem corresponding to the system of singular integral equations considered can be solved effectively and then the methods of reduction, just mentioned, are useful from the practical point of view.

This is, for example, true when the elements of the matrices A and B (or S and D) are rational functions.

Another (more general) case of effective solution was stated in the combined papers by N. P. Vekua and D. A. Kveselava [1], [2].

2°. The method of investigation of a singular equation, stated in § 55 and based on the equivalence theorem by I. N. Vekua, may likewise be generalized to a system of singular equations. This has been done in the interesting paper [2] by N. P. Vekua. In particular, it is to be noted that the reduction of a system of singular equations to an equivalent system of Fredholm equations may be carried out without actually solving the corresponding Hilbert problem, which makes this method of reduction an efficient one.

§ 133. Brief remarks regarding important generalizations and supplements.

In spite of the fact that the theory of systems of singular integral equations has only quite recently reached the closed form in which it has been presented here, there is already a series of important generalizations and supplements to this theory.

1°. A theory of the Hilbert problem for several unknown functions and a theory of systems of singular equations in the case of discontinuous coefficients has been worked out in the two papers [3] and [4] by N. P. Vekua.

2°. A solution of the problem which is itself a generalization of the problem V of § 70, 71 for the case of several unknown functions was given in the paper [4] by B. V. Khvedelidze.

3°. A solution of the same problem, as in 2°, but extended to the case of discontinuous coefficients, was given in the paper [5] by N. P. Vekua.

4°. In § 76, 3° a remark was made with regard to the Poin… problem for an equation of the elliptic type and to a solution of th… problem by B. V. Khvedelidze.

A generalization to the case of a system of equations of the elliptic type, having the form

$$\Delta u_j + \sum_{k=1}^{n}\left[A_{jk}(x, y)\frac{\partial u_k}{\partial x} + B_{jk}(x, y)\frac{\partial u_k}{\partial y} + C_{jk}(x, y)u_k\right] = 0,$$

$$j = 1, 2, \ldots, n,$$

where the $A_{jk}(x, y)$, $B_{jk}(x, y)$, $C_{jk}(x, y)$ are integral functions of the two arguments, was given in the dissertation by A. V. Bitsadze, a brief extract of which was published in his paper [2].

5°. A solution of a system of integro-differential equations of the form

$$A(t_0)\varphi(t_0) + B(t_0)\dot{\varphi}(t_0) + \frac{1}{\pi i}\int_L \frac{P(t_0, t)\varphi(t) + Q(t_0, t)\dot{\varphi}(t)}{t - t_0}\,dt = f(t_0),$$

where A, B, P, Q, are square matrices of the order n, given on L, f is a vector (with n components) given on L, φ is the unknown vector (with n components) and $\dot{\varphi}(t) = d\varphi(t)/dt$, was obtained by L. G. Magnaradze; a short statement of his results was given in his paper [2].

APPENDICES

APPENDIX 1

ON SMOOTH AND PIECEWISE SMOOTH LINES

1°. A definition of a smooth arc or contour L has been given in § 1; the notation of that section will be used here. In particular, the parametric representation of a line L will be written in the form

$$x = \varphi(s), \quad y = \psi(s), \quad s_a \leqq s \leqq s_b, \tag{1}$$

where s is the arc coordinate. The length of the line L will now be denoted by l, so that

$$l = s_b - s_a. \tag{2}$$

Let t_1 and t_2 be any two points on L. Denote by $\sigma(t_1, t_2) = \sigma$ the length of that part of L, included between t_1 and t_2; if L is a contour, the shorter segment will be considered. Thus, for a contour, $0 \leqq \sigma \leqq l/2$, and for an arc, $0 \leqq \sigma \leqq l$. Further, denote by $r(t_1, t_2) = r$ the distance between the points t_1 and t_2. Obviously, r will be a continuous function of the arc coordinates s_1 and s_2 of the points t_1 and t_2. Consider all pairs of points t_1, t_2 such that

$$\sigma(t_1, t_2) \geqq \lambda, \quad 0 < \lambda < l/2, \tag{*}$$

where λ is an arbitrary fixed number in the indicated interval, and put

$$\varrho = \varrho(\lambda) = \min r(t_1, t_2) \tag{**}$$

under the assumption that (*) holds true. This minimum value will be attained for at least one pair of points t_1, t_2. It is easily seen that $\varrho(\lambda) > 0$. In fact, if $\varrho(\lambda)$ would be equal to zero, then the line L would intersect itself, which contradicts the assumption.

Thus a number $\varrho = \varrho(\lambda) > 0$, possessing the following property, corresponds to each λ such that $0 < \lambda < l/2$: if a circle of radius $\varrho_0 < \varrho$ is described about any point t_0 on L, then all the points t of the contour L such that $\sigma(t_0, t) \geqq \lambda$ will lie outside this circle.

2°. Let α_0 be an arbitrary given angle: $0 < \alpha_0 < \dfrac{\pi}{2}$. It follows from the continuity of the change of direction of the tangent to L

that a number $\sigma_0 = \sigma_0(\alpha_0) > 0$ exists, depending only on α_0 and possessing the following property: the angle between the tangents to L at the points t_1 and t_2 does not exceed $\alpha_0/2$, provided only $\sigma(t_1, t_2) \leqq \sigma_0$. In the sequel, it will be assumed that $\sigma_0 < l/2$.

Consider the arc t_1t_2, belonging to L and not exceeding σ_0 in length. It follows from the above that the acute angle between the chord, connecting any two points τ_1, τ_2 of the arc t_1t_2, and the tangent at t_1 (or t_2) does not exceed $\alpha_0/2$; in fact, a point can always be found on an arc $\tau_1\tau_2$ at which the tangent will be parallel to the chord under consideration.

3°. Let t_0 be any fixed point on L. Consider an arc L_0 consisting of those points t of L for which $\sigma_0(t_0, t) \leqq \sigma_0$, where σ_0 is the same as above. Without loss of generality, it can be assumed that in the case where L is a contour the points, corresponding to $s = s_a$ (or s_b) in (1), do not belong to L_0. The point t_0 will divide L_0 into two parts, corresponding to $s > s_0$ and $s < s_0$; an exception will only be made by the case where L is an arc and t_0 coincides with one of its ends. Examine the variation of the distance $r = r(t_0, t)$ for motion of t along L. It is easily seen that

$$\frac{dr}{ds} = \pm \cos \alpha,$$

where α is the acute angle between the chord t_0t and the tangent at the point t; the upper sign refers to the part $s > s_0$, the lower to $s < s_0$. Thus the distance r is seen to be a monotonic function of s on each of these parts, because $\cos \alpha \geqq \cos \alpha_0/2 = k_0$, $0 < k_0 < 1$. On both parts

$$k_0 \mid s - s_0 \mid \leqq r(t_0, t) \leqq \mid s - s_0 \mid. \qquad (***)$$

Now describe about t_0 a circle Γ of radius $R \leqq R_0$, where R_0 is the smaller of the numbers $\varrho(\sigma_0)$ and $k_0 \cdot \sigma_0$; here $\varrho(\sigma_0)$ is the same as $\varrho(\lambda)$ in 1° for $\lambda = \sigma_0$.

It is easily shown that Γ intersects L in exactly two points, except when L is an arc and the distance from t_0 to the nearest end is smaller than R; then Γ intersects L in exactly one point.

In fact, let L first be a contour. When s increases from s_0 to $s_0+\sigma_0$, the distance $r(t_0, t)$ will increase monotonically from 0 to $r_1 \geqq k_0\sigma_0 \geqq R_0$; consequently, the point t will cross the circle Γ exactly once; the same will occur when s decreases from s_0 to $s_0 - \sigma_0$. Other points of intersection between L and Γ will not exist by 1°.

The case of an arc will be considered similarly.

It will be noted that the inequality

$$k_0\sigma(t_1, t_2) \leqq r(t_1, t_2) \leqq \sigma(t_1, t_2), \; 0 < k_0 < 1 \qquad (3)$$

follows for any pair of points on L from the inequality (***), which holds for sufficiently small $\sigma(t_0, t)$; k_0 is a constant for the interval, i.e., a quantity which does not depend on the positions of t_1 and t_2 on L.

5°. As has already been stated at the end of § 1, a simple and continuous arc or contour consisting of a finite number of smooth arcs with contiguous ends will be called a piecewise smooth arc or contour. A piecewise smooth arc is also representable in the form (1), but the derivatives $\varphi'(s)$ and $\psi'(s)$ will have first order discontinuities at the points of contact of the different smooth parts, i.e., at the corner points. If for the passage through a corner point the tangent alters its direction by π, then such a point will be called a cusp.

It will be noted that the inequality (3) will also hold true in the neighbourhood of corner points, other than cusps. Actually, it is sufficient to verify this only for the case where t_1 and t_2 are in the neighbourhood of a corner point c and on different sides of it. An investigation of the triangle ct_1t_2 (Fig. 13) shows that, if r_1 and r_2 are the distances of the point c from t_1 and t_2, then

$$(r_1 + r_2) \sin \frac{\gamma}{2} \leqq r(t_1, t_2) \leqq r_1 + r_2,$$

where γ is the angle at the vertex c. For t_1, t_2, sufficiently close to c, $\sin \gamma/2 \geqq k_0'$, where k_0' is a positive constant. Besides, by application of the inequality (3) to the smooth arcs ct_1 and ct_2, one finds $k_0''(\sigma_1 + \sigma_2) \leqq r_1 + r_2 \leqq \sigma_1 + \sigma_2$, where σ_1 and σ_2 are the lengths of the arcs ct_1 and ct_2, so that $\sigma(t_1, t_2) = \sigma_1 + \sigma_2$, and k_0'' is a constant such that $0 < k_0'' < 1$. Hence the statement follows.

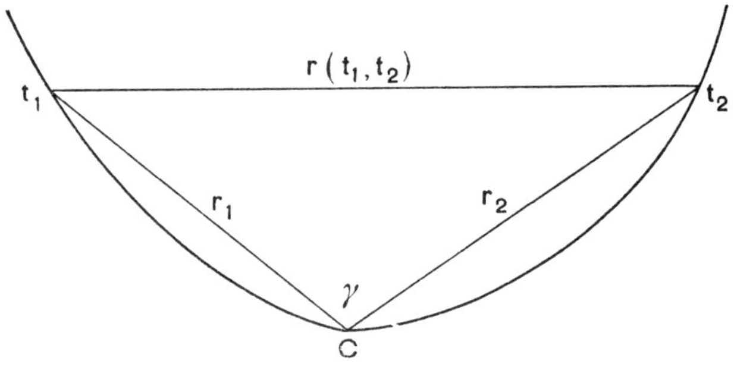

Fig. 13

APPENDIX 2

ON THE BEHAVIOUR OF THE CAUCHY INTEGRAL NEAR CORNER POINTS

1°. Let $\varphi(t)$ be a function, given on a piecewise smooth line L. For the neighbourhoods of corner points, other than cusps, the definition of the $H(\mu)$ condition, given by (3.1), will be retained, i.e.,

$$|\varphi(t_2)-\varphi(t_1)| \leqq A\, r_{12}^{\mu} = A\,|t_2-t_1|^{\mu}, \tag{1}$$

which in this case is equivalent, by what has been said in 5° of Appendix 1, to the definition (3.2), i.e.,

$$|\varphi(t_2)-\varphi(t_1)| \leqq A\, \sigma_{12}^{\mu}; \tag{2}$$

here r_{12} and σ_{12} are the same as $r(t_1, t_2)$ and $\sigma(t_1, t_2)$ of Appendix 1.

However, two different definitions will be given for the neighbourhoods of cusp points. In fact, $\varphi(t)$ will be said to satisfy the $H(\mu)$ condition in the strong form, if (1) holds true, and in the weak form, if (2) holds true.

Since in the neighbourhood of a cusp point r_{12}/σ_{12} is smaller than unity and can be taken as small as one likes, these two conditions are not equivalent; if the condition $H(\mu)$ holds true in the strong form, then the condition $H(\mu)$ also holds true in the weak form, but not conversely.

In the sequel, if it is not clearly stated to the contrary, the condition $H(\mu)$ will always be understood in the weak form, i.e., in the sense of (2).

2°. In § 12, the formula

$$\Phi(t_0) = \frac{1}{2\pi i}\int_L \frac{\varphi(t)dt}{t-t_0} = \tfrac{1}{2}\varphi(t_0) + \frac{\varphi(t_0)}{2\pi i}\log\frac{b-t_0}{a-t_0} + \\ + \frac{1}{2\pi i}\int_L \frac{\varphi(t)-\varphi(t_0)}{t-t_0}\,dt \tag{3}$$

has been introduced for the case in which $L = ab$ is a smooth arc.

This formula obviously also remains in force when L is a piecewise smooth arc and the point t_0 is not a corner point (or end); here and later, $\varphi(t)$ will be assumed to satisfy the H condition on L.

Reasoning just as in § 12, it is easily seen that, if t_0 is a corner point, then (3) will be replaced by

$$\begin{aligned}\Phi(t_0) = \frac{1}{2\pi i}\int_L \frac{\varphi(t)dt}{t-t_0} = \frac{\alpha}{2\pi}\varphi(t_0) + \frac{\varphi(t_0)}{2\pi i}\log\frac{b-t_0}{a-t_0} + \\ + \frac{1}{2\pi i}\int_L \frac{\varphi(t)-\varphi(t_0)}{t-t_0}\,dt,\end{aligned} \tag{4}$$

where α is the positive angle $(0 \leqq \alpha \leqq 2\pi)$ by which an infinitely small vector $\overrightarrow{t_0 t}$ rotates when the point t, remaining to the left of L and revolving about the point t_0, passes from the part $t_0 b$ to the part at_0 (Fig. 14). In the case of an ordinary (not corner) point $\alpha = \pi$, and the formula (3) is again obtained.

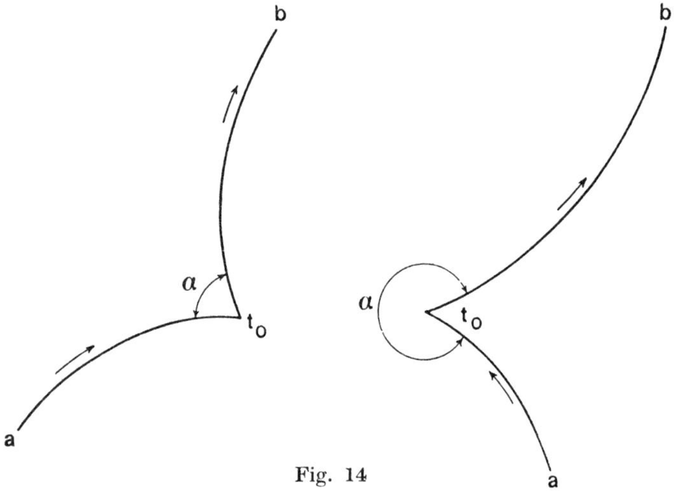

Fig. 14

3°. In § 16, the formulae (16.4), due to Morera, were introduced, where it was assumed that L is a simple smooth contour. Obviously, these formulae will also hold true when L is a simple piecewise smooth contour and t_0 is not a corner point.

It will be shown now that these formulae (and likewise the result regarding the continuity of $\Phi(z)$ on L from the left and from the right) also remain true, when t_0 is a corner point, or, in particular, a cusp.

Let c be one of the corner points. If $\varphi(c) = 0$, then the result becomes obvious, if the integral

$$\Phi(z) = \frac{1}{2\pi i}\int_L \frac{\varphi(t)dt}{t-z} \tag{5}$$

is divided into two integrals, one of which extends from c to some point c' on L, while the other extends from c' to c. Likewise, it is obvious that in this case

$$\Phi^+(c) = \Phi^-(c) = \frac{1}{2\pi i}\int_L \frac{\varphi(t)dt}{t-c}. \tag{6}$$

Now let $\varphi(c) \neq 0$. Consider the integral

$$\Phi_0(z) = \frac{1}{2\pi i}\int_L \frac{\varphi_0(t)dt}{t-z},$$

where $\varphi_0(t) = \varphi(t) - \varphi(c)$. From what has been said above, $\Phi_0(z)$ will be continuous on L from the right and from the left, and the value $\Phi_0^{\pm}(c)$ will be determined by

$$\Phi_0^{\pm}(c) = \Phi_0^{+}(c) = \Phi_0^{-}(c) = \frac{1}{2\pi i}\int_L \frac{\varphi_0(t)dt}{t-c}.$$

But, obviously,

$$\begin{aligned} \Phi(z) &= \varphi(c) + \Phi_0(z) && \text{for } z \in S^+ \\ \Phi(z) &= \Phi_0(z) && \text{for } z \in S^-, \end{aligned} \tag{7}$$

where, as in § 16, S^+ and S^- are to be understood as the finite and infinite parts of the plane, bounded by the contour L, the positive direction of which leaves S^+ on the left. Hence the proposition follows directly.

4°. From the results of 2°, 3° above it is easily seen that in the case, where t_0 is a corner point, the Plemelj formulae (17.2) take the following form:

$$\Phi^+(t_0) = \left(1 - \frac{\alpha}{2\pi}\right)\varphi(t_0) + \frac{1}{2\pi i}\int_L \frac{\varphi(t)dt}{t-t_0}, \tag{8}$$

$$\Phi^-(t_0) = \quad -\frac{\alpha}{2\pi}\varphi(t_0) + \frac{1}{2\pi i}\int_L \frac{\varphi(t)dt}{t-t_0}, \tag{9}$$

where L is an arbitrary piecewise smooth line (not intersecting itself) and α is the angle defined in 2° above.

5°. In § 19, the Plemelj-Privalov Theorem was proved under the assumption that L is a smooth line. It is easily seen that the neighbourhoods of corner points other than cusps do not represent exceptions either in the results or in the proof, given in § 19. This involves again a study of the integral

$$\Psi(t_0) = \frac{1}{2\pi i} \int_L \frac{\varphi(t) - \varphi(t_0)}{t - t_0} dt,$$

of which one is most simply convinced as follows (this method could also have been used in § 19). Without affecting generality, it may be assumed that L is a contour; then, by Morera's formulae,

$$\Phi^+(t_0) = \varphi(t_0) + \Psi(t_0), \quad \Phi^-(t_0) = \Psi(t_0).$$

The investigation of the integral $\Psi(t_0)$, however, may be pursued just as in § 19.

Consideration has still to be given to the neighbourhoods of cusp points. It will be shown that *the theorem, stated in* § 19, *also remains true in the case where* L *is any simple piecewise smooth line (which may have cusps) and where by the* $H(\mu)$ *condition both its weak and its strong form may be understood.*

Simultaneously with this also the theorem of § 22 will be extended to the case of a region bounded by piecewise smooth contours.

In connection with the latter theorem, the following remark will be made. Let ab be a smooth arc and let $\varphi(t)$ satisfy on ab the $H(\mu)$ condition, $\mu < 1$, while $\varphi(a) = 0$. Extend the arc ab beyond a by a segment $a'a$ of the tangent at a; put $\varphi(t) = 0$ on $a'a$ and consider the function

$$\Phi(z) = \frac{1}{2\pi i} \int_{ab} \frac{\varphi(t)dt}{t - z} = \frac{1}{2\pi i} \int_{a'b} \frac{\varphi(t)dt}{t - z}. \tag{10}$$

From the results of § 22 it is easily seen that $\Phi(z)$ satisfies the inequality

$$| \Phi(z_2) - \Phi(z_1) | \leqq C \, | z_2 - z_1 |^{\mu} \tag{11}$$

in the neighbourhood of the arc ab (except possibly near the point b) from the left (right) of $a'b$. This inequality will also hold true for points lying on the arc $a'b$ itself (except possibly near the point b), if by $\Phi(z)$ for z on $a'b$ is understood the boundary value from the left or from the right.

In order to see this, it is sufficient to assume that the arc $a'b$ is completed into any contour and that $\varphi(t) = 0$ on the added portion. (Fig. 15).

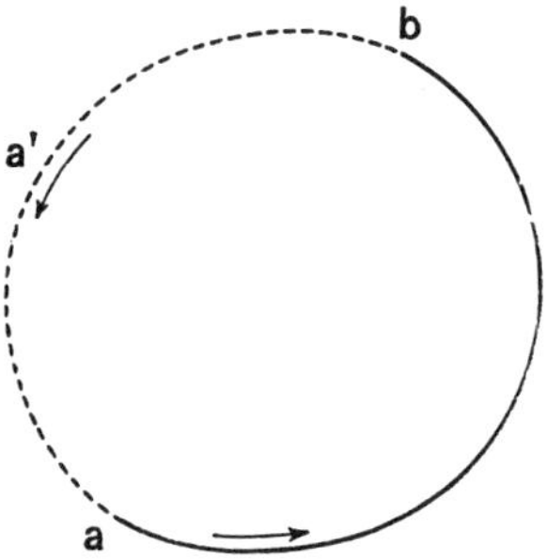

Fig. 15

Now let L be any piecewise smooth contour. Since, obviously, the whole problem reduces to the investigation of the behaviour of $\Phi(z)$ of the formula (5) in a sufficiently small neighbourhood of the corner point, it may be assumed, without affecting generality, that L is a simple contour having a single corner point a. Besides, again without loss of generality, it may be assumed that the angle at a is re-entrant, i.e., that the angle α, shown in Fig. 16, is larger than π; for $\alpha = 2\pi$, this will give a cusp.

Let $\varphi(t)$ satisfy on L the $H(\mu)$ condition (in the weak form). Without loss of generality, it may be assumed that $\varphi(a) = 0$, because in the contrary case one can replace the examination of the integral $\Phi(z)$ by that of the integral

$$\frac{1}{2\pi i}\int\limits_L \frac{\varphi(t)-\varphi(a)}{t-z}\,dt = \begin{cases}\Phi(z)-\varphi(a) & \text{in } S^+,\\ \Phi(z) & \text{in } S^-.\end{cases} \tag{12}$$

Now divide the contour L by the point a and any other point b into the two arcs L' and L''; corresponding to this, the integral $\Phi(z)$ will appear as the sum of two integrals

$$\Phi(z) = \Phi_1(z) + \Phi_2(z), \tag{13}$$

taken over L' and over L'' respectively.

Applying to each of these integrals what has been stated above regarding the behaviour of the integral (10), it is easily concluded that the inequality (11) will hold in S^-, near a, and hence in S^-; in order that the above may become obvious, it is sufficient to extend the smooth arcs, touching at a, by segments of the tangents aa' and aa'' (in the case of a cusp these segments will coincide) and to note that the neighbourhood of S^-, near the point a, lies on one and the same side with respect to the smooth arc baa' as well as with respect to the arc $a''ab$.

In particular, bringing the points z_1 and z_2 of (11) closer to the points t_1 and t_2 of the boundary L, it is seen that the boundary value $\Phi^-(t)$, i.e., from the side of the "indentation", satisfies the $H(\mu)$ condition in the strong form, even if $\varphi(t)$ satisfies the H condition only in the weak form (where it is remembered that the difference between the strong and the weak form of the $H(\mu)$ condition holds true only in the neighbourhood of a cusp).

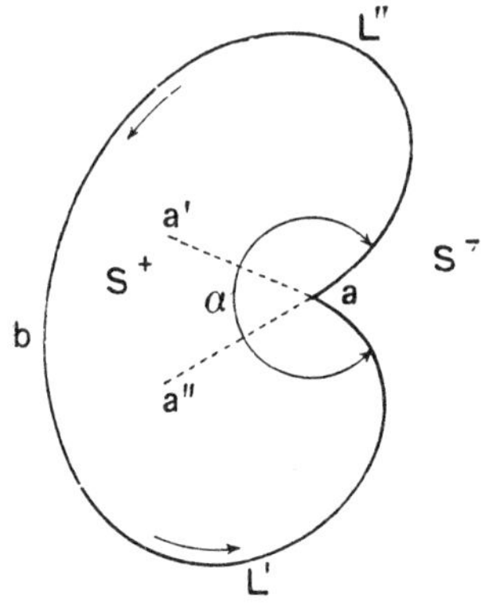

Fig. 16

There remains to consider the behaviour of $\Phi(z)$ in the region S^+, i.e., on the side of the cusp. One will have

$$\Phi^+(t) = \Phi^-(t) + \varphi(t).$$

Since $\varphi(t)$ and $\Phi^-(t)$ satisfy the $H(\mu)$ condition, the same will be true for $\Phi^+(t)$.

Thus the theorem of § 19 (Plemelj-Privalov) *has now been extended to any piecewise smooth line, where cusps are not excluded.*

It will be noted that, if $\varphi(t)$ satisfies the $H(\mu)$ condition in the strong form, then the same will be true for $\Phi^+(t)$, because $\Phi^-(t)$ will satisfy the $H(\mu)$ condition in the strong form.

6°. In 5° the theorem of § 22 was, in passing, extended to the case in which the points z_1 and z_2 lie on the indentation of the neighbourhood of the corner point a of the boundary. There remains the case to be considered, where z_1 and z_2 lie on the side of the cusp, i.e., in the terminology of 5°, in S^+.

It is easily seen that in this case it is insufficient to subject $\varphi(t)$ to the $H(\mu)$ condition in the weak form. Therefore it will now be assumed that $\varphi(t)$ satisfies the $H(\mu)$ condition in the strong form. Then, as has just been shown, the boundary value $\Phi^+(t)$ will satisfy the same condition. After this it is easily seen that the reasoning, introduced in § 22, may also be applied in the present case without any essential alterations. For the consideration of the relation (22.6) only the following has to be kept in mind: firstly, each point z_0 of the region S^+, lying near a, may be connected by a straight line, not intersecting L, either with a or with some point on L near a. In both cases the estimate $|\Phi(z) - \Phi(z_0)| \leqq C |z - z_0|^\mu$ will hold true for each point z on the above lines, as follows from the representation of $\Phi(z)$ in the form (13) or from formula (11).

APPENDIX 3

AN ELEMENTARY PROPOSITION REGARDING BI-ORTHOGONAL SYSTEMS OF FUNCTIONS

Let L be a smooth line in the plane of the complex variable $z = x + iy$ *and let*

$$\varphi_1(t), \ \varphi_2(t), \ \ldots, \ \varphi_n(t)$$

be any system of linearly independent continuous functions of the point $t = x + iy$ on L.

Then it is always possible to select (in an infinite number of ways) *systems of n functions* $\omega_1(t), \ldots, \omega_n(t)$, *satisfying the H condition on L, bi-orthogonal to the system* $\varphi_1(t), \ldots, \varphi_n(t)$ *in the sense that*

$$(\varphi_i\omega_j) = \int_L \varphi_i\omega_j dt = \delta_{ij} = \begin{cases} 1 \text{ for } i = j, \\ 0 \text{ for } i \neq j. \end{cases} \qquad (1)$$

Firstly it will be shown that linear combinations $\psi_i (i = 1, 2, \ldots, n)$ of the functions φ_j exist such that functions χ_j may be selected for them, satisfying the H condition on L, so that $(\psi_i\chi_j) = \delta_{ij}$.

If the requirement that the χ_j should satisfy the H condition is omitted, then the systems φ_i and χ_j may be constructed as follows: orthogonalize in the usual manner the functions φ_i with respect to the arc coordinate s, i.e., form their linear combinations such that

$$\int_L \psi_i \overline{\psi}_j ds = \delta_{ij};$$

then ψ_i and $\chi_j = \overline{\psi}_j \Big/ \dfrac{dt}{ds}$ will satisfy the condition $(\psi_i \chi_j) = \delta_{ij}$.

It will be assumed with regard to the functions, denoted below by ω_j, χ_j, χ'_j, that they satisfy the H condition.

Denote φ_1 by ψ_1 and take any function χ_1 such that $(\psi_1\chi_1) \neq 0$ (where the function $\psi_1 = \varphi_1$ is continuous and not identically zero, since the functions $\varphi_1, \varphi_2, \ldots, \varphi_n$ are linearly independent); there will obviously exist an infinity of such functions. Multiplying χ_1 by a suitable constant, it may obviously be assumed that $(\psi_1\chi_1) = 1$.

Replace the function φ_2 by the function $\psi_2 = \varphi_2 - c\psi_1$, select the constant c such that $(\psi_2\chi_1) = (\varphi_2\chi_1) - c(\psi_1\chi_1) = (\varphi_2\chi_1) - c = 0$. Then $(\psi_1\chi_1) = 1$, $(\psi_2\chi_1) = 0$. Let χ_2' be any function such that $(\psi_2\chi_2') = 1$ (since the functions of the system $\psi_1, \psi_2, \varphi_3, \ldots, \varphi_n$ will be linearly independent, ψ_2 will not be identically zero). Replace the function χ_2' by $\chi_2 = \chi_2' - c\chi_1$, selecting the constant c so that $(\psi_1\chi_2) = (\psi_1\chi_2') - c(\psi_1\chi_1) = (\psi_1\chi_2') - c = 0$. Now there will be the functions $\psi_1, \psi_2, \chi_1, \chi_2$ such that $(\psi_i\chi_j) = \delta_{ij}$ $(i, j = 1, 2)$, while the functions $\psi_1, \psi_2, \varphi_3, \ldots, \varphi_n$ are linearly independent.

Further, replace the function φ_3 by $\psi_3 = \varphi_3 - c_1\psi_1 - c_2\psi_2$, selecting c_1, c_2, so that $(\psi_3\chi_1) = 0$, $(\psi_3\chi_2) = 0$, i.e., that $(\varphi_3\chi_1) - c_1 = 0$, $(\varphi_3\chi_2) - c_2 = 0$. Choose χ_3' such that $(\psi_3\chi_3') = 1$ and replace χ_3' by $\chi_3 = \chi_3' - c_1\chi_1 - c_2\chi_2$ so that $(\psi_1\chi_3) = 0$, $(\psi_2\chi_3) = 0$, etc.

Continuing in this way, n linearly independent functions ψ_i will be obtained, representing linear combinations of the functions φ_i, and n functions χ_j, such that

$$(\psi_i\chi_j) = \delta_{ij}. \tag{2}$$

Since the ψ_j are linearly independent combinations of the φ_i, then, conversely,

$$\varphi_i = \sum_{k=1}^{n} a_{ik}\psi_k,$$

where the a_{ik} are constants such that the determinant of the matrix $A = \| a_{ij} \|$ is not zero.

Now select the constants b_{ij} so that the functions

$$\omega_j + \sum_{l=1}^{n} b_{lj}\chi_l$$

satisfy the condition (1). Expressing this condition, one obtains

$$\delta_{ij} = (\varphi_i\omega_j) = \sum_k \sum_l a_{ik}b_{jl}(\psi_k\chi_l) = \sum_k \sum_l a_{ik}b_{lj}\delta_{kl} = \sum_k a_{ik}b_{kj},$$

and hence it follows that

$$AB = E,$$

where E is the unit matrix and $B = \| b_{ij} \|$. Consequently, the required quantities b_{ij} exist, and, in fact, $B = A^{-1}$. Thus the proposition is proved.

One direct consequence of the statement proved will now be given. *Let $\varphi_1, \varphi_2, \ldots, \varphi_n$ be given, linearly independent, continuous functions of t and let it be known that for any function ω, satisfying* *the H condition on L, there necessarily follows from the rela tion*

$$(\varphi_i\omega) = 0, \ i = 1, 2, \ldots, n$$

that $(\varphi_0\omega) = 0$, *where* φ_0 *is some definite function, continuous on L. Then* φ_0 *is a linear combination of the functions* φ_i.

In fact, if the function φ_0 were linearly independent of $\varphi_1, \varphi_2, \ldots, \varphi_n$, then the functions $\omega_0, \omega_1, \ldots, \omega_n$ would exist such that $(\varphi_i\omega_j) = \delta_{ij}$ $(i, j = 0, 1, \ldots, n)$; in particular,

$$(\varphi_1\omega_0) = 0, \ (\varphi_2\omega_0) = 0, \ \ldots, \ (\varphi_n\omega_0) = 0, \ (\varphi_0\omega_0) = 1,$$

which contradicts the condition of the statement.

Note. The proposition stated at the beginning of this appendix may be considerably generalized, extending the class of functions φ_j and, in contrast, narrowing the class of the functions ω_j.

For example, the statement will remain true, if it is assumed that the functions φ_i may have discontinuities at a finite number of points, remaining absolutely integrable; for this one has to assume identical functions, differing from one another only at the points of discontinuity.

This last condition is essential for the definition of the concept of linear independence; on the strength of this condition, the functions $\varphi_1, \varphi_2, \ldots, \varphi_n$ will be considered to be linearly dependent, if constants $c_1, c_2, \ldots, c_n$, not all zero, exist such that $c_1\varphi_1 + c_2\varphi_2 + \ldots + c_n\varphi_n = 0$ everywhere on L, except possibly at the points of discontinuity.

Further, it is not difficult to prove that, for example, rational functions may be taken in the capacity of the functions $\omega_j(t)$; however, if the line of integration consists only of arcs, then even polynomials may be taken as the functions $\omega_j(t)$. The fact that in the case of contours it is impossible to manage with polynomials alone is obvious: if, for example, the functions φ_i are themselves polynomials, then there would not be any polynomials ω_j such that $(\varphi_i\omega_j) = 0$, and, in particular, $(\varphi_i\omega_i) = 0$, and not 1, as it is required. No time will be spent on this, since it will not be used here.

REFERENCES [1]

The following journals will be referred to by mentioning the relevant number

1 *Doklady*, A.N., S.S.S.R.
(Comptes rendus de l'académie des sciences de l'U.R.S.S.)
2 *Comptes rendus*, Paris.
3 *Bull. des Sciences Math.* 2-e sér.
4 *Ann. de l'Ecole.* Norm. Sup.
5 *Soobshcheniya* A N Gruz. SSR.
6 *Doklady* A N Gruz. SSR.
7 *Arkiv för matematik astronomi och fysik.*
8 *Rendiconti det R. Istituto Lombardo, cl. di. sc. mat. e nat.*
9 *Zeitschr. für Angew. Math. u. Mech.*
10 *Mat. sbornik*, nov. ser.
11 *Izv. Phiziko-matem. obshch. i natushno-issled.* inst. mat. i mech. pri Kazanskom Universitete
12 *Prikladnaya mat. i mech.*
13 *Berichte d. k. Sächs. Ges. d. Wiss. Math.-Phys. Klasse*
14 *Math. Ann.*
15 *Trans. of the American Math Soc.*
16 *Trudy Tbilissk.* Mat. Inst.
17 *Izv. Ross.* Akad. Nauk.
18 *Monatshefte für Math. u. Phys.*
19 *Math. Zeitschr.*
20 *Sitzungsbericht der Akad. d. Wiss. in Wien.* Math.-Nat. Klasse. Abt. II-a.
21 *Sitzungsberichte der Preuss. Akad. d. Wiss.* Phys.-nat. Klasse.
22 *Abhandl. d. Preuss. Akad. d. Wiss.* Math. naturwiss. Klasse.
23 *Trudy Seismolog.* Inst. A.N. SSSR.
24 *Annali di. Mat.*
25 *Annals of Math.*
26 *Duke Math. Jour.*
27 *Publications of the Central Aero-hydro-dynamic Institute of Moscow.*

---

[1] With a few exceptions, work published since spring 1944 has not been included in the list, since the Author had no opportunity to make use of it.

REFERENCES AND AUTHORS INDEX

(Italic type numbers refer to pages of this book)

ABRAMOV, V. M. Абрамов, В. М. [1] The problem of the contact of an elastic half-plane with an absolutely rigid base for the determination of the frictional force. **1.** Vol. XVII no. 4 (1937) pp. 173—178. *287.*

BEGIASHVILI, A. I. Бегиашвили, А. И. [1] Solution of the problem of the pressure of a system of rigid profiles on the straight boundary of an elastic half-plane. **1.** Vol. XXVII no. 9 (1940) pp. 914—916. *292.*

BERTRAND, G. [1] Equations de Fredholm à intégrales principales au sens de Cauchy. **2.** t. 172, (1921) p. 1458—1461. *60.*

——— [2] Le problème de Dirichlet et le potentiel de simple couche. **3.** t. XLVII (1923) pp. 282—288 et 298—307. *33, 60, 185.*

——— [3] La théorie des marées et les équations intégrales. **4.** t. XL. (1923) pp. 151—258. *224.*

BITSADZE, A. V. Бицадзе, А. В. [1] On the local deformations of elastic bodies under compression. **5.** vol. III. no. 5. (1942) pp. 419—424. *306, 309.*

——— [2] Boundary problems for systems of differential equations of the elliptic type. **6.** vol. V. no. 8, (1944) pp. 761—770. (Georgian). *423.*

CARLEMAN, T. [1] Sur la résolution de certaines équations intégrales. **7.** Bd. 16. no. 26 (1922). *92, 124, 155, 237, 335.*

CASORATI, F. [1] Alcuni riflessioni relative alla teorica generale delle funzioni di variabili affatto libere, ossia complesse. **8.** Vol. III. (1866) pp. 337—350. *64.*

FÖPPL, L. [1] Neue Ableitung der Hertzschen Härteformeln für die Walze. **9.** Bd. 16. Heft 3 (1916) S. 165—170. *309.*

Gakhov, F. D. Гахов, Ф. Д. [1] On the Riemann boundary problem. **10.** Vol. 2 (44) no. 4. (1937) pp. 673—683. *87, 92, 237.*

——— [2] Linear boundary problems of the complex function theory. **11.** 3. ser. Vol. X. (1938) pp. 39—79. *87, 92, 204, 404.*

——— [3] Boundary problems of the theory of analytic functions and singular integral equations. Dissertation for the degree of doctor of the phys. math. sciences (defended 26. 10. 1942 in the Math. Inst. Tiflis). *237, 247, 248, 358.*

GALIN, L. A. Галин, Л. А. [1] Mixed problems of the theory of elasticity for the half-plane in the presence of friction. **1.** Vol. XXXIX. no. 3. (1943) pp. 88—93. *299.*

GIRAUD, G. [1] Equations à intégrales principales, étude suivie d'une application. **4.** 3-e sér. t. 51 (1934) pp. 251—372. *60, 161, 162, 421.*

——— [2] Sur une classe d'équations linéaires où figurent des valeurs principales d'intégrales simples. **4.** 3-e sér. t. 56 (1939) pp. 119—172. *61, 161, 162, 415.*

GLAGOLEV, N. I. Глаголев, Н. И. [1] Elastic stresses along the foundations of a dam. **1.** vol. XXXIV no. 7 (1942) pp. 204—209. *288, 299.*

——— [2] Determination of the stresses for the pressure of systems of rigid profiles. **12.** Vol. VII. no. 5. (1943) pp. 383—388. *288.*

Golubev, V. V. Голубев, В. В. [1] Theory of an aircraft wing of finite span. Moscow—Leningrad (1931) **27**. No. 108. *373, 374.*

Harnack, A. [1] Beiträge zur Theorie des Cauchy'schen Integrals. **13**. Bd. 37 (1885) S. 379—398. (reprinted in **14**. Bd. 35 (1899) p. 1—18). *25.*

Hellinger, E. u. Toeplitz, O. [1] Integralgleichungen und Gleichungen mit unendlich vielen Unbekannten. Sonderausgabe aus der Encykl. d. Math. Wiss. (II C 13), 1928. *71, 404.*

Hilbert, D. [1] Über eine Anwendung der Integralgleichungen auf ein Problem der Funktionentheorie. Verhandl. des III. Internal Mathematiker Kongresses, Heidelberg, 1904. *99, 115.*

——— [2] Grundzüge einer allgemeinen Theorie der linearen Integralgleichungen. Leipzig—Berlin 1912 (2nd edition, 1924). *70, 86, 99, 115.*

Hurwitz, W. A. [1] On the pseudo-resolvent to the kernel of an integral equation. **15**. vol. 13 (1912) pp. 405—413. *138.*

Kármán, Th. von and J. M. Burgers [1] General aerodynamic theory — Perfect fluid. Vol. II of Aerodynamic Theory (edited by W. F. Durand) Julius Springer, Berlin 1936 (transl. into Russian 1939 by V. V. Golubev). *373, 374.*

Kartsivadze, I. N. Карцивадзе, И. Н. [1] Fundamental problems of the theory of elasticity for an elastic circular region. **16**. Vol. XII (1943) pp. 95—104 (Georgian with Russian summary). *282, 316, 319.*

Keldysh, M. V. and Sedov, L. I. Келдыш, М. В. Седов, Л. И. [1] Effective solution of certain boundary problems for harmonic functions. **1**. Vol. XVI no. 1 (1937) pp. 7—10. *264, 268, 271, 281, 292.*

Kellog, O. D. [1] Unstetigkeiten in linearen Integralgleichungen. **14**. Bd. 58 (1904) S. 441—456. *70, 71.*

——— [2] Harmonic functions and Green's integral. **15**. vol. 13 (1912) pp. 109—132. *108.*

Kharazov, D. F. Харазов Д. Ф. [1] On a class of singular integral equations, the kernels of which are meromorphic functions of a parameter. **16**. vol. XIII. (1944) pp. 139—152. *162.*

Khvedelidze, B. V. Хведелидзе Б, В. [1] On the Poincaré boundary problem of the theory of the logarithmic potential. **1**. Vol. XXX no. 3 (1941) pp. 195—198. *203, 218, 223, 423.*

——— [2] On the Poincaré boundary problem of the theory of the logarithmic potential for multiply connected regions. **5**. Vol. II no. 7, 10, (1941) pp. 571—578, 865—872. *87, 92, 126, 203, 218, 223.*

——— [3] The Poincaré problem for linear second order differential equations of the elliptic type. **16**. Vol. XII (1943) pp. 47—77. (Georgian with detailed Russian summary). *223, 226.*

——— [4] On a particular linear Riemann boundary problem for systems of analytic functions. **5**. Vol. IV no. 4 (1943) pp. 289—296. *422.*

Kupradze, V. D. Купрадзе, В. Д. [1] The theory of integral equations with integrals in the sense of the Cauchy principal value. **5**. Vol. II, no. **7**. (1941) pp. 587—596. *152, 161.*

——— [2] On the problem of equivalence in the theory of particular integral equations. **5**. Vol. II, no. 9. (1941) pp. 793—798. *161.*

Kveselava, D. A. Квеселава Д. А. [1] Singular integral equations with discontinuous coefficients. **16**. Vol. XIII (1944) pp. 1—27. (Georgian with detailed Russian summary). *247, 324, 355, 358.*

Lavrentef, M. A. Лаврентьев, М. А. [1] On the building up of the flow past an arc of given shape. Moscow 1932. **27**. No. 118. *365*.

Liénard, A. [1] Problème plan de la dérivée oblique dans la théorie du potentiel. Journ. de l'Ecole politechnique IIIe sér. no. 5—7, (1938) pp. 35—158, pp. 177—226. *222*.

Magnaradze, L. G. Магнарадзе, Л. Г. [1] On a new integral equation of the theory of aircraft wings. **5**. Vol. III. no. 6 (1942) pp. 503—508. *373, 374, 377*.

——— [2] On a system of linear singular integro-differential equations and on the linear Riemann boundary problem. **5**. Vol. IV. no. 1. (1943) pp. 3—9. *423*.

——— [3] The theory of a class of linear singular integro-differential equations and its application to the problem of vibration of an aircraft wing of finite span. **5**. Vol. IV. no. 2 (1943) pp. 103—110. *379*.

Mikhlin, S. G. Михлин С. Г. [1] Singular integral equations with two independent variables. **10**. Vol. 1. (43) no. 4. 1936 pp. 535—552; supplement: **10**. no. 6. pp. 963—964. *162*.

——— [2] The problem of equivalence in the theory of singular integral equations. **10**. Vol. 3. (45) no. 1. (1938) pp. 121—141. *162*.

——— [3] On a class of singular integral equations. **1**. Vol. XXIV. no. 4. (1939) pp. 315—317. *124, 146, 149, 162*.

——— [4] On a theorem by F. Noether. **1**. Vol. XLIII no. 4. (1944) pp. 143—145. *162*.

Morera, G. [1] Intorno all'integrale di Cauchy. **8**. ser. 11 Vol. XXII (1889) pp. 191—200. *25, 41, 64*.

Muskhelishvili, N. I. Мусхелишвили, Н. И. [1] Some basic problems of the mathematical theory of elasticity. 2nd ed. A. N. SSSR., 1935. 3rd edition, Edited by J. R. M. Radok, published by P. Noordhoff N.V., Groningen, Holland, 1953. *73, 100, 282, 283, 284, 286, 292, 306, 315, 316, 319, 322*.

——— [2] Application of Cauchy type integrals to a class of singular integral equations. **16**. Vol. X (1941) pp. 1—43. *237, 247, 249, 358, 359, 371*.

——— [3] Systems of singular integral equations with Cauchy type kernels. **5**. Vol. III. no. 10. (1942) pp. 987—994. *179, 415*.

——— [3a] Supplement to [3]. **5**. Vol. IV. no. 2. (1943) pp. 99—101. *144, 415*.

——— [4] On the solution of the Dirichlet problem in the plane. **5**. Vol. I. no. 2. (1940) pp. 99—106. *171*.

——— [5] Notes on the fundamental boundary problems of Potential Theory. **5**. Vol. I. no. 3 (1940) pp. 169—170: Correction of errors. **5**. no. 7. p. 567. *171, 179*.

——— [6] On the fundamental mixed boundary problem of the theory of the logarithmic potential for multiply connected regions. **5**. Vol. II. no. 4. (1941) pp. 309—313. *171*.

——— [7] On the solution of the fundamental boundary problems of the theory of the Newtonian potential. **12**. Vol. IV. no. 4. (1940) pp. 3—26. *171, 179*.

——— [8] The solution of the fundamental mixed problem of the theory of elasticity for the half-plane. **1**. Vol. VIII no. 2. (1935) pp. 51—54. *287*.

——— [9] The fundamental boundary problems of the theory of elasticity for a half-plane. **5**. Vol. II. no. 10. (1941) pp. 873—880. *282, 284, 287, 288, 292*.

——— [10] The fundamental boundary problems of the theory of elasticity for a plane with straight cuts. **5**. Vol. III. no. 2. (1942) pp. 103—110. *282, 309, 311.*

——— [11] On the problem of equilibrium of a rigid stamp on the boundary of an elastic half-plane in the presence of friction. **5**. Vol. III. no. 5. (1942) pp. 413—418. *282, 299.*

——— [12] Sur l'intégration de l'équation biharmonique. **17**. (1919) pp. 663—686. *322.*

——— [13] Applications des intégrales analogues à celles de Cauchy à quelques problèmes de la Physique Mathématique. Tiflis, published by the Tiflis University, 1922. *322.*

Muskhelishvili, N. I. and Kveselava, D. A. Мусхелишвили, Н. И. and Кваселава, Д. А. [1] Singular integral equations with Cauchy type kernels for arcs. **16**. Vol. XI. (1942) pp. 141—172. *73, 227, 235, 237, 247, 323, 358.*

Muskhelishvili, N. I. and Vekua, N. P. Мусхелишвили, Н. И. and Векуа Н. П. [1] The Riemann boundary problem for several unknown functions and its application to systems of singular integral equations. **16**. Vol. XII (1943) pp. 1—46. *128, 381, 422.*

Noether F. [1] Über eine Klasse singulärer Integralgleichungen. **14**. Bd. 82. (1921) S. 42—63. *99, 134, 140, 272.*

Osgood W. F. [1] Lehrbuch der Funktionentheorie. Bd. 1. Leipzig—Berlin, 1912. *34.*

Picard, E. [1] Leçons sur quelques types simples d'équations aux dérivées partielles. Paris. 1927. *33, 45, 87, 92.*

Plemelj, J. [1] Ein Ergänzungssatz zur Cauchyschen Integraldarstellung analytischer Funktionen, Randwerte betreffend. **18**. XIX. Jahrgang, (1908) S. 205—210. *25, 43, 45, 48, 49, 64, 66, 87.*

——— [2] Riemannsche Funktionenscharen mit gegebener Monodromiegruppe **18**. S. 211—245. *247, 384.*

——— [3] Potentialtheoretische Untersuchungen. Leipzig 1911. *165.*

Pogorzelski W. [1] Über die Transformationen einiger iterierten uneigentlichen Integrale und ihre Anwendung zur Poincaréschen Randwertaufgabe. **19**. Bd. 44. (1939) S. 427—444. *202, 224.*

Poincare H. [1] Leçons de Mécanique Céleste. Vol. III, Paris, 1910, chapt. X. *60, 115, 202, 224.*

Privalov, I. I. Привалов, И. И. [1] Introduction to the theory of functions of a complex variable. 6th edition, 1940. *23, 25, 37,*

——— [2] The Cauchy integral. Saratov, 1919 (publ. separately by the Saratov University 1918). *23, 25, 43, 49, 64, 173.*

——— [3] Limiting properties of single-valued analytic functions. Publ. Moscow State University, 1941. *23, 25, 43, 49.*

——— [4] On a boundary problem in the theory of analytic functions. **10**. Vol. 41. no. 4. (1934) pp. 519—526. *92.*

Radon, J. [1] Über die Randwertaufgaben beim logarithmischen Potential. **20**. Bd. 128. Heft. 7. (1919). *176.*

Riemann, B. [1] Grundlagen für eine allgemeine Theorie der Funktionen einer veränderlichen komplexen Grösse. Werke, Leipzig, 1876, S. 3—43. *99.*

Sadowsky, M. [1] Zweidimensionale Probleme der Elastizitätstheorie. **9**. Bd. 8. (1928) S. 107—121. *292.*

Schauder, J. [1] Potentialtheoretische Untersuchungen, Erste Abhandlung. **19**. Bd. 33. (1931) S. 602—640. *169.*

SCHMIDT, H. [1] Strenge Lösungen zur Prandtlschen Theorie der tragenden Linie. **9.** Bd. 17. Heft. 2. (1937) S. 101—116. *379.*

SCHROEDER, K. [1] Über eine Integralgleichung erster Art der Tragflügeltheorie. **21.** XXX, (1938) S. 345—362. *249, 374.*

——— [2] Über die Prandtlsche Integro-differentialgleichung der Tragflügeltheorie. **22.** 1939, No. 16. *274.*

SEDOV, L. I. Седов, Л. И. [1] The theory of the plane flow of an ideal liquid. Moscow—Leningrad 1939 (Published by the Defence Dept.).

SHERMAN, D. I. Шерман, Д. И. [1] The plane problem of the theory of elasticity with mixed boundary conditions. **23.** no. 88, 1938. *323.*

——— [2] The mixed problem of the static theory of elasticity for plane multiply connected regions. **1.** Vol. XXVIII. no. 1. (1940) pp. 29—32. *323.*

——— [3] The elastic plane with straight cuts. **1.** Vol. XXVI. no. 7. (1940) pp. 635—638. *311.*

——— [4] The mixed problem of the potential theory and the theory of elasticity for a plane with a finite number of straight cuts. **1.** Vol. XXVII no. 4. (1940) pp. 330—334. *315.*

SHTAERMAN, I. YA. Штаерман, И. Я. [1] On Hertz's theory of local deformations of elastic bodies under compression. **1.** Vol. XXV. no. 5. (1939) pp. 360—362. *306, 309.*

——— [2] A generalization of Hertz's theory of local deformations of elastic bodies under compression. **1.** Vol. XXIX. no. 3. (1940) pp. 179—181. *306.*

——— [3] Some special cases of the contact problem. **1.** Vol. XXXVIII. no. 7. (1943) pp. 220—224. *306, 309.*

SOBOLEV, S. L. Соболев, С. Л. [1] On a boundary problem of the theory of the logarithmic potential and its application to the reflection of plane elastic waves. **23.** no. 11. (1930) pp. 1—9. *272.*

SIGNORINI, A. [1] Sopra un problema al contorno nella teoria delle funzioni di variabile complessa. **24.** Ser. III. Vol. XXV, (1916) pp. 253—273. *281.*

SÖHNGEN, H. [1] Die Lösungen der Integralgleichung

$$g(x) = \frac{1}{2\pi i}\int_{-a}^{+a} \frac{f(\xi)d\xi}{x-\xi}$$

und deren Anwendung in der Tragflügeltheorie. **19.** Bd. 45.(1939) S. 245—264. *249.*

TAMARKIN, J. D. [1] On Fredholm's integral equations, whose kernels are analytic in a parameter. **25.** 2 Ser. Vol. 28. (1927) pp. 127—152. *162.*

TRICOMI, F. [1] Equazioni integrali contenenti il valor principale di un integrale doppio. **19.** Bd. 27. (1928) pp. 87—133. *162.*

VEKUA, ILYA N. Векуа, Илья, Н. [1] On linear singular integral equations, containing integrals in the sense of the Cauchy principal value. **1.** Vol. XXVI, no. 4. (1940) pp. 335—338. *69, 124, 126, 155, 226, 359, 374.*

——— [2] On a class of singular integral equations with integrals in the sense of the Cauchy principal value. **5.** Vol. II. no. 7. (1941) pp. 579—586. *153, 155, 158, 226.*

——— [3] On the reduction of singular integral equations to Fredholm equations. **5.** Vol. II. no. 8. (1941) pp. 697—700. *155, 160.*

——— [4] Integral equations with special Cauchy type kernels. **16.** Vol. X. (1941) pp. 45—72 *155, 158, 160, 353.*

——— [5] On the theory of singular integral equations. **5.** Vol. III. no. 9 (1942) pp. 869—876. *140, 143, 144, 149, 151, 153, 155, 158, 159, 161, 353*

——— [6] On a linear boundary problem of Riemann. **16.** Vol. XI (1942) pp. 109—139. *100, 192, 194, 202, 204.*

——— [7] On a new integral representation of analytic functions and its application. **5.** Vol. II. no. 6. (1941) pp. 477—484. Supplement to the above, **5.** no. 8 (1941) pp. 701—706. *192, 201, 212, 213, 217.*

——— [8] Boundary problems of the theory of linear elliptic differential equations in two independent variables. **5.** Vol. I. no. 1, 3, 7, (1940) pp. 29—34, 181—186, 497—500. *225.*

——— [9] The complex representation of solutions of elliptic differential equations and its application to boundary problems. **16.** Vol. VII. (1939) pp. 161—253. *225.*

——— [10] Notes on the general representation of solutions of differential equations of the elliptic type. **5.** Vol. IV. no. 5. (1943) pp. 385—392. (Georgian with Russian summary). *225.*

——— [11] On Prandtl's integro-differential equation. **12.** Vol. 9. no. 2. (1945) pp. 143—150. *373, 377, 378.*

——— [12] Allgemeine Darstellung der Lösungen elliptischer Differentialgleichungen in einem mehrfach zusammenhängenden Gebiet. **5.** Vol. 1. no. 5. (1940) pp. 329—335. *226.*

Vekua, Nikolai, P. Векуа, Николай П. [1] On a class of singular integral equations and some boundary problems of Potential Theory. **16.** Vol. X. (1941) pp. 73—92 (Georgian). *260, 422.*

——— [2] On the theory of systems of singular integral equations with Cauchy type kernels. **5.** Vol. IV. no. 3. 1943. pp. 207—214. *415, 422.*

——— [3] The Riemann problem with discontinuous coefficients for several unknown functions. **5.** Vol. V. no. 1. (1944) pp. 1—10 (Georgian). *422.*

——— [4] On the theory of systems of singular integral equations with discontinuous coefficients. **5.** Vol. V. no. 2. (1944) pp. 125—134 (Georgian). *422.*

——— [5] On a linear Riemann boundary problem with discontinuous coefficients for systems of analytic functions. **5.** Vol. V. no. 5. (1944) pp. 473—482 (Georgian). *422.*

Vekua, N. P. and Kveselava, D. A. Векуа, Н. П. and Квеселава, Д. А. [1] On a boundary problem of the theory of functions of a complex variable. **5.** Vol. II. no. 3. (1941) pp. 233—240. *422.*

——— [2] On a boundary problem of the theory of a complex variable and its application to the solution of systems of integral equations. **16.** Vol. IX. (1941) pp. 33—48. *422.*

Walsh, J. L. and Sewell, W. E. [1] Sufficient conditions for various degrees of approximation by polynomials. **26.** Vol. 6. no. 3. (1940) pp. 658—705. *53.*

Warschawski, S. [1] Über das Randverhalten der Ableitung der Abbildungsfunktion bei konformer Abbildung. **19.** Bd. 35. (1932) S. 321—456. *108, 361.*

——— [2] Bemerkung zu [1]. **19.** Bd. 38. (1934) S. 669—683. *53.*

Weissinger, J. [1] Ein Satz über Fourierreihen und seine Anwendung auf die Tragflügeltheorie. **19.** Bd. 47. (1940) S. 16—33. *249, 374.*

INDEX

A CATALOG OF SELECTED

DOVER BOOKS

IN SCIENCE AND MATHEMATICS

CHALLENGING MATHEMATICAL PROBLEMS WITH ELEMENTARY SOLUTIONS, A.M. Yaglom and I.M. Yaglom. Over 170 challenging problems on probability theory, combinatorial analysis, points and lines, topology, convex polygons, many other topics. Solutions. Total of 445pp. 5⅜ × 8½. Two-vol. set.
Vol. I 65536-9 Pa. $7.95
Vol. II 65537-7 Pa. $7.95

FIFTY CHALLENGING PROBLEMS IN PROBABILITY WITH SOLUTIONS, Frederick Mosteller. Remarkable puzzlers, graded in difficulty, illustrate elementary and advanced aspects of probability. Detailed solutions. 88pp. 5⅜ × 8½.
65355-2 Pa. $4.95

EXPERIMENTS IN TOPOLOGY, Stephen Barr. Classic, lively explanation of one of the byways of mathematics. Klein bottles, Moebius strips, projective planes, map coloring, problem of the Koenigsberg bridges, much more, described with clarity and wit. 43 figures. 210pp. 5⅜ × 8½. 25933-1 Pa. $6.95

RELATIVITY IN ILLUSTRATIONS, Jacob T. Schwartz. Clear nontechnical treatment makes relativity more accessible than ever before. Over 60 drawings illustrate concepts more clearly than text alone. Only high school geometry needed. Bibliography. 128pp. 6⅛ × 9¼. 25965-X Pa. $7.95

AN INTRODUCTION TO ORDINARY DIFFERENTIAL EQUATIONS, Earl A. Coddington. A thorough and systematic first course in elementary differential equations for undergraduates in mathematics and science, with many exercises and problems (with answers). Index. 304pp. 5⅜ × 8½. 65942-9 Pa. $8.95

FOURIER SERIES AND ORTHOGONAL FUNCTIONS, Harry F. Davis. An incisive text combining theory and practical example to introduce Fourier series, orthogonal functions and applications of the Fourier method to boundary-value problems. 570 exercises. Answers and notes. 416pp. 5⅜ × 8½. 65973-9 Pa. $11.95

AN INTRODUCTION TO ALGEBRAIC STRUCTURES, Joseph Landin. Superb self-contained text covers "abstract algebra": sets and numbers, theory of groups, theory of rings, much more. Numerous well-chosen examples, exercises. 247pp. 5⅜ × 8½. 65940-2 Pa. $8.95

Prices subject to change without notice.

Available at your book dealer or write for free Mathematics and Science Catalog to Dept. GI, Dover Publications, Inc., 31 East 2nd St., Mineola, N.Y. 11501. Dover publishes more than 175 books each year on science, elementary and advanced mathematics, biology, music, art, literature, history, social sciences and other areas.